AF769411

Encyclopedia of Stoichiometry: Materials Science

Volume I

Encyclopedia of Stoichiometry: Materials Science

Volume I

Edited by **Bruce Hopkin**

NY Research Press
New York

Published by NY Research Press,
23 West, 55th Street, Suite 816,
New York, NY 10019, USA
www.nyresearchpress.com

Encyclopedia of Stoichiometry: Materials Science
Volume I
Edited by Bruce Hopkin

© 2015 NY Research Press

International Standard Book Number: 978-1-63238-170-5 (Hardback)

This book contains information obtained from authentic and highly regarded sources. Copyright for all individual chapters remain with the respective authors as indicated. A wide variety of references are listed. Permission and sources are indicated; for detailed attributions, please refer to the permissions page. Reasonable efforts have been made to publish reliable data and information, but the authors, editors and publisher cannot assume any responsibility for the validity of all materials or the consequences of their use.

The publisher's policy is to use permanent paper from mills that operate a sustainable forestry policy. Furthermore, the publisher ensures that the text paper and cover boards used have met acceptable environmental accreditation standards.

Trademark Notice: Registered trademark of products or corporate names are used only for explanation and identification without intent to infringe.

Printed in China.

Contents

Preface

In my initial years as a student, I used to run to the library at every possible instance to grab a book and learn something new. Books were my primary source of knowledge and I would not have come such a long way without all that I learnt from them. Thus, when I was approached to edit this book; I became understandably nostalgic. It was an absolute honor to be considered worthy of guiding the current generation as well as those to come. I put all my knowledge and hard work into making this book most beneficial for its readers.

This book provides a comprehensive overview of the relevance of stoichiometry in the field of material engineering. It compiles various researches and reviews offering latest information regarding stoichiometry at different levels. The book covers various applications of stoichiometry, such as the role of stoichiometry in nanotechnology, defect chemistry and electrochemical applications. It also discusses intermetallic compounds and metal oxides. The book also elucidates latest methods for solid phase synthesis and the impact of stoichiometry in energy production. This book will be helpful to engineers, academicians and scientists working in the field of materials science.

I wish to thank my publisher for supporting me at every step. I would also like to thank all the authors who have contributed their researches in this book. I hope this book will be a valuable contribution to the progress of the field.

Editor

Part 1

Stoichiometry and Nanotechnology

1

Stoichiometric Boron-Based Nanostructures

Limin Cao[1], Xiangyi Zhang[1], Wenkui Wang[1] and Min Feng[2]
[1]*State Key Laboratory of Metastable Materials, Yanshan University*
[2]*Department of Physics and Astronomy, University of Pittsburgh*
[1]*China*
[2]*USA*

1. Introduction

The elemental boron and boron-containing compounds hold a unique place in chemistry, physics, and materials science (Greenwood, 1973; Donohue, 1974; Matkovich, 1977; Emin, 1987; Albert & Hillebrecht, 2009). In general, they are composed of various multi-atom boron clusters, such as graphitic layer, octahedral B_6 and icosahedral B_{12}, among which the 12-atom B_{12} icosahedral cluster is the most common form (Greenwood, 1973; Donohue, 1974; Matkovich, 1977; Emin, 1987; Albert & Hillebrecht, 2009). The electronic properties of these boron clusters are characterized by the three-center electron-deficient bond, which presents some analogy with metallic bonding because electron deficiency forces electrons to be shared at the cluster surface(Greenwood, 1973; Donohue, 1974; Emin, 1987). Whereas these clusters are linked together by two-center covalent bonds, which lead to the intra-cluster and inter-cluster bonds have comparable strength (Greenwood, 1973; Donohue, 1974; Emin, 1987; Nelmes *et al.*, 1995; Vast *et al.*, 1997; Fujimori *et al.*, 1999). These features make boron-based materials exhibit a wide variety of electronic properties ranging from semiconducting to superconducting. Furthermore, the boron clusters as structural units may be stacked in many different ways. This fact, coupled with the small size and high affinity of boron atoms, make boron-related materials form a unique family with an astonishing number of members.

Boron and borides are widely used in numerous technological applications, particularly in extreme environments where a refractory, light, and hard material is required. Pure boron is, among elemental semiconductors, the least understood as regards its structures and properties. The elemental boron has a low density but a high melting point around 2300°C as well as a hardness close to that of diamond. Moreover, boron is one of the very few elements that can be used in nuclear engineering, high temperature semiconductor devices, thermoelectric power conversion applications, or as a lightweight protective armor for space shuttles (Greenwood, 1973; Donohue, 1974; Matkovich, 1977; Emin, 1987; Albert & Hillebrecht, 2009). The semiconductor boron becomes a superconductor at temperatures 6–12 K under high pressures above 160 GPa (Erements *et al.*, 2001), but the structure and transition mechanism of superconducting boron is still unkown. Magnesium diboride, MgB_2, has the superconductivity at an unexpectedly high temperature of 39 K (Nagamatsu *et al.*, 2001), which is considered as one of the most important discoveries in superconductivity since the high-temperature copper oxide superconductors. Hexagonal boron nitride, h-BN, is a wide band-gap semiconductor with excellent mechanical strength, good thermal conductivity, and strong corrosion resistance properties. Cubic boron nitride, c-BN, is the second hardest material. Boron suboxide, B_6O, is as hard as cubic boron nitride and as tough as diamond (Hubert *et al.*, 1998; He *et al.*, 2002). B_6N, the sub-nitride analogue to B_6O, was reported to be a superhard material with metal-like conductivity (Hubert *et al.*, 1997; Garvie *et al.*, 1997). All of these demonstrate that boron and boron-based compounds constitute one of the most fascinating classes of materials, which are of great scientific and applied importance in terms of their unique chemistry and novel electronic, thermal, and mechanical properties.

Recent interest in low-dimensional nanoscale materials has been motivated by the push for miniaturization of electronic and mechanical devices and a need to understand the fundamentals of nanoscale chemistry and physics. Materials in nanoscale sizes behave very differently from their bulk forms, due to the different ways that electrons interact in three-dimensional (3D), two-dimensional (2D), and one-dimensional (1D) structures. One-dimensional nanostructures afford an ideal system for investigating fundamental phenomena in mesoscopic scales such as size and dimensionality-mediated properties, and exploring applications of these materials in future nanodevices (Dresselhaus *et al.*, 1996; Ajayan & Ebbesen, 1997; Hu *et al.*, 1999; Kuchibhatla *et al.*, 2007). Stimulated by the discovery of fullerenes (Kroto *et al.*, 1985) and carbon nanotubes (Iijima, 1991), as well as their potential fundamental and practical implications, extensive experimental and theoretical studies have been focused on investigating various nanostructures and their applications in developing nanotechnology. Theoretical studies have suggested the existence of novel layered, tubular, fullerene-like, and even quasicrystalline boron solids built from elemental subunits which possess numerous novel structural, electronic, and many other useful properties (Gindulyte *et al.* 1998a, 1998b; Boustani *et al.*, 1999, 2000; Quandt &Boustani, 2005; Szwacki *et al.*, 2007). For example, the proposed boron nanotubes reveal a metallic-like density of states (DOS) (Boustani *et al.*, 1999; Quandt & Boustani, 2005). They may be expected to be very good conductors, much better than carbon nanotubes with potential applications, e.g., in field emission and high-temperature light materials, and in high-temperature electron devices. We reported the first creation of well-aligned, smooth boron nanowires (Cao *et al.*, 2001). Since then, various methods have been developed to synthesize amorphous or crystalline boron nanostructures. Yang and coworkers reported the vapor-liquid-solid (VLS) growth of amorphous boron nanowires using a chemical vapor transport (CVT) process in a sealed quartz ampoule (Wu *et al.*, 2001). Buhro and coworkers synthesized crystalline boron nanowires using a chemical vapor deposition (CVD) method (Otten *et al.*, 2002), they also found that the boron nanowires exhibit the semiconducting electrical properties consistent with those of bulk boron. Subsequently, a laser ablation method was also developed to synthesize boron nanowires and nanobelts by some groups (Zhang *et al.*, 2002; Wang *et al.*, 2003; Meng *et al.*, 2003). Ruoff and coworkers synthesized boron nanoribbons by pyrolysis of diborane at 630-750 °C and ~200 mTorr in a quartz tube furnace (Xu *et al.*, 2004). Yun *et al.* grew inclined boron nanowires bundle arrays in an oxide-assisted vapor-liquid-solid process (Yun *et al.*, 2005). Gao and coworkers have fabricated aligned crystalline boron nanowire arrays using a CVD method, and these boron nanostructures show a good field-emission behavior (Liu *et al.*, 2008). Kirihara *et al.* have measured electrical conductance of single crystalline boron nanobelts fabricated by laser ablation (Kirihara *et al.*, 2006). It is interesting that the pure boron nanobelt is a p-type semiconductor with electrical conductivity on the order of 10^{-3} $(\Omega\,cm)^{-1}$ at room temperature. While doping magnesium into the boron nanobelts does not change the crystalline structure, the electrical conductance increases by a factor of 100–500.

In this chapter we concentrate on our pioneering work on the synthesis of stoichiometric boron-based nanostructures. In Section 2, we describe our first creation of well-aligned smooth boron nanowires using a magnetron sputtering process. The pure boron nanostructures grow vertically on various substrates to form self-assembled arrays over large areas up to several tens of square centimeters without the use of template or catalyst. Highly pure BN nanotubes discussed in Section 3 are synthesized by annealing the pure boron nanowires in N_2 atmosphere at 1500°C. TEM and EELS studies reveal that the products possess a concentric tubular structure and stoichiometric BN composition. Our results illustrate the technological potential of BN nanotubes produced in large quantities be incorporated into future nanocomposites and nanoscale mechanical and electronic devices. In Section 4 we present the creation of B_6N_x/BN coaxial nanowires with radial heterostructures using the simple nitriding processing of pure boron nanowires at 1200°C. The produced nanostructures consist of a core nanowire with rhombohedral structure and stoichiometry of B_6N_x, a metastable high pressure phase, and a hexagonal BN sheath. We proposed a high-pressure-nanocell-assisted growth mechanism for the formation of the B_6N_x core nanostructure and B_6N_x/BN nanoheterostructure. This simple process might enable the studies of high-pressure-induced phase transformation and reaction in nanosystem at ambient pressure, and be extended to bulk fabrication of a wealth of nanoheterostructures and nanocomposites in B-C-N-O system. Finally in Section 5, we briefly summarize our experimental results and also discuss some of the theoretically proposed novel boron-based nanostructures which are waiting for future explorations.

2. Pure boron nanowires

Following the discovery of carbon nanotubes (CNTs) (Iijima, 1991), much interest in one-dimensional (1D) nanostructures has been stimulated greatly due to their potential fundamental and practical implications in areas such as materials science, chemistry, physics and engineering (Dresselhaus *et al.*, 1996; Ajayan & Ebbesen, 1997; Hu *et al.*, 1999; Kuchibhatla *et al.*, 2007). Some studies have focused on the preparation and characterization of new one-dimensional nanometer-sized materials with unique and advanced properties. The others contributed to developing techniques for the manipulation of nanotubes or nanowires with the desirable form of aggregation and dimensionality. Boron is the first group-III element with atomic number 5. It has three valence electrons ($2s^2 2p^1$) but four valence shell orbitals (s, p_x, p_y, p_z). The electron-deficient nature makes boron to form the so-called three-center electron-deficient bonds where the charge accumulation occurs at the center of a triangle formed by three adjacent boron atoms [Fig. 1b], other than to form the conventional covalent two-center bonds [Fig. 1a]. As a consequence, boron holds a special place within chemistry and exhibits the most varied polymorphism of any of the elements (Greenwood, 1973; Donohue, 1974; Matkovich, 1977; Emin, 1987; Albert & Hillebrecht, 2009). The unusual three-center bonding associated with a large variety of uncommon crystal structures of boron and boron-rich borides leads to the formation of a fascinating class of materials with many exceptional and useful properties. Extensive theoretical studies have been focused on investigating the geometrical and electronic structures of boron clusters, and the boron nanostructures possess numerous novel structural, electronic and thermal properties that are not only interesting in theoretical research but also useful in applications (Gindulyte *et al.* 1998a, 1998b; Boustani *et al.*, 1999, 2000; Quandt &Boustani, 2005; Szwacki *et al.*, 2007). The proposed boron nanostructures exhibit nanoscale structural chemistry as abundant and complicated as that of carbon, the most important element in nature and the basis element for living beings. An important motivation for our research is the synthesis of novel boron-based nanostructures and their applications as critical building blocks in the ongoing miniaturization of nanoelectronics and nanocomposites where they may impart stiffness, toughness, and strength.

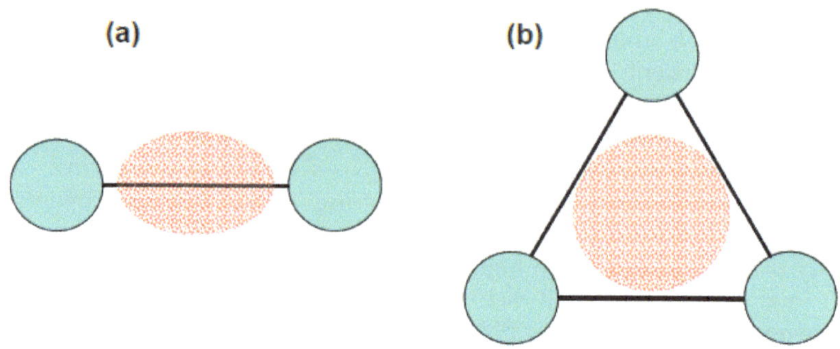

Fig. 1. Schematic of the electronic charge distributions (the dotted regions) of a two-center bond (a) and three-center bond (b).

In this section we describe the first growth of pure boron nanowires and their highly ordered arrays using a magnetron sputtering of high-purity B/B_2O_3 and/or boron targets (Cao *et al.*, 2001, 2002a, 2002b). In a typical experiment, a radio frequency (rf) magnetron sputtering of 80 W power was employed to prepare aligned boron nanowire arrays films. Si(100) substrates were placed on a temperature-controlled heater parallel to the target surface. Prior to sputtering, the vacuum chamber was first pumped to a base pressure better than 4×10^{-5} Pa, highly pure argon (Ar) gas (purity 99.999%) was then introduced. The Si(100) substrate was first heated to 800 °C under the Ar stream, and thereafter growth was initiated at a rf power of 80 W with the total pressure kept at 2 Pa during the process of sputtering. After 6 hours of sputtering deposition, the Si substrate was covered with pitch-black films. Figure 2 shows SEM images of the boron nanowire arrays. The well-aligned boron nanowires grew uniformly on the surface of the substrate over large areas [Fig. 2a]. The largest product, which we obtained, was a uniform film of aligned boron nanowire arrays on a 50-mm-diam Si substrate. The size of the product is limited by the sizes of the sputtering target and the heater in our system. The cross-sectional SEM image [Fig. 2b] shows clearly that the densely aligned boron nanowires grew perpendicular on the substrate surface and they are straight along their axes in the whole length. The clean, smooth, and parallel oriented boron nanowires have uniform diameters of 40-60 nm and a length up to several tens of micrometers [Fig. 2c]. An interesting and unique feature of the boron nanowires is that most of their tips are flat rather than hemispherical in morphology [Fig. 2d]. We found that the formation of boron nanowire arrays is independent from the nature of the substrate. We can obtain well-ordered boron nanowire films with high quality on a set of different substrates, such as SiO_2 wafer, Al_2O_3 wafer, MgO wafer, and many other metal and non-metal plates (Cao *et al.*, 2001, 2002a, 2002b).

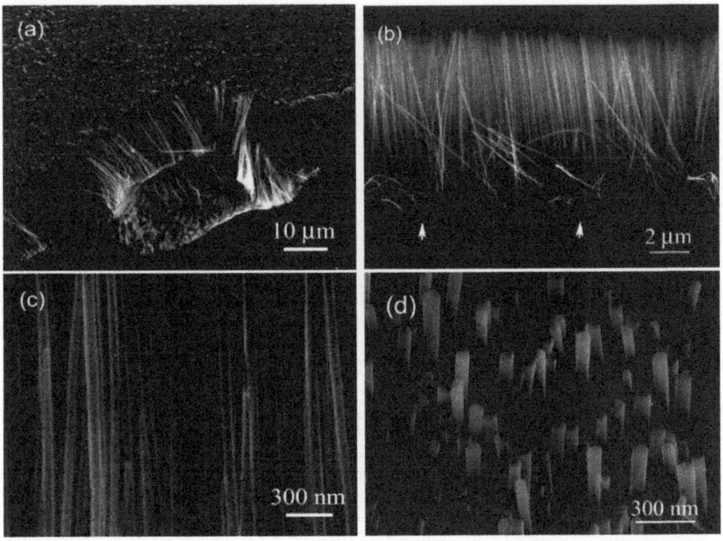

Fig. 2. SEM images of the vertically aligned smooth boron nanowire arrays. (a) Low-magnification side view. (b) Cross-sectional SEM image. (c,d) High-magnification view.

Figure 3 shows a high-resolution transmission electron microscopy (HRTEM) image of a typical boron nanowire. No crystalline fringes can be identified in the HRTEM image at the lattice-resolved scale. This indicates that the boron nanowires are amorphous. No diffraction spots, but some diffuse rings, shown in the selected area electron diffraction (SAED) pattern [Fig. 3, inset] from the boron nanowire, further confirm the amorphous nature of the boron wires. The boron nanowire is sheathed by an amorphous oxide coating that is formed when the boron nanowires are exposed to air after deposition. The chemical characterization of the boron nanowires using EELS shows that the boron nanowire is composed of boron with neglectable traces of oxygen [Fig. 4]. Quantitative EELS studies reveal that the total content of oxygen in the boron wire is less than 5%.

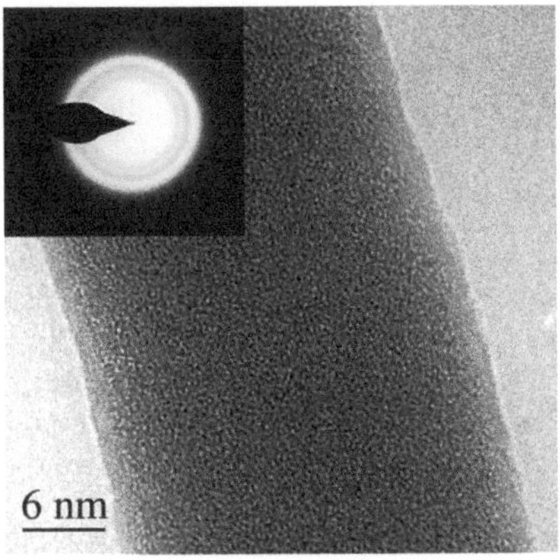

Fig. 3. HRTEM image of a typical boron nanowire. Inset, selected area electron diffraction (SAED) pattern taken from the nanowire.

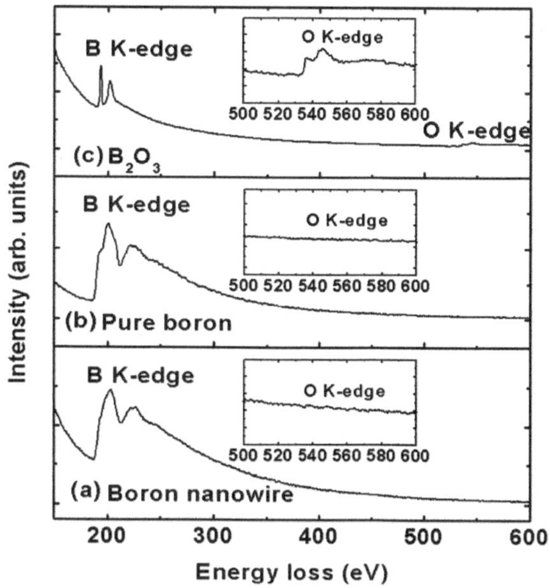

Fig. 4. EELS spectra of (a) a boron nanowire; (b) bulk, pure boron; (c) bulk B_2O_3. The insets are magnified features of the O-K edges. A comparison of the fine structures of the B-K and O-K edges shows that the edges from the boron nanowire and pure boron are very similar, but are obviously different from those of B_2O_3. The distinct O-K edges from B_2O_3 form a sharp contrast to those from pure boron and the boron nanowire.

Energy-filtering imaging offers an effective method to directly image the distribution of particular elements of interest in a relatively large area with high spatial resolution. We mapped the elemental distributions of boron and oxygen in the nanowires using a Zeiss EM912 energy-filtering TEM with an Omega filter. Figure 5 shows the representative bright field image and the corresponding energy-filtered images of a typical boron nanowire. The oxygen elemental distribution profile derived from the line scan analysis across the boron nanowire in the oxygen map [Fig. 5c, inset] shows that the oxygen exists mainly within the coating, which might be the presence of an amorphous boron oxide. We concluded that the oxide outer sheath was formed by exposing the material to air. This is consistent with our HRTEM observations and pervious results reported by others (Morales & Lieber, 1998).

Structural stability of boron nanowires at room temperature was investigated by means of *in situ* high-pressure energy-dispersive x-ray powder diffraction using synchrotron radiation in a diamond anvil cell. Figure 6 shows x-ray diffraction (XRD) pattern of boron nanowire arrays. No diffraction peaks can be observed, also confirming the amorphous nature of boron nanowires as demonstrated by TEM and SAED analysis. *In situ* high-pressure energy-dispersive x-ray powder diffraction (EDXD) measurements were performed at Beijing Synchrotron Radiation Facility (BSRF) using synchrotron radiation and a diamond anvil cell. In the experiments, gold was used as x-ray diffraction reference, and a mixture of methanol, ethanol, and H_2O (16:4:1) served as the pressure medium. The boron nanowire samples peeled off from substrates along with gold powder, and pressure medium was loaded into a hole of 100 μm diameter in a preindented gasket of 42 μm thickness. A microcollimated x-ray beam of 20 × 20 μm^2 was employed for the EDXD studies. All of the EDXD spectra were recorded by a Ge detector at *ED* = 42.5313 keV Å. A series of EDXD spectra [Fig. 7] were recorded at pressures ranging from 0 to 103.5 GPa at room temperature with increasing pressures. Apart from the fluorescence peaks and the diffraction peaks from Au marker, no crystalline Bragg reflections from boron nanowires were observed, as shown in Fig. 7. This suggests that the amorphous structure of boron nanowires is stable up to 103.5GPa, the highest pressure achieved in the study.

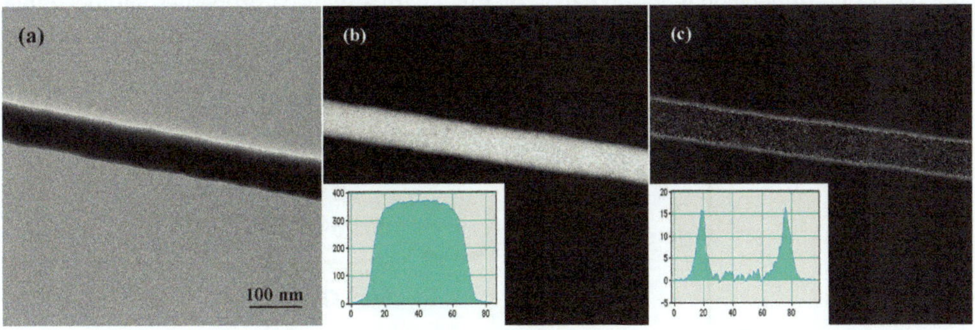

Fig. 5. Bright-field TEM image of a typical boron nanowire (a), and the corresponding boron elemental map (b), and oxygen elemental map (c). Insets show the elemental profiles of boron (b) and oxygen (c) across the boron nanowire, respectively.

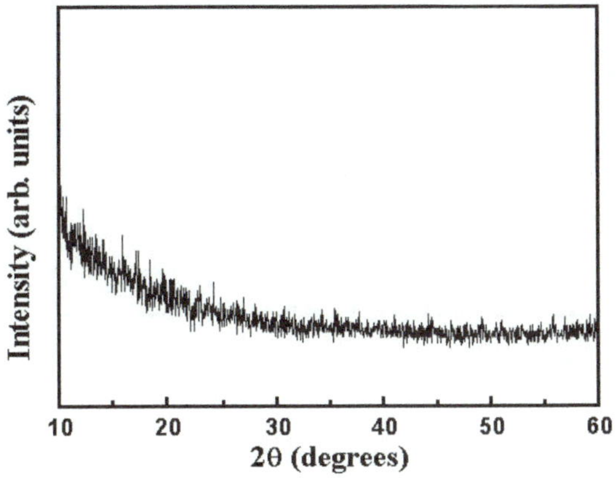

Fig. 6. XRD pattern of the aligned boron nanowire arrays.

From the above observations and discussions, we demonstrated that the nucleation and growth of well-oriented boron nanowire arrays are template-catalyst-independent. The amorphous structure of boron nanowires is stable under high pressure at ambient temperature. The growth of boron nanowires cannot be explained using the classic screw-dislocation model (Frank, 1949) and the vapor-liquid-solid (VLS) model (Wagner & Ellis, 1964; Morales & Lieber, 1998). We suggest that a complicated vapor-solid or vapor-cluster process might govern our system, and the nucleation and growth of the vertically aligned boron nanowire arrays are entirely due to a self-organization process. The successful synthesis of boron nanowires might lead to a deeper understanding of the growth mechanism of one-dimensional nanostructures. Boron nanowires have appeared as a new member in the family of one-dimensional nanostructures. Moreover, large-scale arrays of boron nanostructures provide very exciting opportunities for understanding more about boron, one of the most important and interesting elements. Considering the unique properties of boron-rich solids and other nanostructures, it is reasonable to expect that boron nanowires will display some exceptional properties. A fascinating prospect is that many other one-dimensional boride nanostructures with advanced properties and desirable arrays, as we will show in Section 3 and Section 4, can be fabricated using the boron nanowire arrays as templates or starting materials.

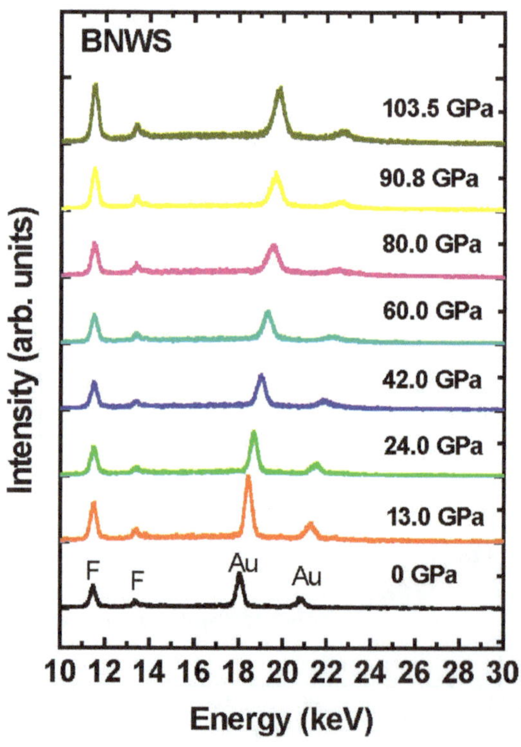

Fig. 7. Synchrotron EDXD spectra of the boron nanowires mixed with an Au pressure standard at various pressures.

3. Stoichiometric boron nitride nanotubes

Carbon and boron nitride (BN) nanotubes are two typical classes of one-dimensional tubular nanostructures, which are of great importance in the fundamental research and in developing nanoscale mechanical and electronic applications (Iijima, 1991; Chopra *et al.*, 1995). They are composed of concentric and seamless cylinders formed by rolling up the planar atomic layers. Carbon nanotubes can be either metals or small gap semiconductors, depending on their diameter and chirality (Hamada *et al.*, 1992; Saito *et al.*, 1992; Dresselhaus *et al.*, 1996). Boron nitride nanotubes, in contrast, are uniformly wide band-gap semiconductors having an energy gap independent of geometrical arrangements and wall number (Blase *et al.*, 1994). Furthermore, BN nanotubes possess a higher resistance to oxidation at high temperature and have thermal conductivity comparable to carbon nanotubes (Xiao *et al.*, 2004; Chang *et al.*, 2005). Theoretical and experimental studies showed that the BN nanotubes possess the elastic modulus of 700-900 GPa, which is among the highest ones in one-dimensional fibers (Chopra & Zettl, 1998; Hernandez *et al.*, 1998; Suryavanshi *et al.*, 2004). These make BN nanostructures superior to carbon species in many mechanical and electronic applications, especially at high temperature.

To realize the practical applications of one-dimensional BN nanostructures in nanoscale electronic and mechanical devices, it is critically important to synthesize them in large quantities. However, so far the large scale synthesis of BN nanotubes and nanowires is still a challenging task. We have synthesized pure boron nanowires in high yield as described in Section 2. Can these novel nanowires be utilized as the starting materials for the creation of other boron-based nanostructures? Our research demonstrates that the answer to this question is an absolute YES. In this section, we describe the large quantity synthesis of pure BN nanotubes using a simple approach of annealing boron nanowire precursors in N_2 atmosphere at 1500°C (Cao *et al.*, 2007). The boron nanowire samples produced by magnetron sputtering were scraped from the substrates, packed into a Ta tube, and then placed in a home-made high temperature annealing furnace. After evacuating the furnace chamber to 10^{-5} Torr, high purity nitrogen gas was introduced into the chamber to a pressure of 150 Torr. The sample was heated to and kept at 1500 °C for 4 hours, and then cooled naturally to room temperature. After nitriding treatment, the nanowire product turns to gray-white from pitch black in color.

SEM studies show that the as-synthesized product consists of one dimensional nanostructures with uniform diameters [Fig. 8]. Energy-dispersive x-ray (EDX) [Fig. 8c] analysis revealed that the resulting nanostructures are composed of boron and nitrogen with B/N atomic ratio of approximately 1. This demonstrates clearly that the pure boron nanostructures have transformed into the corresponding nanoscale BN components by the high-temperature nitriding processing. TEM studies of the product show that the BN nanostructure has the typical tubular structure with a hollow core [Fig. 9]. The nanotube consists of concentric walls with an interlayer spacing of ~0.33 nm, being approximately the hexagonal BN (0002) interplanar distance. Our extensive TEM observations demonstrated that more than 80% of the synthesized BN nano-objects have cylindrical hollow cores with a high crystalline structure (the others are the porous BN nanowires with segmented voids in their interiors). The chemical stoichiometry of the individual nanostructures was further characterized using EELS. A typical EELS spectrum [Fig. 10] exhibits two distinct absorption features beginning at ~188 eV and ~398 eV. These correspond to the characteristic K-shell ionization edges for B and N, respectively. The fine structure of the B and N absorption edges in the spectrum reveals the sp^2 bonding features between boron and nitrogen in the hexagonal BN (Chopra *et al.*, 1995). The quantification analysis of the two edges gives the N/B atomic ratio of about 0.95±0.12, which is consistent with the BN stoichiometry.

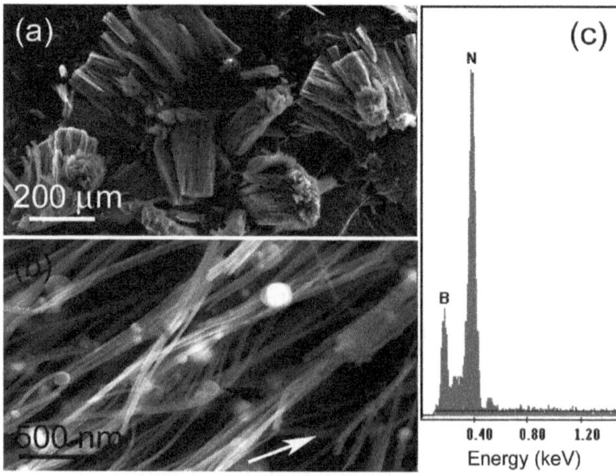

Fig. 8. SEM studies of the as-synthesized BN nanostructures. (a) Low-magnification SEM image of the products. (b) High-magnification SEM image showing the one-dimensional wire-like nanostructures. (C) SEM-EDX spectrum showing that the products are composed of boron and nitrogen.

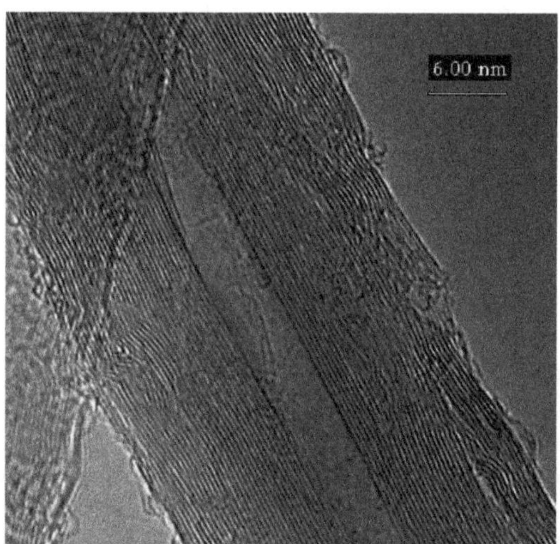

Fig. 9. HRTEM image of the synthesized BN nanotube.

Now we consider the formation mechanism of the hollow BN nanotubes from the solid boron nanowires. Boron is a reactive element at high temperature. High temperature reaction of boron and nitrogen to produce the hexagonal BN component can be expressed as: $2B(solid)+N_2(gas) \rightarrow 2BN(solid)$. In our experiments, the resultant BN nanostructures inherited the one-dimensional features from the starting boron nanowires. We suggest that the formation of the tubular BN nanostructures from the corresponding solid boron nanowires involves the following processes [Fig. 11]. In the heating stage, the nitrogen on the nanowire surface reacts with boron atoms to form hexagonal BN atomic layers [process (a) in Fig. 11]. The BN layers formed will act as a barrier to slower the diffusion of nitrogen into the nanowires. When the temperature is high enough (1500°C in our experiment), the boron atoms, because of its small atomic size, will diffuse outward simultaneously through the BN layers and react with the nitrogen on the outside surface [process (b) in Fig. 11]. Meanwhile, the remnant oxygen (residual in the chamber and desorbed from the chamber's wall) will also react with boron to form boron oxide, which will be evaporated into the vacuum. Since the temperature is high enough, the BN atomic layers formed can rearrange to form concentric cylinders, which makes the structure more energetically stable [process (c) in Fig. 11]. The above reaction, interdiffusion, migration, and rearrangement processes are responsible for the formation of the hollow BN nanotubes. Our model is supported by some recent experimental reports. For example, the Kirkendall effect (Smigelskas & Kirkendall, 1947), which is associated with the non-equilibrium interdiffusion of different components during a solid-state reaction, has been introduced to create a variety of hollow spherical nanocrystals and monocrystalline spinal nanotubes (Yin *et al.*, 2004; Wang et al., 2005; Tu & Gosele, 2005; Fan *et al.*, 2006). A vapor-phase sulphidation process was also reported to synthesize the CdS hemi-cylindrical polycrystalline wires from Cd nanowires (Li & Penner, 2005).

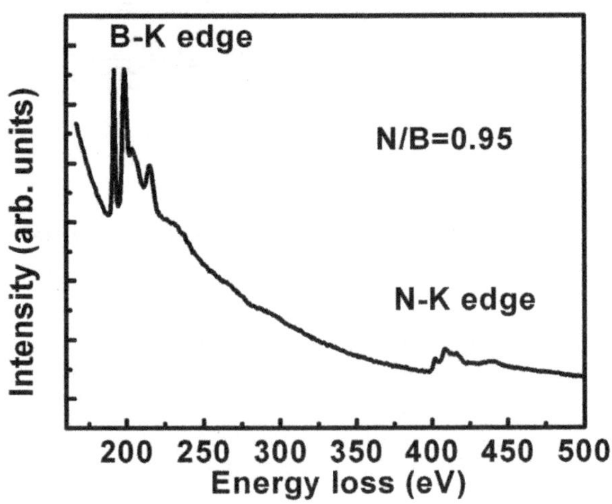

Fig. 10. EELS spectrum of a typical BN nanotube product. The energy-loss fine structures correspond to the sp^2 hybridization of h-BN.

One-dimensional BN nanostructures are of great interest for both fundamental and applied research because of their stable and excellent mechanical and physical properties. Our research has revealed that the high temperature nitriding of pure boron nanowires provides a controllable and simple process to fabricate BN nanotubes. Our results illustrate the technological potential of BN nanotubes synthesized using the simple nitriding process being incorporated into new generation nanoscale mechanical and electronic devices.

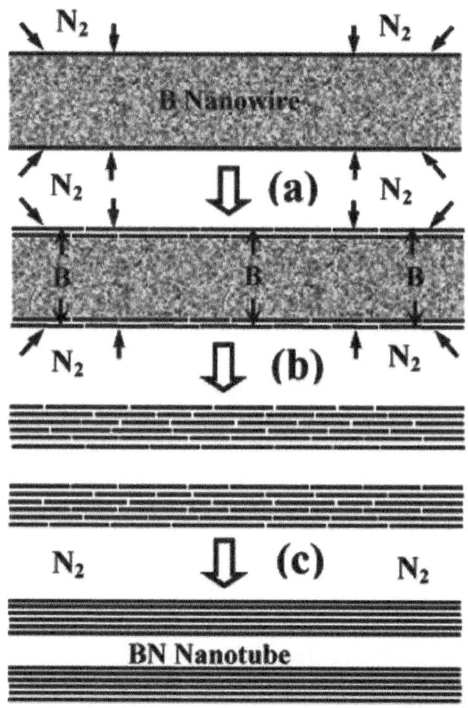

Fig. 11. Schematic diagram of the formation process of hollow BN nanotubes from solid boron nanowires.

4. Boron subnitride nanowires and heterostructured B_6N_x/BN nanocables

One of the substantial challenges in developing nanotechnology is that new nanostructures need to be created and explored to serve as the functional units in the miniaturized nanosystems where high performance and reliability are required. Another practical challenge is that simple processes for reducing cost but improving efficiency are necessitated to mass-fabricate functional nanostructures at commercial scale. Boron and boron-rich borides are unique in terms of their novel structures and properties (Greenwood, 1973; Donohue, 1974; Matkovich, 1977; Emin, 1987; Albert & Hillebrecht, 2009). They provide an ideal platform to explore the fundamental natures in cluster physics and structural chemistry, thanks to the unusual three-center electron-deficient bond and most varied polymorphism existing in boron family. They possess many unrivalled properties ranging from superconducting metals to wide-band semiconductors, and are widely used in numerous technological applications, peculiarly in those where a refractory, light, and hard material is required (Greenwood, 1973; Donohue, 1974; Matkovich, 1977; Emin, 1987; Albert & Hillebrecht, 2009; Erements *et al.*, 2001; Nagamatsu *et al.*, 2001; Garvie *et al.*, 1997; Hubert *et al.*, 1997, 1998; He *et al.*, 2002). For example, B_6O possesses a hardness comparable with that of cubic BN, the second hardest material (Hubert *et al.*, 1998; He *et al.*, 2002); B_6N shows a metal-like character in electron energy loss spectroscopy (EELS) studies (Garvie *et al.*, 1997; Hubert *et al.*, 1997). However, the syntheses of these useful boron-based materials normally require extreme high-pressure high-temperature conditions (Garvie *et al.*, 1997; Hubert *et al.*, 1997, 1998; He *et al.*, 2002). Strategies to obtain these materials in useful shape-designed forms at ambient pressure are of practical importance for their technological applications.

In this section we present the creation of boron subnitride and B_6N_x/BN radial heterostructured nanocables in bulk quantities using a simple process of nitriding the corresponding pure boron nanowires at 1200°C (Cao *et al.*, 2008). The synthesized nanostructures consist of a core B_6N_x nanowire with rhombohedral structure, a metastable high pressure phase, and a shell of hexagonal BN layers. This simple strategy might be used to rationally create other novel boron-based nanoheterostructures and nanocomposites in B-C-N-O system, and in principle, can provide nanoscale platforms to study high-pressure-induced phase transformation and reaction at ambient pressure.

The experimental procedure for the synthesis of B_6N_x nanostructures is similar to that of BN nanotubes discussed in Section 3, except that the nitriding temperature is 1200°C here. SEM and TEM studies of the as-prepared products after nitriding treatment show a smooth one-dimensional wire-like morphology as those of the starting pure boron nanowires. The abrupt interface in the diffraction contrast of the TEM image [Fig. 12a] suggests that the body and skin of the synthesized nanowires are of different phases. Selected area electron diffraction (SAED) technique was applied to identify the phase of nanostructures. SAED patterns shown in Fig. 12b and 12c could be well indexed with the lattice parameters of rhombohedral B_6N (Joint Committee on Powder Diffraction Standards (JCPDS) card: 50-1504; space group $R\bar{3}m$; lattice parameters: a=5.457 Å, c=12.241 Å), in agreement with the chemical analysis. Analysis and diffraction simulation showed that the SAED patterns (Fig. 12b,c) fit precisely to the electron diffractions from [100] and [122] zone axes of rhombohedral B_6N lattice, respectively.

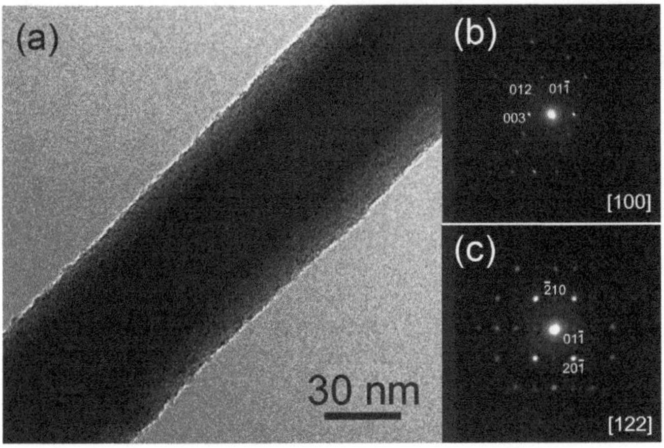

Fig. 12. TEM and SAED analyses of the B_6N_x/BN nanostructures. (a) TEM image showing the coaxial core-shell structure. (b,c) SAED patterns correspond to the electron diffractions from [100] and [122] zone axes of B_6N rhombohedral unit cell, respectively. The weak half-rings (corresponding to a separation of 0.33 nm) in the SAED patterns are from the electron diffraction of hexagonal BN (0002) atomic layers in the nanowire shells.

HRTEM studies highlight that the smooth nanowires consist of a single-crystalline core sheathed with a few graphitic walls with an interlayer spacing of ~0.33 nm [Fig. 13]. The lattice images from the crystalline nanowire cores match the structure of rhombohedral B_6N. The shell layers can be designated to hexagonal BN (0002) planes based on chemical analysis (see below) and TEM studies.

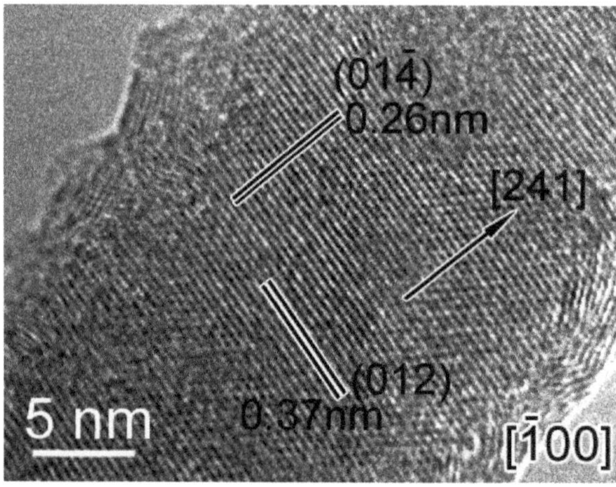

Fig. 13. HRTEM image of the B_6N_x/BN nanostructures. The longitudinal axis of the nanowire orientate roughly along the [241] direction.

We used EDX and EELS to characterize the chemical stoichiometry of the nanostructures. The EDS spectrum revealed that the nanostructures contain boron and nitrogen; the N:B atomic ratio was quantified to be about 1:4.8. Figure 14a shows a representative EELS spectrum taken from the central part of a typical nanowire. Because the shell layers are very thin (2-4 nm), the EELS signals are mainly from the core phase. EELS spectrum presents two sharp threshold peaks beginning at ~185 eV and ~396 eV, corresponding to the characteristic K-shell ionization edges for B and N, respectively. EELS quantification of the spectra gives the N/B atomic ratio of about 0.21±0.034, which is in good agreement with the EDX analysis. We also acquired EELS spectra from pure α-rhombohedral boron and hexagonal BN nanotube for a comparison [see Fig. 14b and 14c]. The fine structures of boron K-edges from the product nanowires show a similarity to those from bulk pure boron (B K-edges typical of icosahedral rhombohedral boron-rich phases), but a distinguishing difference from those of hexagonal BN. Our results are coincident with previous EELS studies of the B_6N and other icosahedral boron-rich phases (Garvie *et al.*, 1997).

We acquired the elemental mapping of B and N in the produced BN_x nanostructures. Figure 15a-c shows the bright field TEM image and the corresponding energy-filtered images of two typical BN_x nanostructures. The B elemental map, shown clearly in Fig. 15b, has uniformly bright contrast in the core throughout the nanowires. Careful observations reveal a reduced intensity at the peripheries of the nanowires in the B map. This suggests that the nanowire shells are boron-deficient compared to the core region where the distribution of B is homogeneous. In contrast, the N elemental map [Fig. 15c] shows a bright contrast sheath and a dark core with much lower brightness, indicating that the coating is nitrogen-rich.

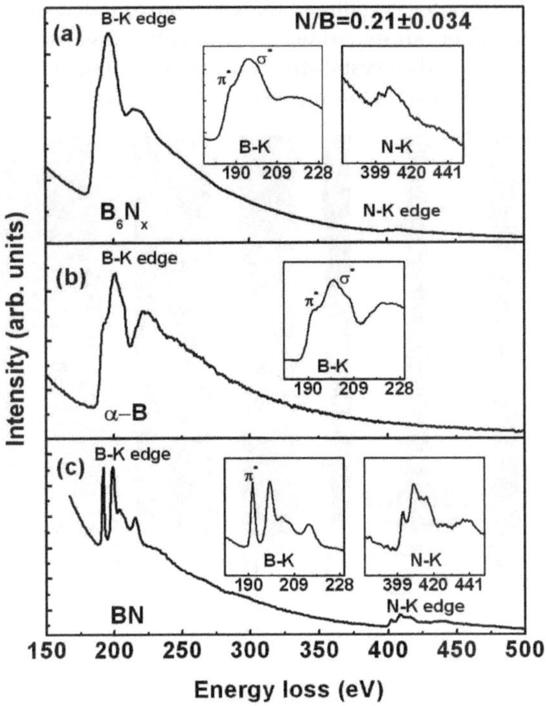

Fig. 14. EELS spectra of B_6N_x nanowire (a), bulk pure α-boron (b), and BN nanotube (c). The B K edge features from B_6N_x nanowire are similar to that from bulk pure boron, but are obviously different from that of BN. The insets show the magnified features of the B and N K-edges from B_6N_x nanostructure (a), pure boron (b), and BN (c).

Linescan analyses of the elemental distributions across the nanowire in the B and N maps show a hill-shaped B concentration profile and valley-like N profile [Fig. 15d,e], confirming that our BN_x nanostructures have a nitrogen-rich shell, and is in good agreement with the conclusion that the nanowire sheath is hexagonal BN. The chemical profiles of B and N, as shown in Figure 15d,e, also highlight two interesting features. First, two shoulder peaks (indicated by the arrows in Fig. 15d) exist in both sides of the B concentration profile. Second, the valley in the intensity profile of the N distribution has a protruding central section (indicated by the arrow in Fig. 15e). Considering that the total N/B atomic ratio in the products is about 1:5, and the BN shell has a higher N/B ratio of 1:1, we can conclude that the core of the synthesized nanostructure has a nominal stoichiometry of B_6N_x (x<1.2). To confirm this assignment, we performed theoretical simulation of the cross-sectional elemental mapping of B and N in the synthetic concentric cylinders of B_6N/BN and B/BN. Figure 15f,g shows the calculated elemental profiles in the B_6N/BN core-shell nanowire with a 40-nm-diameter core and 5-nm-thick shell. The fine features of the calculated B and N mapping cross-sections are in good agreement with the experimental data. In contrast, the theoretical cross-section N profile in the B/BN nanostructure (a 40-nm-diameter B core with a 5-nm-thick BN shell) shows a central dip in the N valley (Fig. 15i), which differs distinctly from the experimental results [Fig. 15e]. Taken together, these structure and composition data show that the produced nanostructures consist of single-crystalline B_6N_x cores surrounded by the hexagonal BN walls. After a careful comparison of our experimental data with those from the known B_xN_y phases (JCPDS Cards), we found that the crystal structure and stoichiometry of our B_6N_x nanowires fit well to that of bulk B_6N crystal, a metastable high pressure phase.

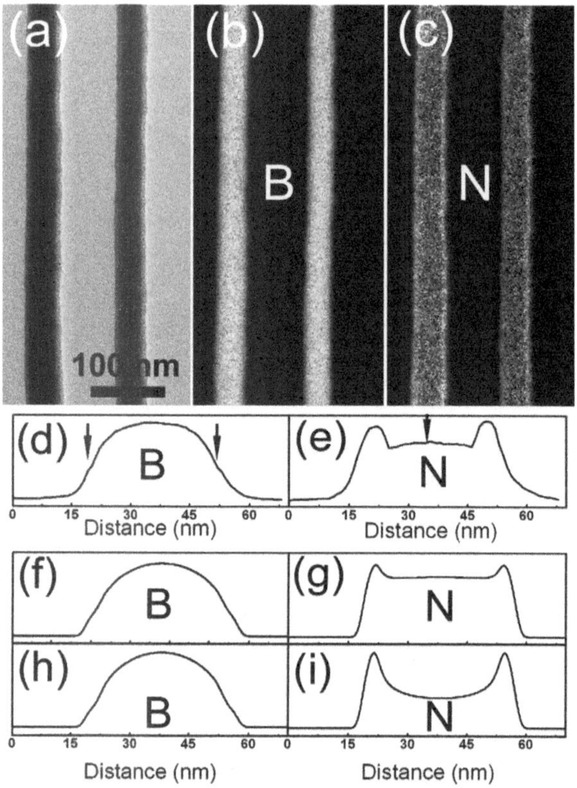

Fig. 15. Chemical analyses of typical B_6N_x/BN nanostructures. (a) Bright-field TEM image. (b) B elemental map. (c) N elemental map. (d,e) Elemental profiles for B and N, respectively, across a typical nanowire. (f,g) Theoretical simulations of the B and N profiles, respectively, on a B_6N/BN core-shell nanocable (40-nm-diameter B_6N core and 5-nm-thick BN shell). (h.i) Elemental mapping cross-sections of B and N, respectively, modelled from a concentric nanowire with B core (40nm diameter) and BN shell (5nm thickness).

Condon *et al.* first showed the existence of B_6N in the direct boron-nitrogen reaction (Condon *et al.*, 1976). By annealing boron powder in N_2, they obtained a new boron subnitride metastable phase with stoichiometry of B_6N and structure resembling to that of B_6O. The microcrystalline B_6N particles were present as the core particles surrounded by hexagonal BN coats. Hubert *et al.* demonstrated the first bulk synthesis of B_6N_{1-x} crystals with sizes up to 1 μm by reacting boron and hexagonal BN at a high pressure (7.5GPa) and high temperature (1700°C) (Hubert *et al.*, 1997). These observations give us a clue to consider the physical origin of the formation mechanism of B_6N_x crystalline core nanowires in our experiments. In the initial stage of the nitriding process, the nitrogen on the nanowire surface reacts with boron atoms to form fragments of BN atomic layers according to the reaction 2B(solid)+N_2(gas)→2BN(solid). As the reaction proceeding, the BN fragments on the outer surface wrap around the host nanowire to form a nano-capsule structure. At high temperature, the B and N atoms in the BN layers can rearrange and the BN fragments interconnect to form perfect concentric cylinders, which makes the structure more energetically stable. Thus the diameter of the BN capsule tends to shrink, and a pressure will build up inside the BN vessel. The BN sheath also separates the residual nanowire from the N_2 atmosphere, and prevents the inward diffusion of N atoms into the nanowire. The BN nano-capsule formed thus acts as a high pressure cell to promote the reaction $12B+N_2 \rightarrow B_6N$. Under these conditions, the residual core nanowires inside the BN shells reacts with N atoms to form B_6N_x nanostructures with the assistance of high-pressure build-up in the nanocells. As a result, the heterostructured B_6N_x/BN nanocables are formed. In this model, the isolated high-pressure BN cell and lower inward migration of nitrogen atoms are responsible for the high-pressure-induced formation of nano-crystalline B_6N_x phase. The temperature also takes a critical role in our experiments. At a higher nitriding temperature of 1500°C, all the pure boron precursors were transformed into BN nanotubes (Cao *et al.*, 2007). The nitriding temperature of 1200°C made for the formation of B_6N_x/BN core-shell nanostructure, suggesting that B_6N_x phase inside the BN nanocell is thermodynamically stable at this temperature. Banhart and Ajayan revealed that carbon onions can act as nanoscopic pressure cells for diamond nucleation and growth in the cores of carbon onions under high temperature and electron irradiation conditions (Banhart & Ajayan, 1996). Golberg and coworkers demonstrated the formation of solid nitrogen inside the BN nanocages, a structural analog of carbon onions, due to the existence of super-high-pressure inside them (Golberg *et al.*, 2002). They also observed the high-pressure-induced phase transformation of hexagonal BN to cubic BN inside the BN high-pressure nanocells. Multiwalled carbon nanotubes as high-pressure cylinders and nanoextruders were also reported to deform, extrude, and break hard nanomaterials that are encapsulated inside the core (Sun *et al.*, 2006). All of these experimental observations support our conclusion on the nanocell-high-pressure-assisted formation mechanism of B_6N_x/BN nanostructures in the nitriding reaction.

It is reasonable to expect that the B_6N nanostructures should share some of the very interesting physical and chemical properties with other boron-rich solids, e.g., low density, high hardness, chemical inertness, unique mechanical, thermal, and electronic properties, etc. The coaxial B_6N/BN nanocable heterostructures are of importance as the promising building blocks in developing nanoscale devices and nano-composites. Our research suggests that the BN nanocells in a similar process can be used as a template to study the pressure-induced phase transformation at nanoscale as well as to synthesize other unique boron-based nanostructures that cannot be formed under normal conditions. This simple synthetic strategy may be used as a general approach for the bulk-fabrication of a variety of intriguing boron-based nanocomposites and nanoheterostructures in B-C-N-O system, through rationally optimizing the synthetic conditions and utilizing the nanocell high-pressure effect.

5. Summary and outlook

In this Chapter, we describe mainly the investigations from our laboratory directed toward the synthesis of stoichiometric boron-based nanostructures. Our goal has been to create new boron and boride nanostructures with advanced properties and desired dimensionality. Through a detailed investigation on the structures and properties of new boron-based nanomaterials, we target at advancing our understanding of the unique boron chemistry and physics in the nanoscale regime. Another goal of our research has been to develop simple and cost-effective methods for mass-fabricating boron-based nanostructures and their highly ordered arrays in order to facilitate their practical applications in nanoscale devices and nano-composites.

We have synthesized large-area arrays of well-aligned boron nanowires using the simple approach of magnetron sputtering. Our research suggests that the magnetron sputtering approach, which is widely used in the preparation of films and multilayers in both industry and academia, can be employed as a general method for synthesis of some nanometer-sized materials with desirable dimensionalities and ordered assemblies via the rational design of the targets and the proper control of the experimental conditions. We have synthesized high-concentration nitrogen-doped carbon nanotube using magnetron sputtering of a mixture of nanometre-sized graphite and nickel powders in a nitrogen atmosphere (Cao *et al.*, 2003).

Using these boron nanostructures as the starting materials, we have created highly pure BN nanotubes in large quantities and B_6N_x/BN radial heterostructured nanocables in high yield by annealing pure boron nanowire precursors in N_2 atmosphere at 1500°C and 1200°C, respectively. Through exploiting the annealing variations in the simple nitriding process, we can expect to tune the properties of the BN nanostructures via the chemical modulation, and to extend this simple process for bulk fabrication of a wealth of boron-based nanoheterostructures and nanocomposites in B-C-N-O system. Our results show promising technological potential of fabricating new and hybrid boron-based nanostructures using simple processes for practical applications.

We believe there is a promising bright future in science and technology for boron and boride nanostructures. Theoretical studies suggest the existence of new boron-based materials showing at least the same variety of nanoscale forms and interesting properties as those known for carbon. Besides the boron nanotubes, theoretical AlB_2 nanotubes show a metallic density of states, independent of their chirality (Quandt *et al.*, 2001); nanotubular B_2O and BeB_2, are new semiconductors and metals in one dimension (Zhang & Crespi, 2002). Theoretical calculations show that these new classes of tubular structures promise to offer properties unavailable to the previously studied C, B, B-C-N, and B-N based systems. As for boron cage-like nanostructures, Szwacki and coworkers predicted using *ab initio* calculations that B_{80}, which is very similar in shape and symmetry to the carbon fullerene C_{60}, is an energetically stable boron fullerene (Szwacki *et al.*, 2007). However, the synthesis of these theoretical novel boron-based nanostructures remains a challenge. Though Ciuparu *et al.* claimed that they have synthesized pure boron single-wall nanotubes (Ciuparu *et al.*, 2004), no any subsequent work has come out following their result since it was published. There remain many fundamental and practical issues about the structures and properties of boron-based nanostructures that deserve future explorations. We expect that the investigations on boron-based nanostructures will provide new insights into boron chemistry and physics, and boron nanostructures will serve as important building blocks for nanotechnology.

6. Acknowledgment

We thank our coworkers who have made contributions to the work described in this chapter, including Ze Zhang, He Tian, Meng He, Chunxiao Gao, Yiqian Wang, Yanchun Li, Jing Liu, Jun Zhang, K. Hahn, and C. Scheu. We would like to acknowledge the Natural Science Foundation of China for financial support.

7. References

Ajayan, P.M. & Ebbesen, T.W. (1997). Nanometre-size Tubes of Carbon. *Reports on Progress in Physics*, Vol. 60, No. 10, (October 1997), pp. 1025-1062, ISSN 0034-4885

Albert, B. & Hillebrecht, H. (2009). Boron: Elementary Challenge for Experimenters and Theoreticians. *Angew. Chem. Int. Ed.*, Vol. 48, No. 46, (November 2009), pp8640–8668, ISSN: 1433-7851

Banhart, F. & Ajayan, P.M. (1996). Carbon Onions as Nanoscopic Pressure Cells for Diamond Formation. *Nature*, Vol. 382, No. 6590, (August 1996), pp. 433-435, ISSN 0028-0836

Blase, X.; Rubio, A.; Louie, S.G. & Cohen, M.L. (1994). Stability and Band Gap Constancy of Boron-Nitride Nanotubes. *Europhysics Letters*, Vol. 28, No. 5, (November 1994), pp. 335-340, ISSN 0295-5075

Boustani, I.; Quandt, A.; Hernandez, E. & Rubio, A. (1999). New Boron Based Nanostructured Materials. *Journal of Chemical Physics*, Vol. 110, No. 6, (February 1999), pp. 3176-3185, ISSN 0021-9606

Boustani, I.; Quandt, A. & Rubio A. (2000). Boron Quasicrystals and Boron Nanotubes: *Ab Initio* Study of Various B_{96} Isomers. *Journal of Solid State Chemistry*, Vol. 154, No. 1, (October 2000), pp. 269-274, ISSN 0022-4596

Cao, L.M.; Zhang, Z.; Sun, L.L.; Gao, C.X.; He, M.; Wang, Y.Q.; Li, Y.C.; Zhang, X.Y.; Li, G.; Zhang, J. & Wang, W.K. (2001). Well-Aligned Boron Nanowire Arrays. *Advanced Materials*, Vol. 13, No. 22, (November 2001), pp. 1701-1704, ISSN 0935-9648

Cao, L.M.; Hahn, K.; Scheu, C.; Ruhle, M.; Wang, Y.Q.; Zhang, Z.; Gao, C.X.; Li, Y.C.; Zhang, X.Y.; He, M.; Sun, L.L. & Wang, W.K. (2002). Template-Catalyst-Free Growth of Highly Ordered Boron Nanowire Arrays. *Applied Physics Letters*, Vol. 80, No. 22, (June 2002), pp. 4226-4228, ISSN 0003-6951

Cao, L.M.; Liu, J.; Gao, C.X.; Li, Y.C.; Li, X.D.; Wang, Y.Q. Wang, Y.Q.; Zhang, Z.; Cui, Q.L.; Zou, G.T.; Sun, L.L. & Wang, W.K. (2002). Synthesis of Well-Aligned Boron Nanowires and Their Structural Stability under High Pressure. *Journal of Physics-Condensed Matter*, Vol. 14, No. 44, (November 2002), pp. 11017-11021, ISSN 0953-8984

Cao, L.M.; Zhang, X.Y.; Gao, C.X.; Wang, W.K.; Zhang, Z.L. & Zhang, Z. (2003). High-Concentration Nitrogen-Doped Carbon Nanotube Arrays. *Nanotechnology*, Vol. 14, No. 8, (August 2003), pp. 931-934, ISSN 0957-4484

Cao, L.M.; Zhang, X.Y.; Tian, H.; Zhang, Z. & Wang, W.K. (2007). Boron Nitride Nanotube Branched Nanojunctions. *Nanotechnology*, Vol. 18, No. 15, (April 2007), pp. 155605(1-4), ISSN: 0957-4484

Cao, L.M.; Tian, H.; Zhang, Z.; Feng, M.; Zhan, Z.J.; Wang, W.K. & Zhang, X.Y. (2008). Heterostructured B_6N_x/BN Nanocable and Nanofeather Nanojunctions. *Crystal Growth & Design*, Vol. 8, No. 12, (December 2008), pp: 4350-4354, ISSN 1528-7483

Chang, C.W.; Han, W.Q. & Zettl, A. (2005). Thermal Conductivity of B-C-N and BN Nanotubes. *Applied Physics Letters*, Vol. 86, No. 17, (April 2005), pp. 173102(1-3), ISSN 0003-6951

Chopra, N.G.; Luyken, R.J.; Cherrey, K.; Crespi, V.H.; Cohen, M.L.; Louie, S.G. &Zettl, A. (1995). Boron-Nitride Nnanotubes. *Science*, Vol. 269, No. 5226, (August 1995), pp. 966-967, ISSN 0036-8075

Chopra, N.G. & Zettl, A. (1998). Measurement of the Elastic Modulus of a Multi-Wall Boron Nitride Nanotube. *Solid State Communications*, Vol. 105, No. 5, (February 1998), pp. 297-300, ISSN 0038-1098

Ciuparu, D.; Klie, R.F.; Zhu, Y.M. & Pfefferle, L. (2004). Synthesis of Pure Boron Single-Wall Nanotubes. *Journal of Physical Chemistry B*, Vol. 108, No. 13, (April 2004), pp. 3967-3969, ISSN 1520-6106

Condon, J.B.; Holcombe, C.E.; Johnson, D.H. & Steckel, L.M. (1976). The Kinetics of the Boron Plus Nitrogen Reaction. *Inorganic Chemistry*, Vol. 15, No. 9, (September 1976), pp. 2173-2179, ISSN 0020-1669

Donohue, J. (1974). *The Structures of the Elements*, pp. 48-83, Wiley, ISBN 0471217883, New York

Dresselhaus, M.S.; Dresselhaus, G. & Ecklund, P.C. (1996). *Science of Fullerenes and Carbon Nanotubes: Their Properties and Applications*, Academic, ISBN 0122218205, New York

Emin, D. (1987). Icosahedral Boron-Rich Solids. *Physics Today*, Vol. 40, No. 1, (January 1987), pp. 55-62, ISSN 0031-9228

Eremets, M.I.; Struzhkin, V.V.; Mao, H.K. & Hemley, R.J. (2001). Superconductivity in Boron. *Science*, Vol. 293, No.5528, (July 2001), pp. 272-274, ISSN 0036-8075

Fan, H.J.; Knez, M.; Scholz, R.; Nielsch, K.; Pippel, E.; Hesse, D.; Zacharias, M. & Gosele, U. (2006). Monocrystalline Spinel Nanotube Fabrication Based on the Kirkendall Effect. *Nature Materials*, Vol. 5, No. 8, (August 2006), pp. 627-631, ISSN 1476-1122

Frank, F.C. (1949). The Infulence of Dislocations on Crystal Growth. *Discussions of the Faraday Society*, No. 5, (1949), pp. 48-54. IDS Number: UL198

Fujimori, M.; Nakata, T.; Nakayama, T.; Nishibori, E.; Kimura, K.; Takata, M. & Sakata, M. (1999). Peculiar Covalent Bonds in a-Rhombohedral Boron. *Physical Review Letters*, Vol. 82, No. 22, (May 1999), pp. 4452-4455, ISSN 0031-9007

Garvie, L.A.J.; Hubert, H.; Petuskey, W.T.; McMillan, P.F. & Buseck, P.R. (1997), High-Pressure, High-Temperature Syntheses in the B-C-N-O System - II. Electron Energy-Loss Spectroscopy (EELS). *Journal of Solid State Chemistry*, Vol. 133, No. 2, (November 1997), pp. 365-371, ISSN 0022-4596

Gindulyte, A.; Lipscomb, W.N. & Massa, L. (1998). Proposed Boron Nanotubes. *Inorganic Chemistry*, Vol. 37, No. 25, (December 1998), pp. 6544-6545, ISSN 0020-1669

Gindulyte, A.; Krishnamachari, N.; Lipscomb, W.N. & Massa, L. (1998). Quantum Chemical Calculations of Proposed Multicage Boron Fullerenes. *Inorganic Chemistry*, Vol. 37, No. 25, (December 1998), pp. 6546-6548, ISSN 0020-1669

Golberg, D.; Bando, Y.; Sato, T.; Grobert, N.; Reyes-Reyes, M.; Terrones, H. & Terrones, M. (2002). Nanocages of Layered BN: Super-High-Pressure Nanocells for Formation of Solid Nitrogen. *Journal of Chemical Physics*, Vol. 116, No. 19, (May 2002), pp. 8523-8532, ISSN 0021-9606

Greenwood, N. N. (1973). In: *Comprehensive Inorganic Chemistry*, J. C. Bailar Jr., H. J. Emeléus, Sir R. Nyholm, A. F. Trotman-Dickenson, (Eds.), Pergamon, Vol. 1, pp. 665-991, ISBN 0080169899, Oxford

Hamada, N.; Sawada, S. & Oshiyama, A. (1992). New One-Dimensional Conductors - Graphitic Microtubules. *Physical Review Letters*, Vol. 68, No. 10, (March 1992), pp. 1579-1581, ISSN 0031-9007

He, D.W.; Zhao, Y.S.; Daemen, L.; Qian, J.; Shen, T.D. & Zerda, T.W. (2002). Boron Suboxide: As Hard as Cubic Boron Nitride. *Applied Physics Letters*, Vol. 81, No. 4, (July 2002), pp. 643-645, ISSN 0003-6951

Hernandez, E.; Goze, C.; Bernier, P. & Rubio, A. (1998). Elastic Properties of C and BxCyNz Composite Nanotubes. *Physical Review Letters*, Vol. 80, No. 20, (May 1998), pp. 4502-4505, ISSN 0031-9007

Hu, J.T.; Odom, T.W. & Lieber, C.M. (1999). Chemistry and Physics in One Dimension: Synthesis and Properties of Nanowires and Nanotubes. *Accounts of Chemical Research*, Vol. 32, No. 5, (May 1999), pp. 435-445, ISSN 0001-4842

Hubert, H.; Garvie, L.A.J.; Buseck, P.R.; Petuskey, W.T. & McMillan, P.F. (1997). High-Pressure, High-Temperature Syntheses in the B-C-N-O System - I. Preparation and Characterization. *Journal of Solid State Chemistry*, Vol. 133, No. 2, (November 1997), pp. 356-364, ISSN 0022-4596

Hubert, H.; Devouard, B.; Garvie, L.A.J.; O'Keeffe, M.; Buseck, P.R.; Petuskey, W.T. & McMillan, P.F. (1998). Icosahedral Packing of B_{12} Icosahedra in Boron Suboxide (B_6O). *Nature*, Vol. 391, No. 6665, (January 1998), pp. 376-378, ISSN 0028-0836

Iijima, S. (1991). Helical Microtubules of Graphitic Carbon. *Nature*, Vol. 354, No. 6348, (November 1991), pp. 56-58, ISSN: 0028-0836

Kirihara, K.; Hyodo, H.; Fujihisa, H.; Wang, Z.; Kawaguchi, K.; Shimizu, Y.; Sasaki, T.; Koshizaki, N.; Soga, K. & Kimura, K. (2006). Mg-Doping Experiment and Electrical Transport Measurement of Noron Nanobelts. *Journal of Solid State Chemistry*, Vol. 179, No. 9, (September 2006), pp. 2799-2804, ISSN: 0022-4596

Kroto, H.W.; Heath, J.R.; O'Brien, S.C.; Curl R.F. & Smalley, R. E. (1985). C_{60}: Buckminsterfullerene. *Nature*, Vol. 318, No. 6042, (November 1985), pp. 162-163, ISSN: 0028-0836

Kuchibhatla, S.V.N.T.; Karakoti, A.S.; Bera, D. & Seal, S. (2007). One Dimensional Nanostructured Materials. *Progress in Materials Science*, Vol. 52, No. 5, (July 2007), pp. 699-913, ISSN 0079-6425

Li, Q.G. & Penner, R.M. (2005). Photoconductive Cadmium Sulfide Hemicylindrical Shell Nanowire Ensembles. *Nano Letters*, Vol. 5, No. 9, (September 2005), pp. 1720-1725, ISSN 1530-6984

Liu, F.; Tian, J.F.; Bao, L.H.; Yang, T.Z.; Shen, C.M.; Lai, X.Y.; Xiao, Z.M.; Xie, W.G.; Deng, S.Z.; Chen, J.; She, J.C.; Xu, N.S. & Gao, H.J. (2008). Fabrication of Vertically Aligned Single-Crystalline Boron Nanowire Arrays and Investigation of Their Field-Emission Behavior. *Advanced Materials*, Vol. 20, No. 13, (July 2008), pp. 2609-2615, ISSN: 0935-9648

Matkovich, V.I. (Ed.). (1977). *Boron and Refractory Borides*, Springer, ISBN 038708181X, New York

Meng, X.M.; Hu, J.Q.; Jiang, Y.; Lee, C.S. & Lee, S.T. (2003). Boron Nanowires Synthesized by Laser Ablation at High Temperature. *Chemical Physics Letters*, Vol. 370, No. 5-6, (May 2003), pp. 825-828, ISSN: 0009-2614

Morales, A.M. & Lieber, C.M. (1998). A laser ablation method for the synthesis of crystalline semiconductor nanowires. *Science*, Vol. 279, No. 5348, (January 1998), pp. 208-211, ISSN 0036-8075

Nagamatsu, J.; Nakagawa, N.; Muranaka, T.; Zenitani, Y. & Akimitsu, J. (2001). Superconductivity at 39 K in Magnesium Diboride. *Nature*, Vol. 410, No. 6824, (March 2001), pp. 63-64, ISSN 0028-0836

Nelmes, R.J.; Loveday, J.S.; Wilson, R.M.; Marshall, W.G.; Besson, J.M.; Klotz, S.; Hamel, G.; Aselage, T.L. & Hull, S. (1995). Observation of Inverted-Molecular Compression in Boron Carbide. *Physical Review Letters*, Vol. 74, No. 12, (March 1995), pp. 2268-2271, ISSN 0031-9007

Otten, C.J.; Lourie, O.R.; Yu, M.F.; Cowley, J.M.; Dyer, M.J.; Ruoff, R.S. & Buhro, W.E. (2002). Crystalline Boron Nanowires. *Journal of the American Chemical Society*, Vol. 124, No. 17, (May 2002), pp. 4564-4565, ISSN 0002-7863

Quandt, A. & Boustani, I. (2005). Boron Nanotubes. *CHEMPHYSCHEM*, Vol. 6, No. 10, (October 2005), pp. 2001-2008, ISSN 1439-4235

Quandt, A.; Liu, A.Y. & Boustani, I. (2001). Density-Functional Calculations for Prototype Metal-Boron Nanotubes. *Physical Review B*, Vol. 64, No. 12, (September 2001), pp. 125422 (1-5), ISSN 1098-0121

Saito, R.; Fujita, M.; Dresselhaus, G.; Dresselhaus, M.S. (1992). Electronic Structure of Graphene Tubules Based C_{60}. *Physical Review B*. Vol. 46, No. 3, (July 1992), pp. 1804-1811, ISSN 0163-1829

Smigelskas, A.D. & Kirkendall, E.O. (1947). Zinc Diffusion in Alpha-Brass. *Transactions of the American Institute of Mining and Metallurgical Engineers*, Vol. 171, (1947), pp. 130-142, IDS No. XR336

Sun, L.; Banhart, F.; Krasheninnikov, A.V.; Rodriguez-Manzo, J.A.; Terrones, M. & Ajayan, P.M. (2006). Carbon Nanotubes as High-Pressure Cylinders and Nanoextruders. *Science*, Vol. 312, No. 5777, (May 2006), pp. 1199-1202, ISSN 0036-8075

Suryavanshi, A.P.; Yu, M.F.; Wen, J.G.; Tang, C.C. & Bando, Y. (2004). Elastic Modulus and Resonance Behavior of Boron Nitride Nanotubes. *Applied Physics Letters*, Vol. 84, No. 14, (April 2004), pp. 2527-2529, ISSN 0003-6951

Szwacki, N.G.; Sadrzadeh, A. & Yakobson, B.I. (2007). B_{80} Fullerene: An ab initio Prediction of Geometry, Stability, and Electronic Structure. *Physical Review Letters*, Vol. 98, No. 16, (April 2007), pp. 166804(1-4), ISSN 0031-9007

Tu, K.N. & Gosele, U. (2005). Hollow Nanostructures Based on the Kirkendall Effect: Design and Stability Considerations. *Applied Physics Letters*, Vol. 86, No. 9, (February 2005), pp. 093111(1-3), ISSN 0003-6951

Vast, N.; Baroni, S.; Zerah, G.; Besson, J.M.; Polian, A.; Grimsditch, M. & Chervin, J.C. (1997). Lattice Dynamics of Icosahedral Alpha-Boron Under Pressure. *Physical Review Letters*, Vol. 78, No. 4, (January 1997), pp. 693-696, ISSN 0031-9007

Wang, Y.L.; Cai, L. & Xia, Y.N. (2005). Monodisperse Spherical Colloids of Pb and Their Use as Chemical Templates to Produce Hollow Particles. *Advanced Materials*, Vol. 17, No. 4, (February 2005), pp. 473-477, ISSN 0935-9648

Wang, Z.K.; Shimizu, Y.; Sasaki, T.; Kawaguchi, K.; Kimura, K. & Koshizaki, N. (2003). Catalyst-Free Fabrication of Single Crystalline Boron Nanobelts by Laser Ablation. *Chemical Physics Letters*, Vol. 368, No. 5-6, (January 2003), pp. 663-667, ISSN: 0009-2614

Wagner, R.S. & Ellis, W. C. (1964). Vapor-Liquid-Solid Mechanism of Single Crystal Growth. *Applied Physics Letters*, Vol. 4, No. 5, (March 1964), pp. 89-90, ISSN 0003-6951

Wu, Y.Y.; Messer, B.& Yang, P.D. (2001). Superconducting MgB_2 Nanowires. *Advanced Materials*, Vol. 13 No. 19, (October 2001), pp. 1487-1489, ISSN: 0935-9648

Xiao, Y.; Yan, X.H.; Cao, J.X.; Ding, J.W.; Mao, Y.L. & Xiang, J. (2004). Specific Heat and Quantized Thermal Conductance of Single-Walled Boron Nitride Nanotubes. *Physical Review B*, Vol. 69, No. 20, (May 2004), pp. 205415(1-5), ISSN 1098-0121

Xu, T.T.; Zheng, J.G.; Wu, N.; Nicholls, A.W.; Roth, J.R.; Dikin, D.A. & Ruoff, R.S. (2004). Crystalline Boron Nanoribbons: Synthesis and Characterization. *Nano Letters*, Vol. 4, No. 5, (April 2004), pp. 963-968, ISSN 1530-6984

Yin, Y.D.; Rioux, R.M.; Erdonmez, C.K.; Hughes, S.; Somorjai, G.A. & Alivisatos, A.P. (2004). Formation of Hollow Nanocrystals through the Nanoscale Kirkendall Effect. *Science*, Vol. 304, No. 5671, (April 2004), pp. 711-714, SSN 0036-8075

Yun, S.H.; Wu, J.Z.; Dibos, A.; Gao, X. & Karlsson, U.O. (2005). Growth of Inclined Boron nanowire Bundle Arrays in an Oxide-Assisted Vapor-Liquid-Solid Process. *Applied Physics Letters*, Vo. 87, No. 11, (september 2005), pp. 113109 (1-3), ISSN 0003-6951

Zhang, P.H. & Crespi, V.H. (2002). Theory of B_2O and BeB_2 Nanotubes: New Semiconductors and Metals in One Dimension. *Physical Review Letters*, Vol. 89, No. 5, (July 2002), pp. 056403(1-4), ISSN 0031-9007

Zhang, Y.J.; Ago, H.; Yumura, M.; Komatsu, T.; Ohshima, S.; Uchida, K. & Iijima, S. (2002). Synthesis of Crystalline Boron Nanowires by Laser Ablation. *Chemical Communications*, No. 23, (October 2002), pp. 2806-2807, ISSN 1359-7345

2

Clay Mineral Nanotubes: Stability, Structure and Properties

Hélio A. Duarte[1], Maicon P. Lourenço[1], Thomas Heine[2] and Luciana Guimarães[3]
[1]*Department of Chemistry, ICEx, Universidade Federal de Minas Gerais, Belo Horizonte, MG,*
[2]*School of Engineering and Science, Jacobs University Bremen, Bremen,*
[3]*Department of Natural Science, Universidade Federal de São João Del Rei, São João Del Rei, MG*
[1,3]*Brazil,*
[2]*Germany*

1. Introduction

Recent developments in nanoscience and nanotechnology opened fundamental and applied new frontiers in science and materials engineering. Advanced materials are being developed with enhanced chemical and physical properties with unique characteristics. The properties of these materials are determined not only by their composition and chemical bonds, but also by size and morphology.

The emerging field of nanotechnology is mostly focused on carbon and inorganic based nanomaterials, such as carbon nanotubes, graphene, transition metal nanotubes and nanowires (Iijima, 1991; Tenne *et al.*, 1992; Endo *et al.*, 1996; Dresselhaus *et al.*, 2001). Systems containing aluminosilicates have been investigated as mesoporous materials in the form of zeolite and alumina. Although they have not yet received as much attention, clay minerals can also form nanostructured layered materials and nanotubes with remarkable geometric properties. Imogolite is the most representative species of this case, since it has been studied in a pre-nano (1970) decade (Cradwick *et al.*, 1972) and has been nearly forgotten until recently. Since 2000 (Bursill *et al.*, 2000; Tamura & Kawamura, 2002; Mukherjee *et al.*, 2005; Nakagaki & Wypych, 2007), these structures gained again prominence in the literature and appear as an emerging field of research. They can be used as nanoreactors for selective catalysts, adsorbent, nanocable, support for the immobilization of metalloporphyrins, encapsulation and ionic conductor (Nakagaki & Wypych, 2007; Kuc & Heine, 2009).

Although the nanotube (NT) term is recent, the idea of a small tubular structure is not new. In 1930, Linus Pauling (1930) proposed the existence of cylindrical structures formed by minerals in nature. Based on asbestos related minerals, Pauling proposed that if two faces of a mineral are not symmetrical, there will be a structural mismatch between the layers leading to its deformation and curvature. Chrysotile, halloysite and imogolite are examples of such structures. Unfortunately, Pauling concluded that layered materials with symmetric structure, such as WS_2 and MoS_2, are not likely to form closed cylindrical structures. It took, however, until 1992 when Tenne, Remskar and others showed that tubular structures are possible from these materials regardless of the missing symmetry (Tenne *et al.*, 1992; Remskar, 2004; Tenne, 2006).

Imogolite, Halloysite, and Chrysotile are examples of naturally occurring nanostructured clay minerals. Imogolite occurs naturally in soils of volcanic origin and is composed of single-walled NTs. The tube walls consist of a curved gibbsite-like sheet ($Al(OH)_3$), where the inner hydroxyl surface of the gibbsite is substituted by $(SiO_3)OH$ groups. This structure possesses a composition of $(HO)_3Al_2O_3SiOH$, which is the sequence of atoms encountered on passing from the outer to the inner surface of the tube (Guimaraes *et al.*, 2007). Halloysite is a clay mineral with stoichiometry $Al_2Si_2O_5(OH)_4 .nH_2O$ that can grow into long tubules and is chemically similar to kaolinite (Giese & Datta, 1973; White *et al.*, 2009). It consists of a gibbsite octahedral sheet ($Al(OH)_3$), which is modified by siloxane groups at the outer surface (Guimaraes *et al.*, 2010). The chrysotile structure is composed of brucite ($Mg(OH)_2$) and tridymite (silicon dioxide, SiO_2) layers. The brucite octahedral sheet forms the outer side of the tube and SiO_4 groups are anchored to the inner side of the tube (Piperno *et al.*, 2007).

The structures of imogolite (Cradwick *et al.*, 1972), halloysite (Bates *et al.*, 1950a) and chrysotile (Bates *et al.*, 1950b) have been identified between the 1950th and 1970ths through spectroscopic methods. However, recently, those clays again became the focus of research and patents (Price & Gaber; Redlinger & Corkery, 2007) due to the great interest in the nanometric structures. Nanostructures (nanotubes and nanospirals) of clay minerals are very versatile systems, and are target materials for applications in catalysis (Imamura *et al.*, 1996), molecular sieves and adsorbents (Ackerman *et al.*, 1993), inorganic support for catalysts (Nakagaki & Wypych, 2007), controlled drug release (Veerabadran *et al.*, 2007), formation of composites, controlled release devices of herbicides, fungicides and insecticides (Lvov *et al.*, 2008) and anti-corrosion agents.

The increasing interest of clay mineral based NTs requires better understanding of their structures and properties. However, in most cases, samples of natural and synthetic compounds present only low crystallinity, leading to low-resolution structural data from X-ray diffraction measurements. Thus, a complementary approach involving spectroscopic methods and computational simulation can help in the interpretation of results and obtained structural data.

In the present chapter, the stability and properties of the nanostructured aluminosilicates will be reviewed and discussed with the focus on the computer modeling of such systems. The first theoretical investigations on the aluminosilicate NTs were mostly based on force fields specially developed for these systems (Tamura & Kawamura, 2002). The size of the unit cell is normally a limitation for using quantum mechanical calculations. Notwithstanding, quantum mechanical methods are being applied to such systems. Density functional theory (DFT), presently the most popular method to perform quantum-mechanical calculations, is the state-of-the-art method to study clay mineral nanotubes with high predictive power. First applications used the approximation to DFT implemented to the SIESTA (Artacho *et al.*, 1999; Soler *et al.*, 2002) code, which uses pseudo potentials and localized numerical atomic-orbital basis sets and it is well parallelized for multicore machines. Recently, the helical symmetry has been implemented in the CRYSTAL (Dovesi *et al.*, 2009) program, reducing significantly the computational costs for treating high-symmetry nanotubes (those at the equilibrium position in case no Peierls distortions are present), and hence making full-electron calculations of these systems feasible. However, if one investigates chemical modification in the NT structure, the use of helical symmetry becomes limited. In the last few years we have used an approximate Density Functional method called Density Functional based Tight Binding with self Consistent Charge corrections (SCC-DFTB) (Elstner *et al.*, 1998) method, as implemented in the deMon-nano (Heine *et al.*, 2009) and DFTB+ programs (Aradi *et al.*, 2007). The SCC-DFTB method, for a recent review see (Oliveira *et al.*, 2009), can lead to results which are nearly equivalent to DFT calculations although some orders of magnitude faster. The SCC-DFTB method uses a non-orthogonal tight-binding approach where all parameters are consistently computed using DFT, together with a minimal valence basis set. This method has been successfully applied to inorganic and carbon NTs (Enyashin & Seifert, 2005; Ivanovskaya *et al.*, 2006; Stefanov *et al.*, 2008; Enyashin *et al.*, 2009; Kuc & Heine, 2009; Rasche *et al.*, 2010). In our laboratory, we have applied successfully the SCC-DFTB method to investigate the stability, electronic and mechanical properties of the nanostructured aluminosilicates (Guimaraes *et al.*, 2007; Kuc & Heine, 2009; Guimaraes *et al.*, 2010).

2. Imogolite-like nanotubes – Gibbsite as a template for new materials

The careful analysis of the imogolite structure is particularly elucidative and can help to envisage strategies to design new materials. It is normally described as a NT where the external part consists of a curved gibbsite-like sheet ($Al(OH)_3$ and in the inner hydroxyls are replaced by $SiO_3(OH)$ groups.

The gibbsite structure (figure 1a) is a layered material with the $Al(OH)_3$ stoichiometry. Normally it crystallizes in hexagonal or prismatic structures with monoclinic symmetry. Each sheet of gibbsite is composed by hexacoordinated aluminum atoms arranged between two layers of hydroxyls. Each hydroxyl bounds to two aluminum centers, resulting in electrically neutral sheets. The layers are kept together through hydrogen bonds.

The hypothetical gibbsite monolayer (Frenzel *et al.*, 2005) and the respective gibbsite NT (Enyashin & Ivanovskii, 2008) have been investigated using DFT and SCC-DFTB calculations. The strain energy, that is, the relative energy with respect to the planar monolayer, depicted in figure 2, does not show a minimum. It presents the same behavior as other inorganic and carbon NTs (Enyashin *et al.*, 2007). However, the hypothetical gibbsite NT is unlikely to be synthesized using conventional synthesis approaches in aqueous solution through hydrolysis, as this is leading to the thermodynamic most stable lamellar structure. It is important to point out that other inorganic and carbon NTs are synthesized in very specific and well controlled experiments and the NTs are the kinetic product of the synthesis. It is well known that graphene is equivalent to a nanotube with infinite diameter and represents the more stable conformation with respect to the carbon NTs.

Figure 1b shows clearly how the fragment SiO_4^{4-} binds to the gibbsite surface to form imogolite. The mismatch of the bond lengths lead to the curvature of the gibbsite layer and to the formation of the imogolite NT. There is an optimal curvature which leads to the minimum strain in the structure. This explains why imogolite is monodisperse with very well-defined geometrical parameters and symmetry.

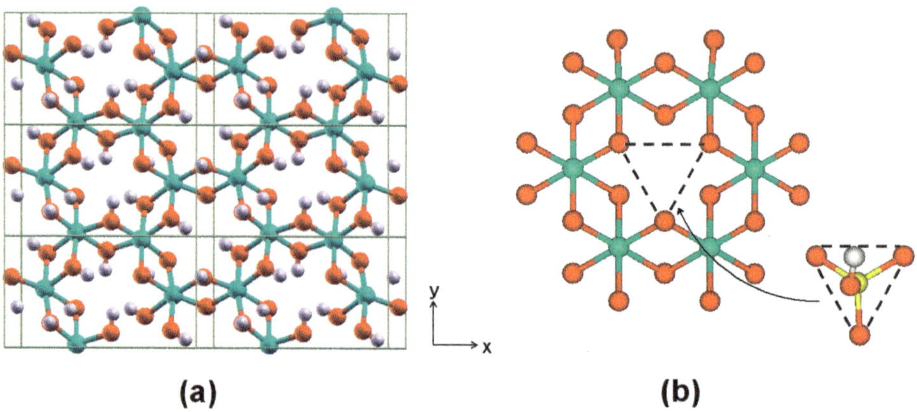

Fig. 1. a) Periodic gibbsite layer model. b) Hexagonal gibbsite ring where silanol is bound.

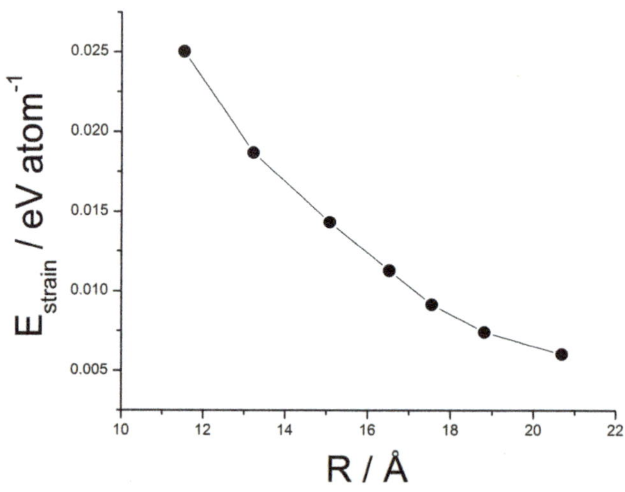

Fig. 2. Calculated strain energies E_{str} as a function of the radius R for zigzag hypothetical gibbsite NTs.

The roll-up process can lead to different symmetries depending on the rolling direction **B** in the 2D lattice (figure 3-a), where $\mathbf{B}=n\mathbf{a}_1+m\mathbf{a}_2$ ($\mathbf{a}_1$, $\mathbf{a}_2$ are lattice vectors of the hexagonal lattice). In principle, three classes of NTs can be constructed: *armchair* (*n,n*), *zigzag* (*n,0*) and *"chiral"* (*n,m*), with $n \neq m$. However, only *zigzag* tubes (figure 3-b) have been experimentally observed.

The synthesis of imogolite occurs in mild conditions and in aqueous solution. However, its mechanism of formation is rather complex and involves self assembly. The Al^{3+} ions in solution rapidly hydrolyze forming polynuclear species (Bi *et al.*, 2004). It has been pointed out that the thermodynamic equilibrium is not achieved rapidly and the kinetics is very slow (Casey, 2006). The silicates in solution are a very complicated system forming many polynuclear intermediates (Exley *et al.*, 2002; Schneider *et al.*, 2004). The imogolite formation mechanism may occur through self assembly, where silicate and aluminate species are combined to form proto-imogolite. It is important to highlight that this process is very sensitive to pH, ionic strength and concentration. The many concurrent reaction channels can be displaced very easily modifying the equilibria and the product. In fact, it is well known that the pH has to be tightly controlled in order to successfully synthesize imogolite. In fact, only recently, it has been shown that the imogolite formation mechanism involves proto-imogolite structures which oligomerize to form the NTs (Doucet *et al.*, 2001; Mukherjee *et al.*, 2005; Yucelen *et al.*, 2011). The fact that the synthesis occurs in aqueous solution means that the pH and, consequently, the involved species acidic constants (pKa=-log(Ka)) are very important and guide the hydrolysis.

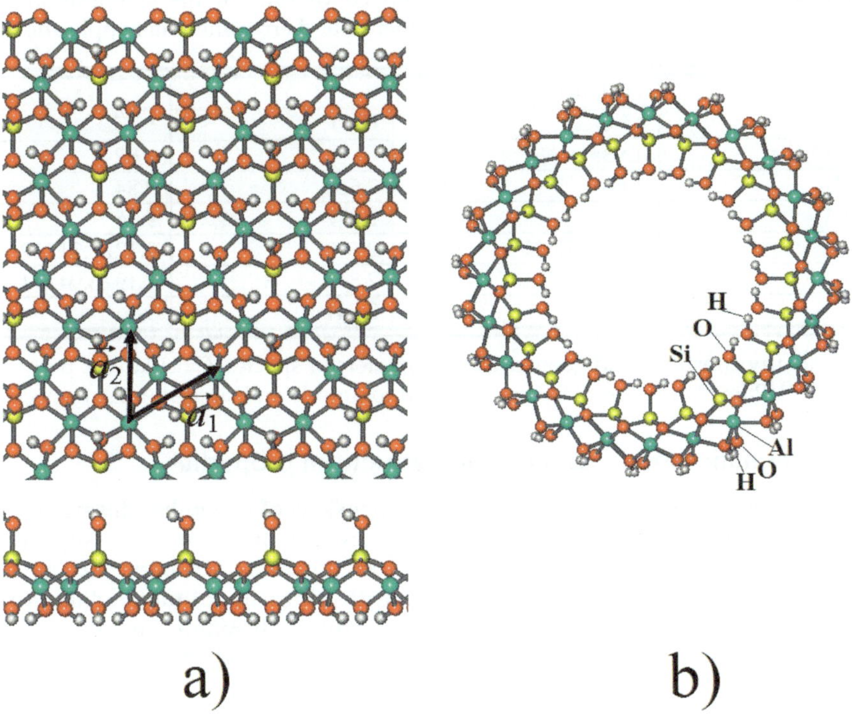

Fig. 3. (a) Hypothetical 2D imogolite layer with vector a_1 and a_2 and (b) *zigzag* (12,0) imogolite NT. White atoms, H; red, O; gray, Al; yellow, Si. Adapted with permission from (Guimaraes *et al.*, 2007). Copyright 2007 American Chemical Society.

Recently, the imogolite-like structure aluminogermanate has been synthesized (Levard *et al.*, 2008; Levard *et al.*, 2010). Here, the SiO_4^{4-} is replaced by GeO_4^{4-} fragments. However, to the best of our knowledge, no other imogolite-like structure except Ge-imogolite has been synthesized so far. Species such as H_3PO_4, H_3AsO_3, H_3AsO_4 are also strong candidates to form imogolite-like structures. However, it seems that their acid/base properties would lead to drastically different experimental conditions in order to perform the synthesis. The experimental conditions for synthesizing other imogolite-like NTs remain to be determined.

In table 1, the pKa of the different species are presented. $Ge(OH)_4$ and $Si(OH)_4$ have similar pKa values, possibly explaining why the alumino-germanate NTs have been synthesized using similar procedures. Comparing the pKa values of the species at table 1, one could argue that aluminoarsenite NTs also could be synthesized in similar experimental conditions of the aluminosilicate NTs, while for NT based on phosphoric and arsenic acid it would be necessary to decrease the pH. Although the synthesis of imogolite-like structures is very challenging, it is an interesting strategy for designing new nanostructured materials. Replacing the $Si(OH)_4$ species in the imogolite structure, one can easily control the diameter and electrostatic potential of the NT inner part.

Finally, gibbsite can be envisaged as a template for developing new nanostructured materials such as imogolite-like NTs. The mild conditions for the synthesis in aqueous solutions make them very attractive for technological and environmental applications.

Species	Distance / Å		pKa
	M-O	O-O	
$[Al(H_2O)_6]^{3+}$	1.934	-	5.52
$Si(OH)_4$	1.663	2.665	9.84
$Ge(OH)_4$	1.799	2.877	9.16
H_3PO_4	1.473/1.641[2]	2.619	2.12/7.21/12.67
H_3AsO_4	1.615/1.811[2]	2.869	2.19/6.94/11.5
H_3AsO_3	1.839	2.841	9.2

Table 1. Acidity constants and geometrical parameters of species. Calculations performed at the PBE/TZVP level of theory; Distances related to double and single bonds, respectively.

3. Imogolite nanotubes – Stability and structural properties

It is still an unsolved problem controlling the dimensions of nanotubes during synthesis in order to produce monodisperse NTs. Several theoretical studies on NTs, such as C, BN, MoS_2, TiO_2 (Hernandez *et al.*, 1998; Seifert *et al.*, 2000; Enyashin & Seifert, 2005) have shown that the strain energy decreases monotonically with increasing of tube radius. No energy minimum is observed in the strain energy curve. Therefore, these NTs are not thermodynamical products and they must be seen as kinetic products.

However, as shown elsewhere (Mukherjee *et al.*, 2005; Yucelen *et al.*, 2011), dealing with a number of experimental conditions (e.g., reactant composition, concentration, pH, temperature and time) it is possible to control structure, dimensions and composition of aluminosilicate (imogolite) and aluminogermanate NTs. Imogolite NTs are single walled and present well defined structure and dimensions. The external and internal diameters of imogolite NTs are estimated to be 2.3 and 1.0 nm, respectively, with average length of 100 nm.

At present, the stability of imogolite NTs is well investigated. Several theoretical studies (Tamura & Kawamura, 2002; Konduri *et al.*, 2006; Alvarez-Ramirez, 2007; Guimaraes *et al.*, 2007; Zhao *et al.*, 2009; Demichelis *et al.*, 2010; Lee *et al.*, 2011) using different methodologies indicated that there is clearly a minimum in the strain energy curve of the imogolite. However, the minimum value is still a matter of controversy. In 1972, based on X-ray and electron diffraction analyses, Cradwick et al. (1972) first reported that the circumference of natural imogolite NT is composed by 10 hexagonal gibbsite rings. Few years later, Farmer et al. (1977) have synthesized the first imogolite nanotube which contained 12 hexagonal gibbsite rings around its circumference, figure 3.

The first theoretical assessment on NT stability was carried out in the framework of molecular dynamics simulation using a classical many-body potential (Tamura & Kawamura, 2002) with specific parameters for imogolite. The total energy obtained with this method has the minimum strain energy per atom around a tube diameter of 2.6-2.9 nm, which means 16 gibbsite units around the circumference. Konduri et al. carried out molecular dynamics simulations for imogolite NTs employing the CLAYFF force field (Konduri *et al.*, 2006). According to this work, the force field accurately reproduced the properties of aluminosilicate minerals including gibbsite, and the CLAYFF simulations (Konduri *et al.*, 2006) reproduced the experimental findings of Farmer et al. (1977) with 12 gibbsite units around the tube.

The *zigzag* and *armchair* imogolite NTs stabilities have been studied within SCC-DFTB by (Guimaraes *et al.*, 2007). The calculated strain energy per atom for both chiralities have shown the same behavior, although *zigzag* NTs are more stable than *armchair* ones and have a minimum with 12 gibbsite units around circumference, i.e., (12,0) (figure 4).

The NT stability can also be explained in the framework of a model based on the classical theory of elasticity. For several NTs, including C, BN, MoS_2, TiO_2 (Hernandez *et al.*, 1998; Seifert *et al.*, 2000; Enyashin & Seifert, 2005) the tube's strain energy E_{str} per atom can be related to the elastic modulus Y, the thickness h of monolayer and by the tube radius R:

$$E_{str} = \frac{a}{R^2} \sim \frac{Yh^3}{R^2} \tag{1}$$

The strain energy per atom follows the general trend $1/R^2$ for all known NTs except for imogolite. When the tube is formed by a symmetric layer, equation 1 is valid. Imogolite is composed of nonsymmetrical aluminosilicate layer and a difference in the surface tensions $\Delta\sigma$ of outer and inner tube surfaces must be taken into account. As a result, an additional contribution is included to strain energy as can be seen in equation 2 and 3.

$$E_{str} = \frac{a}{R^2} + \frac{b}{R} \sim \frac{Yh^3}{R^2} + \frac{\Delta\sigma \cdot h}{R} \tag{2}$$

$$E_{str} = \frac{5.2}{R^2} - \frac{1.1}{R} \tag{3}$$

In which E_{str} is given in eV atom^{-1}, R in Å, $\boldsymbol{a}$ in eV atom^{-1} Å^2, and $\boldsymbol{b}$ in eV atom^{-1} Å. The surface energy $\Delta\sigma$ supports a negative curvature, which decreases the strain energy and introduces a minimum into the $E_{str}(R)$ curve. The fit of the obtained E_{str} and R values for imogolite NTs using equation 2 describes the change of the strain energy in the wide range of radii quite well (figure 4).

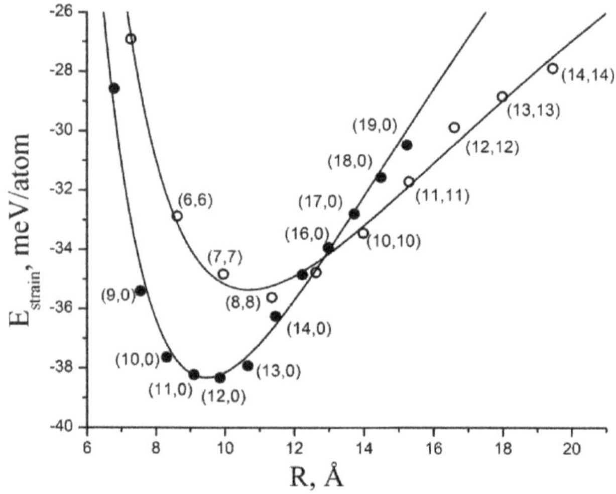

Fig. 4. Calculated strain energies E_{str} as a function of the radius R for *zigzag* (closed circles) and *armchair* (*open circles*) imogolite NTs. Reprinted with permission from (Guimaraes *et al.*, 2007).Copyright 2007 American Chemical Society.

First-principles calculations based on density functional theory (DFT) have been performed to study the energetics of imogolite NT as a function of tube diameter (Zhao *et al.*, 2009). A localized linear combination of numerical atomic-orbital basis sets has been used for the valence electrons and nonlocal pseudopotentials have been adopted for the atomic core. The DFT strain energy curve for imogolite NTs indicates an energy minimum for (9,0) structure. Furthermore, there is a local energy minimum for (12,0) nanotube, being 0.14 kJ mol^{-1} less stable than (9,0) structure. The authors assign both global and local energy minima as the natural and synthetic imogolite NTs. According to them, due to the curvature effect of the NTs, the energy minimum arises from shortening of Al-O and Si-O in the inner wall and increase of Al-O bonds in the outer wall.

Recently, first-principles calculations based on DFT have also been performed in order to study the origin of the strain energy minimum in imogolite NTs (Lee *et al.*, 2011). Although the same methodology (DFT), functional (PBE), local basis and program (SIESTA) had been used as the previous discussed work (Zhao *et al.*, 2009), the strain energy curve profile and minimum for imogolite are different. Lee et al. (2011) have found a minimum at (8,0) and the strain energy curve does not present any local minimum, in contrast to Zhao et al. (2009) that found the most stable structure at (9,0) and a local minimum at (12,0).

Demichelis *et al.* (2010) also contributed to the imogolite energy minimum topic. The authors explored the structure and energetics of imogolite NTs in the framework of full electron DFT. In contrast to the previously discussed works mentioned so far, Demichelis *et al.* (2010) have used a hybrid functional (B3LYP) in the CRYSTAL program, without the usage of parameterized pseudo potentials (Demichelis *et al.*, 2010). The obtained total energy curve presents a well defined minimum at (10,0) for zigzag NTs and (8,8) for armchair. In order to closely compare the results with ones obtained by Zhao *et al.* (2009), Demichelis *et al.* (2010) have optimized the most stable imogolite structures (n=8-13) using the PBE exchange-correlation functional. The total energy curve presents a minimum at (9,0), in contrast to (10,0) from B3LYP, although the absolute energy difference is only 0.4 kJ mol^{-1} per formula unit. Besides, Demichelis *et al.* (2010) have assigned the reason imogolite *zigzag* NTs (*n*,0) are more stable than *armchair* (*n*,*n*), which are mainly related to the geometrical setting of the inner wall. According to Demichelis *et al.* (2010), oxygen atoms from neighboring SiO_4 present shorter distances for (n,0) tubes compared to (*n*,*n*). Moreover, the presence of hydrogen bonds chains in the inner wall of the zigzag tubes allows stabilization of the curled structure in comparison to the armchair one. Lee *et al.* (2011) also presented evidences that the unique arrangement of inner silanol groups (Si-OH) and the hydrogen network are the origin of the strain energy minimum and are the reason for preference of the zigzag chirality. According to those authors, depending on the rolling direction, inner silanol OH groups produce distinct hydrogen bond (HB) networks, e.g., for *zigzag* tubes occurs disk inner HB because inner OH groups are aligned with zigzag like rolling direction in parallel and helix-like inner HB networks occurs to armchair. The *zigzag* NTs can effectively construct inner HB networks. In order to evaluate the *zigzag* preference, Lee *et al.* (2011) have investigated the structural relaxation of hydrogen saturated curved gibbsite-like imogolite, i.e., a piece of gibbsite like with armchair configuration. The obtained results have shown the curved gibbsite-like tubes spontaneously change the chirality from armchair to zigzag by shortening inner HB distances and changing the rolling direction. However, it is important to note that for all discussed works the calculations have been performed in the gas phase and it does not take into account the water solvent and the rather large interaction of the protons with the solvent. Furthermore, the synthesis of imogolite is carried out in aqueous solution and the water must play an important role in the HB network formed inside and outside the imogolite NT.

Besides the structural properties, the electronic and mechanical properties of imogolite NTs have also been calculated. For instance, from SCC-DFTB (Guimaraes *et al.*, 2007) estimates, imogolite is insulator with high band gap value. The calculated Young's moduli for imogolite lies in the range of 175-390 GPa, similar to the other inorganic NTs such as MoS_2 (230 GPa) and GaS (270 GPa). The electrostatic field based on the SCC-DFTB charges is shown at figure 5. Imogolite presents negative charges at the inner walls and positive charges at the outer walls. However, it is important to note that these are gas phase calculations and in the aqueous solution the acidity of the hydroxyl groups can change the charge distribution along the structure.

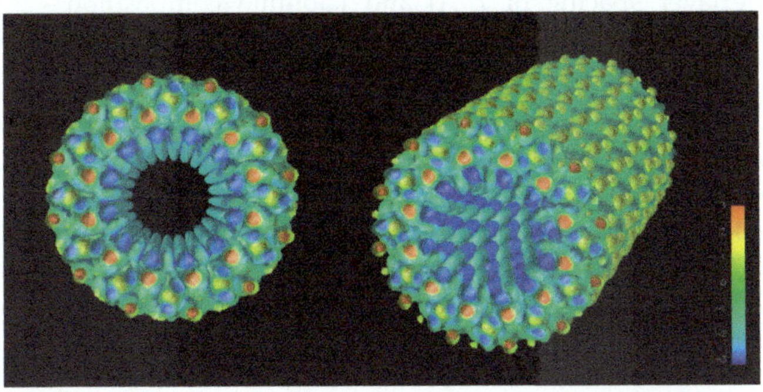

Fig. 5. Electrostatic field of the imogolite (12,0). Adaptedwith permission from (Guimaraes *et al.*, 2007). Copyright 2007 American Chemical Society.

4. Halloysite nanotubes – Stability and structural properties

Halloysite is a clay mineral normally described as a gibbsite octahedral sheet ($Al(OH)_3$), which is modified by siloxane groups at the outer surface (figure 6), and has a 1:1 Al:Si ratio and stoichiometry $Al_2Si_2O_5(OH)_4.nH_2O$ (Guimaraes *et al.*, 2010). Halloysite exhibits a range of morphologies, and according to Joussein *et al.* (2005) the structure will depend on crystallization conditions and geological occurrences. Various morphologies are reported in the literature, as platy and spheroidal crystals, scroll, glomerular or 'onion-like' and the hollow tubular structure, which is the most common one. The size of halloysite tubes varies from 500-1000 nm in length, 15-100 nm in inner diameter, depending on the substrate (Guimaraes *et al.*, 2010).

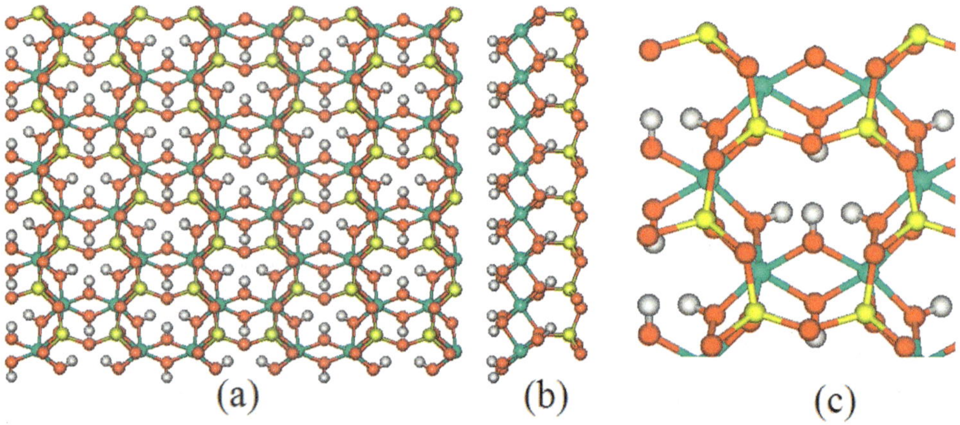

Fig. 6. Halloysite layer formed by gibbsite octahedral sheet and siloxane groups. (a) top view and (b) side view. (c) Detail of the top view. White atoms are H, red - O, blue - Al, yellow - Si.

Halloysite has the same stoichiometrical composition of kaolinite, except for its water content. Layered halloysite occurs mainly in two different polymorphs, the hydrated form (with interlayer spacing of 10 Å) with the formula $Al_2Si_2O_5(OH)_4.2H_2O$ and the anhydrous form (with interlayer spacing of 7 Å) and kaolinite composition - $Al_2Si_2O_5(OH)_4$. The intercalated water is weakly bound and can be readily and irreversibly removed (Joussein *et al.*, 2005).

According to Lvov *et al.* (2008) the reason why planar kaolinite rolls into a tube remains unclear. In the review article of Joussein *et al.* (2005) some questions are pointed out. Dixon and Mckee (Dixon & McKee, 1974) proposed the tubes are formed by layer rolling, caused by dimensional mismatch between the octahedral and tetrahedral layers and weak interaction bonds. In the hydrated halloysite, the rolling leaves a small space between the adjacent layers, although the dehydration does not change the structure. As reported by Bailey (1990) the dimensional mismatch between the octahedral and tetrahedral layers also occurs to kaolinite. However, the mismatch is corrected by rotation of alternate tetrahedral in opposite directions, while in halloysite the rotation is blocked by interlayer water molecules.

Halloysite NTs are attractive materials due to availability and vast range of applications. Besides, in contrast to other nanomaterials, naturally occurring halloysite is easily obtained and an inexpensive nanoscale container. For instance, halloysite is a viable nanocage for inclusion of biologically active molecules with specific sizes due to the empty space inside the NT (Price & Gaber; Price *et al.*, 2001). It has been used as support for immobilization of catalysts such as metallocomplexes (Nakagaki & Wypych, 2007; Machado *et al.*, 2008) and for the controlled release of anti-corrosion agents, herbicides, fungicides (Price & Gaber; Shchukin *et al.*, 2006; Shchukin & Mohwald, 2007). It exhibits interesting features and offers potential application as entrapment of hydrophilic and lipophilic active agents, as enzymatic nanoscale reactor (Shchukin *et al.*, 2005); as sustained release of drugs (Price *et al.*, 2001; Levis & Deasy, 2003; Kelly *et al.*, 2004; Veerabadran *et al.*, 2007); as adsorbing agent for dye removal (Liu *et al.*, 2011). It can be employed to improve mechanical performance of cements and polymers (Hedicke-Höchstötter *et al.*, 2009).

Imogolite and halloysite have the same gibbsite layer composition but differ in the arrangement of silicate atoms and in the Al:Si ratio, 2:1 and 1:1, respectively. The way silicon atoms are bonded to gibbsite octahedral rings is also different. In imogolite NT, $(SiO_3)OH$ groups are anchored to the inner side of the tube at gibbsite octahedral rings (figure 7a), while in halloysite siloxane groups are bonded via only one oxygen atom to gibbsite octahedral rings at the outer part (figure 7b), and the apical oxygen of tetrahedra becomes the vertices of octahedra.

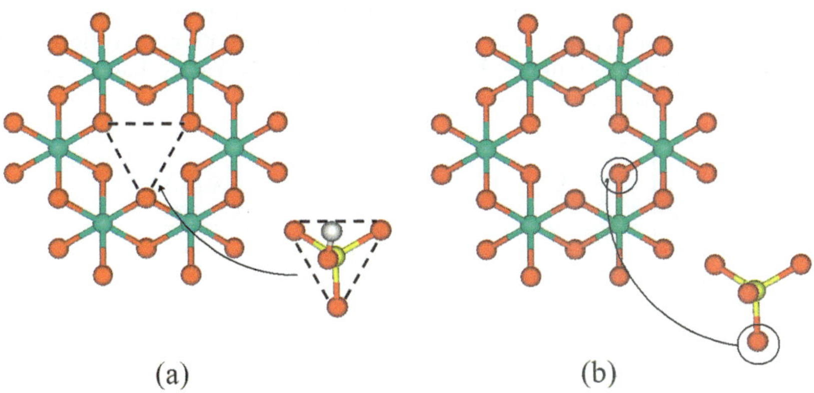

Fig. 7. Scheme presenting the different way silicon atoms are bonded to gibbsite octahedral ring at (a) imogolite and (b) halloysite.

As discussed earlier, the strain energy of imogolite NTs is an apparent exception, once instead of decreasing monotonically this function presents a minimum. At a first glance, the strain energy per atom for halloysite NTs (figure 8) decreases with increasing tube radius (R) and converges approximately as $1/R^2$, as demonstrated with SCC-DFTB calculations (Guimaraes *et al.*, 2010). However, a detailed look at the calculated values E_{str} shows that they can be better fitted by the following equation (Eq. 4):

$$E_{str} = \frac{49.0}{R^2} - \frac{3.0}{R} \qquad (4)$$

In which E_{str} is given in eV $atom^{-1}$ and R in Å. The values of 49.0 and 3.0 are given in eV $atom^{-1}$ $Å^2$ and eV $atom^{-1}$ Å, respectively. For a wide region between 24 and 54 Å of the extrapolated curve, halloysite NTs have slightly negative values for strain energies and are more stable than the respective monolayer. Thus, halloysite NTs are described by a similar equation used to fit the strain energies of imogolite NTs (Guimaraes *et al.*, 2007). It is not an unexpected result, since halloysite NTs are composed of an asymmetrical aluminosilicate layers and should have different tension promoting the formation of a curved structure.

The minimum of E_{str} curve for halloysite NTs is much less pronounced compared to that of imogolite NTs, the minimum is only 7 meV/atom below the energy of the layer, which is 5-6 times smaller than the corresponding values for imogolite. This explains the morphological distinction between experimental observations on halloysite and imogolite, that exist as multi-walled and single-walled NTs, respectively. The strain energy difference between halloysite NTs is small enough to explain the existence of a set of multi-walled NTs with large radii distribution. In contrast, imogolite NTs are strongly monodisperse.

Halloysite is an aluminosilicate which has two different basal faces. The first one consists of a tetrahedral silicate surface Si-O-Si while the other basal surface has gibbsite octahedral layer ($Al(OH)_3$). In principle, both faces are – as ideal structures in theory - electrically neutral. The charges inside and outside halloysite NTs are related to their structure and adsorption properties. The charges obtained with SCC-DFTB calculations (Guimaraes *et al.*, 2010) have been used to get the electrostatic potential map of some halloysite NTs, as shown in figure 9. As it can be seen, the inner wall of tube is mainly positively charged, while the outer surface has a weakly negative charge, in good agreement with observations by Lvov *et al.* (2008). According to these authors, below pH 8.5 the tube cavity has a positive inner surface and negatively charged outer surface.

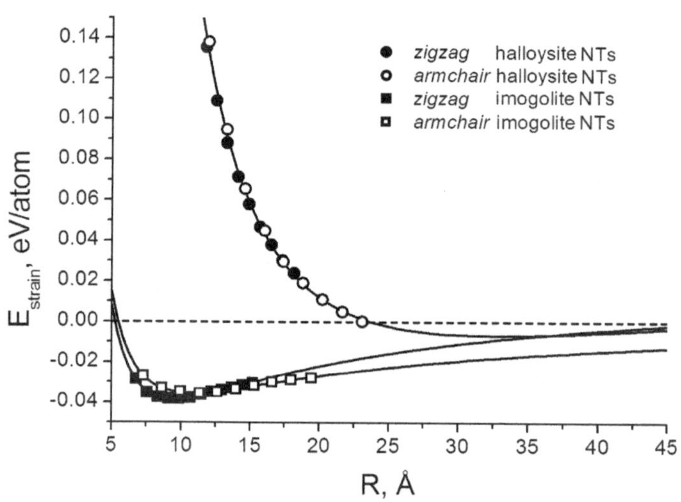

Fig. 8. Strain energy as a function of tube radius for (*n,0*) (closed circles) and (*n,n*) (open circles) single walled halloysite NTs and (*n,0*) (closed squares) and (*n,n*) (open squares) single walled imogolite NTs. Reprinted with permission from (Guimaraes *et al.*, 2010) Copyright 2010 American Chemical Society.

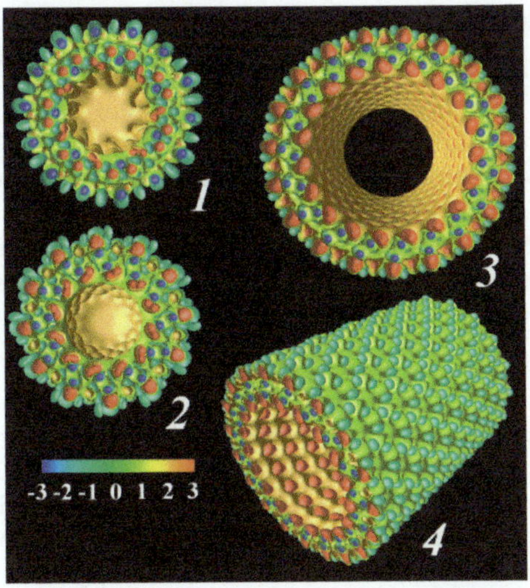

Fig. 9. Electrostatic field nearby halloysite NTs of different chiralities: *1* – *armchair* (7,7), *2* – *zigzag* (12,0), *3* – *zigzag* (19,0) (the views along the tubes' axis are shown) and *4* – a diagonal view for *zigzag* (19,0) NT. Different colors show equipotential surfaces: -3.0, -2.0, -1.0, 1.0, 2.0 and 3.0 e/Å. Reprinted with permission from(Guimaraes *et al.*, 2010). Copyright 2010 American Chemical Society.

Experimental results from natural samples indicate that the halloysite structure at the edge is disrupted, and the surface groups can be protonated or deprotonated originating variable charge (Theng *et al.*, 1982). For instance, halloysite presents negative charge at pH higher than 3 (Theng *et al.*, 1982), and its isoelectric point is around pH 3. In this way, the edges are considered to be positively charged at low pH, neutral at isoeletric point and negatively charged at higher pH (Braggs *et al.*, 1994). The negative charge can be ascribed to the deprotonation of water and hydroxyl groups bound to aluminum and silicon at the edges (Theng *et al.*, 1982), and the hydroxyl groups are considered to be the principal reactive sites. Furthermore, Machado *et al.* (2008) have shown the immobilization of anionic and cationic metalloporphyrins into halloysite NTs occurs at high rates while for neutral metalloporphyrins the immobilization was not observed. The cationic immobilization can occur via SiO^- groups, while anionic immobilization may occur through aluminol groups at halloysite edges.

5. Chrysotile nanotubes – Structural properties

Chrysotile and lizardite are fibrous natural phylosilicate minerals which belong to the serpentine group and present 1:1 structure. They have the same empirical formula $Mg_3Si_2O_5(OH)_4$ (Falini *et al.*, 2004; Anbalagan *et al.*, 2010), as can be seen in figure 10. Chrysotile constitutes approximately 95% percent of the manufactured asbestos and presents three polytypes: clinochrysotile (Whittaker, 1956a), orthochrysotile (Whittaker, 1956b) and parachrysotile (Whittaker, 1956c). Clinochrysotile is the most common one. While lizardite, more abundant than chrysotile, presents a planar shape, chrysotile presents a tubular form. Chrysotile and lizardite are composed by octahedral sheet, brucite (magnesium dihydroxide, $Mg(OH)_2$) and tetrahedral layer tridymite (silicon dioxide, SiO_2), figure 10. The outer part of chrysotile is formed by brucite and the inner part by tridymite.

Figure 10 shows the structures of tridymite, brucite, lizardite and chrysotile. The superposition of the tetrahedral and octahedral layers results in 1:1 lizardite which has the hexagons formed by Mg-O bounds (from brucite) located on the center of the hexagon formed by Si-O bounds (from tridymite). The connections of brucite and tridymite to form lizardite occur via the apical oxygen of the SiO_4 layer which are connected directly with the Mg atoms of brucite. The connection of brucite and tridymite layers occurs in the same way as in chrysotile NTs.

Chrysotile is a nanosized and tube-shaped material with lower mechanical strength and it is always uncapped. Chrysotile (Piperno *et al.*, 2007; Anbalagan *et al.*, 2010) can be synthesized in aqueous solution under mild conditions, easily modified (Wypych *et al.*, 2004; Wang *et al.*, 2006; Wang *et al.*, 2009) and functionalized (Nakagaki & Wypych, 2007). Therefore chrysotile is an interesting target material to be used as component of hybrid materials, support for catalysis, ionic channels, molecular sieving, for gas storage (Halma *et al.*, 2006; Nakagaki *et al.*, 2006; Nakagaki & Wypych, 2007) and other applications in nanotechnology. Stoichiometric chrysotile has been synthesized and characterized by structural and spectroscopy analyses (Falini *et al.*, 2002; Falini *et al.*, 2004). Chrysotile is found as multiwalled nanotubes with inner diameter around 1-10 nm, outer diameter around 10-50 nm and the size can reach the millimeter range (Falini *et al.*, 2004). Chrysotile can also be found in spiral form (Yada, 1967, 1971).

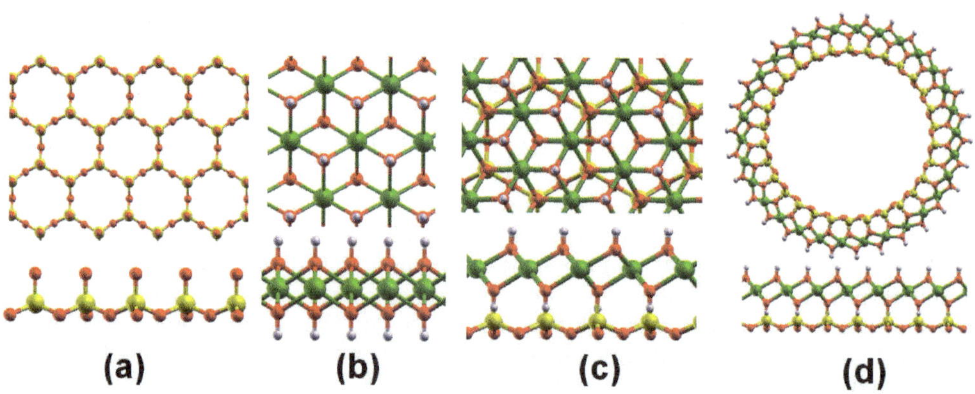

Fig. 10. Top view and side view of (a) tridymite (SiO_2), (b) brucite ($Mg(OH)_2$), c) lizardite ($Mg_3Si_2O_5(OH)_4$) layers and (d) chrysotile NT. Atoms label: Si, yellow; O, red; H, white; Mg, green.

Chrysotile NTs were synthesized and characterized by Piperno and co-workers (2007) using atomic force microscopy and transmission electron Microscopy (TEM). The results have shown that chrysotile NTs exhibit elastic behavior at small deformation. The chrysotile Young's modulus evaluated by (Piperno *et al.*, 2007) are 159 ± 125 GPa. The stoichiometric chrysotile fibers demonstrate a hollow structure with quite uniform outer diameter around 35 nm and inner diameter about 7-8 nm. The NTs are open ended with several hundred nanometers in length.

Only few theoretical studies concerning chrysotile NTs have been carried out. The chrysotile unit cell is composed by hundreds up to thousands of atoms and, therefore, DFT or *ab initio* calculations on such systems are computationally time consuming. D'Arco *et al.* (2009) have studied the stability and structural properties of some armchair chrysotile NTs using the DFT method and helical symmetry approach as it is implemented in the CRYSTAL program (Dovesi *et al.*, 2009). The structural results are in good agreement with the experimental data for NTs and lizardite monolayer. Preliminary results of the strain energy curve of chrysotile calculated using the SCC-DFTB method decreases monotonically with the increase of the radii indicating the monolayer is more stable than the NTs. The chirality does not affect the relative stability of the NTs, i.e., strain energy profile for *zigzag* and *armchair* NTs present the same pattern. In spite of the polydispersity of the chrysotile NTs and the environmental concern of asbestos, many attempts for modifying and functionalizing chrysotile NTs have been reported. Chrysotile has been studied in many fields such as support for immobilization of metalloporphyrins, oxidation catalysts, fixation of CO_2 by chrysotile under low-pressure (Larachi *et al.*, 2010), modification of chrysotile surface by organosilanes, functionalization of single layers and nanofibers to produce polymer nanocomposites (Wang *et al.*, 2006; Nakagaki & Wypych, 2007; Wang *et al.*, 2009) and to produce self-assembled systems (De Luca *et al.*, 2009). Furthermore, many studies have reported the partial or total substitutions of magnesium atoms at chrysotile sites for different atoms as Fe and Ni (Bloise *et al.*, 2010). The substitution of Mg atoms at chrysotile by Ni results in another nanotubular material called pecorite (Faust *et al.*, 1969) with empirical formula $Ni_3Si_2O_5(OH)_4$ similar to that of chrysotile. Pecorite and its planar form (called nepouite) can be found in nature (Faust *et al.*, 1969) or synthesized (McDonald *et al.*, 2009; Bloise *et al.*, 2010). Since nickel atoms are usually applied in catalysis, Ni-containing phyllosilicates (Ni-lizardite or nepouite) have been used as catalysts precursors for carbon dioxide reforming of methane (Sivaiah *et al.*, 2011).

The acid leaching of chrysotile is a process used to synthesize SiO_2 nano tubular structure which has been reported recently (Wang *et al.*, 2006). The process occurs by leaching of brucite layers and the reminiscent product is an amorphous material called nano-fibriform silica (Wang *et al.*, 2006) which presents tubular shape and the diameter around 20-30 nm. SCC-DFTB calculations of SiO_2 NTs indicate that these structures are not stable and may easily collapse to the silica structure. However, it opens an interesting opportunity to functionalize the NT surface and eventually create a carbon based structure surrounding the tridymite, SiO_2, structure. Actually, Wang *et al.* (2009) have been able to modify the outer surface of the nano-fibriform silica with dimethyldichorosilane. Theoretical investigations of these recently synthesized systems can bring important insights about their structural and mechanical properties and eventually indicate the possibility to design materials with enhanced properties.

6. Final remarks

Nanostructured aluminosilicates are becoming the target for new advanced materials. Their availability, the syntheses in mild conditions and their well defined structures are very attractive characteristics. They are easily functionalized and much effort has been devoted to modify their structures and to enhance their physical and chemical properties. Particularly, the aluminosilicate nanostructure can be envisaged for the development of nanoreactors, controlled release devices, ion conductors for batteries, gas storage and separation systems. They are insulator and the stiffness of the NT is similar to other inorganic NTs and comparable to steel. Much progress in characterizing and developing new materials based on clay mineral NTs has been obtained in the last few years. The modification (Kang *et al.*, 2010) and the functionalization (Kang *et al.*, 2011) of the imogolite NTs inner walls are recent notable achievements that open new perspectives on the field. Understanding the formation mechanism of such nanostructured clay minerals is also an important achievement broadening the fundamental knowledge about clay mineral NTs. The synthesis of new imogolite-like structures is an important issue and deserves more attention. Actually, the aluminogermanate NTs (Levard *et al.*, 2008; Levard *et al.*, 2010) are an important example of the feasibility of this task and more effort in this direction must be done. In fact, lamellar gibbsite can be seen as a template for modeling and synthesizing new nanostructured imogolite-like structures. Actually, the use of clay NTs for developing new advanced materials has not yet received much attention commensurate with their potential for technological application.

7. Acknowledgments

Support from the agencies Fundação de Amparo à Pesquisa de Minas Gerais (FAPEMIG), Coordenação de Aperfeiçoamento de Pessoal de Nível Superior (CAPES), and Conselho Nacional para o Desenvolvimento Científico e Tecnológico (CNPq), Deutsche Forschungsgemeinschaft (DFG) and Deutscher Akademischer Auslandsdienst (DAAD) are gratefully acknowledged. This work has also been supported by the Brazilian Initiative National Institute of Science and Technology for Mineral Resources, Water and Biodiversity, INCT-ACQUA.

8. References

Ackerman, W.C., Smith, D.M., Huling, J.C., Kim, Y.W., Bailey, J.K. & Brinker, C.J. (1993) Gas vapor adsorption in imogolite - a microporous tubular aluminosilicate. *Langmuir*, 9, 1051-1057.

Alvarez-Ramirez, F. (2007) Ab initio simulation of the structural and electronic properties of aluminosilicate and aluminogermanate natotubes with imogolite-like structure. *Physical Review B*, 76.

Anbalagan, G., Sivakumar, G., Prabakaran, A.R. & Gunasekaran, S. (2010) Spectroscopic characterization of natural chrysotile. *Vibrational Spectroscopy*, 52, 122-127.

Aradi, B., Hourahine, B. & Frauenheim, T. (2007) Dftb+, a sparse matrix-based implementation of the dftb method. *Journal of Physical Chemistry A*, 111, 5678-5684.

Artacho, E., Sanchez-Portal, D., Ordejon, P., Garcia, A. & Soler, J.M. (1999) Linear-scaling ab-initio calculations for large and complex systems. *Physica Status Solidi B-Basic Research*, 215, 809-817.

Bates, T.F., Hildebrand, F.A. & Swineford, A. (1950a) Morphology and structure of endellite and halloysite. *American Mineralogist*, 35, 463-484.

Bates, T.F., Sand, L.B. & Mink, J.F. (1950b) Tubular crystals of chrysotile asbestos. *Science*, 111, 512-513.

Bi, S.P., Wang, C.Y., Cao, Q. & Zhang, C.H. (2004) Studies on the mechanism of hydrolysis and polymerization of aluminum salts in aqueous solution: Correlations between the "core-links" model and "cage-like" keggin-al-13 model. *Coordination Chemistry Reviews*, 248, 441-455.

Bloise, A., Belluso, E., Fornero, E., Rinaudo, C., Barrese, E. & Capella, S. (2010) Influence of synthesis conditions on growth of ni-doped chrysotile. *Microporous and Mesoporous Materials*, 132, 239-245.

Braggs, B., Fornasiero, D., Ralston, J. & Stsmart, R. (1994) The effect of surface modification by an organosilane on the electrochemical properties of kaolinite. *Clays and Clay Minerals*, 42, 123-136.

Bursill, L.A., Peng, J.L. & Bourgeois, L.N. (2000) Imogolite: An aluminosilicate nanotube material. *Philosophical Magazine A, Physics of Condensed Matter Structure Defects and Mechanical Properties*, 80, 105-117.

Casey, W.H. (2006) Large aqueous aluminum hydroxide molecules. *Chemical Reviews*, 106, 1-16.

Cradwick, P.D., Wada, K., Russell, J.D., Yoshinag.N, Masson, C.R. & Farmer, V.C. (1972) Imogolite, a hydrated aluminum silicate of tubular structure. *Nature-Physical Science*, 240, 187-&.

D'Arco, P., Noel, Y., Demichelis, R. & Dovesi, R. (2009) Single-layered chrysotile nanotubes: A quantum mechanical ab initio simulation. *Journal of Chemical Physics*, 131.

De Luca, G., Romeo, A., Villari, V., Micali, N., Foltran, I., Foresti, E., Lesci, I.G., Roveri, N., Zuccheri, T. & Scolaro, L.M. (2009) Self-organizing functional materials via ionic self assembly: Porphyrins h- and j-aggregates on synthetic chrysotile nanotubes. *Journal of the American Chemical Society*, 131, 6920-6921.

Demichelis, R., Noel, Y., D'Arco, P., Maschio, L., Orlando, R. & Dovesi, R. (2010) Structure and energetics of imogolite: A quantum mechanical ab initio study with b3lyp hybrid functional. *Journal of Materials Chemistry*, 20, 10417-10425.

Dixon, J.B. & McKee, T.R. (1974) Internal and external morphology of tubular and spheroidal halloysite particles. *Clays and Clay Minerals*, 22, 127-137.

Doucet, F.J., Schneider, C., Bones, S.J., Kretchmer, A., Moss, I., Tekely, P. & Exley, C. (2001) The formation of hydroxyaluminosilicates of geochemical and biological significance. *Geochimica Et Cosmochimica Acta*, 65, 2461-2467.

Dovesi, R., Saunders, V.R., Roetti, C., Orlando, R., Zicovich-Wilson, C.M., Pascale, F., Civalleri, B., Doll, K., Harrison, N.M., Bush, I.J., D'Arco, P. & Llunell, M. (2009) Crystal09 user's manual. Pp.,University of Torino, Torino.

Dresselhaus, M.S., Dresselhaus, G. & Avouris, P. (2001) Carbon nanotubes: Synthesis, structure, properties and applications. Pp. *Topics in applied physics*, 80,Springer-Verlag, Berlin Heidelberg.

Elstner, M., Porezag, D., Jungnickel, G., Elsner, J., Haugk, M., Frauenheim, T., Suhai, S. & Seifert, G. (1998) Self-consistent-charge density-functional tight-binding method for simulations of complex materials properties. *Physical Review B*, 58, 7260-7268.

Endo, M., Iijima, S. & Dresselhaus, M.S. (1996) Carbon nanotubes. Pp.,Pergamon.

Enyashin, A.N. & Seifert, G. (2005) Structure, stability and electronic properties of TiO_2 nanostructures. *Physica Status Solidi B-Basic Solid State Physics*, 242, 1361-1370.

Enyashin, A.N., Gemming, S. & Seifert, G. (2007) Simulation of inorganic nanotubes. Pp. 33-57. In S. Gemming, M. Schreiber, and J.-B. Suck, Eds. *Materials for tomorrow*, 93,Springer Berlin Heidelberg.

Enyashin, A.N. & Ivanovskii, A.L. (2008) Theoretical prediction of al(oh)(3) nanotubes and their properties. *Physica E-Low-Dimensional Systems & Nanostructures*, 41, 320-323.

Enyashin, A.N., Popov, I. & Seifert, G. (2009) Stability and electronic properties of rhenium sulfide nanotubes. *Physica Status Solidi B-Basic Solid State Physics*, 246, 114-118.

Exley, C., Schneider, C. & Doucet, F.J. (2002) The reaction of aluminium with silicic acid in acidic solution: An important mechanism in controlling the biological availability of aluminium? *Coordination Chemistry Reviews*, 228, 127-135.

Falini, G., Foresti, E., Gazzano, M., Gualtieri, A.E., Leoni, M., Lesci, I.G. & Roveri, N. (2004) Tubular-shaped stoichiometric chrysotile nanocrystals. *Chemistry-A European Journal*, 10, 3043-3049.

Falini, G., Foresti, E., Lesci, G. & Roveri, N. (2002) Structural and morphological characterization of synthetic chrysotile single crystals. *Chemical Communications*, 1512-1513.

Faust, G.T., Fahey, J.J., Mason, B. & Dwornik, E.J. (1969) Pecoraite, ni6si4o10(oh)8, nickel analog of clinochrysotile, formed in the wolf creek meteorite. *Science*, 165, 59-60.

Frenzel, J., Oliveira, A.F., Duarte, H.A., Heine, T. & Seifert, G. (2005) Structural and electronic properties of bulk gibbsite and gibbsite surfaces. *Zeitschrift Fur Anorganische Und Allgemeine Chemie*, 631, 1267-1271.

Giese, R.F. & Datta, P. (1973) Hydroxyl orientation in kaolinite, dickite, and nacrite *American Mineralogist*, 58, 471-479.

Guimaraes, L., Enyashin, A.N., Frenzel, J., Heine, T., Duarte, H.A. & Seifert, G. (2007) Imogolite nanotubes: Stability, electronic, and mechanical properties. *ACS Nano*, 1, 362-368.

Guimaraes, L., Enyashin, A.N., Seifert, G. & Duarte, H.A. (2010) Structural, electronic, and mechanical properties of single-walled halloysite nanotube models. *Journal of Physical Chemistry C*, 114, 11358-11363.

Halma, M., Bail, A., Wypych, F. & Nakagaki, S. (2006) Catalytic activity of anionic iron(iii) porphyrins immobilized on grafted disordered silica obtained from acidic leached chrysotile. *Journal of Molecular Catalysis A-Chemical*, 243, 44-51.

Hedicke-Höchstötter, K., Lim, G.T. & Altstädt, V. (2009) Novel polyamide nanocomposites based on silicate nanotubes of the mineral halloysite. *Composites Science and Technology*, 69, 330-334.

Heine, T., Rapacioli, M., Patchkovskii, S., Frenzel, J., Koester, A.M., Calaminici, P., Escalante, S., Duarte, H.A., Flores, R., Geudtner, G., Goursot, A., Reveles, J.U., Vela, A. & Salahub, D.R. (2009) Demon, demon-nano edn. Pp.,deMon-Sofware, Mexico, Bremen.

Hernandez, E., Goze, C., Bernier, P. & Rubio, A. (1998) Elastic properties of c and bxcynz composite nanotubes. *Physical Review Letters*, 80, 4502-4505.

Iijima, S. (1991) Helical microtubules of graphitic carbon. *Nature (London)*, 354 56.

Imamura, S., Kokubu, T., Yamashita, T., Okamoto, Y., Kajiwara, K. & Kanai, H. (1996) Shape-selective copper-loaded imogolite catalyst. *Journal of Catalysis*, 160, 137-139.

Ivanovskaya, V.V., Heine, T., Gemming, S. & Seifert, G. (2006) Structure, stability and electronic properties of composite mo1-xnbxs2 nanotubes. *Physica Status Solidi B-Basic Solid State Physics*, 243, 1757-1764.

Joussein, E., Petit, S., Churchman, J., Theng, B., Righi, D. & Delvaux, B. (2005) Halloysite clay minerals - a review. *Clay Minerals*, 40, 383-426.

Kang, D.Y., Zang, J., Jones, C.W. & Nair, S. (2011) Single-walled aluminosilicate nanotubes with organic-modified interiors. *Journal of Physical Chemistry C*, 115, 7676-7685.

Kang, D.Y., Zang, J., Wright, E.R., McCanna, A.L., Jones, C.W. & Nair, S. (2010) Dehydration, dehydroxylation, and rehydroxylation of single-walled aluminosilicate nanotubes. *Acs Nano*, 4, 4897-4907.

Kelly, H.M., Deasy, P.B., Ziaka, E. & Claffey, N. (2004) Formulation and preliminary in vivo dog studies of a novel drug delivery system for the treatment of periodontitis. *International Journal of Pharmaceutics*, 274, 167-183.

Konduri, S., Mukherjee, S. & Nair, S. (2006) Strain energy minimum and vibrational properties of single-walled aluminosilicate nanotubes. *Physical Review B*, 74, 033401.

Kuc, A. & Heine, T. (2009) Shielding nanowires and nanotubes with imogolite: A route to nanocables. *Advanced Materials*, 21, 4353-+.

Larachi, F., Daldoul, I. & Beaudoin, G. (2010) Fixation of co2 by chrysotile in low-pressure dry and moist carbonation: Ex-situ and in-situ characterizations. *Geochimica et Cosmochimica Acta*, 74, 3051-3075.

Lee, S.U., Choi, Y.C., Youm, S.G. & Sohn, D. (2011) Origin of the strain energy minimum in imogolite nanotubes. *Journal of Physical Chemistry C*, 115, 5226-5231.

Levard, C., Rose, J., Masion, A., Doelsch, E., Borschneck, D., Olivi, L., Dominici, C., Grauby, O., Woicik, J.C. & Bottero, J.-Y. (2008) Synthesis of large quantities of single-walled aluminogermanate nanotube. *Journal of the American Chemical Society*, 130, 5862-+.

Levard, C., Rose, J., Thill, A., Masion, A., Doelsch, E., Maillet, P., Spalla, O., Olivi, L., Cognigni, A., Ziarelli, F. & Bottero, J.Y. (2010) Formation and growth mechanisms of imogolite-like aluminogermanate nanotubes. *Chemistry Of Materials*, 22, 2466-2473.

Levis, S.R. & Deasy, P.B. (2003) Use of coated microtubular halloysite for the sustained release of diltiazem hydrochloride and propranolol hydrochloride. *International Journal of Pharmaceutics*, 253, 145-157.

Liu, R., Zhang, B., Mei, D., Zhang, H. & Liu, J. (2011) Adsorption of methyl violet from aqueous solution by halloysite nanotubes. *Desalination*, 268, 111-116.

Lvov, Y.M., Shchukin, D.G., Mohwald, H. & Price, R.R. (2008) Halloysite clay nanotubes for controlled release of protective agents. *Acs Nano*, 2, 814-820.

Machado, G.S., Castro, K., Wypych, F. & Nakagaki, S. (2008) Immobilization of metalloporphyrins into nanotubes of natural halloysite toward selective catalysts for oxidation reactions. *Journal of Molecular Catalysis a-Chemical*, 283, 99-107.

McDonald, A., Scott, B. & Villemure, G. (2009) Hydrothermal preparation of nanotubular particles of a 1:1 nickel phyllosilicate. *Microporous and Mesoporous Materials*, 120, 263-266.

Mukherjee, S., Bartlow, V.A. & Nair, S. (2005) Phenomenology of the growth of single-walled aluminosilicate and aluminogermanate nanotubes of precise dimensions. *Chemistry Of Materials*, 17, 4900-4909.

Nakagaki, S., Castro, K., Machado, G.S., Halma, M., Drechsel, S.M. & Wypych, F. (2006) Catalytic activity in oxidation reactions of anionic iron(iii) porphyrins immobilized on raw and grafted chrysotile. *Journal of the Brazilian Chemical Society*, 17, 1672-1678.

Nakagaki, S. & Wypych, F. (2007) Nanofibrous and nanotubular supports for the immobilization of metalloporphyrins as oxidation catalysts. *Journal of Colloid and Interface Science*, 315, 142-157.

Oliveira, A.F., Seifert, G., Heine, T. & Duarte, H.A. (2009) Density-functional based tight-binding: An approximate dft method. *Journal of the Brazilian Chemical Society*, 20, 1193-1205.

Pauling, L. (1930) The structure of the chlorites. *Proceedings of the National Academy of Sciences of the United States of America*, 16, 578-582.

Piperno, S., Kaplan-Ashiri, I., Cohen, S.R., Popovitz-Biro, R., Wagner, H.D., Tenne, R., Foresti, E., Lesci, I.G. & Roveri, N. (2007) Characterization of geoinspired and synthetic chrysotile nanotubes by atomic force microscopy and transmission electron microscopy. *Advanced Functional Materials*, 17, 3332-3338.

Price, R. & Gaber, B. Controled release of active agents using inorganic tubules. Pp.,5651976, US Patent.

Price, R.R., Gaber, B.P. & Lvov, Y.M. (2001) In-vitro release characteristics of tetracycline hcl, khellin and nicotinamide adenine dineculeotide from halloysite; a cylindrical mineral. *Journal of Microencapsulation*, 18, 713.

Rasche, B., Seifert, G. & Enyashin, A. (2010) Stability and electronic properties of bismuth nanotubes. *Journal of Physical Chemistry C*, 114, 22092-22097.

Redlinger, M. & Corkery, R. (2007) Cosmetic skincare applications employing mineral derived tubules for controlled release. Pp.,0202061, US Patent.

Remskar, M. (2004) Inorganic nanotubes. *Advanced Materials*, 16, 1497-1504.

Schneider, C., Doucet, F., Strekopytov, S. & Exley, C. (2004) The solubility of an hydroxyaluminosilicate. *Polyhedron*, 23, 3185-3191.

Seifert, G., Terrones, H., Terrones, M., Jungnickel, G. & Frauenheim, T. (2000) Structure and electronic properties of mos2 nanotubes. *Physical Review Letters*, 85, 146-149.

Shchukin, D.G. & Mohwald, H. (2007) Surface-engineered nanocontainers for entrapment of corrosion inhibitors. *Advanced Functional Materials*, 17, 1451-1458.

Shchukin, D.G., Sukhorukov, G.B., Price, R.R. & Lvov, Y.M. (2005) Halloysite nanotubes as biomimetic nanoreactors. *Small*, 1, 510-513.

Shchukin, D.G., Zheludkevich, M., Yasakau, K., Lamaka, S., Ferreira, M.G.S. & Mohwald, H. (2006) Layer-by-layer assembled nanocontainers for self-healing corrosion protection. *Advanced Materials*, 18, 1672-+.

Sivaiah, M.V., Petit, S., Beaufort, M.F., Eyidi, D., Barrault, J., Batiot-Dupeyrat, C. & Valange, S. (2011) Nickel based catalysts derived from hydrothermally synthesized 1:1 and 2:1 phyllosilicates as precursors for carbon dioxide reforming of methane. *Microporous and Mesoporous Materials*, 140, 69-80.

Soler, J.M., Artacho, E., Gale, J.D., Garcia, A., Junquera, J., Ordejon, P. & Sanchez-Portal, D. (2002) The siesta method for ab initio order-n materials simulation. *Journal of Physics-Condensed Matter*, 14, 2745-2779.

Stefanov, M., Enyashin, A.N., Heine, T. & Seifert, G. (2008) Nanolubrication: How do mos2-based nanostructures lubricate? *Journal of Physical Chemistry C*, 112, 17764-17767.

Tamura, K. & Kawamura, K. (2002) Molecular dynamics modeling of tubular aluminum silicate: Imogolite. *Journal of Physical Chemisty B*, 106, 271-278.

Tenne, R. (2006) Inorganic nanotubes and fullerene-like nanoparticles. *Nature Nanotechnology*, 1, 103-111.

Tenne, R., Margulis, L., Genut, M. & Hodes, G. (1992) Polyhedral and cylindrical structures of tungsten disulfide. *Nature*, 360, 444-446.

Theng, B.K.G., Russell, M., Churchman, G.J. & Parfitt, R.L. (1982) Surface-properties of allophane, halloysite, and imogolite. *Clays and Clay Minerals*, 30, 143-149.

Veerabadran, N.G., Price, R.R. & Lvov, Y.M. (2007) Clay nanotubes for encapsulation and sustained release of drugs. *Nano*, 2, 115-120.

Wang, L.J., Lu, A.H., Wang, C.Q., Zheng, X.S., Zhao, D.J. & Liu, R. (2006) Nano-fibriform production of silica from natural chrysotile. *Journal of Colloid and Interface Science*, 295, 436-439.

Wang, L.J., Lu, A.H., Xiao, Z.Y., Ma, J.H. & Li, Y.Y. (2009) Modification of nano-fibriform silica by dimethyldichlorosilane. *Applied Surface Science*, 255, 7542-7546.

White, C.E., Provis, J.L., Riley, D.P., Kearley, G.J. & van Deventer, J.S.J. (2009) What is the structure of kaolinite? Reconciling theory and experiment. *Journal of Physical Chemistry B*, 113, 6756-6765.

Whittaker, E.J.W. (1956a) The structure of chrysotile .2. Clino-chrysotile. *Acta Crystallographica*, 9, 855-862.

Whittaker, E.J.W. (1956b) The structure of chrysotile .3. Ortho-chrysotile. *Acta Crystallographica*, 9, 862-864.

Whittaker, E.J.W. (1956c) The structure of chrysotile .4. Para-chrysotile. *Acta Crystallographica*, 9, 865-867.

Wypych, F., Schreiner, W.H. & Richard, E. (2004) Grafting of phenylarsonic and 2-nitrophenol-4-arsonic acid onto disordered silica obtained by selective leaching of brucite-like sheet from chrysotile structure. *Journal of Colloid and Interface Science*, 276, 167-173.

Yada, K. (1967) Study of chrysotile asbestos by a high resolution electron microscope. *Acta Crystallographica*, 23, 704-&.

Yada, K. (1971) Study of microstructure of chrysotile asbestos by high resolution electron microscopy. *Acta Crystallographica Section a-Crystal Physics Diffraction Theoretical and General Crystallography*, A 27, 659-&.

Yucelen, G.I., Choudhury, R.P., Vyalikh, A., Scheler, U., Beckham, H.W. & Nair, S. (2011) Formation of single-walled aluminosilicate nanotubes from molecular precursors and curved nanoscale intermediates. *Journal of the American Chemical Society*, 133, 5397-5412.

Zhao, M.W., Xia, Y.Y. & Mei, L.M. (2009) Energetic minimum structures of imogolite nanotubes: A first-principles prediction. *Journal of Physical Chemistry C*, 113, 14834-14837.

Part 2

Defect Chemistry: Stoichiometry and Surface Structures

3

Structure, Morphology, and Stoichiometry of GaN(0001) Surfaces Through Various Cleaning Procedures

Azusa N. Hattori[1] and Katsuyoshi Endo[2]

[1]*Nanoscience and Nanotechnology Center, The Institute of Scientific and Industrial Research, Osaka University*

[2]*Research Center for Ultra-Precision Science and Technology, Graduate School of Engineering, Osaka University*

Japan

1. Introduction

Perfect surfaces are expected to upgrade the quality of films grown on them. Generally, the surface plays a crucial role for a thin-film system, because of the large contribution of the interface and surface regions. The structures of clean surfaces are of particular importance since knowledge of the structures is the first step in understanding the fundamental issues of contact formation, chemical reactivity, growth processes, and so on.

Gallium nitride (GaN) has excellent properties such as a direct and wide band-gap energy, and high electron mobility, and has thus attracted much attention owing to its application in a wide range of electronic devices. For practical device fabrication, various materials have been integrated on GaN substrate surfaces, where the initial GaN surfaces often play a crucial role in the device operation. A number of groups have investigated GaN cleaning procedures for device fabrication. However, there is no standard method of preparing GaN substrates using chemical solutions or by ultra high-vacuum (UHV) treatment, and the results obtained in previous studies were not in conformity with one another. Through the various processes of cleaning GaN(0001) surfaces, we found a strong interrelation between surface morphology and stoichiometry; rough surfaces have broken stoichiometry while flat surfaces retain their stoichiometry (Hattori et al., 2009; 2010).

In this chapter we summarize the surface structures, morphology, and stoichiometry of GaN surfaces treated by annealing in UHV, sputtering, and chemical solutions. For the UHV-treated GaN surfaces, there were structure differences among samples subjected to the same treatment, and we clarify the effect of crystal quality, such as dislocations, the concentration of hydrogen impurities, and the residual reactant molecules in GaN films, on the surface structure. In the case of performing wet etching on a GaN system, selecting an etchant solution with a certain pH and oxide-reduction potential and controlling the etching time are important to obtain an oxide-free and balanced-stoichiometry surface. Finally, the reactions of GaN in solutions will be explained on the basis of the theoretical potential-pH equilibrium diagram. At first, the backgrounds of GaN crystal, and GaN(0001) surface structures are introduced in section 2 and 3, respectively. Surface structure, morphology, and stoichiometry through each treatment are reported in section 4. Two different hydride vapor phase epitaxy (HVPE) freestanding GaN(0001) single crystal substrates were prepared; HVPE1 and HVPE2 which were purchased from different suppliers. The details of experimental condition and all references were written in Ref.(Hattori et al., 2009; 2010).

2. GaN crystal

GaN has excellent properties such as a direct and wide band-gap energy of 3.4 eV at room temperature (RT). It also has high electron mobility, and has thus attracted much attention for its potential use in a wide range of electronic devices. Usually GaN crystal (Fig. 1) has a hexagonal phase (wurtzite, a = 0.319 nm and c = 0.519 nm) (Edgar, 1994; Strite & Morkoc, 1992).

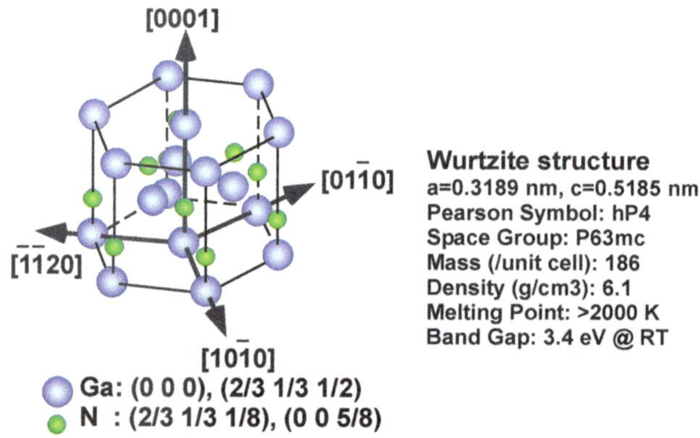

Fig. 1. Crystal structure of wurtzite GaN.

Figure 2 shows the atomic structures of each GaN face. The most common face of GaN is Ga-polar GaN(0001) (c-face), and recently non polar (11$\bar{2}$0) (a-face) and (10$\bar{1}$0) (m-face) GaN have been used for applications. The c-axis-oriented optoelectronic devices in particular suffer from undesirable spontaneous and piezoelectric polarization effects. The aim to eliminate such detrimental effects has led to renewed interest in non polar (a- and m-plane) GaN (Waltereit et al., 2000).

2.1 HVPE

For commercially available samples, the major fabrication method is hydride vapor phase epitaxy (HVPE). In the HVPE method, GaCl and NH_3 react to produce GaN, with H_2 and HCl gases as by-products: $GaCl + NH_3 \rightarrow GaN + H_2 + HCl$ (Maruska & Tietjen, 1969) (Fig. 3).

The crystallinity (such as defect density) and/or properties of GaN strongly depend on the growth conditions. HVPE GaN has a threading dislocation density of about 10^6-10^7 cm^{-2}, although GaN crystal quality has been improved in recent years and is expected to be improved, further. HVPE growth (Maruska & Tietjen, 1969) is a very popular method of fabricating both crystalline substrates and homoepitaxial layers because of the high

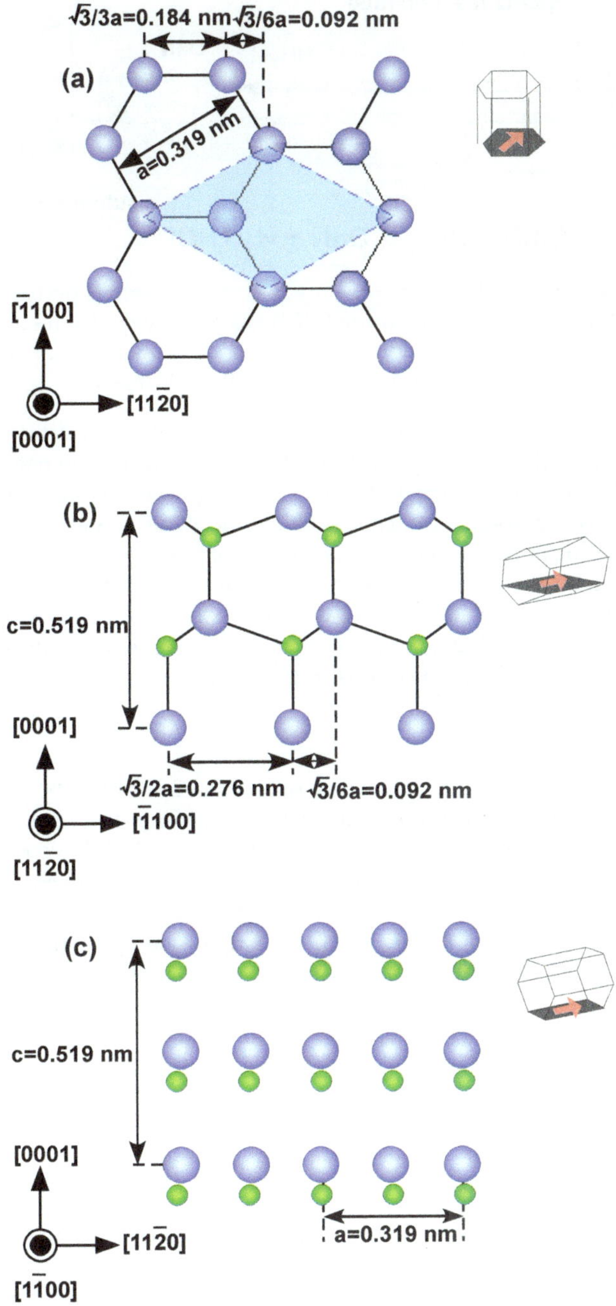

Fig. 2. Atomic arrangement of (a) GaN(0001) (*c*-face), (b) GaN(11$\bar{2}$0) (*a*-face), and (c)(10$\bar{1}$0) (*m*-face). The atomic arrangement is shown in the direction of each arrow in the inset crystals.

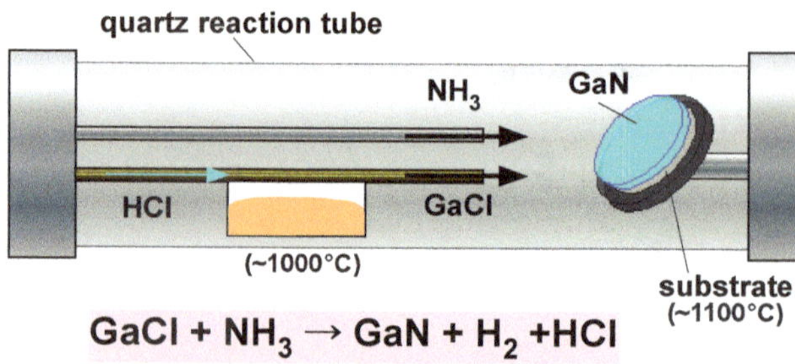

Fig. 3. Schematic illustration of HVPE growth method. The growth is performed in a quartz tube reactor by bubbling a hydrogen chloride (HCl) gas flow over a Ga melt, which results in the formation of gaseous gallium chloride (GaCl). The GaCl flows through the nozzle towards the sample, where it reacts with ammonia (NH_3) to form GaN.

growth rates of around 100 μm/hr. Also, many methods of stably fabricating larger and higher-quality GaN crystals at a low price by HVPE growth have been patented. Of course, dislocations severely influence device performance, though GaN crystals have been realized as products in the laser market. To achieve electronic devices with excellent properties, for instance, carrier transport, we should improve the crystal quality and surface-finishing methods and understand the local (microscopic) surface structure that becomes the interface under electrode materials. A dislocation density of less than 10^3 cm^{-2} is required for laser diode applications to obtain a higher yield and cost-effectiveness. Many researchers have tried to fabricate GaN crystals with lower dislocation densities.

2.2 Hydrogen impurity

The crystal properties and surface structures of GaN are affected by impurities in the crystals; in particular, the presence of hydrogen as an impurity has been reported (Nickel, 1999). It was shown that H in GaN exhibits unique features which have not been observed in more traditional semiconductors such as Si and GaAs (Nickel, 1999), for example, the formation of an acceptor-hydrogen complex, especially in GaN:Mg, with p-type conductivity. Also, the presence of ambient H changes the formation energy of the GaN surface and causes the formation of reconstruction structures. The hydrogen impurity sometimes arises from an impurity component of the source gases, especially ammonia. A hydrogen concentration of 10^{19}-10^{23} cm^{-3} was measured for molecular beam epitaxy (MBE)-GaN by nuclear reaction analysis, and a strong correlation between hydrogen concentration and crystal strain (dislocations) was observed. This correlation leads to a higher density of the hydrogen impurity in metal-organic chemical vapor deposition (MOCVD) films than in HVPE substrates (Hattori et al., 2009), affecting the dislocation density, when the hydrogen impurity is supplied from the source gases.

Figure 4(a) shows the typical major temperature programmed desorption (TPD) curves of an as-received HVPE1 sample. The temperature was ramped up at a rate of 0.5 K/s. Similar TPD curves to those in Fig. 4(a) were observed for the HVPE2 sample. As shown in the figure, the major detected values of m/e are 2, 12, 14, 16, 17, 18, 28, 35, 36, 69, and 104 in atomic units (AMU), which can be assigned to H_2^+, C^+, N^+/CH_2^+, O^+/NH_2^+, OH^+/NH_3^+, H_2O^+,

N_2^+/CO^+, Cl^+, HCl^+, Ga^+, and $GaCl^+$, respectively. Thus, the main desorption species are considered to be H_2, CH_4, NH_3, H_2O, N_2, CO, (H)Cl, Ga, and GaCl.

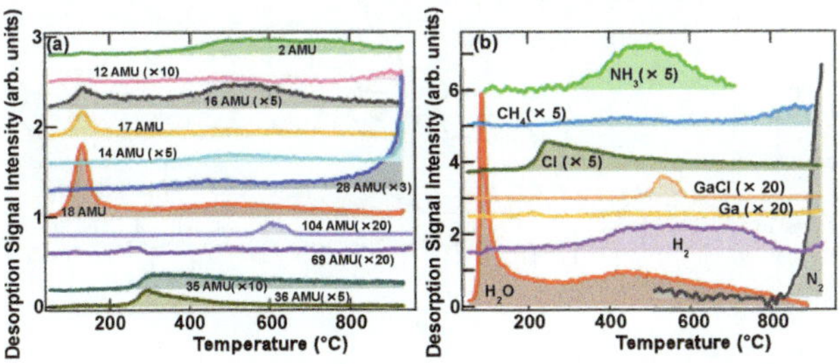

Fig. 4. (a) TPD curves of major mass numbers detected for HVPE1 samples. (b) TPD spectra of desorption species estimated from (a) considering the cracking elements and ratio. The temperature was ramped up at a rate of 0.5 K/s. The TPD curves for HVPE2 samples were similar to those for HVPE1 samples.

Figure 4(b) shows the desorption curves of these species estimated from Fig. 4(a). H_2O peaks below 200°C, N_2 peaks above 800°C, and broad H_2 peaks from 400 to 800°C can be seen for both samples. The origin of the H_2O desorption should be physical adsorption on the surface from humidity in the air. The m/e = 35 (Cl^+), 36 (HCl^+), and 104 ($GaCl^+$) peaks were clearly observed. In the HVPE method (Fig. 3), GaCl and NH_3 react to produce GaN, with H_2 and HCl gases as by-products: GaCl + NH_3 $\rightarrow$ GaN + H_2 + HCl (Maruska & Tietjen, 1969). The desorption of GaCl, (H)Cl, NH_3, and H_2 in HVPE-GaN samples strongly indicates that they originated from unreacted source materials and/or residual product materials. This implies that the HVPE growth conditions are not yet optimized. In the TPD of MOCVD samples, no GaCl desorption and little Cl desorption were detected (Hattori et al., 2009). In the MOCVD process, the gallium and nitrogen sources are usually $Ga(CH_3)_3$ and NH_3, respectively, and the products are GaN and CH_4: $Ga(CH_3)_3$ + NH_3 $\rightarrow$ GaN + $3CH_4$ (Amano et al., 1986). It was reported that intense CH_4 desorption occurs from MOCVD samples rather than HVPE samples (Hattori et al., 2009); the CH_4 species is one of the by-product elements in the MOCVD film and/or from surface carbon contamination.

3. GaN(0001) surface structures

3.1 GaN(0001)2×2 surface structures

The structures of clean surfaces are of particular importance since knowledge of the structures is the first step in understanding the fundamental issues of contact formation, chemical reactivity, growth processes, and so on. The surface structures of as-grown samples of Ga-polar GaN(0001) grown by MBE and MOCVD on substrates of Si-polar SiC(0001), sapphire(0001), Si(111), and so on have been studied *in-situ*. For the MBE growth of a GaN(0001) surface under a Ga-rich condition, 2×2, 2×3, 3×2, 3×4, 4×4, 5×5, 10×10, $5\sqrt{3}\times2\sqrt{13}$, 5×2.5, $\sqrt{7}\times\sqrt{7}$, and $\sqrt{3}\times\sqrt{3}$ reconstruction structures have been reported. In general, the unintentional presence of arsenic on the surface leads to these MBE reconstruction structures.

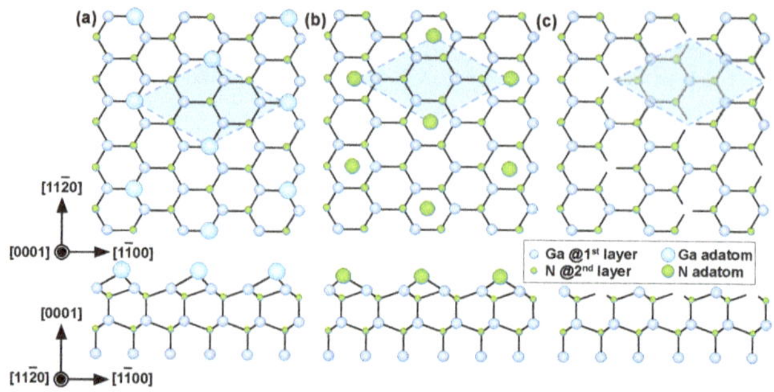

Fig. 5. GaN(0001)2×2 models: (a) Ga adatom, (b) N adatom, and (c) vacancy.

Ordered 1×1 and 2×2 reconstruction structures are formed on GaN(0001) surfaces. On samples grown by MOCVD, an ordered 1×1 surface with a nearly one-to-one stoichiometric composition has been reported. Figure 5 shows the 2×2 reconstruction models. Some theoretical studies have predicted that a GaN(0001) surface exhibits a 2×2 reconstruction structure under a Ga-rich condition: GaN(0001)2×2-Ga and its minimum-energy configuration have been proposed. Reconstructed GaN(0001)2×2 surfaces (GaN(0001)2×2-N), however, have been reported as a result of experiments on samples grown by MBE under nitrogen-rich growth conditions. These studies were performed *in-situ*; films were fabricated by controlling the stoichiometry in UHV, where the surfaces were subsequently observed. These *in-situ* methods cannot be universally applied to commercial sample surfaces.

3.2 As-received HVPE-GaN(0001) surface structures

Figure 6 shows low-energy electron diffraction (LEED) and reflection high-energy electron diffraction (RHEED) patterns of the two different as-received commercial HVPE1 and HVPE2 samples. LEED patterns of the HVPE1 sample (Fig. 6(a)) and HVPE2 sample (Fig. 6(b)) show almost no specific features. Note that LEED in this E_p range reflects the ordering of a few surface layers. It was reported that an MOCVD-grown GaN(0001) surface is terminated by hydrogen, and that MOCVD-grown GaN(0001) show clear 1×1 spots even after exposure to air (Hattori et al., 2009). That is why the survival of the 1×1 ordering on MOCVD samples is caused by an inert H-termination cap, similar to the case of Si(001)-H surfaces.

In general, commercially available GaN sample surfaces are surface-lapped (mechanically polished) and finished by chemical mechanical polishing (CMP), dry etching, and so on, after fabrication and subsequent exposure to air. The surface figuring treatments are performed to obtain mirror-polished smooth surfaces, where the presence of a surface oxide layer was reported. Thus, no specific LEED spots indicates that surface crystallinity and ordering were clearly destroyed in at least a few surface layers by the surface-finishing treatments.

In the RHEED patterns for the HVPE1 (Fig. 6(c)) and HVPE2 (Fig. 6(d)) samples, where the electron mean free path at this E_p is ~20 nm (Ichimiya & Cohen, 2004), we observed weak transmission diffraction spots (cyan circles) in addition to clear 1×1 surface spots (dashed arcs, corresponding to Laue zones). The transmission spots indicate the existence of 3D islands of

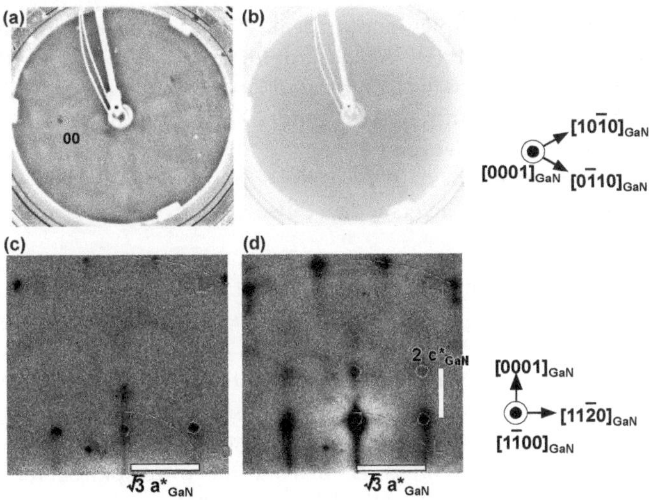

Fig. 6. LEED and RHEED patterns of as-received samples: (a) and (c) HVPE1, and (b) and (d) HVPE2 samples, respectively. Note that the 00 beam in LEED was not at the screen center. The incident electron direction was $[1\bar{1}00]$ in RHEED. E_p in LEED was 95 eV. In RHEED, Laue zones (L_0 and L_1) and transmission spots are indicated by pink dashed arcs and cyan circles, respectively. The scales of a^* and c^* are the reciprocal lattice units corresponding to a and c, respectively.

GaN(0001) on the as-received GaN(0001) surfaces. The main purpose of the surface-finishing treatments is to remove damage by lapping and to achieve surface flattening without damage. However, no effective polishing method or etchant has yet been confirmed for GaN substrates, although macroscopic surface roughness has been improved. Murata *et al.* evaluated the thickness of damage (depth profile) in commercial HVPE GaN substrates treated by CMP by measuring photocurrent density. They concluded that commercial HVPE GaN substrates have a damaged layer with a thickness of a few hundred nm and many scratches on the surface (Murata et al., 2009).

Figures 7(a) and (b) show the core-level spectra of Ga $3d_{5/2,3/2}$ and N 1s, respectively, for the as-received HVPE1 sample obtained by X-ray photoelectron spectroscopy (XPS). The Ga 3d spectra are asymmetric and can be fitted with four symmetric Voigt components with different chemical shifts corresponding to Ga bonding to nitrogen: Ga_N (binding energy (BE) = 19.2-20.3 eV), Ga bonding to oxygen: Ga_{O_x} (BE = 19.6-21.0 eV), Ga bonding to N-H: Ga_{H-N} (BE = 19.21 eV), and metal-Ga (BE = 18.4-18.49 eV). It is known that hydrogen is unintentionally doped in GaN, and hydrogen behavior in GaN films has been investigated by Fourier transform infrared (FTIR) absorption, electron-energy-loss spectroscopy, and so on. Kong *et al.* confirmed the stretch mode of the N-H bond by FTIR and denoted the hydrogen-related complex as Ga$\cdots$H-N. They also assigned a Ga$\cdots$H-N peak in the Ga 3d spectra in XPS. Note that the energy difference between Ga $3d_{3/2}$ and Ga $3d_{5/2}$ is less than 0.1 eV, negligible compared with our resolution. Upon comparing GaN samples with and without CMP treatment after film fabrication, we found a small Ga_{O_x} peak and negligible metal-Ga peaks in the no-CMP samples (Hattori et al., 2009). For the enhancement of the GaO_x and metal-Ga peaks in the as-received HVPE1 sample, we consider that CMP treatment after film(crystal) fabrication destroys the surface termination, promotes oxidation, and induces the formation of Ga metal impurities and/or segregation. The surface oxidation and the existence of metal impurities after CMP have been pointed out to be general phenomena. These results also indicate that CMP treatment has a severe effect on the surface structures and condition. Note that the existence of Ga oxide layers with a thickness of ~ 1.5 nm was reported for HPVE samples.

The N 1s spectra can be fitted with five components: N 1s core electrons bonding to hydrogen: N_{H_3} (BE = 405.6-406.2 eV) and N_{H_2} (BE = 397.7-399.72 eV), an N 1s core-electron bonding to gallium: N_{Ga} (BE = 396.2-397.86 eV), and two Ga LMM Auger electrons with satellites. There is no significant difference of the N 1s spectra between the HVPE and MOCVD samples. The impurity components of O 1s, C 1s, and Cl 2p were observed by XPS. From the intensities of these components, we can estimate their atomic percentages (atomic%) in the surface region using the established photoelectron cross sections and mean free paths. Table 1 summarizes the atomic% of each component for the HVPE1 and HVPE2 samples. Here, the sum of Ga, N, O, C, and Cl atomic% is fixed to 100%. We can confirm that the as-received HVPE1 and HVPE2 samples include non-negligible amounts of O, C, and Cl as impurities.

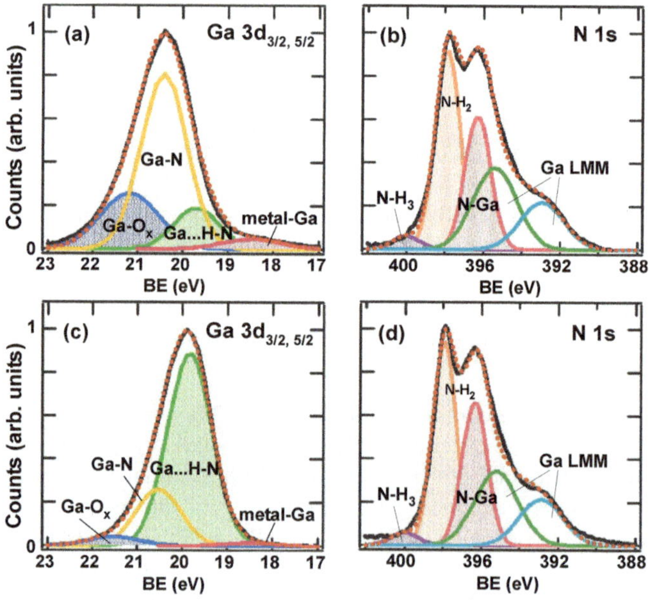

Fig. 7. Core-level XPS spectra of (a) and (c) gallium 3d, and (b) and (d) nitrogen 1s: (a) and (b) for an as-received (non etched) HVPE1 sample, and (c) and (d) for a 100 s HF-etched HVPE1 sample. Measured Ga 3d and N 1s spectra (black curves) are fitted with Voigt functions of four components (Ga_N, Ga_{O_x}, Ga_{H-N}, and metal-Ga) and five components (N_{Ga}, N_{H_2}, N_{H_3}, and two Ga LMM Auger satellites), respectively. The red dashed curves indicate the sum of the component peaks.

4. Surface stoichiometry through treatments

4.1 Annealing in UHV

Figure 8 shows schematic phase diagrams of (a) HVPE substrates (HVPE1 and HVPE2) after annealing, determined by LEED, RHEED, XPS, Auger electron spectroscopy (AES), TPD, and scanning tunneling microscope (STM) measurements. Relatively clean surfaces were obtained by annealing the HVPE samples; 1×1 structures exist on the HVPE surfaces after annealing at approximately 500-600°C. At a higher annealing temperature (≥500-600°C), nitrogen sublimation in the form of ammonia started to occur and 3D islands with facets formed. Finally, samples were damaged after annealing above 800°C.

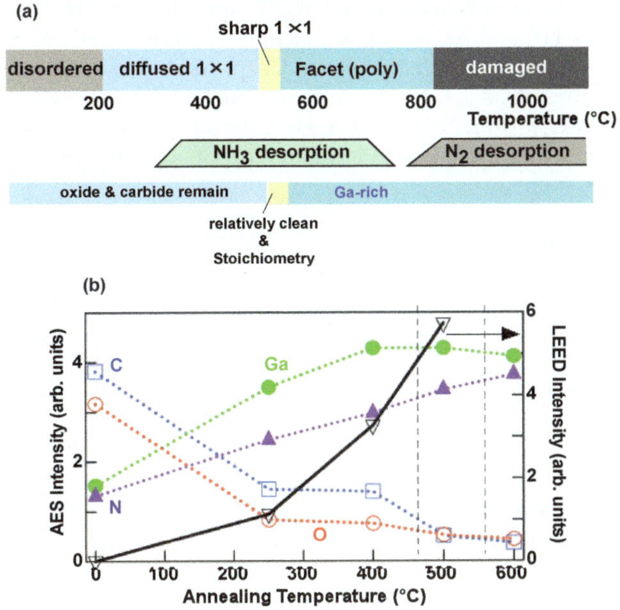

Fig. 8. Schematic phase diagrams for HVPE1 and HVPE2 samples treated by annealing in UHV. Nitrogen sublimates in two forms: as ammonia at around 300-700°C and as molecular nitrogen above 800°C. The phases were observed at RT after annealing. (b) Intensity change of (10) LEED spots as a function of annealing temperature (open black inverted triangles). In addition, the AES intensities of C (277 eV, open blue squares), N (392 eV, filled violet triangles), O (525 eV, open red circles), and Ga (1098 eV, filled green circles) are plotted against annealing temperature. In (b) annealing was performed at 250°C for 12 hrs., 400°C for 1 hr, 500°C for 10 min, and 600°C for 10 min. The incident E_p for LEED and AES were 95 eV and 2 keV, respectively.

We found that three-step annealing results in relatively clean and flat surfaces with balanced stoichiometry. The established conditions for each step were (1) 200°C for 12 hrs., (2) 400°C for 1 hr, and (3) 500°C for 5 min. AES signals (Fig. 8(b)) showed the reduction of oxide and carbide from the surface upon annealing at ~200 and 500°C, which is consistent with the TPD results (Fig. 4). Figure 9(a) shows an LEED pattern and a typical STM image of an HVPE1 sample annealed (degassed) at 200°C for 12 hrs. The LEED pattern showed a diffused 1×1 structure and STM indicated that many clusters remained on the surface. These clusters are ascribed to the formation of oxide/carbide layers. The peak-to-valley (PV) height and root-mean-square (RMS) roughness were 2.5 and 0.86 nm, respectively. After another two steps, the sharpest 1×1 LEED patterns appeared and STM images showed wide and flat terraces of ~10 nm size and steps, as shown in Fig. 9(b). PV and RMS were 1.7 and 0.39 nm, respectively. Sharp 1×1 spots without any transmission spots are seen in the RHEED pattern, as shown in Fig. 9(c). On relatively clean surfaces, the AES intensities of C and O decreased to ~20% of those for the as-received surfaces, but no ordered atomic arrangements were observed anywhere.

Table 1. Atomic percentages of surface species, Ga, N, O, C, and Cl for HVPE1 and HVPE2 samples with different treatments (as-received, 600°C annealing in UHV, Ar^+ sputtering, and 100 sec-HF, 500 sec-HF, and 700 sec-HNO_3 etching), determined by XPS intensities and sensitivity factors for Ga 3d, N 1s, O 1s, C 1s, and Cl 2p. As shown in Figs. 2, we can estimate the component intensities of Ga_N, Ga_{O_x}, Ga_{H-N}, and metal-Ga from Ga 3d spectra, and those of N_{Ga}, N_{H_2}, and N_{H_3} from N 1s spectra. Here the sum of Ga, N, O, C, and Cl atomic-% is fixed to be 100%. Note that these values have an error of approximately ±10%, e.g., 16% is 16±2%.

Sample	Treatment	Ga [%]					N [%]				O	C	Cl
		Ga_N	Ga_{H-N}	Ga_{O_x}	metal	sum	N_{Ga}	N_{H_2}	N_{H_3}	sum	[%]	[%]	[%]
HVPE1	received	16	9	13	2	**40**	15	22	6	**43**	**6**	**11**	<1
	sputtering	26	20	–	–	**46**	22	25	4	**51**	**2**	–	<1
	annealing	18	15	8	1	**42**	21	14	3	**38**	**11**	**8**	<1
	100 sec-HF	9	30	2	<1	**41**	19	26	2	**47**	**3**	**9**	<1
	500 sec-HF	7	26	2	<1	**35**	24	26	5	**55**	**3**	**7**	<1
	700 sec-HNO_3	8	27	2	<1	**37**	24	26	2	**52**	**3**	**8**	<1
HVPE2	received	13	19	3	2	**37**	20	20	3	**43**	**11**	**8**	<1

In the HVPE2 samples, similar dependence of the behavior of surface structures on the annealing temperature to that shown in the HVPE1 sample was confirmed. However, the microscopic structures observed by STM were quite different from those of the HVPE1 sample. Figure 10 shows an RHEED pattern and typical STM image of an HVPE2 sample after three-step annealing (200°C for 12 hrs. + 400°C for 1 hr. + 500°C for 5 min). Sharp 1×1 spots without any transmission spots are seen in the RHEED pattern, as shown in Fig. 10(a). The LEED pattern also showed sharp 1×1 spots. The STM image indicated a flat surface with numerous grains of a few nm in size as shown in Fig. 10(b). Some grains gathered to form petal-like shapes (indicated by a diamond in the inset in Fig. 10(b)). On less than a few percent of the surface area, the petal-like structures coalesced with each other (indicated by diamonds in Fig. 10(c)), but these structures are scattered. The side length of the diamonds is ~2 nm; thus, the area of one petal-like structure corresponds to $\approx 6a \times 6a$. Note that the directions of the sides of the diamonds are parallel to the $\langle 11\bar{2}0 \rangle$ directions. These petal-like structures resemble those in the STM image in Ref. (Feenstra et al., 2000), where a 12×12-reconstructed structure, which had not been previously observed on (0001) or (000$\bar{1}$) surfaces. Feenstra *et al.* (Feenstra et al., 2000) associated this novel structure with the "1×1" Ga adlayers on GaN(0001) layers, implying the existence of an inversion domain immediately below the novel reconstruction area of the surface. On HVPE2 sample surfaces, petal-like structures were reproducibly observed but neither pits nor holes were observed. They were present in at

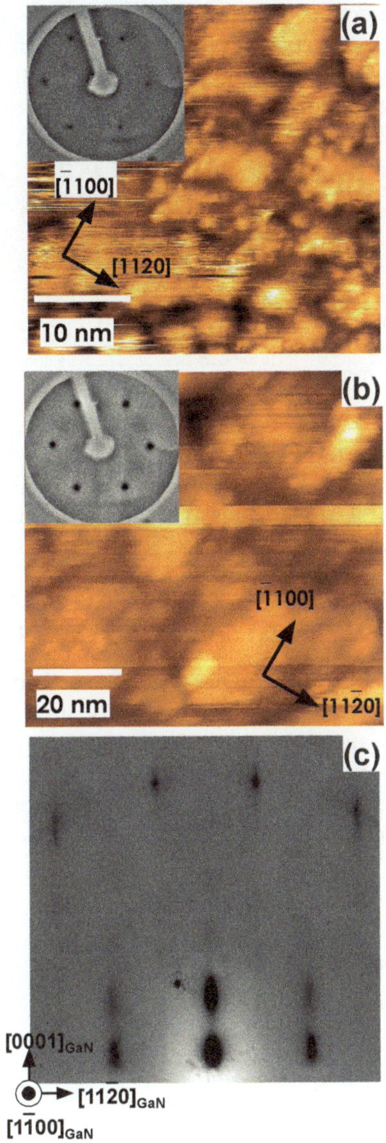

Fig. 9. (a) Typical STM image and LEED pattern after degassing at 200°C for 12 hrs. (b) Typical STM images and LEED pattern (in inset), and (c) RHEED pattern after three-step annealing (200°C for 12 hrs., 400°C for 1 hr, and 500°C for 10 min). The observed conditions were (a) $E_p = 85$ eV, $V_s = -4.8$ V, $I_t = 0.3$ nA; (b) $E_p = 85$ eV, $V_s = -5.0$ V, $I_t = 0.1$ nA; and (c) [1$\bar{1}$00] incident, 15 kV.

most a few % in the whole observed area. The HVPE2 sample was grown by multistep lateral epitaxial overgrowth, and there are defects distributed inside this crystal. Thus, we associate the petal-like structure on the surface with an area with high defect density immediately below the inside of the crystal, and conclude that the petal-like structure in this study is correlated with the novel reconstruction in the prior work (Feenstra et al., 2000). The petal-like structures observed on HVPE2 samples were not observed on HVPE1 samples.

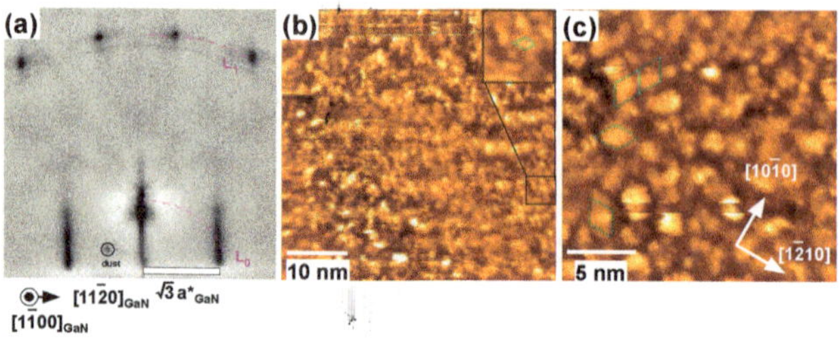

Fig. 10. (a) Typical STM image and LEED pattern (inset) after degassing at 200°C for 12 hrs. (b) Cross-sectional profile through the dashed line in (a); PV and RMS were 2.5 and 0.86 nm, respectively. (c) and (e) Typical STM images and LEED pattern (in inset of (c)) after three-step annealing (200°C for 12 hrs., 400°C for 1 hr, and 500°C for 10 min). (d) and (f) Cross-sectional profiles through the dashed lines in (c) and (e), respectively. PV and RMS in (d) were 1.7 and 0.39 nm, respectively. The observed conditions were (a) $E_p = 85$ eV, $V_s = -4.8$ V, $I_t = 0.3$ nA; (c) $E_p = 85$ eV, $V_s = -5.0$ V, $I_t = 0.1$ nA; and (e) $V_s = +4.2$ V, $I_t = 0.5$ nA.

Annealing in UHV, which is one of the general methods of obtaining clean and well-defined surfaces, could not produce the expected clean GaN surfaces; only postannealing below ~550°C could reduce (but not eliminate) surface contamination while maintaining the stoichiometry; however, it did not induce any surface reconstructions. The annealing treatment at high temperature removed surface contamination completely, but facets started to appear on the GaN surface above 550°C and the surface become Ga-rich with unbalanced stoichiometry (Table 1).

Figure 11(a) shows a typical RHEED pattern of a HVPE1 sample annealed at 600°C for 20 min after the three-step annealing. We can see characteristic "chevrons" in the RHEED patterns, corresponding to facets, in addition to transmission spots from 3D GaN islands. The transmission spot intensities were weak, indicating a small number of epitaxial islands. The chevron patterns were broad, indicating small domains. Although the chevron directions are $\langle 11\bar{2}4\rangle$ at the $[1\bar{1}00]$ incidence (Fig. 6(a)), we consider that these are part of $\langle 101\bar{2}\rangle$ reciprocal rods, which correspond to $\{101\bar{2}\}$ facets, from the LEED results of Fig. 11(b). Figures 11(b) and (c) respectively show an LEED pattern and a typical STM image of a HVPE1 sample annealed at 600°C for 5 min after the three-step annealing. We can see characteristic spots in the LEED pattern (some of them are indicated by arrows); the motion of these spots with changing E_p was different from that of the fundamental spots, reflecting the inclined reciprocal rods of the facets. The motion of the facet spots was restricted in the directions from the 00 spots to the 10, 01, $\bar{1}$1, $\bar{1}$0, 0$\bar{1}$, and 1$\bar{1}$ spots, implying that the inclined reciprocal rods are oriented in the $\langle 10\bar{1}n\rangle$ directions. Since the projection of the RHEED chevrons with the $\langle 11\bar{2}4\rangle$ direction (Fig. 11(a)) in the $\langle 10\bar{1}n\rangle$ direction gives the $\langle 10\bar{1}2\rangle$ direction, we consider the existence of six fold facets of $\{10\bar{1}2\}$. This facet plane is the so-called R-plane. There have been many reports about facets formed on GaN surfaces. Most of them were formed during chemical etching and at the beginning of crystal fabrication, where facets with a size of few tens of μm, for example, $\{10\bar{1}\bar{1}\}$ facets, were formed. The $\{10\bar{1}\bar{2}\}$ orientation of the facet planes on GaN(0001) films has also been reported. For MOCVD GaN(0001) samples annealed at around 800-900°C in UHV, facet LEED patterns were reported in some references.

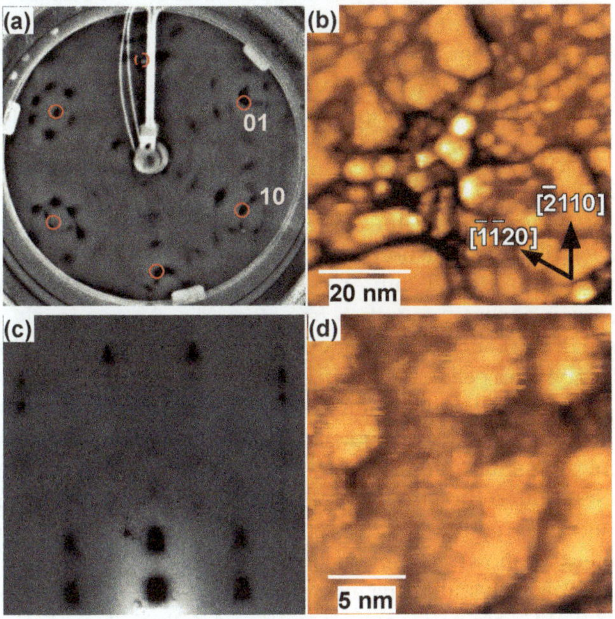

Fig. 11. (a) LEED pattern (E_p = 68 eV) of HVPE1 sample annealed at 600°C for 5 min after three-step annealing. Fundamental spots are marked by red circles. The other spots correspond to facets, some of them indicated by arrows. (b) Typical topographic STM image ($V_s = +2.0$ V, I_t = 0.2 nA) of a facet surface. Numerous protrusions ascribed to facets with a size of a few nm were observed.

According to STM images, many hillocks (a few tens of nm in diameter and a few nm in height) appeared on the surface, as shown in Fig. 11(c). Although the shapes of the facet islands were not clear in the STM images, these hillocks should be the origin of the facet patterns in the LEED and RHEED patterns. From the results of XPS, we found that the facet surface was Ga-rich, corresponding to the higher desorption of NH_3 than Ga and GaCl below approximately 600°C in TPD (Fig. 4). At present, although the mechanism of facet formation by annealing in UHV is not clear, facets are thought to grow mainly by the sublimation of nitrogen; the enhancement of grooves on the surfaces in the sublimation produces isolated 3D islands with facets. Instead of GaN facets, the segregated Ga droplets was also suggested. Anyway, the stoichiometry is broken on the faceting rough surfaces, and these unbalanced rough surfaces could not be recovered by additional treatments.

4.2 Sputtering

Ar ion sputtering is commonly used to prepare clean surfaces and has been used for Ga compound III-V semiconductor surfaces such as GaP, GaAs, GaSb, and GaN. For GaN, it was reported that nitrogen ions are more effective than Ar and Xe ions in removing C and O, and subsequently the annealed surfaces exhibited greater ordering. In previous studies, the reduction of surface contamination and the appearance of 1×1 surface diffraction patterns were revealed. Here, we present the surface structures and compositions obtained through Ar^+ sputtering and subsequent annealing, and discuss their microscopic phenomena. We found that Ar^+ sputtering of the GaN surface is very effective for removing oxide and carbide from the surface, but the enhancement of surface roughness caused by the sputtering could not be recovered by postannealing.

The as-received GaN surfaces have carbon and oxygen as impurities (Table 1 and Fig. 8(b)). Ar ion sputtering is effective for removing C and O surface contamination. Table 1 and Fig. 12(a) show a significant reduction of C 1s, O 1s, and Ga 3d (Ga_{O_x}) components with increasing sputtering time. After 20 min Ar^+ sputtering, the C 1s peak was not detectable and the surface oxygen component was only a few atomic%. As previously mentioned, HVPE GaN (HVPE1 and HVPE2) samples include Cl as an impurity. In XPS spectra, small but obvious Cl 2s, 2p, and Auger peaks (<1 atomic%) always appeared, even after Ar^+ sputtering. This implies the existence of Cl in the bulk of the HVPE HVPE1 samples, as mentioned in subsection 3.3. Figure 12(a) shows a higher rate of increase of Ga 3d (Ga_N) than that of N 1s (N_{Ga}) with increasing sputtering time; the intensity of Ga_N increases over tenfold while that of N_{Ga} increases around threefold after 20 min sputtering. The Ga_N and N_{Ga} intensities against all elements observed on the surface increased from 16 to 26 atomic% and 15 to 22 atomic%, respectively, as shown in Table 1. The higher rate of increase of Ga_N is thought to be caused by the difference in sputtering efficiency between Ga and N. A molecular dynamics simulation showed that Ga atoms are always sputtered with N atoms in pairs, while N atoms are mostly sputtered alone. This simulation also showed that in Ar^+ sputtering at an incident energy of 500 eV, the nitrogen sputtering yield is about five times higher than that of gallium. This indicates that the Ar^+-sputtered GaN surface does not maintain its stoichiometry and becomes Ga-rich at the near-surface. Under our conditions (20 min sputtering), however, the stoichiometry did not change greatly (Table 1), despite the increase of the Ga_{H-N} and N_{H_2} components. The hydrogen included in the samples is expected to play a role in stabilizing the stoichiometry against sputtering.

Figure 12(b) shows a typical RHEED pattern of an HVPE1 sample after Ar^+ sputtering for 20 min and subsequent three-step annealing (200°C for 12 hrs. + 400°C for 1 hr. + 500°C for 5 min). Compared with Fig. 9(c), the 1×1 spots are more diffused and the background level is higher. Faint transmission spots (cyan circles) are seen. The transmission spots indicate the existence of three-dimensional (3D) islands on the surface. Indeed, we observed many islands in STM images as shown in Fig. 12(c). In the RHEED patterns some faint transmission diffraction patterns and chevron patterns corresponding to facets were also observed at different incident azimuth angles. The chevron patterns, implying the presence of facets, were observed after 600°C annealing for both 10 min and 12 hrs. The existence of epitaxial 3D GaN islands implies that the recovery from an Ar^+-sputtered surface is not sufficient to form a flat surface; postannealing cannot produce a flat surface. In contrast, without Ar^+

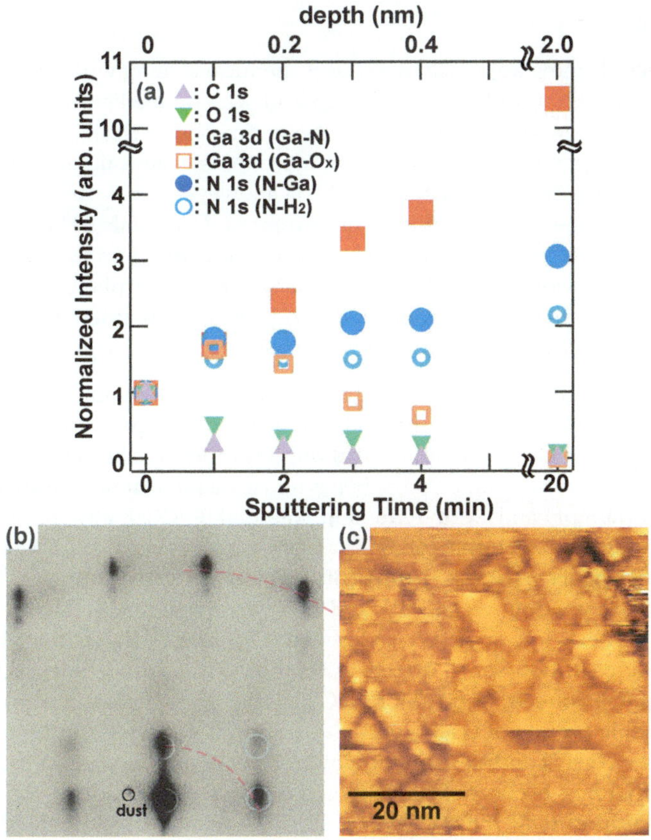

Fig. 12. (a) Normalized XPS intensities of components of C 1s (purple triangles), O 1s (green inverted triangles), Ga $3d_{(Ga_N)}$ (red solid squares), Ga $3d_{(GaO_x)}$ (pink open squares), N $1s_{(N_{Ga})}$ (blue solid circles), and N $1s_{(N_{H_2})}$ (cyan open circles) as functions of Ar^+ sputtering time for HVPE1 samples. Here, each intensity was normalized to the as-received (no-treated) value. The Ar^+ sputtering rate was ~0.1 nm/min. (b) RHEED pattern and (c) STM image for HVPE1 sample after Ar^+ sputtering and subsequent three-step annealing (200°C for 12 hrs., 400°C for 1 hr. and 500°C for 10 min). The incident electron direction was $[1\bar{1}00]$.

sputtering, the intensity of the transmission spots is strong, indicating a large number of epitaxial islands. For HVPE surfaces without Ar^+ sputtering, 3D islands with facets were also formed by annealing above 550 and 600°C. For Ar^+-sputtered HVPE1 samples, the formation of 3D islands started at a slightly lower annealing temperature (500°C). This would imply that the sputtering-induced surface damage leads to easier NH_3 sublimation, thus producing 3D islands. Metallic Ga (α, β, γ, δ, ϵ-Ga, Ga(II), or Ga(III)) in the liquid phase shows halo patterns in diffraction patterns, however, we could not observe any halo patterns in RHEED patterns. According to the STM observation, many small clusters and grains appeared to form.

4.3 Wet treatment

As another surface cleaning approach, we expect chemical etching with subsequent annealing to be also a possible method of obtaining clean and well-defined surfaces. For instance, for Si surfaces chemically etched by a wet process, we can obtain clean surfaces without oxide layers but with hydrogen termination. In general, GaN is chemically stable and expected to be hardly etched (Pankove & Moustukas., 1999). However, GaN can be etched in various chemicals because of the low crystallinity of current GaN crystals. Thus, it is necessary to appropriately control the etching conditions so that oxide layers on the surface are removed while avoiding the erosion of bulk GaN. The GaN surface morphology and stoichiometry depend on the different reactivity between GaN and Ga oxide in different etchants. Therefore, we should predict the reaction in solutions on the basis of a potential-pH equilibrium diagram.

A potential-pH equilibrium diagram, which maps out possible stable phases of an aqueous electrochemical system for a particular metal, can be used as a guide to explain the oxidation and etching reactions of target materials. Although it indicates the condition of Ga metal in solution, it can help us to speculate on the behavior of Ga ions (atoms) in solid GaN bulk in a solution with a certain pH and oxide-reduction potential (ORP). Figure 13 shows a simplified potential-pH equilibrium diagram for the gallium-water system at 25°C (Pourbaix, 1974) showing the four etchant conditions. Figure 13(a) indicates regions of "immunity", "corrosion", and "passivation", depending on the pH and OPR of the solution. The diagram indicates the stability of Ga metal in a specific environment. Immunity shows that the Ga metal has not reacted, while corrosion shows that a general oxidation-reduction reaction occurs and uncharged Ga atoms convert to Ga^{3+} or GaO_3^{3-} ions. Passivation occurs when the metal forms a highly stable coating of Ga_2O_3 on the surface. This theoretical pH-potential diagram is valid under the defined condition, under which no other reactions of Ga^{3+} and GaO_3^{3-} ions, or Ga_2O_3 with other ions occur in the solutions (Pourbaix, 1974). Indeed, for example, Ga_2O_3 dissolves into a solution at all values of pH, as shown in Fig. 13(b). Note that the pH-potential diagram is only valid when the dissolution of all substances is excluded. Thus, when the Ga atoms (ions) of a GaN crystal come in contact with a solution they react to form Ga^{3+}, GaO_3^{3-}, or Ga_2O_3, depending the pH and OPR conditions of the solution, as shown in Fig. 13(a). To obtain an oxide-free balanced-stoichiometry GaN surface, the reaction in the solution should stop before GaN itself starts to be etched by controlling the dipping time.

The solubility of Ga_2O_3 depends on the pH of the solution (Pourbaix, 1974), as shown in Fig. 13(b); the solubility of Ga_2O_3 increases in stronger acidic and basic solutions, that is, at a lower pH and higher pH, respectively. The diagram of Fig. 13(a) predicts the passivation phase, i.e., a Ga_2O_3 coating on a GaN surface, in HF solution, and Fig. 13(b) indicates the etching of Ga_2O_3 in HF solution. Since the reaction speed of the etching is faster than that of the oxidation (Pourbaix, 1974), the rate-determining factor is the oxidation in this case.

The appropriate dipping time in each solution was determined experimentally. Figure 14(a) shows the LEED intensities of the (10) spot (filled circles) as a function of dipping time in 0.5 wt% HF solution for HVPE1 samples without annealing. The figure indicates that dipping for 100 s led to the strongest intensity. The surface stoichiometry was almost retained after 100 s-HF treatment (Ga/N $\sim$0.93 $\rightarrow\sim$0.87), while the surface became considerably Ga-defective/N-rich (Ga/N $\sim$0.64) after 500 s-HF etching. Regarding the roughness obtained

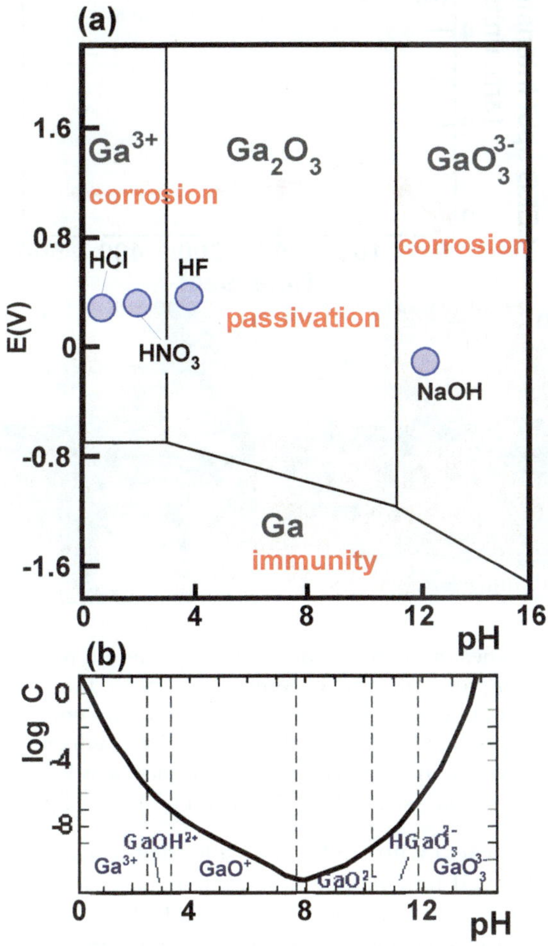

Fig. 13. (a) Potential-pH equilibrium diagram for the gallium-water system at 25°C, quoted from Ref. (Pourbaix, 1974). Predominant ion boundaries are represented by lines. The diagram represents the theoretical conditions (regions) of the "corrosion", "immunity", and "passivation" of gallium. The diagram indicates the stability of Ga metal in a specific environment. Immunity shows that the metal has not reacted, while corrosion shows that a general oxidation-reduction reaction occurs and uncharged Ga atoms convert to Ga^{3+} or GaO_3^{3-} ions. Passivation occurs when the metal forms a highly stable coating of Ga_2O_3 on the surface. The circles represent etchant conditions in this study. (b) Dependence of solubility of Ga_2O_3 on pH in the theoretical equilibrium condition (Pourbaix, 1974). C is the concentration of gallium in solution in all its dissolved forms (g-at Ga/l). The gallium, oxide, and hydroxide ions represented at the bottom are the possible dissolved substances at the pH ranges partitioned by the solid lines.

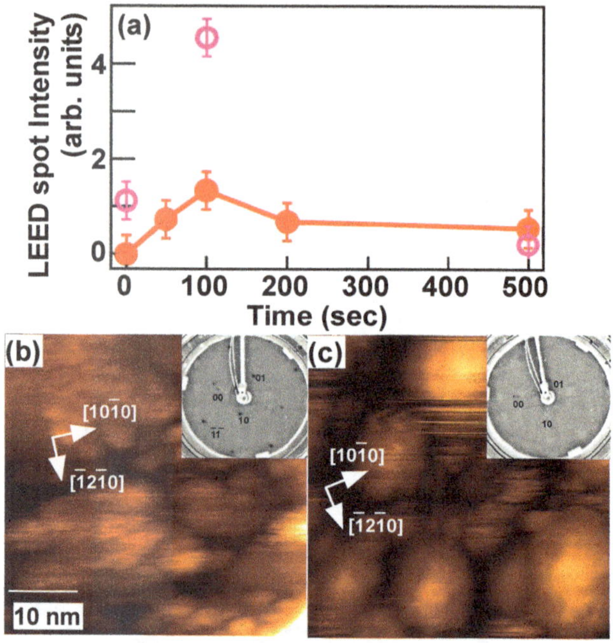

Fig. 14. (a) LEED intensities of the fundamental (10) spot as a function of the HF dipping time for HVPE-GaN samples. Solid circles represent as-treated samples, while open ones represent three-step postannealed samples after the wet treatment. STM images of HVPE-GaN samples dipped in HF for (b) 100 s and (c) 500 s, which were subsequently degassed at 200°C for 12 hrs. in UHV. The LEED condition was E_p = 95 eV in (a)-(c). The STM conditions were (b) $V_s = +4.2$ V, I_t = 0.5 nA and (c) $V_s = -2.8$ V, I_t = 0.5 nA.

from STM, the surface treated with HF for 100 s after degassing had a PV value of $\approx$ 1.8 nm and an RMS value of $\approx$ 0.51 nm, making it flatter than an as-received surface (PV $\approx$ 2.5 nm and RMS $\approx$ 0.86 nm)(Fig. 14(b)). Reconstructed GaN(0001)2×2 surfaces (GaN(0001)2×2-N) appear on the 100 s-HF treated GaN surfaces by three-step annealing. As shown in Fig. 15(b), the STM image revealed that the surface was very flat with certain atomic defects (PV and RMS were 0.86 and 0.17 nm, respectively). The stoichiometry was almost retained, but, there was a tendency of slightly a Ga-defective/N-rich surface. The surface containing local Ga-defective/N-rich area may explain the formation of GaN(0001)2×2-N on part of the surface after postannealing.

Appropriate etching conditions can lead to clean and reconstructed GaN surfaces. However, inappropriate etching conditions result in rough surfaces with unbalanced stoichiometry. When the dipping time in HF solution exceeded 100 s, many protrusions of $\sim$10 nm diameter and a few nm in height appeared on the surface, as shown in Fig. 14(c). The surface flatness and stoichiometry could not be recovered by any treatment. This result means that Ga atoms in the solid GaN crystal are oxidized to Ga_2O_3 and subsequently dissolve, assuming the inert property of the paired N atoms, resulting in a rough surface with unbalanced stoichiometry (N-rich/Ga-defective) when the dipping time exceeds 100 s.

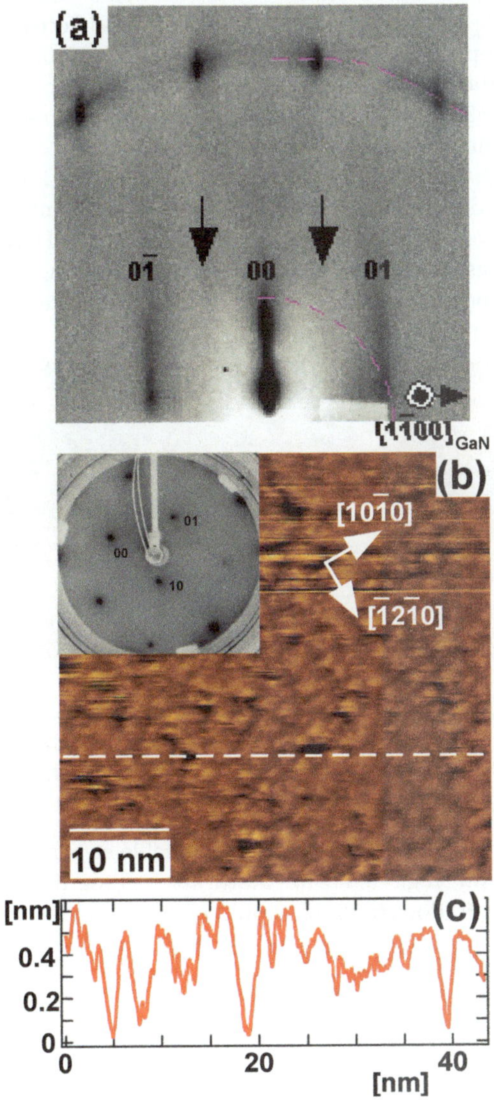

Fig. 15. (a) RHEED pattern, (b) STM images with inset of LEED image for HVPE1-GaN samples after 100 s-HF treatment and three-step postannealing in UHV. In the RHEED patterns, faint 2×2 surface reconstruction streaks (indicated by arrows) can be seen, in addition to 1×1 surface streaks on the integral Laue zones (L_0 and L_1). (c) Cross-sectional profile through the dashed lines in (b). PV and RMS were 0.86 and 0.17 nm. The incident electron energies for RHEED and LEED were 15 keV and 95 eV, respectively. The STM condition were $V_s = -2.5$ V, $I_t = 0.5$ nA.

In the case of HCl, HNO_3, and NaOH solutions which are in the "corrosion" region, the etching treatment led to an oxide-free surface, but unbalanced Ga-defective/N-rich stoichiometry (Table 1), even for appropriate etching time (Hattori et al., 2010). The pH-E diagram of Fig. 13(a) predicts Ga elution as Ga^{3+} or GaO_3^{3-} in the corrosion phases in HCl, HNO_3, and NaOH solutions. The unbalanced Ga-defective/N-rich stoichiometry after etching means in addition to Ga_2O_3 etching, Ga elusion on GaN surface was occurred. STM images after postannealing revealed that the HNO_3 (HCl)-treated sample had a rough surface and the NaOH-treated sample had many small protrusions. The different types of corrosion may induce different morphologies: huge protrusions are formed by HNO_3 and HCl and small protrusions by NaOH (Figs. 16(a) and (b)). For the surface cleaning procedures of GaAs and GaP, only HF solution produced oxide-free and stoichiometry-balanced surfaces, while HCl solution created an As-rich GaAs(001) surface.

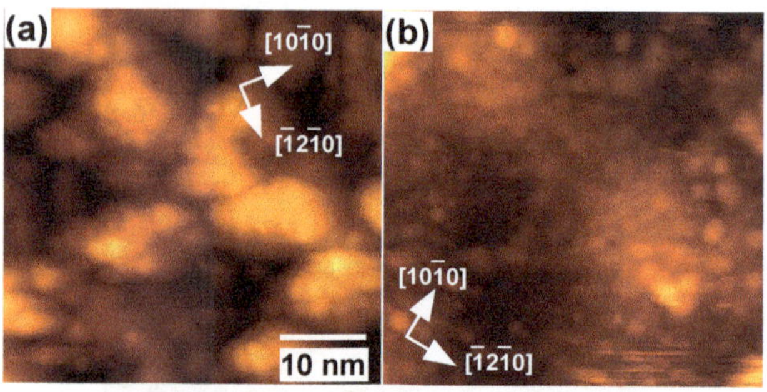

Fig. 16. Typical STM images after three-step annealing of (a) HNO_3- and (b) NaOH-treated HVPE-GaN samples. The STM image of a HCl-treated and postannealed surface was similar to that (a). The STM conditions were (a) $V_s = -4.5$ V, $I_t = 0.3$ nA and (b) $V_s = +4.5$ V, $I_t = 0.4$ nA.

The important points for wet treatment are to select suitable etchant solutions whose pH and ORP are in the "passivation" region and to control the etching time in those solutions. At the first trial, lower concentration solutions should be used because it is easier to control the reactions.

5. Summary

We summarize the surface structures, morphology, and stoichiometry of GaN(0001) surfaces treated by annealing in UHV, sputtering, and chemical solutions. For the UHV-treated GaN surfaces, there were structure differences among samples under the same treatment, and we clarified the effect of crystal quality, such as dislocations, the concentration of hydrogen impurities, and the residual reactant molecules in GaN films, on the surface structure. Because of hydrogen impurities, nitrogen desorbs as ammonia upon annealing above a temperature of 600°C, and the clean surfaces with balanced stoichiometry could not be produced. The sputtering treatment was effective in removing C and O surface contamination, but the increased surface roughness as a result of Ar^+ sputtering could not be recovered by postannealing, and the formation of 3D islands with facets was enhanced on the sputtered surfaces. Because the nitrogen sputtering yield is about five times higher than that of gallium, the Ar^+-sputtered GaN surface does not maintain its stoichiometry and becomes Ga-rich at the near-surface. In the case of performing wet etching on a GaN system, selecting an etchant solution with a certain pH and oxide-reduction potential and controlling the etching time are important to obtain an oxide-free and balanced-stoichiometry surface.

For GaN(0001) surfaces, an unbalanced surface stoichiometry typically leads to a very rough Ga-rich surfaces(Figs 11(b), 12(c), 16(a), and 16(b)) or N-rich surfaces ((Figs 11(d) and 14(c)). For the binary compound GaN system, Ga and N have different reactions to annealing, sputtering (gas reaction), and wet etching (wet chemical reaction, and the stoichiometry is easily broken. The surface morphology showed close correlation with the surface stoichiometry. The structures of clean and flat surfaces are of particular importance since knowledge of the structures is the first step in understanding the fundamental issues of contact formation, chemical reactivity, growth processes, and so on. To obtain clean and flat GaN(0001) surfaces, we should understand the reaction differences between Ga and N, and maintain their stoichiometry.

6. Acknowledgements

The authors acknowledge helpful discussions with Prof. Hiroshi Daimon and Prof. Ken Hattori. ANH also expresses many thanks to her beloved husband Ken for useful advices and constant encouragement throughout this investigation.

7. References

Hattori A. N., Endo K., Hattori K. & Daimon H. (2009) GaN(0001) surfaces on commercial hydride vapor phase epitaxy and metal-organic chemical vapor deposition substrates in ultra high vacuum *Applied Surface Science*, Vol. 256, pp 4745-4756.

Hattori A. N., Kawamura F., Yoshimura M., Kitaoka Y., Mori Y., Hattori K., Daimon H. & Endo K. (2010) Chemical etchant dependence of surface structure and morphology on GaN(0001) substrates *Surface Science*, Vol. 604, pp 1247-1253.

Properties of Group III Nitrides, edited by Edgar J. H.(1994) INSPEC, London.

Strite S. & Morkoc H. (1992) GaN, AlN, and InN: A review *Journal of Vacuum Science and Technology B*, Vol. 10, pp 1237-1266.

Waltereit P.,Brandt O. , Trampert A., Grahn H. T., Menniger J., Ramsteiner M., Reiche M., & Ploog K. H. (2000) Nitride semiconductors free of electrostatic fields for efficient white light-emitting diodes *Nature*, Vol. 406, pp 865-868.

Maruska H. P. & Tietjen J. J. (1969) The preparation and properties of vapor-deposited single-crystal-line GaN *Applied Physics Letter* Vol. 15, pp 327-329.

Hydrogen in Semiconductors II, Nickel N. H. (1999) Academic Press, Berlin.

Amano H., Sawaki N., Akasaki I., $ Toyoda Y. (1986) Metalorganic vapor phase epitaxial growth of a high quality GaN film using an AlN buffer layer *Applied Physics Letters* Vol. 48, pp 353-355.

Reflection High Energy Electron Diffraction, Ichimiya A. & Cohen P. I. (2004) Cambridge University Press, Cambridge.

Murata J., Sadakuni S., Yagi K., Sano Y., Okamoto T., Arima K., Hattori ĄA. N., Mimura H. & Yamauchi K. (2009) Planarization of GaN(0001) Surfaces by Photo-Electrochemical

Method with Solid Acidic or Basic Catalyst *Japanese Journal of Applied Physics*, Vol. 48, pp 121001-1-4.
Feenstra R. M., Chen H., Ramachandran V., Smith A. R. & Greve D. W. (2000) Reconstructions of GaN and InGaN surfaces *Applied Surface Science*, Vol. 166, pp165-172.
Gallium Nitride Pankove J. I. & Moustukas T. D. (1999) Academic Press, San Diego.
Atlas of Electrochemical Equibilia in Solutions Pourbaix M. (1974) NACE, Houston.

4

Nonstoichiometry and Properties of SnTe Semiconductor Phase of Variable Composition

Elena Rogacheva
National Technical University "Kharkov Polytechnic Institute"
Ukraine

1. Introduction

According to the modern views, all semiconductor compounds have a homogeneity region (HR), i.e. represent phases of variable composition. The existence of thermodynamically equilibrium deviation from stoichiometry in a chemical compound is connected with a decrease in the crystal free energy under introduction of nonstoichiometric defects (NSD) whose appearance leads to a growth in configurational entropy. The location and size of HR depend on the ratio of the formation energies for different types of defects. The stoichiometric composition may lie inside or outside HR (which corresponds to two-sided and one-sided HR, respectively). Changing properties through deviation from stoichiometry is one of the most important methods of controlling properties of semiconductors.

A large group of semiconductor phases of variable composition are so-called heavily nonstoichiometric (NS) phases with wide HR and high concentrations of NSD. With regard to these phases, there arise a number of questions related to the defect interaction and its effect on the crystal structure, energy band spectrum, and physical properties. The fundamental concepts of statistical thermodynamics of point defects are valid on condition that defects can be considered as noninteracting particles and their distribution has a purely statistical character (Kröger, 1973). That is why the Wagner-Schottky theory (Wagner & Schottky, 1930) can provide satisfactory description only for the compounds with a very narrow HR, whereas for phases with a wide HR it is necessary to take into consideration the defect interaction (Coulomb, deformational, etc.), which can lead to the defect ordering. At present, there is extensive experimental material confirming the possibility of a short- and long-range ordering of NSD in wide HR (Collongues, 1974, 1993; Rabenau, 1975). An intermediate character of chemical bond in semiconductors and the appearance of free charge carriers induced by nonstoichiometry determine a great variety of possible mechanisms of nonstoichiometry. However, problems related to properties of heavily NS phases practically have not been studied. Physical phenomena accompanying significant deviations from stoichiometry are similar to phenomena taking place under heavy doping; the difference consists in the fact that under deviation from stoichiometry, the role of impurity atoms (IA) is played by NSD.

IV-VI binary semiconductors are very convenient objects for studying fundamental problems of heavily NS phases because they include compounds with wide HR (for example, SnTe). Besides, these compounds are of practical importance too, as they are widely applied in thermoelectricity, optoelectronics, IR-devices and other fields of science and technology (Nimtz & Schlicht, 1985; Gornik et al., 1982; Ravich et al, 1970).

In this chapter, an overview and analysis of our experimental and theoretical data on the crystal structure, mechanical, thermal, galvanomagnetic and thermoelectric properties vs. composition of the one-sided binary semiconductor phase SnTe are given together with the results obtained by other authors. The chapter consists of three sections. Section 1 contains the results of the studies of the SnTe HR boundaries at different temperatures and the concentration dependences of properties determined by the crystal lattice dynamics and electronic subsystem of the crystal in the SnTe HR. We were first who observed peculiarities in the property–composition dependences at the composition corresponding to the maximum in the melting curves. Using the percolation theory approach, we attributed the peculiarities to the transition from weak to heavy self-doping within the HR and determined compositions optimal for long-range ordering. We proposed a principally new model of the SnTe band structure, which takes into consideration a high concentration of NSD. In section 2, the results of the study of phase transitions in SnTe bulk crystals and thin films are given. New phase transitions at a high degree of deviation from stoichiometry were detected and attributed to the processes of vacancies redistribution. The influence of kinetic factors on the temperature dependences of the properties and the character of relaxation processes was considered. Section 3 deals with the peculiarities of SnTe doping. Some new physical phenomena connected with simultaneous presence of NSD and IA are established. It is shown that the presence of cation vacancies causes a new mechanism of impurity solubility associated with filling of NS vacancies. The influence of IA on equilibrium of NSD in SnTe is considered. Recharging of IA with variable valence in the presence of the NS vacancies is revealed. It is suggested that the above mentioned phenomena are common for heavily NS phases and should be taken into account when developing semiconductor materials for different applications.

2. Concentration dependences of properties in the SnTe homogeneity region

SnTe is a congruently melting semiconductor compound formed in the Sn-Te system at 1063 K (Fig. 1,a), which can crystallize in two structures: cubic β (space group Fm3m) and rhombohedral α (R3m) (Abrikosov & Shelimova, 1975; Merzhanov, 1988). SnTe undergoes a ferroelectric phase transition (FPT) β→α at a temperature $T_C \sim 100$ K (Littlewood, 1982,1984; Suski, 1985). Chemical bonds in SnTe are of ionic-covalent-metallic character. It is assumed that the dominant role in the formation of covalent bonds is played by atomic p-orbitals. The difference in electronegativity of Sn and Te atoms determines the appearance of ionic component of chemical bond, and a high concentration of charge carriers due to deviation from stoichiometry leads to the existence of a metallic component (Littlewood, 1982,1984).

2.1 Homogeneity region and defect structure of SnTe

The HR of SnTe was studied in a number of works (Brebrick, 1963,1964,1967,1971; Maselsky&Lubell, 1963; Shelimova& Abrikosov, 1964). The maximum extension of the HR (50,1-50,9 at.% Te) is at 673 K; the maximum in the melting curves corresponds to 50.4 at.% Te ($Sn_{0.984}Te$). It was found that the unit cell parameter decreases linearly with increasing Te content and that the predominant defects are cation vacancies. SnTe always exhibits *p*-type conductivity and has a high hole concentration *p*, which increases with increasing Te content from $p = 2 \cdot 10^{26}$ m^{-3} up to $p = 1.5 \cdot 10^{27}$ m^{-3} at 300 K. Most experimental data and calculations

(Akchurin& Ufimtsev, 1979; Lin et al., 1986; Kattner et al., 1985; Logachev & Moyzhes, 1970) on the SnTe HR are related to the temperatures higher than the eutectic temperature in the Sn-Te system (~ 678 K).

We determined the boundaries of the SnTe HR at 450 – 820 K (Rogacheva et al, 1986,1991a) using a complex of different methods, in particular microstructural and X-ray diffraction (XRD) methods, measurements of microhardness *H*, Seebeck coefficient *S*, electrical conductivity σ and other properties. The samples were subjected to different types of heat treatment (HT), specifically, annealed at 820 K for 500 hours (HT1), at 670 K for 500 hours (HT2), at 570 K for 2100 hours (HT3), at 450 K for 4200 hours (HT4), and then quenched in water. It is seen from Fig. 2,a that the unit cell parameter *a* decreases linearly within the HR, and as the temperature of annealing decreases, the SnTe HR narrows. In Fig. 1,b, the HR of SnTe is plotted on the basis of our data and data of other authors. It is seen that with decreasing temperature, the boundaries of the SnTe HR are shifted to the composition of $Sn_{0,984}Te$. This is rather interesting fact, as usually in NS phases at $T \to 0$ the phase composition corresponds to the stoichiometric one. One can suggest that SnTe remains one-sided phase in the entire temperature range. Let us call such phase "genuine nonstoichiometric". The existence of such phases can be connected with a decrease in the free energy due to the formation of charge carriers (Korzhuev, 1985) or due to the formation of a superstructure of defects.

Under introduction of excess Te into the SnTe crystal the following types of the defect formation are possible: 1) Te atoms are built into the anion sublattice causing the appearance of the cation vacancies; 2) Te atoms are embedded in the tetrahedral interstitial sites; the probability of this process is very low because it leads to a significant deformation of the crystal lattice; 3) Te atoms occupy positions of Sn atoms (antistructural defects) creating vacancies in the anion sublattice or partially filling the existing cation vacancies. In IV-VI compounds, the energy of the anion vacancy formation significantly exceeds that for the cation vacancy. That is why the predominant defects in SnTe are cation vacancies.

The presence of the maxima in liquidus and solidus curves at 50.4 at.% Te indicates the existence of a composition with maximum thermal and thermodynamic stability within the SnTe HR. One can expect that in the concentration dependences of properties there must be peculiarities corresponding to the maxima in the melting curves. To verify this suggestion, we carried out detailed studies of the concentration dependences of properties within the SnTe HR.

2.2 Static distortions of the crystal lattice caused by nonstoichiometric vacancies

The presence of point defects causes geometrical distortion of the crystal lattice as a result of displacement of the atoms (ions) surrounding a defect. A rigorous solution of the problem of the determination of atom displacements in the vicinity of a defect is complicated (Maradudin, 1968; Lanno & Burguen, 1984; Burguen&Lanno, 1985). Under the assumption of elastic continuous medium, the displacement field around a vacancy is spherically symmetrical, and the displacement diminishes proportionally to r^{-3}, where *r* is the distance from the defect. As a result of the displacement of Te atoms surrounding a vacancy, the distance between them decreases, which leads to a decrease in the SnTe unit cell parameter. From the concentration dependence of *a* in the SnTe HR, one can estimate the crystalochemical radius of vacancies: r_V=0.06±0.005 nm. It is seen that the vacancy radius is 0.5–0.6 of the ionic radius of Sn, and the vacancy volume is 0.1-0.3 of the cation volume. It should be noted that in diamond-like lattices the volume of vacancies amounts to ~0,8 of the atom volume and in FCC-lattices with metallic bonds it is 0,5-0,6 of the atom volume (Gorelik & Dashevskii, 1988). Thus, in comparison with vacancies of the above mentioned types, NS vacancies have much smaller volume, which indicates significant static displacements of Te atoms near a vacancy. Such considerable distortions in the crystal lattice in the vicinity of vacancies under the condition of their random distribution must lead to an increase in the degree of anharmonicity of atom vibrations and other effects. This, in turn, can stimulate redistribution of vacancies corresponding to the minimum of free energy of the system.

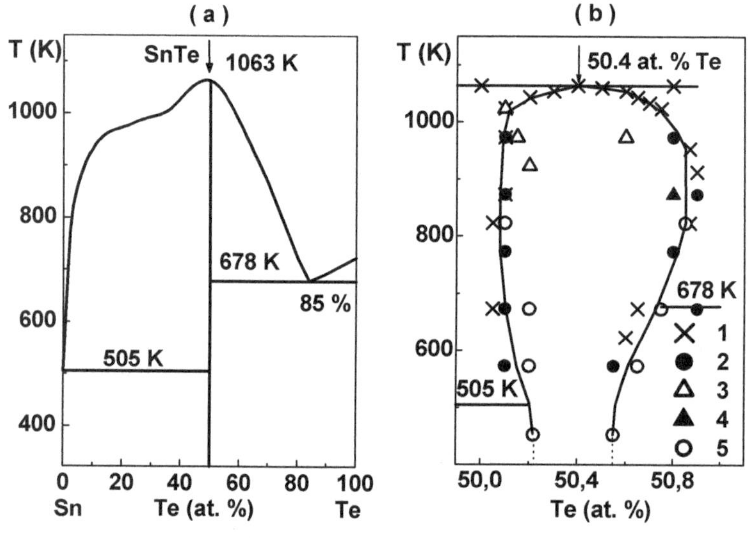

Fig. 1. Sn-Te system phase diagram (a) and homogeneity region of SnTe compound (b).
b: 1 - (Brebrick, 1963, 1971; Brebrick & Strauss, 1964), 2 - (Shelimova & Abrikosov, 1964), 3 - (Tairov et al., 1970), 4 - (Mazelsky & Lubell, 1963), 5 - (Rogacheva et al., 1991a).

The influence of defects on mechanical properties of crystals is well known. That is why one can expect that crystal disordering connected with deviation from stoichiometry must noticeably affect such parameters as ultimate strength, Youngs' modulus, microhardness *H*, etc. H represents a generalized characteristic of resistance to plastic deformation under contact compression (Grigorovich, 1976). The main mechanism of plastic deformation is dislocation movement which is impeded primarily by other dislocations and point defects. Doping usually leads to deformation hardening of the crystal as a result of elastic and electrostatic interactions between dislocations and IA (Suzuki et al., 1991). Thus, one can suggest that under introduction of NSD, *H* must increase. According to the literature data (Abrikosov & Shelimova, 1975), *H* increases linearly with increasing Te content within the SnTe HR. However, our detailed study (Rogacheva et al., 1986,1991a) showed that the *H*-curves exhibit an anomalous behavior in the vicinity of 50.4 at.% Te (Fig. 2,b-e).

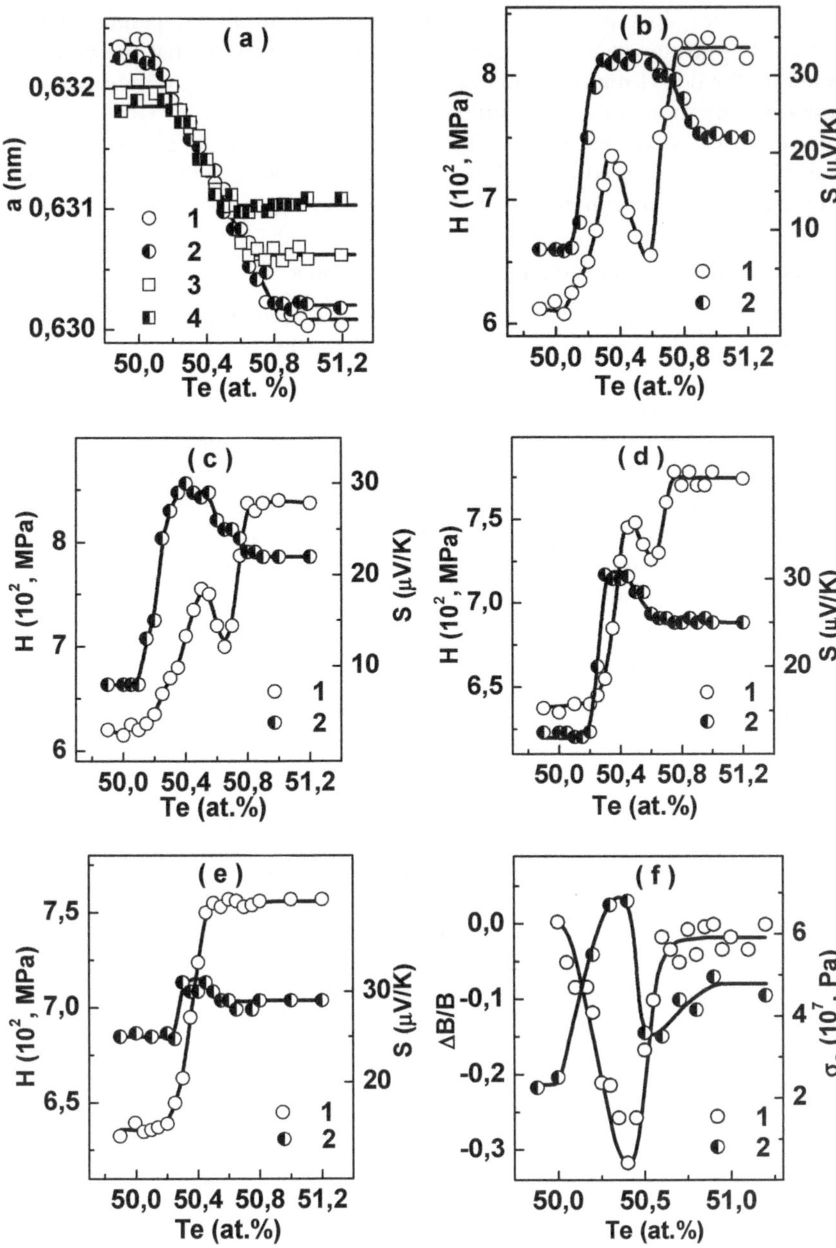

Fig. 2. Concentration dependences of the unit cell parameter *a* (a), microhardness *H* (b-e), Seebeck coefficient *S* (b-e), X-Ray linewidth *ΔB/B* (f), and ultimate compressive strength σ_c (f) in the SnTe homogeneity region after different heat treatments : a: 1 – HT1, 2 – HT2, 3 – HT3, 4- HT4; b-e: 1 – H, 2 – S; b - HT1, c - HT2, d - HT3, e - HT4 (e); f: 1 - *ΔB/B*, 2 - σ_c (Rogacheva et al., 1986, 1991a).

The inflection near 50,4 at.% Te was also observed in the dependence of ultimate compressive strength on Te content (Fig.2,f). It can be suggested that the behavior of H is determined by the defect interaction which starts to manifest itself at their sufficiently high concentration. This suggestion is additionally supported by the fact that at 50.4 at.% Te, the minimum of the X-ray line width β is observed (Fig. 2,f) (Rogacheva et al., 1986; 2003a). As all the samples were prepared using the same technique, the smallest value of β at 50,4 at.% Te may indicate the maximum degree of structure ordering.

2.3 Nonstoichiometric vacancies and lattice dynamics

Defects affect the character of lattice vibrations and can significantly change the phonon spectrum of crystal. Among the properties determined by the crystal lattice dynamics are heat capacity, the Debye temperature θ_D, thermal expansion coefficient α, the probability of Mössbauer effect f', phonon thermal conductivity λ_p. One can expect that the presence of NS vacancies in the crystal will lead to an increase in the anharmonicity of atom vibrations leading to a growth in α, decrease in λ_p, θ_D and f'. However, the situation can change under NSD ordering, when defects form a periodic structure and in the harmonic approximation do not scatter phonons. That is why from the dependences of dynamic properties on the vacancy concentration one can indirectly judge about the character of the vacancy distribution in the cation sublattice.

2.3.1 Phonon spectrum of SnTe

The Debye approximation assumes a single atom in a unit cell and only acoustic branches of vibrations (Maradudin, 1968). The existence of a $\theta_D(T)$ dependence signals about non-applicability of the Debye approximation. According to (Gul'tyaev & Petrov, 1951) at sufficiently high temperatures, θ_D is constant, and its value for SnTe (139±3 K) is close to the thermodynamic θ_D (130 K). In (Bukchpan, 1968), θ_D did not change in the range 85-250 K and equaled (132±3) K. Analysis of the data reported in (Brukhanov et al., 1964; Varnek et al., 1980; Bukchpan, 1968) shows that θ_D determined from f' at 290 K (132-139 K) practically coincides with the value of θ_D obtained from heat capacity measurements (Gul'tyaev & Petrov, 1951). Because of the closeness of its vibrational spectrum to the Debye one, SnTe is a convenient object for studying the influence of NS vacancies on the phonon spectrum and dynamic properties of the crystal.

2.3.2 Nuclear gamma resonance

The influence of various factors on phonon spectra can be studied by measuring the Mössbauer spectrum parameters (f', isomer shift δ, linewidth Γ, absorption area S). The introduction of cation vacancies must lead to a disturbance of the phonon spectrum of the crystal (Maradudin, 1968) and, hence, to a change in f', S, and θ_D. According to (Bekker et al., 1973), in the Sn-Te system, the value of δ practically does not depend on the ratio of constituents, whereas according to (Varnek et al., 1980), δ changes under increasing nonstoichiometry. In (Bekker et al., 1973), at 300 K an increase in Γ with increasing Te content was observed, while in (Varnek et al., 1980) no additional broadening of lines was registered. According to (Bekker et al., 1973), the area S, which is directly proportional to f', and calculated on its basis values of effective θ_D somewhat increase with increasing degree of deviation from stoichiometry, at the same time in (Varnek et al., 1980; Bekker et al., 1973) it is reported that θ_D does not practically depend on composition. Thus, the data reported by different authors disagree. That is why we carried out a detailed study on the influence of deviation from stoichiometry on the Mössbauer parameters of SnTe (Baltrunas et al., 1986) at 300 K on bulk samples annealed at 820 K for 300 hours. A solid solution of Sn^{119} in Pd served as γ-source. The area S was calculated as $S = \pi \cdot f' \cdot \Gamma/2$. The spectra of ^{119}Sn atoms represented singlet lines, whose width is typical for highly symmetrical environment of Sn atoms. The values of δ were close to the literature data on SnTe (Flinn, 1978) and corresponded to Sn^{2+}. It was found that an increase in Te content is accompanied by a decrease in δ (Fig. 3,a). The concentration dependences of Γ and S are nonmonotonic: in the Γ curve, a slight maximum at 50.4 at.% Te is seen, and in the S curve, there are two maxima: a slight maximum at 50.4 at.% Te and a distinct one at 50.8 at.% Te. It is known (Flinn, 1978) that isomeric shift in Sn is determined by effective numbers of valence 5s and 5p electrons (n_s and n_p): $\delta = -0.41 + 3.10\, n_s - 0.20\, n_s^2 - 0.17\, n_s n_p$. The chemical bond in SnTe is realized mainly by p-electrons with a small contribution of valence s-states. The appearance of holes leads to a deficit of valence electrons, which explains the decrease in δ with increasing hole concentration. The Mössbauer coefficient f' is one of the parameters related to θ_D which is determined by the mean-square displacement of atom under thermal vibrations $\langle x^2 \rangle$:

$$f' = \exp\left[-\left(4\pi^2 \langle x^2 \rangle / \lambda^2\right)\right]$$

and can be calculated using the Debye model (Wertkheim, 1966):

$$f' = \exp\left[-\frac{E^2}{2Mc^2 k_0 \theta_D}\left(\frac{3}{2} + \frac{\pi^2 T^2}{\theta_D^2}\right)\right] \quad (1)$$

where E is the transition energy, M is the nucleus mass, κ_0 is the Boltzmann constant. It follows from general considerations that the introduction of defects must result in a local weakening of bonds between particles, a decrease in f' and θ_D. For example, a decrease in θ_D was registered under introduction of impurities into Ge: for pure Ge, $\theta_D = 374 \pm 2$ K, while for Ge containing 5.4 10^{19} cm^{-3} Ga, $\theta_D = 362 \pm 2$ K (Zhdanova & Kontorova, 1965). Growth instead of drop in θ_D under deviation from stoichiometry in the Sn-Te system despite the introduction of charged vacancies can result from the vacancy ordering. The maximum of S at 50.4 at.% Te supports the idea about the vacancy ordering, showing simultaneously a high sensitivity of the Mössbauer method to the processes of point defect redistribution. The sharp growth in S at 50.75 at.% Te may be an evidence of the formation of another ordered phase with a higher strength of chemical bonds and higher θ_D, and existing in a narrow range of concentrations near the boundary of the SnTe HR. As the distance between vacancies corresponding to 50.8 at.% Te is $d=2a$, the ordering process is very likely. The increase in vacancy concentration leads to an increase of the distribution asymmetry of the electric charge around the Mössbauer nucleus and to the broadening of linewidth (Fig. 3,a). The decrease of Γ after 50.4 at.% Te is caused apparently by the redistribution of vacancies which results in their symmetrization around the Mössbauer nuclei.

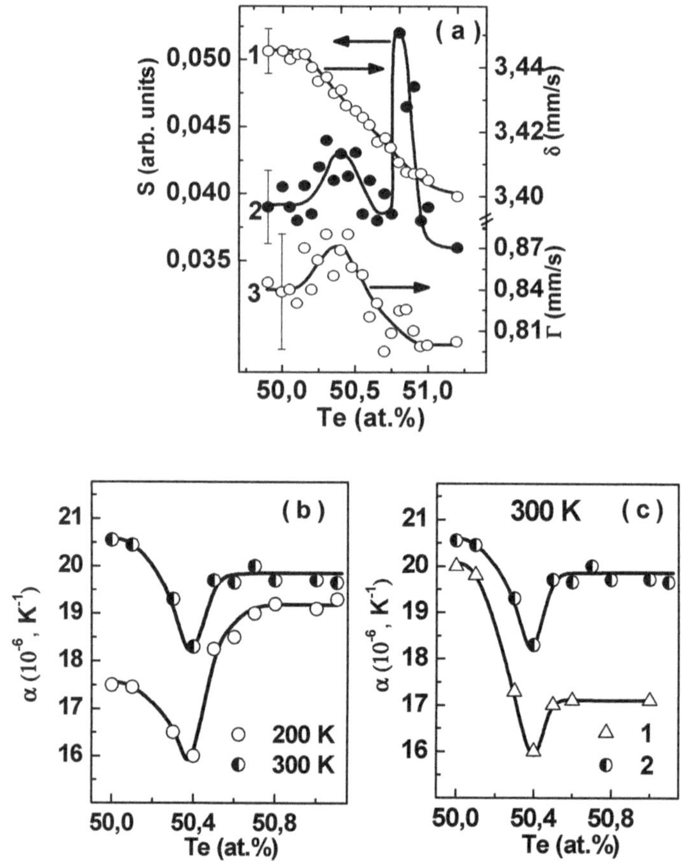

Fig. 3. The isomer shift δ, the area under the Mössbauer spectrum line S, the linewidth Γ (a), and thermal expansion coefficient α (b,c) vs Te concentration in the Sn-Te system. δ is recalculated relative to the $Ba^{119}SnO_3$ source (Baltrunas et al., 1986). a: 1 - δ, 2 - S, 3 - Γ; b: dynamic regime; c: 1 - stationary regime; 2 - dynamic regime (Nashchekina et al., 2008).

2.3.3 Phonon thermal conductivity

Measurement of phonon thermal conductivity λ_p is one of the tools for studying dynamic irregularity caused by the presence of point defects. Anharmonicity of the lattice vibrations leads to scattering of phonons by phonons, which determines a finite thermal conductivity even in case of undoped crystal. The introduction of point defects results in additional scattering of phonons by point defects and can intensify phonon-phonon scattering due to the increase in cubic anharmonicity of lattice vibration. The authors of (Damon, 1966) who measured the $\lambda_p(T)$ dependences of SnTe monocrystalline samples with various p in the range of 100-500 K concluded that along with the processes of three-phonon scattering, scattering of phonons by vacancies takes place. It was pointed out that this scattering is rather strong and for the most part is caused by fields of deformations created by vacancies.

The authors of (Erasova & Kaydanov, 1968) arrived at the conclusion that λ_p of SnTe is practically independent of the cation vacancies concentration. We conducted a detailed study of λ_p of SnTe as a function of the cation vacancy concentration on samples with 49,9-51 at.% Te (Rogacheva et al., 1995). In Fig. 4,a,b the T-dependences of total λ, electronic λ_e and phonon λ_p thermal conductivity and thermal resistance $1/\lambda_p$ in the range of 300-650 K are presented, and in Fig. 4,c,d , the isotherms of these parameters are given. Analysis of the $1/\lambda_p$ (T) dependences showed that for all samples a linear dependence $1/\lambda_p = AT + B$ is observed. The value of A characterizing the slope of the straight lines remains constant, whereas the value of B corresponding to additional thermal resistance caused by vacancies increases with increasing Te content. The straight line drawn through the origin of coordinates parallel to the obtained lines corresponds to the temperature dependence of $1/\lambda_p$ of a hypothetical defectless SnTe. It is known (Berman, 1979; Mogilevskii & Chudnovskii, 1972; Drabl & Goldsmit, 1963) that if thermal conductivity is associated only with three-phonon processes, the Eiken law is fulfilled: $1/\lambda_p = kT/\theta_D^3$, where k is the coefficient taking into account the degree of anharmonicity of crystal lattice vibrations. According to the obtained results, the introduction of vacancies does not lead to breaking the Eiken law, and the linear behavior of the temperature dependences of $1/\lambda_p$ indicates that thermal resistance caused by vacancies does not depend on temperature and is determined only by scattering by defects. This result is in good agreement with the theory: at $T > \theta_D$ thermal resistance caused by defects does not depend on temperature. SnTe satisfies these conditions, as $\theta_D \cong 130$ K. The constancy of the $1/\lambda_p$ straight lines slope shows that the coefficient of proportionality in the Eiken law does not virtually change under increasing degree of deviation from stoichiometry, and this fact is another argument in favor of the suggestion about the vacancy ordering. It follows from Fig. 4 that for the defectless SnTe $\lambda_p T = 1600$ W/m. Theoretical calculations of this parameter performed in different works yield rather close values: 1620 W/m (Julian C.L., 1965), 1300 W/m (Keyes R.W., 1959, Klemens P.G., 1955). From $1/\lambda_p$ dependences on the IA concentration one can estimate thermal resistance produced by each IA and the cross-section of phonon scattering σ, using the equation (Ioffe, 1954):

$$\frac{\lambda}{\lambda_0} = 1 + \frac{N}{N_0} \cdot \Phi \cdot \frac{l_0}{a}, \tag{2}$$

where N is the IA concentration; N_0 is the number of atoms in 1 cm^3 of the crystal; a is the distance between neighboring atoms; l_0 is the phonon mean free path; Φ is the factor in the equation $\sigma = \Phi a^2$; λ and λ_0 are the λ_p of the crystal lattice with and without impurities. In our case, cation vacancies play the role of IA. The mean free path l_0 is calculated using the Debye equation: $\lambda_0 = (C_v \cdot l_0 \cdot V_{gr})/3$, where C_v is the heat capacity per 1 cm^3 and V_{gr} is the mean group velocity ($V_{gr} = (E/\rho)^{1/2}$, where E is the elastic modulus and ρ is the density). From equation (2) we obtain $\sigma \cong 10\ a^2$ and $\Phi = 10$. For comparison, values of Φ under doping of PbTe with I, Cl, Br are $\Phi = 3\text{-}3.7$ (Ioffe, 1954). For isovalent IA in PbTe, σ is much smaller than for I, Cl, Br (by a factor of ~5). It was established (Drabl & Goldsmit, 1963) that for I and Cl impurities in Bi_2Te_3, Φ is 13, and the large value of Φ was attributed to the fact that I and Cl occupy interstitial sites and strongly scatter phonons. It follows from our data that cross-section of phonon scattering by NS vacancies is sufficiently large.

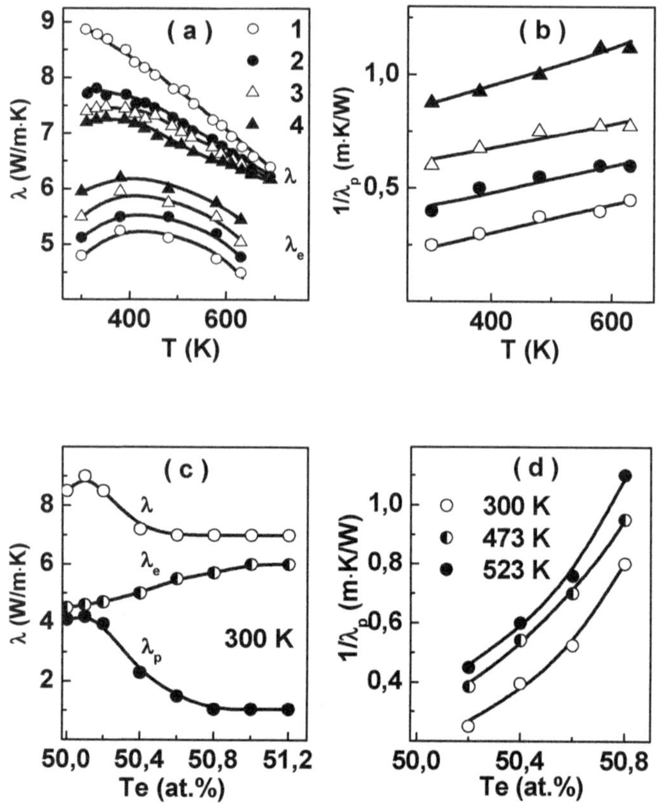

Fig. 4. Temperature (a,b) and concentration (c,d) dependences of total λ (a,c), electronic λ_e (a,c) and phonon (c) thermal conductivities and thermal resistance $1/\lambda_P$ (b,d) of SnTe. a,b: 1 - 50.2 at.% Te, 2 - 50.4 at.% Te, 3 – 50.6 at.% Te , 4 – 50.8 at.% Te (Rogacheva et. al, 1995).

In accordance with the Klemens theory (Klemens, 1955) scattering of phonons by defects can be described taking into account contributions of three factors: local change of mass $\Delta M/M$, strength of atomic bonding $\Delta G/G$, elastic distortion of the crystal caused by the difference in effective sizes of the atom (ion) and the substituting defect $\Delta R/R$. The total scattering cross-section according to Klemens is as follows

$$\sigma = \frac{V_0^2 \omega^4}{4\pi V^4}\left\{\left(\frac{\Delta M}{M}\right)^2 + 2\left[\frac{\Delta G}{G} - \frac{b}{\sqrt{5}}\cdot\gamma\cdot\frac{\Delta R}{R}\right]^2\right\}, \tag{3}$$

where V_0 is the volume per atom, ΔG is the local change in the elastic modulus, ΔM is the local change of mass, γ is the Gruneisen parameter, b = 14.2 (Berman, 1979). Assuming $\omega=\omega_{max}=k_0\theta_D/h$, θ_D =130K, γ=2, $\Delta R/R$=0.13, V_0=31 10^{-30} m^3, we obtain $\sigma \cong 10$ a^2, which practically coincides with σ determined on the basis of the experimental data.

2.3.4 Coefficient of thermal expansion

Change in the crystal volume under changing temperature is caused by an asymmetry of the resulting interaction potential in the crystal lattice, manifesting itself as the interaction of vibrations of the phonon spectrum. The degree of the volume change is characterized by the coefficient of volume expansion β (Novikova, 1974):

$$\beta = \frac{1}{V}\left(\frac{\partial V}{\partial T}\right)_P = \frac{1}{V}\left(\frac{\partial^2 F}{\partial P \cdot \partial T}\right), \ \beta \cong 3\alpha, \tag{4}$$

where F is the free energy, α is the coefficient of linear expansion. As F is an additive function, and can be considered for solids as the sum of free energies of the crystal lattice, electron gas, magnons and so on, β is also an additive function: $\beta = \beta_L + \beta_l + \beta_m +$, where L, l, m denote lattice, electronic and magnetic contributions respectively. As a rule, at $T>0.1\theta_D$ the lattice contribution plays the main role. It is known that the introduction of IA usually leads to an increase in α of the crystal (Novikova, 1974; Zhdanova & Kontorova, 1963; Abrikosov et.al., 1981, 1982). This is in good agreement with the theoretical calculations performed for diluted solid solutions (Varisov et.al, 1966; Timmesfeld & Elliot, 1970) according to which the increase in α is connected with a growth in configurational entropy of the ideal crystal under introduction of IA as well as with the effect of free charge carriers on heat capacity and the Gruneisen parameter. In accordance with (Timmesfeld & Elliot, 1970), the change in α under introduction of substitutional point defects occurs due to: i) a change in elastic constants in the vicinity of the defect, ii) a change in anharmonicity near the defect, iii) a change in the normal model of a crystal with defects. These three factors can partially compensate each other, so the resulting change in α can be positive or negative. In (Masharov, 1963), a dynamic model of the crystal with a small concentration of vacancies or IA is suggested. According to this model, "defect" crystal can be considered as an ideal one with the phonon energy depending on the defect concentration. Using this model, the authors of (Varisov et al., 1966) calculated corrections in α caused by vacancies and associated with the change in configurational potential energy.

The effect of deviation from stoichiometry on thermal expansion of phases of variable composition has not been investigated in detail (Novikova & Shelimova, 1965,1967; Belson & Houston, 1970). That is why we have conducted studies on the influence of deviation from stoichiometry, temperature, and kinetic factors on α of SnTe crystals (Rogacheva et al., 1993; Rogacheva, 2003b; Nashchekina et al., 2008). The $\alpha(T)$ dependences in the range 5-300 K or 77-300 K were obtained under two regimes. The stationary regime consisted in cooling samples down to the liquid helium temperature slowly and then heating up to room temperature, keeping at the measurement temperatures until complete temperature stabilization is reached. Under the dynamic regime samples were quickly quenched in liquid nitrogen, and the $\alpha(T)$ curves were measured under subsequent rapid heating. The first regime was close to quasi-equilibrium conditions, whereas the other one corresponded to apparently non-equilibrium conditions of heating. For both regimes the minimum value of α within the SnTe HR was observed at 50.4 at.% Te (Fig. 3, b,c). It can be suggested that an increase in the dynamic stability of the SnTe crystal lattice under increasing concentration of NSD is connected with ordering of vacancies. It is known that $\alpha = \beta \cdot k_0/\varepsilon^2 \cdot a$, where ε is the coefficient of quasi-elastic bond, β is the coefficient of anharmonicity, a is the distance between atoms; k_0 is the Boltzmann constant. Although the equation is rather rough and corresponds to a two-atom model, under the condition of preservation of the crystal structure, on the basis of α values one can make comparative estimates of the degree of crystal anharmonicity and values of average displacements of atoms from the equilibrium positions: $x = \alpha \cdot a \cdot T$ - proportional to temperature. The calculation shows that the value of x for SnTe is ~0,005 a at 300 K. In comparison with SnTe, in $Sn_{0,984}Te$ the number of vacancies in the cation sublattice increases from 0,4 to 1,6 %, and α, β/ε^2, x decrease by ~6 %. There is a simple relationship between α and θ_D (Novikova, 1974): $\theta_D = \sqrt{A/(\alpha V^{2/3} M)}$, where M is the molecular weight; A is the constant depending on the type of chemical bond. The estimate shows that as α changes by ~6 %, θ_D increases by ~5 %. The calculation of the change in θ_D within the SnTe HR performed on the basis of the Mössbauer data in (Bekker, 1973) demonstrated that for 300 K the change in θ_D amounts to ~ 8%, which is in good agreement with the result of our calculation of the change in θ_D based on the data on α.

2.4 Electronic properties and band structure

2.4.1 Electronic properties

Deviation from stoichiometry in semiconductor compounds leads to perturbation not only in the crystal lattice but also in the electronic subsystem of crystal. In Figs. 5,6 the temperature and concentration dependences of S, σ, R_H, p, Hall mobility μ_H, and Nernst-Ettinsgausen coefficient $Q_\perp$ for SnTe samples with different degrees of deviation from stoichiometry are presented (Laptev & Rogacheva, 1986, 1988). With increasing Te content, σ and p increases, μ_H decreases. The S - composition dependence has a maximum at ~50.4 at.% (Fig. 2,b-e), and this fact is usually explained on the basis of the model of the valence band which consists of two subbands with different densities of states. From the concentration dependences of S, σ, R_H and $Q_\perp$ in a weak magnetic field assuming single type of carriers, isotropic case, strong degeneration, elastic scattering, and absence of phonon drag, we estimated scattering parameter t and effective mass of density of states at the Fermi level m^*_1 (Fig. 6). The calculations showed that t = 0, which corresponds to scattering by close-range components of vacancy potential in the low temperature range and to acoustic scattering at high temperatures. The m^*_1 (p) dependence is described by the equation: $m^*_1 = 0.28 m_0 (p/p_O)^{2/3}$, where p_0 is the minimum p in the SnTe HR. Such dependence indicates band spectra nonparabolicity, which does not correspond to the dispersion laws (Kane, 1957) or (Cohen, 1961). The T- dependences of kinetic coefficients are characterized by maxima in the $R_H(T)$ and $S(T)$ curves, a drop in $R_H(T)$ and $S(T)$ at high temperatures, a sharp growth in S at T~300-500 K for the samples with a low p. The attempts to explain the anomalies using existing models (Andreev, 1967; Efimova et al., 1965; Kaydanov et al., 1967), i.e. by the increasing contribution of intrinsic conductivity to kinetic effects at T>600 K fail, because within those models the temperature of maxima in the $R_H(T)$ and $S(T)$ curves is estimated at more than 1000 K.

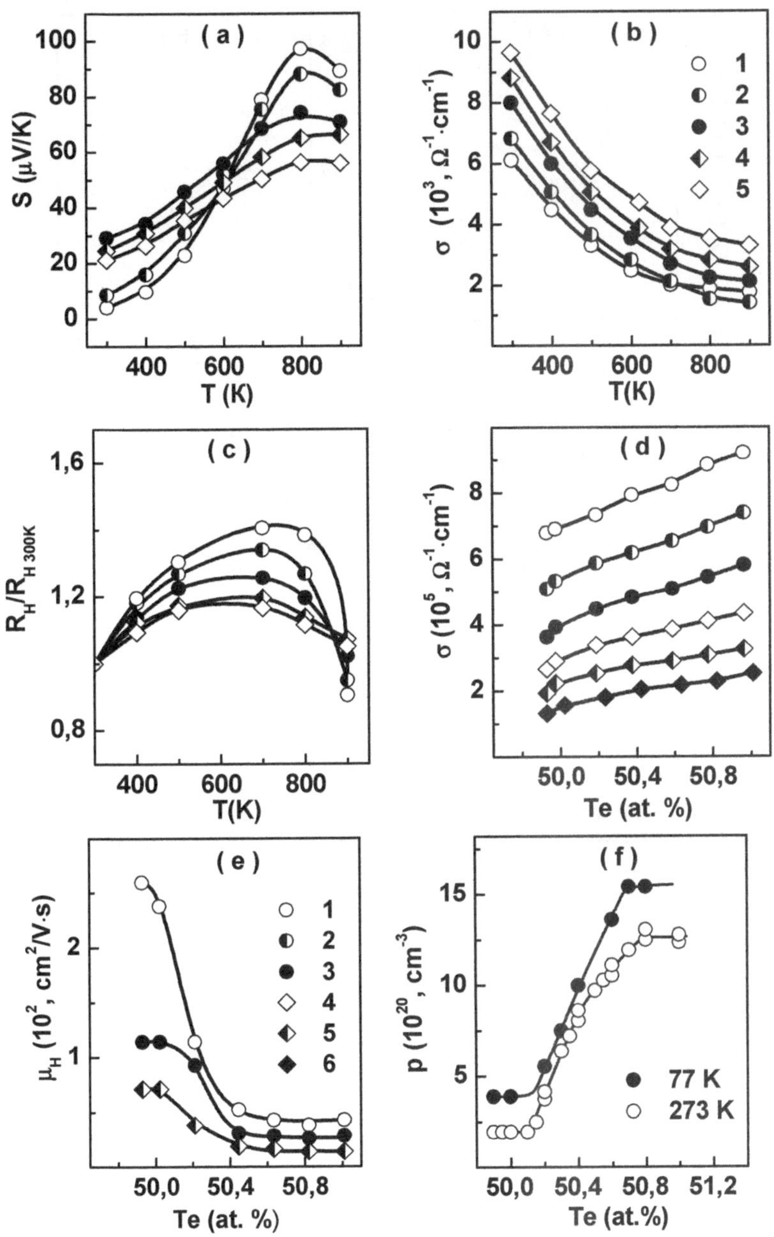

Fig. 5. Temperature (a-c) and concentration (d-f) dependences of the Seebeck coefficient S (a), electrical conductivity (b, d), Hall coefficient (c), Hall charge carrier mobility μ_H (e), and hole concentration p (f). a,b,c: 1 – 50.0 at.% Te, 2 – 50.2 at.% Te, 3 – 50.4 at.% Te, 4 – 50.6 at.% Te, 5 – 50.8 at.% Te; d, e, f: 1 – 273 K, 2 – 373 K, 3 – 473 K, 5 – 573 K, 6 – 773 K (Laptev &Rogacheva, 1986, 1988).

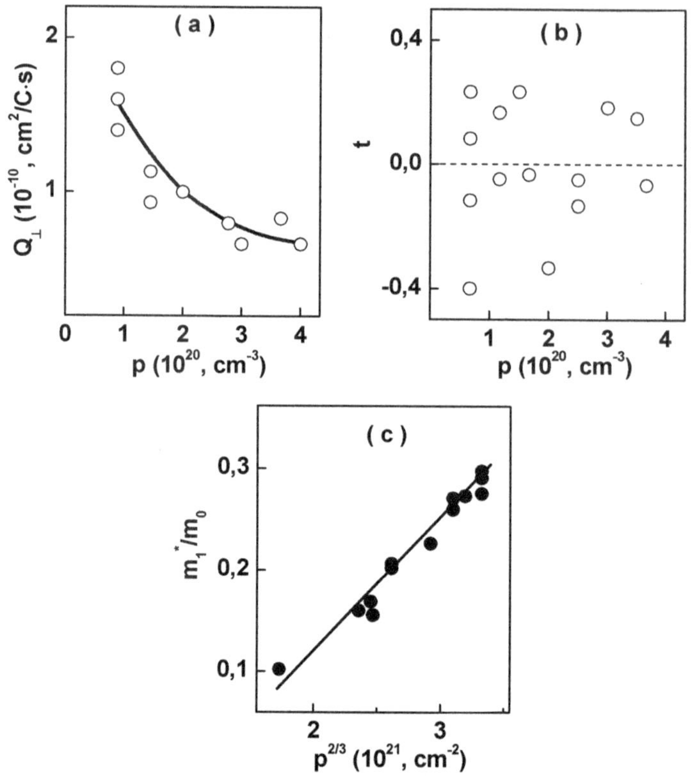

Fig. 6. Dependences of the transverse Nernst-Ettinsgausen coefficient $Q_{\perp}$ (a), scattering parameter *t*, and effective mass of the density of states m_1^*/m_0 (c) on the hole concentration *p* in SnTe (Laptev &Rogacheva, 1986, 1988).

2.4.2 Hall-factor in SnTe

The number of carriers supplied by each Sn vacancy to the SnTe valence band is of fundamental importance in studies of the energy structure of this compound. It is related to the value of the Hall factor *r* defined by the relationship $p = r/R_H{\cdot}e$.

Assuming $r = 1$, the authors of (Brebrick, 1963) conclude that each Sn vacancy is a doubly ionized acceptor and supplies two carriers. According to (Houston et al., 1963), this conclusion is correct if one assumes that $r = 0.6$. For determining *r* we measured the dependences of longitudinal ρ_{zz} and transverse ρ_{xx} magnetoresistance and R_H on magnetic field *H* for SnTe samples with $p = (0.75\text{-}2.5){\cdot}10^{26}$ m^{-3} (Kolomoets et al., 1986) (Fig. 7,a,b). All the dependences exhibited oscillations whose amplitude increased with increasing *p* at constant *H* (Fig. 7,b.). It was found that $r = 0.9\text{-}1.0$. Comparison of this value of *r* with the dependence of *p* on the concentration of Te in the Sn-Te system (Fig. 7,c) led to the conclusion that each vacancy supplies four holes to the valence band. This conclusion was used to forecast the effect of vacancies on the energy spectrum of SnTe (Laptev & Rogacheva, 1986, 1988).

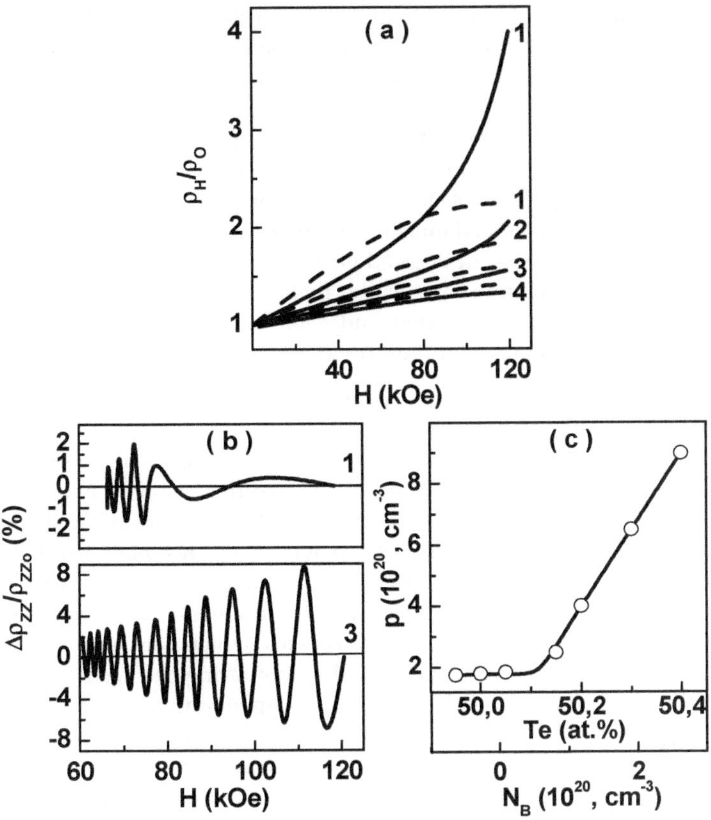

Fig. 7. Dependences of the longitudinal ρ_{XX} magnetoresistance on the magnetic field: a) monotonic part; b) oscillatory part. The dashed curves represent ρ_{ZZ}. c) dependence of the hole concentration *p* in SnTe on the Te content and on the concentration of cation vacancies (Kolomoets et.al, 1986).

2.4.3 New model of the valence band

The transport properties of SnTe exhibit anomalies which are common for all IV-VI compounds: a sharp change of R_H under increasing temperature, nonmonotonic character of the *S(p)* dependence, anomalous concentration and temperature dependences of hole mobility and so on. On the basis of measurements of R_H and *S* in the range of 4,2-300 K the authors of (Allgaier & Sheie, 1961) suggested a model of two valence bands for SnTe, which was further developed in (Brebrick et al.,1962; Sagar & Miller, 1962; Brebrick, 1963; Brebrick & Strauss, 1963; Kafalas et al.,1964; Andreev, 1967; Kaydanov et al., 1967; Rogers, 1968; Rabii, 1969). Analyzing the literature data on the SnTe band structure one can note two circumstances. On the one hand, none of the models describes all experimental results using a single set of parameters of the energy spectrum and hole scattering, nor can explain a number of the facts, in particular, the character of non-parabolicity of the main band extremum, the origin of "heavy" extremum, the concentration and temperature dependences of the transport properties and so on. On the other hand, none of the models takes into account high concentration of NSD, although it must significantly affect the electronic spectrum. In (Nimtz G., Schlicht, 1985; Heinrich, 1979; Kaydanov & Ravich, 1985; Sobolev, 1981) no impurity levels or bands in the SnTe energy gap were detected. According to the theoretical estimations (Kaydanov & Ravich, 1985), the defect potential in narrow-gap semiconductors like SnTe must be short-range due to interband screening, significant for a small energy gap, and a large value of static dielectric constant connected with the presence of soft modes. Highly localized potential must result in the formation of deep levels, the impurity state energy can fall within a zone of allowed energies creating a resonant (quasi-local) state. Disturbing potential of vacancies is one of the highest among defects. If a vacancy is not charged, this potential is highly localized and rapidly decreases in comparison with the potential created by a donor or acceptor impurity (Parada & Pratt, 1969).

We proposed a model of the SnTe energy band structure taking into account a high level of self-doping in this compound (Laptev & Rogacheva, 1986, 1988), which allowed us to interpret the temperature and concentration dependences of transport properties satisfactorily. According to the model, in the valence band, a resonant band of deep vacancy levels playing the role of additional "heavy" extremum, is formed (Fig.8, a). The band width, number of states and carrier dispersion within the resonant band depend on the vacancy concentration. For description of conductivity over the resonant band, a model of narrow impurity band was applied using approximation of strong coupling. We made assumption about an ordered distribution of vacancies, which was supported by the results of studies of the properties within the SnTe HR. Comparing the results of the theoretical calculations of the concentration and temperature dependences of the kinetic coefficients based on the proposed model with the experimental data we determined a set of parameters for SnTe. At 300 K, ΔE_g = 0.2 eV; the distance from the edge of the valence band to the vacancy band is equal to ΔE_V = 0.42 eV and does not depend on temperature. Within the HR, the vacancy band width changes from 0,038 to 0,19 eV, and the effective mass - from 2,08 to 1,53 m_0. This model lends itself for the interpretation of a wide range of phenomena, not enough satisfactorily explained by the other models of the SnTe band structure.

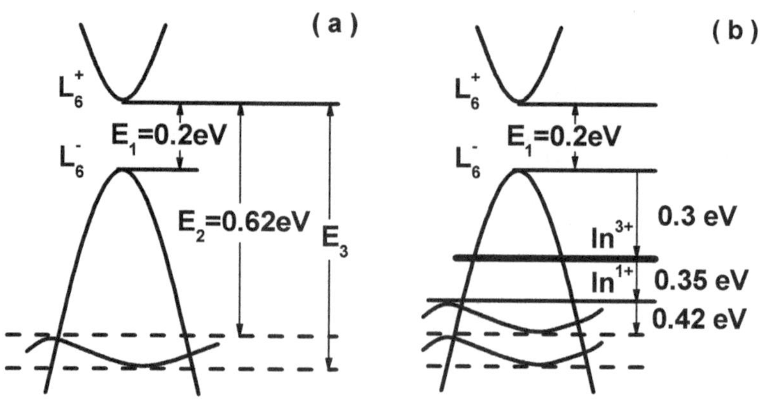

Fig. 8. Models of the band structures of SnTe (Laptev &Rogacheva, 1986, 1988) (a) and SnTe doped with In (Rogacheva & Laptev 1999) (b).

It follows from the said above that when studying the energy spectrum of heavily NS semiconductor phases, it is necessary to take into consideration the possibility of the formation of "defect" bands. The specificity of the "defect" bands consists in the dependence of its parameters on the concentration of defects. The proposed method determines a new approach to study of the band structure of IV-VI compounds, which can be extended to other phases.

2.5 Discussion

It is seen from the results described in Section 2 that in the SnTe HR, the concentration dependences of different physical properties, which characterize both the electronic and lattice subsystems of the crystal, exhibit nonmonotonic behavior: practically in all dependences there are peculiarities (extrema, inflection points) in the vicinity of the composition 50.4 at.% Te corresponding to the maximum in the melting curves.

As was mentioned above, point defects can be considered as noninteracting structural elements only at sufficiently small concentrations. Under increasing concentration, their interaction stimulates the processes of structure ordering. According to the recent theoretical studies (Mycielski, 1986), the Coulomb interaction between impurity centers leads to a uniform distribution of impurity and the formation of superstructure of ionized defects in semiconductors. At present, it is well known that in case of strongly NS phases, wide HR often split into a number of ordered phases with a more restricted nonstoichiometry (Collongues,1974, 1993). Analysis of possible variants of ordered distribution of NS cation Sn vacancies at different their concentrations showed (Rogacheva et al., 1986,1991a) that in the range of the SnTe HR the following types of ordering are theoretically feasible: 1) the location of vacancies in the sites of a primitive unit cell with a parameter $a = 4a_0$, where a_o is the unit cell parameter of SnTe with random distribution of vacancies, the average distance between vacancies d corresponding to $d = a = 4a_0$. This variant of ordering can be realized at 50.1 at.% Te near the boundary of the HR on the Sn side; 2) the location of vacancies in the sites of a fcc lattice with $a = 4a_0$, $d = 2.8a_0$ can occur at 50.4 at.% Te; 3) the distribution of vacancies over the sites of a primitive unit cell with a unit cell parameter $a = 2a_0$ and d=$2a_0$ can take place at 50.8 at.% Te, i.e. near the HR boundary on the Te side. The results of the analysis allows us to suggest that the maximum within the SnTe HR thermal stability of the sample containing 50.4 at.% Te is connected with the formation of a fcc lattice of vacancies and leads to the appearance of the anomalies in the concentration dependences of the properties. The results of the Mössbauer studies (see 2.3.2) indicate the possibility of the formation of another ordered phase at 50.8 at.% Te.

Earlier in a large number of solid solutions we observed concentration anomalies of the properties in the range of small impurity concentrations (~ 1 at.%) (Rogacheva, 1989, 2003; Rogacheva & Krivulkin, 2001, 2002; Rogacheva et.al., 1998) and suggested that there exists a phase transition from diluted to concentrated solid solutions connected with the formation of percolation channels (Rogacheva, 1993). Reaching the percolation threshold like any critical phenomenon usually manifests itself through the appearance of anomalies in the concentration dependences of properties. If one treats NSD as IA, then again there must exist a critical concentration corresponding to the transition from the defect discontinuum to defect continuum. In terms of this assumption the peculiarities of properties in the vicinity of 50,4 at.% Te can be attributed to reaching the percolation threshold in the vacancy subsystem, when the process of vacancy interaction spreads over the entire system stimulating long- or short-range ordering. Considering the observed critical phenomenon in the framework of the percolation theory (Stauffer & Aharony, 1992) and assuming that peculiarities in the concentration dependences of properties within HR correspond to the percolation threshold, it is possible to calculate radius of the NSD "action sphere" R_o from the composition, at which peculiarities are observed, using the equation $4/3\ \pi\ N_c(2R_o)^3 \approx 2.7$, where N_c is the average number of sphere centers per volume unit. The calculation shows that for SnTe $R_0 \approx 1.4\ a_0$, where a_0 is the unit cell parameter. The formation of continuous chains of NSD upon reaching the percolation threshold can stimulate such redistribution of NSD in the crystal lattice, which would lead to the realization of their configuration corresponding to a minimum of the thermodynamic potential. The fact that at 50.4 at.% Te, we observed minima of the $\Delta B/B$, α, and a maximum of f' is an evidence in favor of NSD (vacancies) ordering in SnTe.

3. Temperature phase transitions in SnTe

The existence of the peculiarities in the concentration dependences of the properties within the SnTe HR indicates the presence of phase transitions (PT) connected with the defect redistribution, whose character must depend on composition and temperature. That is why we carried out a complex of works related to detecting possible temperature PT associated with the processes of NSD self-organization. It was mentioned above that SnTe undergoes a ferroelectric phase transition (FPT) at a temperature T_c close to 100 K. It is known that T_C decreases with increasing hole concentration p, and FPT is not observed for p_{77}=(7-10) 10^{26}m^{-3}, i.e. starting from the composition ~ 50.4 at.% Te (Kobayashi et al., 1976; Sugai et al., 1977; Brilson et al, 1974; Iizumi et al., 1975). The latter circumstance explains the significant discrepancy in T_c values reported in different works. Some authors (Fano et al., 1977; Bashkirov et al., 1985; Grassie et al., 1979) reported a much higher value of T_C (160 K and 140-145 K) and presumed the existence of at least two PT in SnTe. The structural instability of SnTe is usually attributed to the softening of the transverse optical mode with zero momentum as a result of the interband electron-phonon interaction (Littlewood,1982; Suski, 1985). The authors of (Khandozhko et al.) observed steps at T_n= 376/n (n=1-6) in the T-dependences of resonant fields of ^{119}Sn nuclei in NMR experiments and suggested that the registered effects are associated with the movement of the crystal lattice defects under a change in temperature.

3.1 Ferroelectric phase transition

For studying the structural instability of a ferroelectric type in SnTe crystals and thin films, we used dilatometry, diffractometry, and measurements of transport properties (Rogacheva et al., 1993; Nashchekina et.al., 1995, 1998, 1999a,1999b, 2008). The studies were conducted on SnTe samples with 50.0-50.8 at.% Te. For the stoichiometric SnTe, in the $\alpha(T)$ curve near 90-100 K (Rogacheva et al., 1993; Nashchekina et.al., 1999b), a distinct λ-anomaly typical of the second-order PT (Fig. 9,a) was observed. In the $a(T)$ dependence in the range 90-100 K (Fig. 9,b), there was also registered an anomalous section (Nashchekina et.al., 1999c). In Fig.

10, a,b,c, the *σ(T,)* $R_H(T)$, and $\mu_H(T)$ dependences are shown (Nashchekina et.al., 1999b). For all samples, σ decreases with growing temperature, and for the stoichiometric sample, in the *σ(T)* curve in the vicinity of 90-95 K there is observed an anomaly manifesting itself through additional drop in *σ*. In the samples with 50.4 and 50.6 at.% Te no anomalies are detected. As temperature increases up to 100-150 K, R_H practically does not change (which is characteristic of semiconductors with a high degree of degeneration), but then R_H grows monotonically, which is usually interpreted in terms of the two-band model of the valence band (Nimtz & Schlicht, 1985; Ravich et al,1970). No temperature anomalies of R_H were observed in the vicinity of the FPT for all bulk samples investigated. This indicates that FPT in SnTe is not accompanied by any noticeable change in *p*, but additional scattering of carriers takes place leading to additional decrease in mobility $\Delta\mu_H$. On the basis of the experimental data, $\Delta\mu_H$ was calculated. Its maximum value corresponds to T_C = 93 K and reaches ~8 % (see the insert of Fig. 10,c). Similarly, additional electrical resistance $\Delta\rho$ as a result of FPT was estimated and compared with the results reported in (Kobayashi et al., 1975) and (Grassie et al., 1979), where the *ρ(T)* dependences were studied for SnTe single crystals with p_{H77} = 1.2 × 10^{20} cm^{-3} and polycrystalline films with p_{H77} = 0.76 × 10^{26} m^{-3}, respectively. In Fig. 10,d, the $\Delta\rho(T)$ dependences obtained in (Nashchekina et al., 1999a) and (Kobayashi et al., 1975) are given. The additional enhancement in *ρ* at T_C amounted to ~2 % in (Kobayashi et al., 1975), ~8 % in (Nashchekina et al., 1999a), and ~14 % in (Grassie et al., 1979). The $\Delta\rho$ value increases as we move from single crystals (Kobayashi et al., 1975) to polycrystals (Nashchekina et al., 1999a), and then to polycrystalline films (Grassie et al., 1979). It can be suggested that imperfection of structure induces additional mechanisms of electron-phonon interaction leading to Δρ increase. The growth in *ρ* near the FPT was explained quantitatively by two alternative mechanisms. The authors of (Kobayashi et al., 1975; Katayama et al., 1976).), who attributed the anomalous behavior of ρ in the range of the FPT to interband scattering of holes by transverse optical phonons near *L*-point of the Brillouin zone, obtained the following equation for Δρ:

$$\Delta\rho = T / \gamma(T - T_C) \quad \text{at } T > T_C \qquad\qquad \Delta\rho = T / 2\gamma(T - T_C) \quad \text{at } T < T_C \tag{5}$$

where *γ* is the coefficient independent of temperature. For the explanation of the anomaly in *ρ*, the authors of (Minemura & Morita, 1978) used the model of band-to-band electron-phonon interaction, assuming that the interband interaction through deformation potential is weak. In this model, Δρ changes with temperature according to the law: $\Delta\rho \sim T^2$. In Fig. 10,e the dependence of $\Delta\rho$ on $T/(T-T_C)$, where T_C = 93 K, is presented. One can notice that there is a good agreement between the experimental data and theoretical predictions (Kobayashi et al., 1975; Katayama et al., 1976).): the slope of the straight line at $T \ll T_C$ is almost twice as large as that at $T > T_C$; in the temperature range near the FPT, there is a significant deviation from the theoretical dependence. Studying the *σ(T)* dependences of the thin films showed that for the SnTe films with p_H = (1.0-2.5)·10^{26} m^{-3} near 100 K in the *σ(T)* curves there is an anomaly associated with the FPT, although less pronounced than for bulk samples. At higher values of p_H, the anomaly is not detected (Nashchekina et al., 1998). Thus, the results of our studies convincingly confirm the existence of the FPT in the stoichiometric SnTe, the determined value of T_C being in good agreement with the results of other authors.

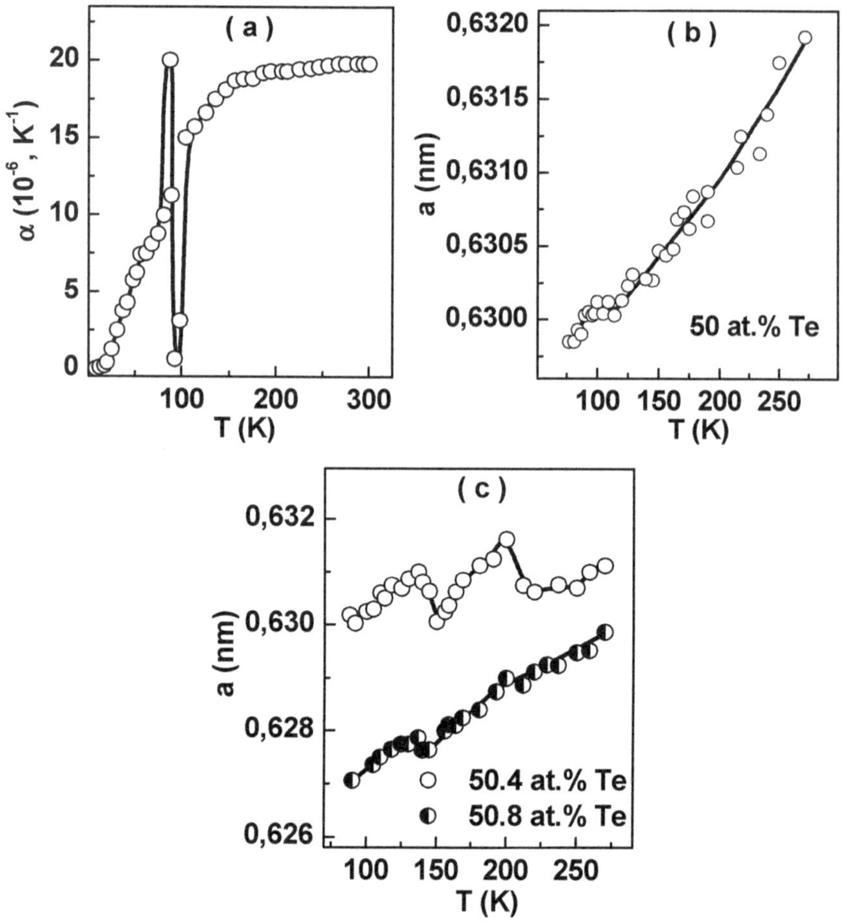

Fig. 9. Temperature dependences of the coefficient of linear expansion for stoichiometric SnTe (a) and unit cell parameter for stoichiometric (b) and nonstoichiometric (c) SnTe

3.2 Other phase transitions in SnTe

3.2.1 X-Ray diffraction studies

In (Nashchekina et al., 1999b), SnTe with various degrees of deviation from stoichiometry (50.0, 50.4, and 50.8 at.%Te) was studied by using XRD analysis, and the *a(T)* dependences were measured during heating in the temperature range 80-290 K (Fig 9,b,c). It has been mentioned above that the *a(T)* curve for the sample of the stoichiometric composition exhibits the anomaly in the range 90-100 K (Fig. 9a), which is apparently associated with the known FPT. For 50.4 at.% Te, well-pronounced jumps in the unit cell parameter ($\Delta a/a \cong 0.015$) are observed in the intervals 135-150K and 200-215 K corresponding to a negative *a*. At 50.8 at.% Te these effects become less pronounced (Fig.9,c). The presence of the anomalies in the *a(T)* curves indicates the instability of the crystal lattice in certain temperature intervals. Taking into consideration what was said above regarding the composition of 50.4 at.% Te (see 2.5), it is natural to assume that the emergence of new PT in SnTe can be due to a change in the equilibrium conditions and the type of polymorphism in the presence of vacancies as well as due to redistribution of intrinsic defects as a result of their interaction.

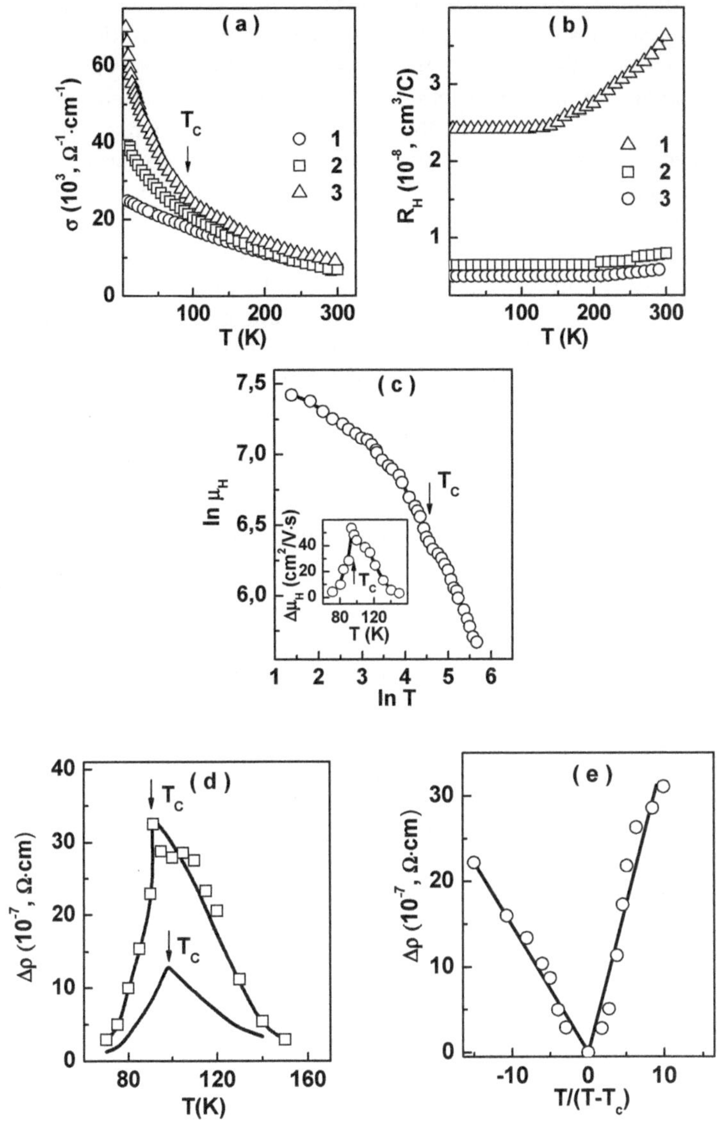

Fig. 10. Temperature dependences of the electrical conductivity σ (a), Hall coefficient R_H (b), charge carrier mobility μ_H (c), additional decrease in carrier mobility $\Delta\mu_H$ (c) and electrical resistance $\Delta\rho$ at FPT (d,e) for SnTe with different stoichiometry. a,b: 1 - 50 at.% Te, 2 – 50.4 at.% Te, 3 – 50.6 at.% Te; c,d,e: 50.0 at.% Te (Nashchekina et al., 1999b).

There may arise a question whether the ordering processes are possible at such low temperatures. The redistribution of defects is usually associated with migration of atoms and follows the diffusion mechanism. The rate of diffusion in normal crystals at temperatures from 80 to 300 K is low, and such transition must be hampered. However, if diffusion occurs through the vacancy mechanism (which is the most probable in the given case), the self-diffusion coefficient D and the vacancy diffusion coefficient D_i are connected via the relation $D = f \cdot A \cdot D_i$, where f is the correlation factor and A is the relative fraction of vacancies (Bokshtein et al., 1974; Show, 1975). Hence, an increase in the vacancy concentration leads to an increase in the self-diffusion rate. Such a sharp increase in the diffusion coefficient upon the introduction of structural vacancies caused by deviation from stoichiometry is confirmed experimentally (Show, 1975). Since the concentration of structural vacancies in SnTe at low temperatures (~ 1-2 %) is much higher than the concentration of thermal vacancies responsible for the vacancy mechanism of diffusion in most crystals (10^{-14}-10^{-18}% at 300 K if we assume that the energy of the vacancy formation is E_v= 1-1.5 eV), we can expect that the diffusion rate in SnTe is much higher than in stoichiometric crystals. The presence of NSD of various types formed as a result of rapid cooling also stimulates the increase in the rate of diffusion processes (Bokshtein et al., 1974; Show, 1975). It should also be kept in mind that the samples were subjected to the long (200 h) homogenizing annealing at 820 K, resulting in a high degree of crystal homogeneity and a quite uniform distribution of vacancies over the volume. Under such conditions, small displacements of atoms are sufficient for the formation of a long-range order at low temperatures. On the other hand, the possibility of defect ordering at temperatures above 300 K should not be ruled out either.

The observed anomalies are characterized by a considerable variation of a for the same symmetry and by the negative sign of α. This suggests that such a strong manifestation of PT is determined not only by the thermodynamic but also kinetic factors. The $a(T)$ curves were recorded during heating of the samples cooled preliminary to 80 K at a rate of ~20K/min. It can be assumed in this connection that a certain nonequilibrium state with a higher than in the equilibrium state is established in the sample with 50.4 at.% Te as a result of a rapid cooling and incompleteness of the diffusion processes whose rate is small at such low temperature. In other words, the structural instability of the samples with 50.4 and 50.8 at.% Te in certain temperature intervals is caused by the presence of an equilibrium PT associated with a redistribution of NSD, but the sharp change in a accompanying these PT in the process of sample heating is apparently because of the nonequilibrium nature of the initial structure formed as a result of rapid cooling from room temperature. The diffusion rate increases due to heating, and the system abruptly reaches the equilibrium state corresponding to a given temperature at the point where an equilibrium PT must take place. It should be noted that the authors of (Fukui et al., 1984) obtained amorphous SnTe films by condensation on substrates at T=4.2 and 77 K, and the resistivity of such films decreased in a jump-like manner upon heating at 180 K, i.e., at the temperature close to the jump in a in the $a(T)$ curve (Fig. 9,c). One should also note that the positions of two steps in the temperature dependence of the resonant field of ^{119}Sn nucleus (Khandozhko et al., 1978) coincide with the positions of the steps in the a(T) curve obtained in our studies. Thus, the existence of the PT was confirmed by two different methods.

3.2.2 Thermal expansion coefficient

In connection with the results of the XRD studies, it seems important to clarify how kinetic factors can affect the manifestation of the anomalies in the *T*-dependences of other properties. Besides, the high concentration of cation vacancies must lead to the specificity of relaxation processes, which occur in the system deviated from the equilibrium state. To additionally confirm the existence of new PT and to establish the influence of the NSD concentration on the character of the relaxation processes, we measured the $\alpha(T)$ dependences using two regimes described in 2.3.4. (Nashchekina et al, 2008). For the samples measured under the stationary regime, an increase in α with increasing temperature continued up to ~ 130-140 K (the Debye temperature of SnTe), and the α(T) curves exhibited anomalies. In the $\alpha(T)$ curve of the sample with 50.4 at.% Te (Fig. 11,a), an anomalous behavior is observed in the temperature ranges 140-150 K and 190-200 K. The $\alpha(T)$ curves for the samples with 50.7 and 51.1 at.% Te have plateaus at 180-190 K (Fig. 11,b). When measurements were carried out under the dynamic regime, the behavior of the $\alpha(T)$ curves changed significantly for all studied compositions (Fig. 11). In contrast with the stationary regime, a sharp initial increase in α was observed, and under the subsequent heating, the $\alpha(T)$ dependences exhibited either an oscillatory behavior (50.4 - 50.7 at.% Te) or a monotonic decrease in α with increasing temperature (51.1 at.% Te). The oscillatory behavior was most pronounced in the sample with 50.4 at.% Te (Fig. 11,a). In the sample with the composition 51.1 at.% Te, which is outside the HR, approaching the equilibrium state occurred in a monotonic way. For all studied compositions, starting from 220-240 K, the curves obtained in the stationary and dynamic regimes virtually coincided (Fig. 11).

Analyzing the obtained data, one can suggest that due to quick cooling of the samples in liquid nitrogen, they get oversaturated with defects and the processes of defect redistribution in the samples are not finished. A significant deviation from the equilibrium is confirmed by the sharp increase in α measured in the dynamic regime compare to the values of α obtained in the stationary regime, which was clearly observed in all samples (Fig. 11). With increasing Te concentration, a growth in α caused by immersing the sample in liquid nitrogen becomes more significant, which indicates that the degree of disordering in the system increases. Under heating of such nonequilibrium system, due to an increase in the diffusion rate, which leads to the defect movement and partial annihilation of defects, complex processes must take place. These processes are caused, on the one hand, by the relaxation of the system towards the equilibrium state and decrease in α, and on the other hand, by phase transformations occurring in a system whose state is far from equilibrium. Such processes can lead to a gigantic lattice instability in the vicinity of the temperature of a PT (equilibrium or nonequilibrium). When returning to the equilibrium state, the system can pass through a number of intermediate states corresponding to metastable equilibria at given temperatures. As a result, the character of the $\alpha(T)$ dependences becomes more sophisticated. The fact that values of α measured under the two above described regimes coincide at temperatures above ~ 240 K shows that in the vicinity of 240 K the system returns to the equilibrium and that the state of the defect subsystem is identical for both regimes. As is seen from Fig. 11, the character of the relaxation processes and kinetics of the process of reaching equilibrium in the crystal defect subsystem depend on the cation vacancy concentration. The *α(T)* dependences for most of the studied samples exhibit an oscillatory behavior, although locations, amplitudes and nature of oscillations vary. The oscillations are most pronounced for the sample with 50.4 at.% Te, and with increasing Te concentration these oscillations become less distinct (Fig. 11). One can suggest that these oscillations are associated mainly with the defect redistribution processes (PT) taking place under nonequilibrium conditions. That is why the oscillations observed in the α(T) curve of the sample with 50.4 at.% Te measured under the dynamic regime (Fig. 11) can be connected with a high probability of the redistribution of NSD which takes place in a system whose state is far from the thermodynamic equilibrium. A larger magnitude of the anomaly at 190-200 K observed in the sample with 50.4 at.% Te under the dynamic regime compare to the stationary regime is apparently caused by such redistribution. For the samples with 50.5-50.8 at.% Te, the *α(T)* dependences are similar but the manifestation of the anomalies is less pronounced. The most probable reason for that is deviation from the optimum ordering composition and, as a result, diminishing stimuli for ordering. However, the increase in *α* in comparison with its equilibrium value becomes more noticeable, and an oscillatory character of the relaxation processes manifests itself more distinctly. To all appearances, the oscillatory character of the relaxation processes is connected to a great extent with a high concentration of NSD, whose presence leads to loosening of the crystal lattice, slackening of bonds and decrease in the period of oscillations, thus making them observable.

3.2.3 Electronic properties

The temperature dependences of R_H, μ_H, σ in the range 77 - 300 K were obtained for SnTe epitaxial and polycrystalline thin films with different thicknesses *d* (0.2 – 2.0 μm) and different carrier concentrations: p_{H77} = 3.5·10^{26}, 4.5 × 10^{26} m^{-3}, and p_{H77} = 1 × 10^{27} m^{-3} (Nashchekina & Rogacheva, 1998). Techniques were developed for the preparation of mono- and polycrystalline SnTe films with controlled concentration of NSD and high μ_H (Nashchekina et al, 1999a; Nashchekina et al, 1995). The best results were attained using the hot wall method, which permits to vary the cation vacancy concentration in films, and thus their properties, in a wide range. Analysis of the temperature dependences of kinetic characteristics showed that the values of mobility in mono- and polycrystalline films are rather close and comparable to those in bulk crystals. It was established that the exponent in the $\mu_H(T)$ relationship is determined by the hole concentration, temperature, as well as by the degree of the film structure perfection, and its values lie in the range from 0 to 0.7. The highest values of the exponent are found in monocrystalline films with low *p* at low temperatures.

For the epitaxial SnTe films with p_{H77} = 3.5·10^{26} m^{-3} and thicknesses *d* = 0.2 and 0.4 μm in the temperature range 180- 200 K, an increase in R_H and a pronounced slowing down of the drop in μ_H are observed (Fig 12). For the sample with *d* = 0.6 μm, in the vicinity of 200 K, pronounced anomalies of R_H and μ_H are registered representing evidence of the qualitative changes in the electron and phonon spectra. The values of μ_H for this sample are higher than those for the other two. One can suggest that the anomalies in the R_H(T) and μ_H (T) dependences in the vicinity of 200 K are caused by the PT connected with a redistribution of NSD. The high mobility and the high, for such films, value of *b* below 200

K indicate the possibility of the formation of an ordered structure in which electron scattering by defects becomes insignificant due to their regular distribution over the crystal lattice. The sharp character of transition in the sample with d = 0.6 μm and the gradual one in the samples with d = 0.2 and 0.4 μm can be associated with kinetic factors determining scale and degree of ordering. At small d, the ordering processes can be hampered due to a considerable contribution of the surface energy. For the investigated films, the thickness of 0.6 μm seems to be optimal for ordering. One can assume that the increase in R_H above 200 K, interpreted usually in the framework of the two-band model, is connected with a disordered distribution of NSD and/or the appearance of defects of a new type. In Fig. 12,a, the $R_H(T)$ and $\mu_H(T)$ dependences for the sample with $p = 4.5 \times 10^{26}$ m^{-3} are shown. As is seen, there are two distinct anomalies in the temperature ranges 135-150 and 200-215 K. In Fig. 12, the $\rho(T)$ dependences for polycrystalline SnTe films (d=0.27 and 2.0 μm) with $p=(3\text{-}4)\cdot 10^{27}$ m^{-3} deposited on glass substrates at room temperature are presented. In the film with d = 0.27 μm, ρ increases practically linearly with increasing temperature, and in the region 135-145 K a horizontal section is observed, whereas in the 2 μm thick film there are two anomalous sections in the temperature ranges 145-155 K and 185-200 K. It follows from the obtained data that the temperature intervals in which the anomalies of ρ, R_H and μ_H are observed are practically the same (135-150 K, 185-215 K) for both mono- and polycrystalline films. It is seen that the reconstruction of the defect subsystem is accompanied not only by a sharp (~ 30%) drop in mobility but also by a noticeable change in R_H (15-20 %). Let us note that under the FPT no change in R_H was observed, i.e. insignificant changes in the crystal lattice symmetry do not practically affect the electronic subsystem. On the other hand, the processes of the defect redistribution accompanied by a change in the electronic structure of defects, in the character of chemical bonds in the vicinity of a defect, which determine the number of charge carriers and their mobility, cause changes not only in μ_H but also in R_H. The fact that under certain conditions the ordering processes are more easily realized in thin films is by no means exceptional. In a large number of III-V alloys with isovalent substitution, the phenomenon of atomic scale ordering during film growth, which is not thermodynamically stable in the bulk alloys, has been revealed. The probability of cation vacancy ordering in SnTe films increases due to the effect of such factors as the Coulomb repulsion between charged defects and a large deformation energy connected with their formation. As is seen from the results obtained, the realization of these processes is controlled by kinetic factors. In bulk crystals the diffusion processes are inhibited, which prevents the defect redistribution and formation of a configuration corresponding to the minimum of free energy. A detailed study of the T- dependences of ρ and S for the SnTe polycrystalline films with $p=(3\text{– }5)\cdot 10^{27}$ m^{-3} was conducted in the range of 80–300 K (Rogacheva et al, 2003c). In the $\rho(T)$ and $S(T)$ curves within the range 80–150 K a number of peculiarities in the form of steps and plateaus were exhibited most distinctly. The possible reasons for the appearance of the temperature peculiarities are the system's passing through different quantum states; the processes of self-organization taking place in an open system; microdomain structure of thin films; relaxation processes. The pronounced anomalies observed in the ranges of 135-150 and 190–200 K are attributed to PT caused by the redistribution of NSD.

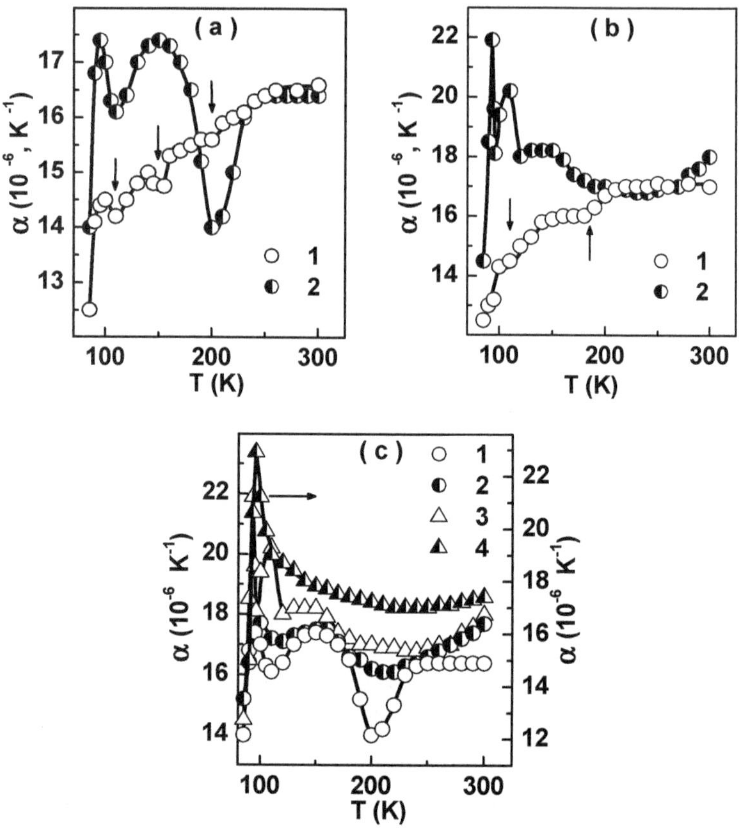

Fig. 11. Temperature dependences of the thermal expansion coefficient α for SnTe with 50.4 (a) and 50.7 (b) at.% Te (1 – stationary regime; 2 – dynamic regime) and $\alpha(T)$ dependences for SnTe with different stoichiometry obtained under a dynamic regime (1 - 50.4 at.% Te, 2 - 50.5 at.% Te , 3 - 50.7 at.% Te, 4 - 51.1 at.% Te) (Nashchekina et al, 2008).

4. Problems of doping of non-stoichiometric SnTe

4.1 Mechanism of impurity dissolution in nonstoichiometric SnTe

One of the basic ways of controlled changing of properties of semiconductors is doping, which is accompanied by a partial disordering of the structure by IA. It is impossible to develop physical principles for controlled doping of NS phases without the knowledge of the specific physical effects connected with the simultaneous introduction of IA and NSD in a crystal. IA can be introduced into a NS compound in the elementary form or in combination with other components of the host compound, for example, by cation substitution. In the Sn-Te-X (X is the impurity component) concentration triangle, different schemes of the IA introduction correspond to changing the alloy composition along a certain section passing through SnTe compound. We will limit our consideration to two doping schemes - the introduction of elementary impurities and the introduction of IA by cation substitution.

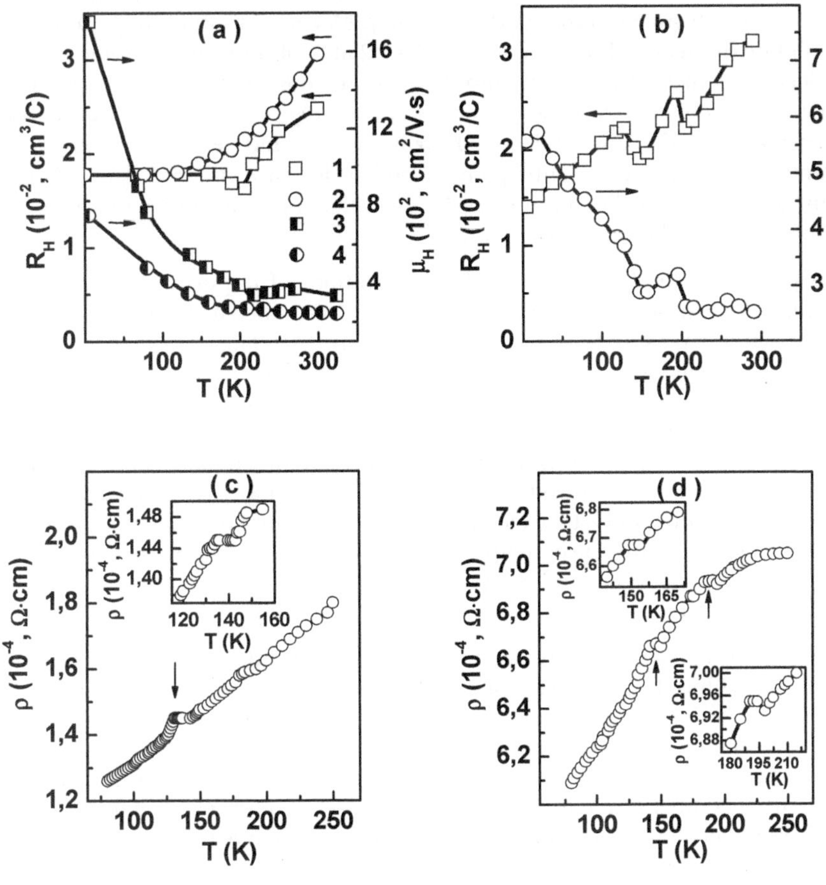

Fig. 12. Temperature dependences of Hall coefficient R_H (a,b), charge carrier mobility μ_H (a,b), and electrical resistance ρ (c, d) for SnTe thin films deposited on KCl at 520 K (a, b) or glass at 300 K (c, d) substrates with carrier concentration $p_{H77} = 3.5 \cdot 10^{26}$ m^{-3} (a), $4.5 \cdot 10^{26}$ m^{-3} (b), $4.0 \cdot 10^{27}$ m^{-3} (c), $3.0 \cdot 10^{27}$ m^{-3} (d) and with different thicknesses d (Nashchekina et al.,1999a, 1995; Nashchekina & Rogacheva, 1998). a: 1, 3-*d* =0.6 μm, 2, 4-*d*=0.2 μm; b: *d* =0.4 μm; c: *d* = 0.27 μm; d: *d* =2 μm

The basic mechanisms of the impurity dissolution in a stoichiometric semiconductor compound are the following: i) IA enter interstices and ii) IA fill the basic structural sites in one sublattice at the simultaneous formation of vacancies in the other one. To establish the mechanisms of IA dissolution in a crystal with NS vacancies, we performed a complex of studies (Dzyubenko et al., 1981,1983; Rogacheva & Dzyubenko, 1986; Rogacheva et al., 1985,1988) of the dependences of properties on the IA concentration at a fixed content of intrinsic defects. $Sn_{0.984}Te$ was chosen as the initial composition, and thus the crystal lattice contained ~1.6 at.% of vacant cation sites. The analysis of the concentration dependences of properties (Fig. 13) shows that there are two types (I and II) of impurities. For impurities I

(In, Ga, Cd, etc.), a nonmonotonic behavior of the concentration dependences of properties is observed, which suggests that under changing IA concentration different mechanisms of IA dissolution are realized (Fig. 13, a). At 0.1 – 0.2 at.% of impurity, extreme points or bends are observed, which divide HR into two subregions with different character of the concentration dependences of properties. In the subregion of small IA concentrations, the sharp increase in *H*, *p*, *a*, and the decrease in μ_H and λ_p occur with increasing IA concentration. In the second subregion, *H* and *p* decrease, μ_H and λ_p increase as the IA concentration increases. For impurities II, the behavior characteristic only of the second subregion is observed (Fig. 13, b).

The sharp increase in *H* and *a* at small concentrations of impurity atoms evidences that a significant deformation of the crystal lattice takes place, whereas the decrease in μ_H and λ_p is connected with increasing hole and phonon scattering by IA. The extreme character of the concentration dependence of *p* can be attributed to the change in the electrical action of IA from acceptor to donor under increasing IA concentration, which indicates a reconstruction of chemical bonds in the vicinity of IA. In this connection one can suggest that the introduction of the first portions of IA into the crystal lattice takes place by the interstitial mechanism, and IA are localized in the tetrahedral interstices of the close packing of Te atoms. In semiconductor crystals, which usually have relatively small coordination numbers and rather loose structure, the enthalpy of the IA formation is comparable with the enthalpy of the vacancy formation, and that is why interstitials play more important role in semiconductors than in metals.

The behavior of the properties in the range of IA concentrations greater than 0.1-0.2 at.% for impurity I or within the entire HR for impurity II suggests that the main mechanism of the IA dissolution is the filling of cation vacancies. The decrease in *a* observed in SnTe under increasing deviation from stoichiometry (Fig. 13,a) indicates the compression of the crystal lattice. A vacancy creates tensile strains around the defect, and the introduction of an IA leads to partial compensation of these strains. The decrease in *H* shows that the overall level of strains in the crystal lattice decreases. The main parameters that determine the degree of change in *H* under the introduction of point defects are the energies of elastic and electrical interaction between the defect and dislocation. Since the energy of elastic interaction is proportionate to the change in volume under the substitution of the host atom by the defect, the rate of change in *H*, occurring when the IA fills the vacancy, must depend somehow on the difference between the cation and vacancy radii. The comparison of the rate of the decrease in *H* when IA fill vacancies for impurities with the same valence showed that with decreasing difference in the radii of the impurity cation and cation vacancy this rate decreases (Dzubenko et al, 1983). IA localized in vacancies give their valence electrons to Te thus reducing the overall concentration of holes. The reduction in the crystal lattice distortions under the filling of vacancies leads to a reduction in phonon and electron scattering and consequently to an increase in μ_H and λ_p. It is established that all elementary impurities introduced into $Sn_{0.984}Te$ reduce *p*, simultaneously reducing *S*. The exception is In, which exhibits variable valence. In case of In impurity, *S* increases with decreasing *p*.

A question arises whether the observed effect related to a change in the localization of IA (from interstitials to cation vacancies) is an equilibrium one or corresponds to a metastable state produced by incomplete diffusion. In this connection we investigated the effect of thermal treatment on the concentration dependence of *H* and *a* in SnTe doped with Cd, In, and Sb (Rogacheva et al., 1985). In particular we applied annealing at 820K for 120 h (HT1) and for 420 h (HT2) and aging at room temperature for 2 years (HT3). In all samples subjected to HT1, a sharp increase in *H* and *a* was observed under the introduction of impurity up to 0.1-0.2 at. %, and then for *a* the rate of the increase decreased and for *H* it sharply dropped (Fig. 14). After HT2 in the systems doped with Sb, *H* decreased monotonically in the entire solubility region, and *a* increased at a lower rate than after HT1. In the systems doped with Cd and In, the general behavior of the dependences of *H* on composition did not change, although the relative increase in *H* and *a* at small concentrations decreased. HT3 caused practically no qualitative changes in the concentration dependences of the properties. Thus, at small IA content, the character of the property-composition dependences is controlled to a great extent by diffusion processes, and in a number of cases, the fixation of nonequilibrium states is possible. The stability of these states depends upon the IA diffusion coefficients. Thus, the presence of NSD in SnTe leads to the appearance of a new mechanism of IA dissolution consisting in the occupation of vacancies by IA. In case of some impurities (In, Ga, etc.), at small concentrations (< 0.2 at. %), IA enter tetrahedral interstices, and the mechanism is controlled by diffusion processes.

4.2 Solubility limits of impurities in SnTe

The limits of solubility for different elements in SnTe at fixed concentrations of intrinsic defects were determined on the basis of the results of the microstructure studies and measurements of the concentration dependences of properties after homogenizing annealing of samples at 820 K for 300 hours. $Sn_{0.984}Te$ was chosen as the initial composition. The analysis of the obtained results shows that the solubility limits (C_{lim}) for different elements differ substantially. When vacancies are occupied by IA, defects of a new type – impurity substitution defects – are created in the crystal lattice in addition to vacancies, which causes the appearance of elastic stress fields because of the difference in the cation sizes and violates the periodicity in the distribution of the crystal lattice potential because of the difference in the electronic configurations of Sn and impurity atoms. It is therefore natural to expect that C_{lim} depends both on the crystallochemical radii of Sn and IA and on their valence states. Fig. 15,a shows the dependences of C_{lim} on the relative difference ($\Delta r/r_{Sn} = (r_{imp}-r_{Sn})/r_{Sn}$) of the ionic radii of the impurity (r_{imp}) and Sn (r_{Sn}) atoms (Rogacheva & Dzubenko, 1986). It is evident from Fig. 15 that there is a correlation between C_{lim} and $\Delta r/r_{Sn}$ both for positive and for negative values of the size factor: when $\Delta r/r_{Sn}$ increases, the solubility of the impurity decreases. The different slopes of the straight lines for positive and negative values of $\Delta r/r_{Sn}$ for SnTe show that the compression and tensile deformations, appearing when the vacancies are filled by cations larger or smaller than Sn, affect the stability of the crystal lattice in different ways. For the same relative difference in the ionic radius, when Sn ions are replaced by smaller cations, the value of C_{lim} is higher than in case of substitution by larger cations. The influence of the vacancy concentration in SnTe on the solubility of IA was investigated in (Rogacheva et al, 1988; Vodoriz, 2009). It was established that the solubility of Sb in SnTe increases linearly with an increase in the concentration of cation vacancies up to ~ 1 at.% (Rogacheva et al, 1988) (Fig. 15,b). When IA are introduced in a NS phase with a vacancy concentration exceeding ~1 at. %, chemical interaction between impurity and host atoms in the solid solution becomes an additional factor affecting the solubility limit and character of the change in properties (Rogacheva et al., 1988). Similar results were obtained for SnTe doped with Cu (Vodoriz et al., 2009).

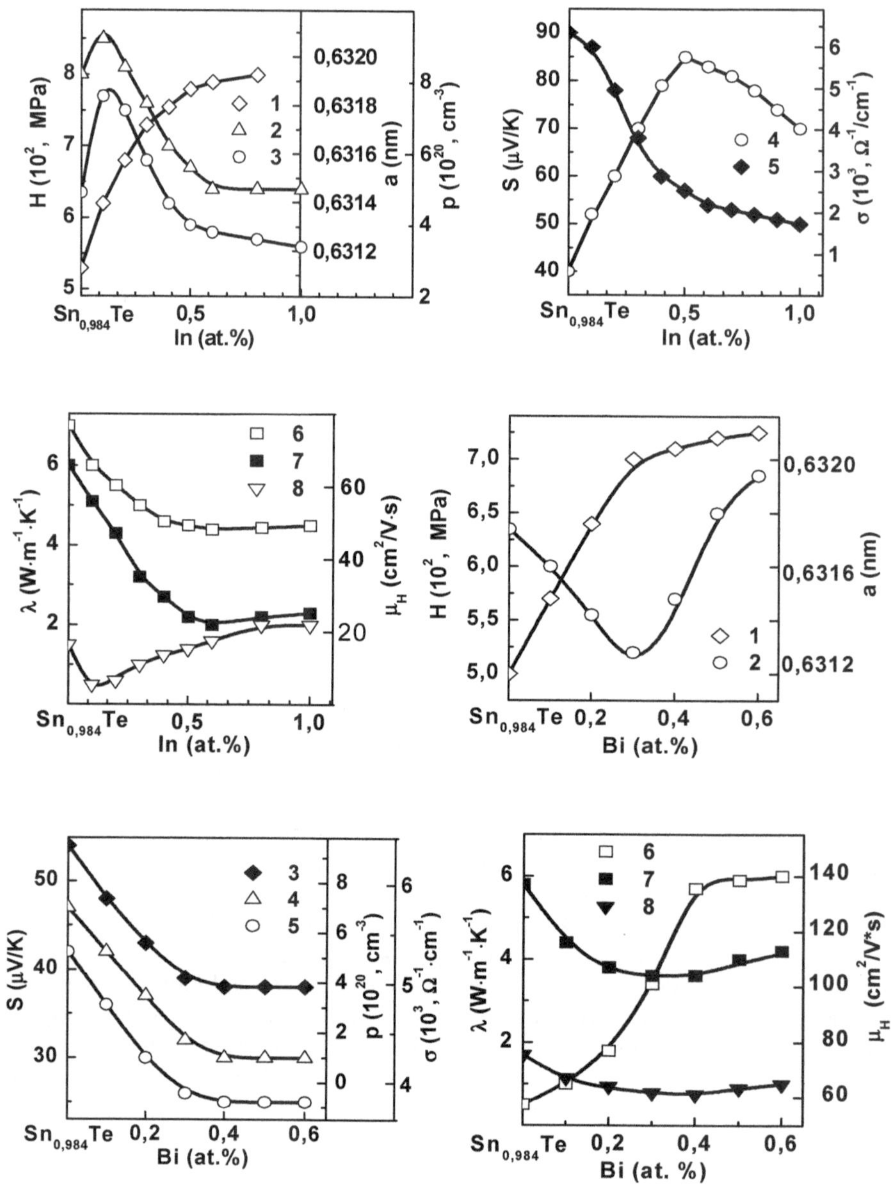

Fig. 13. Unit-cell parameter *a*, microhardness *H*, carrier concentration *p*, Seebeck coefficient *S*, electrical conductivity *σ*, total thermal conductivity *λ*, lattice thermal conductivity λ_p, and Hall carrier mobility μ_H of SnTe as a function of In and Bi content. In: 1 – a; 2 – p; 3 – H; 4 – S; 5 – σ; 6 - μ_H; 7 - λ; 8 - λ_p; Bi: 1 – a; 2 – H; 3 – σ; 4 – p; 5 – S; 6 - μ_H; 7 - λ; 8 - λ_p. (Dzubenko et al., 1981,1983)

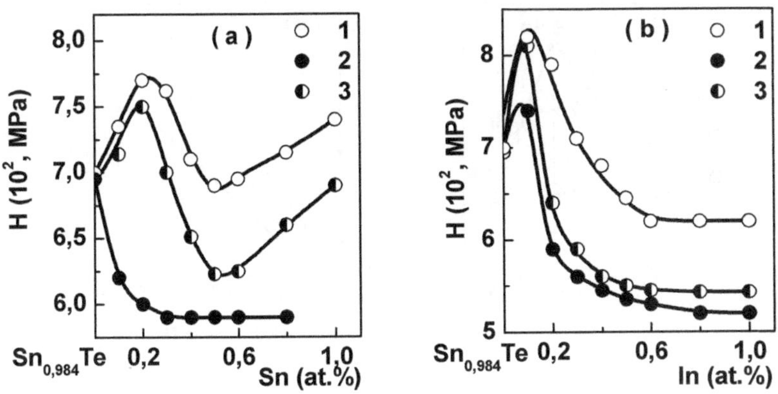

Fig. 14. The dependences of microhardness *H* on Sb (a) and In (b) concentration in $Sn_{0.984}Te$ after annealing at 820 K for 120 h (1), 420 h (2) and aging at room temperature for 2 years (3) (Rogacheva et al., 1985).

4.3 The influence of a third component on the equilibrium of intrinsic defects in SnTe

The introduction of a third component X in a binary compound must lead to a change in the thermodynamic equilibrium conditions and to a change in the intrinsic defect concentration (Kröger, 1973). One can judge about the character of this change by the position of the boundaries of HR of a SnTe-based phase in a ternary Sn-X-Te system. At a fixed T, such HR represents a two-dimensional region within the solubility isotherm. We plotted the solubility isotherms using a wide array of experimental methods (Rogacheva, 2005). It was established that the solubility isotherms have a pronounced singular character and the direction of maximum solubility corresponds to the cross-section passing through the stable compound formed in an X-Te system. In the systems with Cu and Ag, the isotherms are oriented in the direction of Cu_2Te and Ag_2Te compounds, respectively, while in the systems with Sb and Bi - in the direction of Sb_2Te_3 and Bi_2Te_3 (Fig. 16). The observed regularities can be qualitatively predicted on the basis of the active mass law (Kröger, 1973; Glazov & Pavlova, 1988): there is a dynamic equilibrium between the concentrations of ionized intrinsic and impurity defects, the presence of charged impurity centers leads to an increase in the concentration of intrinsic defects of the opposite sign. The analysis of the solubility isotherms allowed us to reveal another interesting effect: in all isotherms in the range of small IA concentrations (up to ~ 0.5-1.0 at.%), a shift of the SnTe HR boundary in the direction of increasing degree of deviation from stoichiometry is observed. The formation of additional vacancies in the diluted solid solutions can be stimulated by such factors as a decrease in the energy of vacancy formation in a strongly distorted crystal, a larger increase in the configuration entropy under introduction of IA into a crystal with vacancies, and other factors. We attribute the bents in the solubility isotherms in the range of small IA concentrations to a transition from a weak doping when the interaction between IA can be neglected to a heavy doping when this interaction becomes cooperative.

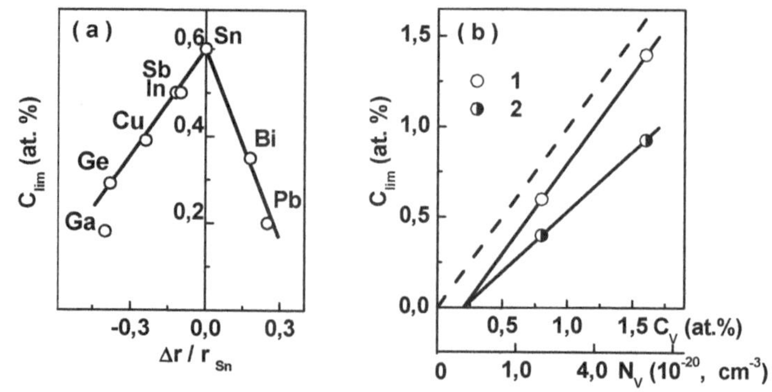

Fig. 15. The dependences of the solubility limit (C_{lim}) of impurities in SnTe on the relative difference in the ionic radii of impurity and Sn atoms ($\Delta r/r_{Sn} = (r_{imp} - r_{Sn})/r_{Sn}$) (a) (Rogacheva&Dzubenko, 1986) and the dependence of the solubility limit of Sn (1) and Sb (2) on the concentration of cation vacancies in SnTe (b) (Rogacheva et al., 1988).

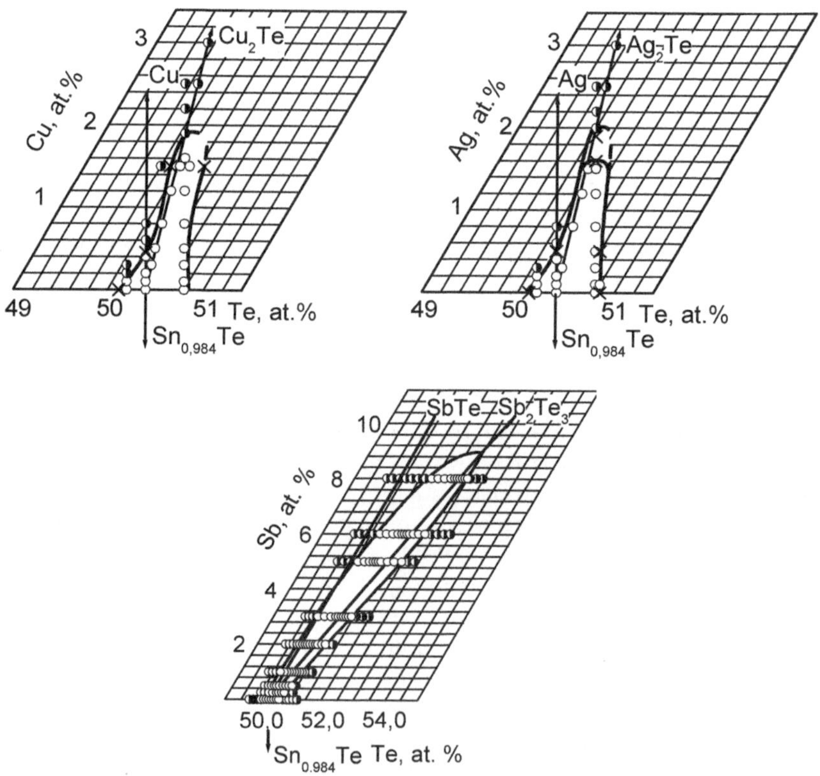

Fig. 16. Isotherms of solubility at 520 K in the Sn-Cu-Te, Sn-Ag-Te, and Sn-Sb-Te ternary systems (Rogacheva, 2005).

4.4 Impurities with variable valence and nonstoichiometric vacancies in SnTe

Under the introduction of cations with variable valence into SnTe, an unusual effect connected with the possibility of the formation of chemical bonds of different types in SnTe-based solid solutions was detected (Rogacheva et al., 1983,1991,2003d; Baltrunas et al., 1988; Gorne et al., 1987). The effect consists in an anomalous influence of deviation from stoichiometry on properties of SnTe doped with In. The movement along In isoconcentrates in the Sn-Te- In ternary system in the direction of increasing Te concentration leads to the recharging of In cations ($In^{1+} \rightarrow In^{3+}$), compensation of the acceptor action of vacancies by the donor action of In^{3+} and a decrease in the hole concentration, which continues up to the intersection with the SnTe - In_2Te_3 section. It was shown that the change of the charge state of In atoms is associated with a change in their localization in the crystal lattice. The model of energy spectrum of SnTe doped with In was proposed (Rogacheva & Laptev, 1999). According to the model, the valence band contains two resonance bands corresponding to In^{1+} and In^{3+}. The vacancy band is the lowest at the energy scale of the valence band, followed by In^{1+} and In^{3+} bands (Fig. 8,b). The width of the V_0 or In^{1+} subbands and effective mass of holes are determined by the distance between defects. In (Rogacheva et al, 1997), on the basis of the Mössbauer measurements it was shown that introducing Ag_2Te into $Sn_{0.984}Te$ results in partial change of Sn valence from Sn^{2+} to Sn^{4+}.

5. Conclusions

The specificity of chemical bonds and defect structure of semiconductors in comparison with metals determines the necessity of further development of the concept of a nonstoichiometric phase. Semiconductor compounds with one-sided wide homogeneity region represent a special group of nonstoichiometric phases. We discussed the specificity of properties of such phases using SnTe as an example. The conducted studies yielded the following results.

- It was established that as temperature decreases, the SnTe HR boundaries shift towards the composition 50.4 at.% Te, corresponding to the maximum in melting curves. This composition is marked with special points in the concentration dependences of different properties. The possibility of the formation of vacancies superstructure at 50.4 at.% Te is suggested.
- It was shown that it is fruitful to use the percolation theory approach for studying properties of phases with wide HR.
- A new approach to studying the genesis of energy band spectrum of semiconductor phases with wide HR, taking into account a high concentration of NS defects, was proposed, and a principally new model of energy band spectrum was developed for SnTe and for SnTe doped with indium.
- It was shown that the presence of NS vacancies in the host compound leads to the appearance of a new mechanism of impurity dissolution consisting in the occupation of vacancies by impurity atoms. The solubility limits for different elements in SnTe at fixed concentrations of NS cation vacancies were determined. At vacancy concentrations less than ~1 at.%, there is correlation between solubility limit and relative difference in ionic radii of Sn atoms and those of the impurity. The solubility limit increases with increasing concentration of NS vacancies in the host material.

- It was revealed that the introduction of cations with variable valence (In, Cu) leads to a specific effect caused by the presence of NS vacancies and consisting in the recharging of cations under changing concentration of impurity atoms or deviation from stoichiometry.
- For SnTe, new temperature phase transitions associated with redistribution of NS defects were revealed.
- There was proposed a classification of one-sided phases, the composition of a phase at T=0 being chosen as a main criterion. "Genuine NS phases" remain one-sided in the entire temperature range, even at $T \rightarrow 0$. The formation of such phases can be connected with a decrease in the crystal free energy under deviation from stoichiometry due to an increase in the contribution of the electron subsystem to the crystal total energy and defect ordering.

6. Acknowledgments

The author thanks Dr. Olga Nashchekina for fruitful and stimulating discussions.

7. References

Abrikosov, N.Kh., & Shelimova, L.E. (1975). *Semiconducting materials based on A^4B^6 compounds*, Nauka, Moscow.

Allgaier, R.S. & Sheie, P.O. (1961). Electrical properties of p-type SnTe. *Bull. Am. Phys. Soc.*, vol. 6, No. 2, (February, 1961), pp. 436-437, ISSN 0003-0503.

Akchurin, R.Kh., & Ufimtsev, V.B. (1979) Calculation of lead and tin tellurides homogeneity regions boundaries. *Zhurnal fizich. khimii,* Vol. 53, No. 6, (June, 1979), pp. 1441-1445, ISSN 0044-4537.

Andreev, A.A. (1967). Parameters of the two valence bands model for SnTe from Hall effect measuring at high temperature. *Fizika tverdogo tela,* Vol. 9, No. 5, (May, 1967), pp. 1560-1561, ISSN 0367-3294.

Baltrunas, D., Motiejunas, S., & Rogacheva, E.I. (1986). Effect of the Deviation from Stoichiometry on the Mossbauer Parameters of SnTe. *Phys. Stat. Solidi. A,* Vol. 97, No. 2, (February, 1986), pp. K131-K133, ISSN 1862-6319.

Baltrunas, D.I., Gorne, G.V., Laptev, S.A., Moteyunas, S.V., & Rogacheva, E.I. (1988). Nature of SnTe-based solid solutions in the system SnTe-InTe. *Izv. AN SSSR. Neorgan. Material,* Vol. 24, No 10, (1988), pp. 1398-1401.

Bashkirov, Sh.Sh., Dobrjakov, I.A., Liberman, A.B., & Tsarevskiy, S.S. (1985). Mossbauer and X-ray investigations of the high-temperature phase transition in SnTe. *Kristalographiya,* Vol. 30, No. 6, , pp. 1016-1017, ISSN 0023-4761.

Bekker, A.A., Zhukov, Yu.M., Kaukis, A.A., & Nesmeyanov, A.N. (1973). Tin telluride study using gamma-resonance spectroscopy method. *Vestnik Moskovskogo Univ., Ser. Khimiya,* No. 4, (1973), pp. 434-436, ISSN 0579-9384.

Belson, H.S., & Houston, B. (1970). Thermal expansion of tin telluride. *J. Appl. Phys.*, Vol. 41, No. 1, (January- February, 1970), pp. 422-424, ISSN 0021-8979.

Berman, R. (1979). *Thermal Conduction in Solids,* Mir, ISBN 1704060000, Moscow, (in Russian).

Bokii, G.B., (1971). *Crystal Chemistry,* Nauka, Moscow, (in Russian).

Brebrick, R.F. (1967). Non-stoichiometry in Binary Semiconductor Compounds. *Progress Solid State Chem.*, Vol. 3, No 3, (1967), pp. 213-264, ISSN 0079-6786.

Brebrick, R.F. (1963). Deviations from stoichiometry and Electrical Properties in SnTe. *J. Phys. Chem. Solids,* Vol. 24, No 1, (January, 1963), pp. 27-36, ISSN 0022-3697.

Brebrick, R.F., & Strauss, A.I. (1964) Partial Pressures in Equilibrium with Group IV Tellurides. II. Tin Telluride. *J. Chem. Phys.*, Vol. 41, No 1, (January, 1964), pp. 197-205, ISSN 0021-9606.

Brebrick, R.F. (1971). Composition stability limits for the Rocksalt-Structure Phase $(Pb_{1-y}Sn_y)_{1-x}Te_x$ from Lattice Parameter Measurements. *J. Phys. Chem. Solids.*, Vol. 32, No 3, (March, 1971), pp. 551-562, ISSN 0022-3697.

Brebrick, R.F., Harman, T.C., & Strauss, A.J. (1962). Experimental investigation of band structure of SnTe. *Bull. Am. Phys. Soc. ser. II.*, Vol. 7, No. 2, (February, 1962), pp. 203, ISSN 0003-0503.

Brebrick, R.F., & Strauss, A.J. (1963). Anomalous thermoelectric power as evidence for two-valence bands in SnTe. *Phys. Rev.*, Vol. 131, No. 1, (July , 1963), pp. 104-110, ISSN 1943-2879.

Bokshtein, B.S., Bokshtein, S.Z, & Zhukhovitskii, A.A. (1974). *Thermodynamics and kinetics of diffusion in solids*, Metalurgiya, Moscow, (in Russian).

Brilson, L.J., Burstein, E., & Muldawer, L. (1974). Raman observation of the ferroelectric phase transition in SnTe. *Phys. Rev. B*, Vol. 9, No. 4, (February, 1974), pp.1547-1551, ISSN 1098-0121.

Brukhanov, V.A., Delyagin, N.N., Kuzmin, R.I. & Shpinel', V.S. (1964). Mossbauer effect in the binary compounds of tin. *JETF*, Vol. 46, No. 6, (June, 1964), pp. 1996-2001, ISSN 0044-4510.

Bukchpan, S. (1968). Determination of the Debye temperature of SnTe using the Mossbauer effect in ^{119}Sn and ^{125}Te. *Solid State Commun*, Vol. 6, No 7, (July , 1968), pp. 477-478, ISSN 0038-1098.

Burguen, Zh., & Lanno, M. (1985). *Point defects in Semiconductors. Experimental aspects*, Mir, Moscow, (in Russian).

Cohen, M.L. (1961). Energy bands in the Bismuth structure. 1 A nonellipsoidal model for electrons in Bi. *Phys. Rev.*, vol. 121, No. 2, (January, 1961), pp. 387-395, ISSN 1943-2879.

Collongues, R. (1974). *Nonstoichiometry. Inorganic materials of variable composition*, Mir, Moscow, (in Russian).

Collongues, R. (1993). Nonstoichiometry in ternary oxides and materials science. *Jpn. J. Appl. Phys.*, Vol. 32, No. 3, (March, 1993), pp. 442-447, ISSN 0021-4922.

Damon, D.H. (1966). Thermal Conductivity of SnTe between 100° and 500°K. *J. Appl. Phys.*, Vol. 37, No. 8, (July, 1966), pp. 3181-3190, ISSN 0021-8979.

Drabl, J. & Goldsmit, G. (1963) *Thermal conductivity of semiconductors*, Izd. Inostr. Lit., Moscow, (in Russian)

Dzyubenko, N.I., Rogacheva, E.I., Kosevich, V.M., Laptev, S.A., & Arinkin, A.V. (1983). The influence of indium, gallium, antimony, and bismuth on the properties of tin telluride. *Izv AN SSSR, Neorgan Mater*, Vol. 19, No. 9, (September, 1983), pp. 1457-1461.

Dzyubenko, N.I., Kosevich, V.M., Ob'edkov, A.G., & Rogacheva, E.I. (1981). The behavior of the IV group elements in tin monotelluride and alloys on its basis. *Izv AN SSSR, Neorgan Mater.*, Vol. 17, No. 1, (January, 1981), pp. 34-38.

Efimova, B.A., Kaydanov, V.I., Moyzhes B.Ya. & Chernik, I.A. (1965). About SnTe band tructure. *Fizika tverdogo tela*, vol. 7, No. 10, (October, 1965), pp. 2524-2527, ISSN 0367-3294.

Erasova, N.A., & Kaydanov, V.I. (1968). Investigation of the tin telluride thermal conductivity. *Fizika i Tekhnika Poluprovodnikov*, Vol. 2, No. 1, (January, 1968), pp. 34-36.

Fano, V., Fedeli, G., & Ortalli, J. (1977). Phase transition in SnTe by Mossbauer spectroscopy. *Sol. St. Communs.*, Vol. 22, No. 7, (May, 1977), pp. 467-470, ISSN 0038-1098.

Flinn, P.A. (1978). Tin isomer shifts, In: *Mossbauer Isomer Shifts*, Shenoy, G.K., Wagner, F.E., North Holland, Amsterdam.

Fukui, K., Inoguchi, K., Kondo, S., & Tatsukawa, T. (1984). Amorphous-crystalline transformation of SnTe thin films. *Jpn. J. Appl. Phys.*, vol. 23, No. 8, (August, 1984), pp. 1141-1143, ISSN 0021-4922.

Houston, B.B., Allgaier, R.S., Babiskin, J., & Siebenmann, P.G. (1964). Relationship between real and nominal carrier concentrations in p-type SnTe. *Bull. Am. Phys. Soc.*, Vol. 9, No. 1, 2, (January, 1964), pp. 60, ISSN 0003-0503.

Glazov, L.M., & Pavlova, L.M. (1988). *Chemical Thermodynamics and Phase Equilibrium*. Mettalurgia, Moscow, (in Russian).

Grassie, A.D. C., Agapito, J.A., & Gonzalez, P. (1979). Anomalous resistivity at the structural phase transition of polycrystalline SnTe. *J. Phys. C*, Vol. 12, No. 24, (December, 1979), pp. L925-L927, ISSN 0022-3719.

Gorelik, S.S. & Dashevskii, M.Ya. (1988). *Materials science of semiconductors and isolators*, Metalurgiya, Moscow, (in Russian).

Gorne, G.V., Zhigareva, N.K., Ivanova, A.B., & Rogacheva, E.I. (1987). Chemical interaction in SnTe-based solid solutions in the Sn-In-Te system, *Izv. AN SSSR. Neorgan.Mater.*, Vol. 23, No.8, (August, 1987), pp. 1152-1156.

Gornik, E., Heinrich, H., & Palmetshofer L. (Eds.) (1982). *Physics of narrow gap semiconductors*, Springer, ISBN-10: 3540111913, Berlin.

Grigorovich, V.K. (1976). *Hardness and microhardness of metals*, Nauka, Moscow, (in Russian).

Gul'tyaev, P.V., & Petrov, A.V. (1951). Heat capacity of some $A^{IV}B^{VI}$ semiconductors. *Fizika tverdogo tela*, Vol. 1, No. 3, (March, 1951), pp. 1368-1370, ISSN 0367-3294.

Ioffe, A.V., & Ioffe, A.F. (1954). Some regularities in the values of semiconductor thermal conductivity. *Dokl. Akad. Nauk SSSR. Ser. Neorgan. Mater.*, Vol. 97, No. 5, (May, 1954), pp. 821-822.

Iisumi, M., Hamaguchi, Y., Komatsubara, K.P., & Kato, Y. (1975). Phase transition in SnTe with low carrier concentration. *J. Phys Soc. Jap.*, Vol. 38, No. 2, (February, 1975), pp. 443-449, ISSN 0031-9015.

Heinrich, H. (1980). Defects in IV-VI compounds. *Lecture Notes in Physics*, vol. 133, (1980), pp. 407-426, ISSN 0075-8450.

Julian, C.L. (1965). Theory of Heat Conduction in Rare-Gas Crystals. *Phys. Rev. A*, Vol. 137, No. 1A, (January, 965), pp. A128-A137 , ISSN 1050-2947.

Kafalas, J.A., Brebrick, H.F., & Strauss, A.J. (1964). Evidence that SnTe is a Semiconductors. *Appl. Phys. Letters.*, vol. 4, No. 5, (March, 1964), pp. 93-94, ISSN 0003-6951.

Kane, E.O. (1957). Band structure in indium antimonide. *J. Phys. Chem. Solids*, vol. 1, No. 4, (March, 1957), pp. 249-261.

Kaydanov, V.I., Chernik, I.A., & Efimova, B.A. (1967). Investigation of band structure and mechanism of charge carrier scattering in SnTe. *Fizika i tekhnika poluprovodnikov*, Vol. 1, No. 6, (June, 1967), pp. 869-379.

Kaydanov, V.I., & Ravich, Yu.L., (1985). Deep and resonant states in $A^{IV}B^{VI}$ semiconductors. *Physics-Uspekhi (Advances in Physical Sciences)*, Vol. 145, No. 1, (January, 1985), pp. 51-86, ISSN 0042-1294.

Katayama, S. (1976). Anomalous resistivity in structural phase transition of IV-VI compounds. *Solid State Commun.*, Vol. 19, No. 4, (June,1976), pp. 381-383, ISSN 0038-1098.

Kattner, U., Lukas, H.L., & Petzow, G. (1985). Optimization and Calculation of the Sn-Te system. *J. Less-Common Metals.*, Vol. 114, No. 1, (December, 1985), pp. 129-144, ISSN 0925-8388.

Khandozhko, A.G., Slinko, E.I., Letjuchenko, S.D., & Tovstjuk, K.D. (1978). NMR temperature instability in SnTe. *Ukrain. Phys. J.*, Vol. 23, (1978), pp. 1747-1749, ISSN 2071-0186.

Klemens, P.G. (1955). The Scattering of Low Frequency Lattice Waves by Static Imperfections. *Proc. Phys. Soc.*, Vol. A68, No. 12A, (December, 1955), pp. 1113-1128, ISSN 0370-1328.

Kobayashi, K.L.I., Kato, Y., Katayama, Y., & Komatsubara, K.P. (1975). Resistance anomaly due to displacive phase transition in SnTe. *Solid State Commun.*, Vol. 17, No. 7, (October, 1975), pp. 875-878, ISSN 0038-1098.

Kobayashi, K.L., Kato, Y., Katayama, Y., & Komatsubara, K.F. (1976). Carrier-concentration-dependent phase transition in SnTe. *Phys. Rev. Letters*, Vol. 37, No. 12, (September, 1976), pp. 772-774, ISSN 0031-9007.

Kolomoets, N.V., Laptev, S.A., & Rogacheva, E.I. (1986). Hall factor of SnTe. *Sov. Phys. Semicond.*, Vol. 20, No. 3, (March, 1986), pp. 283-285, ISSN 0038-5697.

Korzhuev, M.A. (1986). *Germanium telluride and its physical properties*, Nauka, Moscow, (in Russian).

Kröger, F.A. (1973). *Chemistry of Imperfect Crystals*, North-Holland, ISBN-13: 9780444105615, Amsterdam.

Lanno, M. & Burguen, Zh. (1984). *Point defects in semiconductors. Theory*, Mir, Moscow, (in Russian).

Laptev, S.A. & Rogacheva, E.I. (1986). Resonant band of tin vacancies and kinetic phenomena in SnTe. *Fiz. Elektronika*, Vol. 32, (1986), pp. 56-62.

Laptev, S.A. &Rogacheva, E.I. (1988). Vacancy energy state in SnTe. *Fiz. Elektronika*, Vol. 36, (1988), pp. 40-45.

Lin, J.-Ch., Ngai, V., & Chang, Y.A. (1986). Thermodynamic Properties and Defect Structure of Semiconducting Compound Phases: Tin Telluride. *Metallurgical Transactions, A.*, Vol. 17 A, No. 7, (Jule, 1986), pp. 1241-1245, ISSN 1073-5623.

Littlewood, P.B. (1982). Phase transitions and optical properties of IV-VI compounds. *Lect. Notes Phys.*, Vol. 152, (1982), pp. 238-246, ISSN 0075-8450.

Littlewood, P.B. (1984). Structure and bonding in narrow gap semiconductors. *Crit. Revs. Solid State and Mater. Sci.*, Vol.11, No. 3, (1984), pp. 225-285, ISSN 1040-8436.

Logachev, Yu.A., & Moyzhes, B.Ya. (1970). About deviation from stoichiometry in PbTe, SnTe and GeTe. *Izv. AN SSSR, Neorgan.Mater.*, Vol. 6, No. 10, (October, 1970), pp. 1792-1797.

Maradudin, A. (1968). *Defects and the vibrational spectrum of crystals*, Mir, Moscow, (in Russian).

Maselsky, R., & Lubell, M.S. (1963). Nonstoichiometry in some Group IV Tellurides. *Advanc. Chem. Ser.*, Vol. 39, No. 2, (January, 1963), pp. 210-217, ISSN 0065-2393.

Masharov, S.I. (1963). Frequency spectrum of acoustic vibrations of crystals with vacancies. *Izvestiya vuzov*, No. 5, (May, 1963), pp. 39-42 , ISSN 1573-9228.

Merzhanov, L.A. (1988). Optimization of data on thermodynamic properties and tin-tellurium phase diagram. *Vestnik Moskovskogo Univ., Ser.Khimiya*, Vol. 29, No. 5, (September-October, 1988), pp. 471-474, ISSN 0579-9384.

Minemura, C., & Morita, A. (1978). Electrical resistivity anomaly in the vicinity of structural phase transition of p-SnTe. *Solid State Commun.*, Vol. 28, No 3, (October, 1978), pp. 273-275, ISSN 0038-1098.

Mogilevskii, V.M., & Chudnovskii, A.F. (1972). *Thermal conductivity of semiconductors*, Nauka, Moscow, (in Russian).

Mycielski, T. (1986). Formation of supperlattice of ionized resonant donors or acceptors in semiconductors. *Solid State Commun.*, Vol. 60, No. 2, (October, 1986), pp. 165-168, ISSN 0038-1098.

Nashchekina, O.N., Rogacheva, E.I., Shpakovskaya, L.P., Pinegin, V.I., & Fedorenko, A.I. (1995). Nonstoichiometric defects in semiconductor SnTe thin films. *Mat. Res. Soc. Symp. Proc.*, Vol. 378, No. 2, pp. 255-260, ISSN 0272-9172.

Nashchekina, O.N., & Rogacheva, E.I. (1998). Temperature anomalies of transport properties in SnTe epitaxial thin films. *Phys. Stat. Sol. A*, Vol. 169A, No. 2, (October, 1998), pp. 235-238, ISSN 1521-3951.

Nashchekina, O.N., Rogacheva, E.I., Fedorenko, A.I., & Shpakovskaya, L.P. (1999a). Preparation and study of SnTe thin films. *Functional Materials*, Vol. 6, No. 2, (March-April, 1999), pp. 374-379, ISSN 1027-5495.

Nashchekina, O.N., Rogacheva, E.I., & Fedorenko, A.I. (1999b). Phase transition in SnTe and temperature anomalies of kinetic properties. *Functional Materials*, Vol. 6, No. 4, (July-August, 1999), pp. 653-657, ISSN 1027-5495.

Nashchekina, O.N., Rogacheva, E.I., Fedorenko, A.I, Isakina., A.P, & Prokhvatilov, A.I. (1999c). Low-temperature lattice instability in SnTe. *Low Temperature Physics*, Vol. 25, No. 4, (April, 1999), pp. 390-395, ISSN 0022-2291.

Nashchekina, O.N., Rogacheva, E.I., & Popov, V.P. (2008). Nonstoichiometry and thermal expansion of SnTe. *J. Phys. and Chem. Solids*, Vol. 69, Nos 2-3, (February-March, 2008), pp.273-277, ISSN 0022-3697.

Nimtz, G., & Schlicht, B. (Eds). (1985). *Narrow-gap semiconductors*, Springer, ISBN-13: 9780387120911, Berlin.

Novikova, S.I. (1974). *Thermal expansion of solids*, Nauka, Moscow (in Russian).

Novikova, S.I. & Shelimova, L.E. (1965). Phase transition in SnTe. *Fizika tverd. tela*, vol. 7, No. 8, (August, 1965), pp. 2544-2545, ISSN 0367-3294.

Novikova, S.I. & Shelimova, L.E. (1967). Low-temperature phase transition in SnTe. *Fizika tverd. tela*, vol. 9, No. 5, (May, 1967), pp. 1336-1338, ISSN 0367-3294.

Parada, N.J., & Pratt, G.W. (1969). New Model for vacancy states in PbTe. *Phys. Rev. Lett.* vol. 22, No. 5, (February, 1969), pp. 180-182, ISSN 0031-9007.

Rabenau, A. (Ed.). (1975). *Problems of nonstoichiometry*, Metallurgy, Moscow, (in Russian).

Rabii, S. (1969). Energy-band structure and electronic properties of SnTe. *Phys. Rev.*, vol. 182, No. 3, (June, 1969), pp. 821-828, ISSN 1943-2879.

Ravich, Yu.I., Efimova, B.A., & Smirnov, I.A. (1970). Semiconducting Lead Chalcogenides, Plenum Press, ISBN-13: 9780306304262, New York.

Rogacheva, E.I., Dzubenko, N.I., Laptev, S.A., Kosevich, V.M., & Ob'edkov, A.G. (1983). Effect of indium on properties of SnTe with different degree of deviation from stoichiometry. *Izv. AN SSSR. Neorgan. Mater.*, Vol. 19, No. 4, (March, 1983), pp. 573-577.

Rogacheva, E.I., Ivanova, A.B., Zhigareva, N. K., & Ob'edkov, A.G. (1985). Effect of thermal treatment on mechanism of impurities dissolving in compounds of the $A^{IV}B^{VI}$ type. *Izv. AN SSSR. Neorgan. Mater.*, Vol. 21, No. 12, (December, 1985), pp. 1762-1764.

Rogacheva, E.I., Gorne, G.V., Laptev, S.A., Arinkin, A.V., & Vesene, T.B. (1986). Concentration dependences of properties in the SnTe homogeneity region. *Izv. AN SSSR. Neorgan. Mater.*, Vol. 22, No. 1, (January, 1986), pp. 34-37.

Rogacheva, E.I., & Dzyubenko, N.I. (1986). Solubility of impurities in nonstoichiometric SnTe. *Izv. AN SSSR. Neorgan. Mater.*, Vol. 22, No. 5, (May, 1986), pp.760-761.

Rogacheva, E.I., Gorne, G. V., Laptev, S.A., Baltrunas, D.I., Moteyunas, S.V. & Vesene, T.B. (1988). Effect of deviation from stoichiometry on the solubility of antimony in tin monotelluride. *Izv. AN SSSR. Neorgan. Mater.*, Vol. 24, No. 6, (June 1988), pp. 786-789.

Rogacheva, E.I. (1989). Concentration-Dependent microhardness in semiconductor solid solutions. *Izv. AN SSSR. Neorgan. Mater.*, Vol. 25, No. 5, (May, 1989), pp. 643-646.

Rogacheva, E.I., Gorne, G.V., Zhigareva, N.K., & Ivanova, A.B. (1991a). Homogeneity region of tin monotelluride. *Inorganic Materials*, Vol. 27, No. 2, (February, 1991), pp.194-197, ISSN 0020-1685.

Rogacheva, E.I., Gorne, G.V., Zhigareva, N.K., & Ivanova, A.B. (1991b). Influence of thermal treatment on the solubility of indium in tin monotelluride. *Inorganic Mater.*, Vol. 27, No. 2, (February 1991), pp. 198-202, ISSN 0020-1685.

Rogacheva, E.I., Sinelnik, N.A., Nashchekina, O.N., Popov, V.P., Lobkovskaya, T.A. (1993). Defects of non-stoichiometry and dynamic stability of SnTe crystal lattice. *Acta Physica Polonica A*, Vol. 84, No. 4, (April, 1993), pp. 733-736, ISSN 0587-4246.

Rogacheva, E.I. (1993). Critical Phenomena in Heavily-Doped Semiconducting Compounds. *Jpn. J. Appl. Phys.*, vol. 32, No. 3, (March, 1993), pp. 775-777, ISSN 0021-4922.

Rogacheva, E.I., Gorne, G.V., & Nashchekina, O.N. (1995). Deviation from stoichiometry and lattice properties of semiconducting SnTe phase. *Mat. Res. Soc. Symp. Proc.*, Vol. 378, No. 2, (1995), pp. 107-112, ISSN 0272-9172.

Rogacheva, E.I., Seregin, P.P., & Nashchekina, O.N. (1997). Charge state of tin atoms in the ternary semiconductor phases of $Sn_{1-x}Ag_{2x}Te$. *Proceedings of the 11th. Intern. Conf. on Ternary and Multinary Compounds, ICMTC-11*, University of Salford, 8-12 September, 1997.

Rogacheva, E.I., Sologubenko, A.S., & Krivulkin, I.M. (1998). Microhardness of $Pb_{1-x}Mn_xTe$ semimagnetic solid solutions. *Inorg. Materials*, vol. 34, No. 6, (June 1998), pp. 545-549, ISSN 0020-1685.

Rogacheva, E.I., & Laptev, S.A. (1999). Electronic properties and energy spectrum of SnTe doped with In, *Proceeding of Fourth International conference on Material Science and Material Properties for Infrared Optoelectronics*, ISBN-13: 9780819434913, Kiev (Ukraine), 29 September - 2 October 1998.

Rogacheva, E.I., & Krivulkin, I.M. (2001). Isotherms of thermal conductivity in PbTe-MnTe solid solutions. *Fiz. Tverd.Tela*, vol. 43, No. 6, (June, 2001), pp. 1000-1003, ISSN 0367-3294.

Rogacheva, E.I., & Krivulkin, I.M. (2002). The temperature and concentration dependences of the charge carrier mobility in PbTe-MnTe solid solutions, *Semiconductors*, vol. 36, No. 9, (September, 2002), pp. 966-970, ISSN 1063-7826.

Rogacheva, E.I. (2003a). Self-organization processes in impurity subsystem of solid solutions. *J. Phys. Chem. Solids*, vol. 64, Nos. 9-10, (September, 2003), pp. 1579-1583, ISSN 0022-3697.

Rogacheva, E.I. (2003b). Non-Stoichiometry and Problem of Heavy Doping in Semiconductor Phases. *Materials Science in Semiconductor Processing*, Vol. 6, Nos. 5-6, (October - December, 2003), pp. 491-496, ISSN 1369-8001.

Rogacheva, E.I., Nashchekina, O.N., Tavrina, T.V., Vekhov, Ye.O., Sipatov, A.Yu., Dresselhaus, M.S. (2003c). Non-Stoichiometry in SnTe Thin Films and Temperature Instabilities of Thermoelectric Properties. *Materials Science in Semiconductor Processing*, Vol. 6, Nos. 5-6, (October-December, 2003), pp. 497-501, ISSN 1369-8001.

Rogacheva, E.I., Nashchekina, O.N., & Dresselhaus, M.S. (2003d). The Optimization of Thermoelectric Parameters when introducing impurities with variable valence, *Proceedings of 22 Int. Conf. On Thermoelectrics*, La Grande Motte, 17-21 August 2003.

Rogacheva, E.I. (2005). Non-stoichiometry and properties of ternary semiconductor phases of variable composition based on IV-VI compounds. *Journal of Physics and Chemistry of Solids, Special issue*, vol. 66, No. 11 (November, 2005), pp. 2104-2111: Proc. Of The 14th International Conference on Ternary and Multinary Compounds, Denver, USA,

Rogers, L.M. (1968). Valence band structure of SnTe. *J. Phys. D: Appl. Phys*, Vol. 1, No. 7, (July, 1968), pp. 845-852, ISSN 0003-0503.

Shelimova, L.E. & Abrikosov, N.Kh. (1964). Sn-Te system in the region of SnTe compound. *Zhurnal neorgan. khimii*, Vol. 9, No. 8, (August, 1964), pp. 1879-1883, ISSN 0036-0236 (in Russian).

Sagar, A, & Miller, R.C. (1962). Experimental investigation of band structure of tin telluride. *Proceeding of International Conference on the Physics of Semiconductors*, London, July, 1962.

Show, D. (Ed.) (1975). *Atomic diffusion in semiconductors*, Mir, Moscow, (in Russian).

Sobolev, V.V. (1981). *Intrinsic energy levels in $A^{IV}B^{VI}$ compounds*, Shtiintsa, Kishinev.

Stauffer, A D., & Aharony, A. (1992). *Introduction to percolation theory*, Taylor & Fransis, ISBN-13: 978-0748402533, London and Washington D.C.

Sugai, S., Murase, K., & Kawamura, H. (1977). Observation of soft to-phonon in SnTe by Raman scattering. *Sol. St. Communs*, Vol. 23, No. 2, (July, 1977), pp. 127-129, ISSN 0038-1098.

Sugawara, K., Huang, C.-Y., Inoue, M., Yagi, H., & Tatsukawa, T. (1979). Ferroelectric semiconductor SnTe: Mn EPR as a probe of phase transition. *J. Phys. Soc. Japan*, Vol. 47, No. 5, (November, 1979), pp. 1739-1740, ISSN 0031-9015.

Suski, T. (1985). Phase transitions and resistivity anomalies in (Pb, Sn, Ge)Te compounds. *Mater. Sci.*, Vol. 11, No. 1, (January, 1985), pp. 3-53, ISSN 0022-2461.

Suzuki, T., Yoshinaga, H., & Takeuchi, S. (1991). *Dislocation Dynamics and Plasticity*, Springer, Berlin.

Tairov, S.M., Ormont, B.F., & Sostak, N.O. (1970). Study of the Pb-Sn-Te system near PbTe-SnTe quasibinary section. *Izv. Akad. Nauk SSSR. Ser. Neorgan. Mater.*, Vol. 6, No. 9, (September, 1970), pp. 1584-1588.

Timmesfeld, K.H., & Elliot, R.J. (1970). Thermal expansion and volume change of crystals containing substitutional point imperfections. *Physica Status Solidi B*, Vol. 42, No. 2, (March, 1970), pp. 859-869, ISSN 1521-3951.

Varisov, A.Z., Rezanov, A.I., & Sabitov, R.M. (1966). Anharmonic effects in crystals with vacancies. *Izv. Vyssh. Uchebn. Zaved. Fiz.*, Vol. 9, No. 5, (May, 1966), pp. 147-150, ISSN 0021-3411.

Varnek, V.A., Mazalov, L.N., Sidorov, Yu.G., Kanter, Yu.O., & Mashchanov, V.I. (1980). Peculiarities of Mossbauer effect in SnTe and $Pb_{0,9}Sn_{0,1}Te$. *Izv. Akad. Nauk SSSR. Ser. Neorgan. Mater.*, Vol. 16, No. 5, (May, 1980), pp. 931-932.

Vodoriz, O.S., Dzubenko, N.I., & Rogacheva, E.I. (2009). Influence of copper and nonstoichiometry on SnTe structure and electrophysical properties. *Visnik Lvivskogo university, Ser. Fizuchna*, Vol. 44, (2009), pp. 31-38, ISSN 1024-588X.

Wagner, C., & Schottky, W. (1930). Theorie der geordneten Mischphasen. *Z. Phys. Chem.*, Vol. 11, No 1, (January, 1930), pp. 163-210, ISSN 0942-9352.

Wertkheim, G. (1966). *The Mossbauer effect*, Mir, Moscow, (in Russian).

Zhdanova, V.V. & Kontorova, T.A. (1965). Thermal expansion of doped germanium. *Fizika tverdogo tela*, Vol. 7, No. 11, (November, 1965), pp.3331-3338, ISSN 0367-3294.

5

Ellipsometry and Its Applications in Stoichiometry

Yu-Xiang Zheng, Rong-Jun Zhang and Liang-Yao Chen
Department of Optical Science and Engineering,
School of Information Science and Engineering, Fudan University, Shanghai,
China

1. Introduction

Ellipsometry is a powerful tool to gain the optical properties of materials though measuring the change of polarization state of the probe light after interaction with the sample. It offers a sensitive, nondestructive and comprehensive way to accurately determine film thickness and optical constants of extensive materials, such as metals, ceramics, glasses, semiconductors, and its compounds and composites. These materials can be liquid phase or even gaseous phase, can be isotropic or anisotropic, and can be bulk materials or multi-layer thin films.

Actually, the principle of ellipsometry was established one hundred years ago, but ellipsometry was developing slowly over long time. During the past decades, the ellipsometry techniques have developed rapidly, benefit from the advances in computer science and technology.

In this chapter, we will provide an overview of principles, measurement techniques, data analysis procedures for ellipsometry, and introduce the related applications of ellipsometry, especially in the field of stoichiometry.

2. What is ellipsometry?

2.1 Measurement principles of ellipsometry

As Fig.1 shows, one beam of linearly polarized light with known wavelength, incident onto the surface of an isotropy sample, the polarization state of the incident light turns to elliptical polarization. This variation in polarization state depends on the parameters that related to the sample and light, such as the incident angle of light, the refractive index of the sample. If the sample is multilayered thin film system, the parameters will include the optical constants and thickness of each layer, etc.

For convenience, the ratio ρ of the Fresnel amplitude reflection coefficients for p- and s-polarized light is introduced and given by

$$\rho = \frac{r_p}{r_s} \tag{1}$$

Where, r_p and r_s represent the amplitude reflection coefficients for p- and s- wave respectively. Both r_p and r_s are complex, so is the ratio ρ, which is usually expressed in terms of the two ellipsometry angles ψ and Δ $(0\leq\psi\leq90^\circ, 0\leq\Delta\leq360^\circ)$ as follows

$$\rho = \tan\psi \exp(i\Delta) \tag{2}$$

Where, ψ and Δ are also called ellipsometry parameters. The ratio of the modulus of the amplitude reflection coefficients is given by $\tan\psi$, and the phase difference between p- and s-polarized reflected light is given by Δ. The two ellipsometry parameters ψ and Δ are obtained directly from the measurement of ellipsometry.

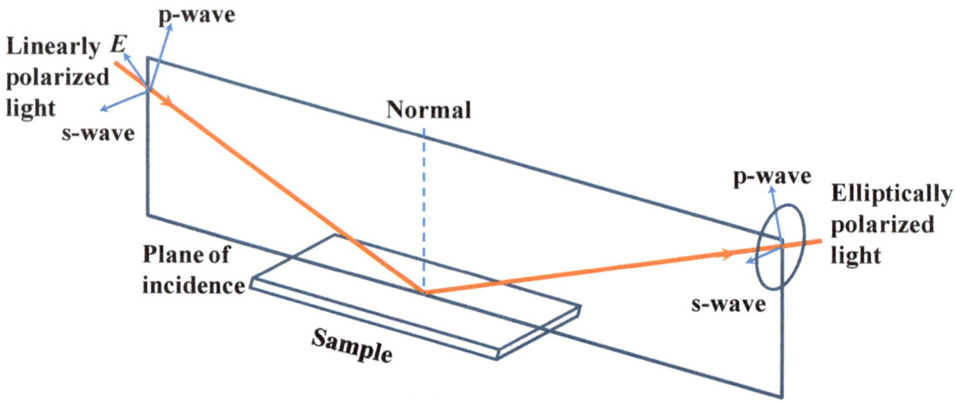

Fig. 1. Reflection of p- and s-polarized light

Usually the samples are multilayered thin films, and the purpose of ellipsometry measurement is to get the film structure and optical constants from the measured ellipsometry parameters, such as the refractive index, extinction coefficient, and thickness of each layer, etc. for the sample.

Fig.2 shows the process of ellipsometry data fitting and model analysis, and it includes the following steps:

1. To collect the ellipsometry data of the sample from the ellipsometry measurement.
2. To build a suitable optical model for the sample with parameters $n_i(\lambda), k_i(\lambda), d_i$, i means the ith layer of the film. The film may consist of only one kind of material or consist of several kinds of materials.
3. To select suitable dispersion model, such Sellimeier model, Cauchy model, Lorentz model, Drude model, effective medium approximation (EMA) model etc., for each layer. Which dispersion model should be selected for a certain layer depends on the specific type of the film and we will discuss later in detail. In the model, some parameters are known and the others are unknown. The unknown parameters will be determined through mathematical inversion method.
4. To calculate the theoretical values of ψ and Δ using the selected model and compare them with the measured values. The following function is usually employed for the comparison:

$$\text{RMSE} = \sqrt{\frac{1}{2n-m-1}\sum_{i=1}^{n}\left[(\Psi_i^{calc} - \Psi_i^{meas})^2 + (\Delta_i^{calc} - \Delta_i^{meas})^2\right]} \tag{3}$$

Where, the superscripts "meas" and "calc" represent the measured and calculated ellipsometry parameters respectively. In Eq. (3), n and m are the numbers of the measured data points and the analytical parameters, respectively. The unknown parameters in the optical model, such as film thickness or optical constants, are varied and try to produce a "best fit" to the experimental data. The best fitting results will lead to minimum of the RMSE value or a value small enough, and then the physical parameters are obtained once a good fit is achieved.

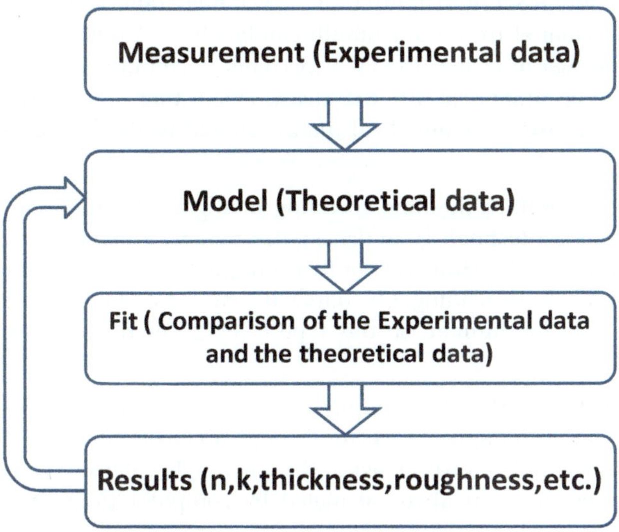

Fig. 2. Process of ellipsometry data fitting and model analysis

From the above introduction, one may find that ellipsometry is an indirect way to get the optical constants and thicknesses of each layer for a sample. The precision of the extracted parameters by the ellipsometry measurement depends on two aspects:

1. The measurement accuracy of ψ and Δ.
2. The right model and suitable fitting method.

The general iterative algorithm which applied the most is least square method. However, the algorithm must have a suitable initial estimate, or we can not get the reasonable numerical solution. In order to determine the film parameters more quickly and accurately, genetic algorithms, simulated annealing algorithm, artificial neural network algorithm and other optimization algorithms have been gradually applied to the ellipsometric data processing. It has greatly improved the convergence speed of the fitting process.

Here, we only provide a very simple introduction about the measurement principle of ellipsometry. As for the detailed knowledge on this topic, please refer to some excellent books or references [1-7].

2.2 Types of ellipsometry

2.2.1 Null ellipsometer

In the 19th century and early 20th century, as the only light detector, human eyes can only detect the presence or disappearance of the signal light from a quantitative point of view, so the ellipsometers were based on null ellipsometry at that time [8-10]. The null ellipsometry is usually of PCSA structure, P, C, S and A are the polarizer, the phase compensator, the sample, and the analyzer device respectively. The order of letters shows the order components arranged in the light path along the propagation of light. The operation of null ellipsometry is to rotate the P, C and A, let the incident light intensity on the detector to the smallest, and identify a set of azimuth (P, C, A) combination. The ellipsometry parameters ψ and Δ are calculated from the azimuth combination. Automation technology had not been developed at that time, and the PCA mechanical azimuth adjustment was all manual. In order to obtain accurate azimuth, the measurement time can be up to several minutes, after the azimuth measured, manual calculation should be done based on the principles of polarized light.

In 1960s, in order to reduce the time spending in angle calibration, H. Takasaki [11] used automation servomotor technology to drive polarizer and compensator and set the azimuth of polarizers automatically. However, in this configuration, it was still unavoidable to use human eye in the azimuth reading. Of course, the subsequent ellipsometry data processing was still involved in the human. So this type of ellipsometer was only a semi-automatic ellipsometer.

In 1967, J. L. Ord *et al.* applied the computer to send pulses to the stepper motors, read the light intensity signals from optical detector through an analog-digital (AD) signal converter, and precisely control and record the azimuths when light intensity comes to zero [12]. The ellipsometry parameters were then calculated by computer program from the measured azimuths.

During the operation of null ellipsometer, it is necessary to change the azimuths of the polarizers or compensator through mechanical rotating of optical devices. Although the application of automation technology has improved the rotary speed, but still subject to the rotation limit of the mechanical components. Besides the mechanical method of changing the polarization state, there is a phase modulated ellipsometry that add an electro-optic modulated device to the polarized device, changing the output light's polarization state by phase modulation. The speed of changing polarization state depends on the electro-optical effect (about several kHz), which is much higher than that of the mechanical rotation (usually 1 ~ 100Hz). In 1960s, an ADP crystal with a large electro-optic coefficient was used for phase modulation [13, 14]. In the same period, Winterbottom [15] proposed and proved a new method to achieve phase modulation by using magneto-optical Faraday effect to change the polarization state of the output light, and in 1970s H. J. Mathieu made this method come true [16]. With the support of automatic circuit control system, the ellipsometer using Faraday-effect cell obtained higher measurement speed compared with the stepper motor-driven automation null ellipsometer. In addition, in 2004, K. Postava *et al.* [17] applied the photoelastic modulation (PEM) crystal with photoelastic effect, which have usually been used in the phase modulation photometric ellipsometer, to the null ellipsometer with PCSA structure, and produced an automatic phase modulation null ellipsometer with high signal to noise ratio, high precision and high sensitivity but also fast measurement speed. However, due to the electro-optic effect and photoelastic effect of the modulator crystal are very sensitive to temperature, the phase-modulated ellipsometer must work under the conditions of constant temperature and the calibration process of coefficient of the crystal phase modulation is relatively complicated.

2.2.2 Photometric ellipsometer

In 1937, C. V. Kent and J. Lawson [18] first reported the photometric instrument with the polarizer and compensator adjusted to produce circularly polarized light on reflection from the sample. The analyzer was rotated at 40 Hz, and the pseudonull condition detected by amplifying the output of the photocell detector and listening to the 80 Hz components on a pair of headphones.

The modern type photometric ellipsometry technique, with the application of the stepper motor, photodetector and computer control system, appeared in early 1970s [19, 20]. This type of ellipsometry technique no longer needs to adjust the polarizers in the light path to a specific azimuth and make the light reach the detector for a particular polarization. One may use computer to precisely set the azimuths of polarizer, analyzer and compensator, and make the polarization state of light reaching the detector changing continuously. The signal of intensity of light with different polarization was changed to experimental data through a linear AD converter and collected by computer. The data were then analyzed based on principles of ellipsometry by computer. The ellipsometry parameters were obtained from the Fourier analysis during the data processing. All the operation (from measurement to data processing) can be performed by computer. Therefore, this photometric ellipsometer represents a fast and fully automatic measurement.

Photometric ellipsometer can be divided into phase modulation type and mechanical rotation type according to the method of continuously changing the light polarization. The phase modulated ellipsometer (PME) [21, 22] is similar to the phase modulated null ellipsometer in principle, but unlike the latter, PME works without mechanical rotation of optical devices. PME has a very high measurement speed with kHz modulation frequency. However, PME also has a fatal flaw same to phase modulated null ellipsometer: as a key component, the modulator crystal is very sensitive to temperature drift, measurement error will significantly increase if not used in constant temperature, and the entire build process is complicated, so the cost is relatively high. Mechanically rotating photometric ellipsometer can divided into rotating polarizer ellipsometer (RPE), rotating analyzer ellipsometer (RAE) [23, 24], rotating polarizer and analyzer ellipsometer (RPAE) [25, 26] and rotating compensator ellipsometer (RCE) [27]. In 1975, P. S. Hauge *et al.* applied rotating compensator into RCE for the first time. In the situation of fixing the polarizer and analyzer, one can measure all four Stokes vector components through rotating compensator, and this method can be used to measure all polarization states, including the complete polarization and partial polarization state. If the dual rotating compensator [28] structure is used, all sixteen Mueller matrix elements can be measured. Compensator of RCE is a quarter-wave plate, strictly speaking, valid only for a single wavelength, making it restricted in the field of multi-spectral applications once.

RPE, RAE and RPAE belong to the same kind of photometric ellipsometer, and they have common advantages, such as simple structure, mature, low cost, *etc*. But they have an obvious disadvantage: when the ellipsometry parameter Δ approaching 0°or 180°, the error of measurement would be significant. Most of the early photometric ellipsometers only rotate one of the three polarizers. RPAE is the latest invention. P and A can rotate at different speed at the same time. The rotation ratio of P to A is flexible and can be 1:2 [25, 26], 1:1 [29], *etc*. The advantage RPAE has but the other two do not have is that it is not necessary to know the background light intensity. So RPAE can obtain two sets of ellipsometry parameters for the absolute calibration self-consistency, which is unique for RPAE. Because the rotating polarizer is relatively far away from the detector in RPE, any small deviation in installation of polarizer will bring the light spot reaching the detector rotate around a non-central point, tend to increase measurement error. Except the application in parallel measurement, the RPE is not as useful as RAE. In addition to C. V. Kent [18], the first RAE was built in 1962 by W. Budde [23]. In 1969, B. D. Cahan developed the first automated photometric ellipsometer using RAE-based structure [24].

2.2.3 Spectroscopic ellipsometer

It can't determine the thickness of each layer with ellipsometry parameters of single wavelength for multilayer samples. In order to determine the optical constant exactly, and to get the structure of multilayer films with enough data by ellipsometry, the spectroscopic ellipsometer (SE) has been developed. In 1975, D. E. Aspnes *et al*. reported an SE system based on RAE for the first time [30]. A Xe lamp and Czerny-Turner grating monochrometer were employed in the SE system. The spectral resolution was 0.2 nm. For single wavelength λ = 400 nm, the total acquisition time was 7 seconds with cycles averaged per data point =1000.

In 1984, R. H. Muller *et al*. developed an SE system based on Faraday Cell phase self-compensation technique [31]. A special filter with uneven thickness was employed to choose wavelength at different position when white light used as light source. According to interference theory, different point at the filter with different thickness corresponds to different central wavelength of transmittance. This method has low quality of monochromatic light, but it has a high rate for changing wavelength with speed as high as 114 nm/s, and it can take 400 sets of ellipsometry parameters for different wavelength from 370 nm to 720 nm in 3 seconds.

In 1990, Y. T. Kim *et al*. realized a real time spectrum measurement with the combination of RPE, prism spectrometer, and optical multichannel analyzer (OMA) [32]. It took 40 ms to finish the measurement of 128 sets of ellipsometry parameters over the whole spectrum. In 2003, this group developed a generalized ellipsometer with multichannel detecting using combination of RCE and OMA [33]. The measurement time depends on the rotating frequency of compensation device, and it can take 150 sets of ellipsometry parameters in the energy range of 2 - 5 eV in 0.25 seconds.

The spectral measurement technology has a new development with the combination of multiple gratings and two dimensional array detectors. With the development of densely folded spectral images of a charge-coupled device (CCD) spectrometer [34, 35], it is possible to obtain the real time spectrum with high accuracy and resolution in the collecting time of CCD without moving of the mechanical devices. H. Y. You *et al.* realized a fast measurement of ellipsometry parameters in visible light range with the combination of array grating spectrometer and RPAE [36]. Recently, P. H. Mao *et al.* reported a new type of ellipsometer using an integrated analyzer composed of 12 sub-analyzers with different azimuth angles [37]. The new method having the merits of high speed and reliability in the optical data measurement can be potentially used in the fields of real time process monitoring.

Up to date, diverse of ellipsometry has been developed, such as imaging ellipsometry [38-40], infrared spectroscopic ellipsometry [41-50], Mueller matrix ellipsometry [51-63], *etc.* Here, we will not make a full introduction about the developments of ellipsometry. As an example, a rotating-polarizer-analyzer ellipsometer (RPAE) will be introduced as follows.

2.3 Example: A rotating-polarizer-analyzer ellipsometer (RPAE)

Here, we take the RPAE as an example to show how an ellipsometer works. An improved spectroscopic ellipsometer by synchronous rotation of the polarizer and analyzer (RPA) was suggested, designed, and constructed many years ago [64, 65]. The configuration of RPAE is shown in Fig. 3, light goes through a fixed polarizer, a rotating polarizer, sample and a rotating analyzer before entering the detector.

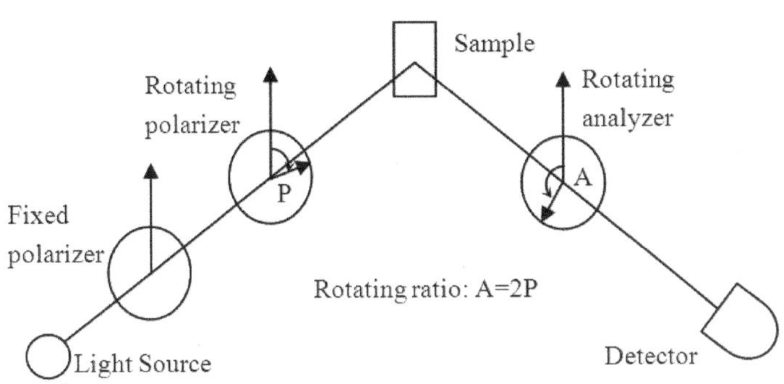

Fig. 3. Optical configuration of the RPAE

For a convenient mathematical presentation, assuming that the coordinates s and p are perpendicular and parallel, respectively, to the incident plane, the azimuthal angles of both P and A are related to the s axis. Then the electric field that finally emerges from the analyzer is

$$E_f = \begin{bmatrix} 1 & 0 \end{bmatrix} \begin{bmatrix} \cos A & \sin A \\ -\sin A & \cos A \end{bmatrix} \begin{bmatrix} r_s & 0 \\ 0 & r_p \end{bmatrix} \begin{bmatrix} \cos P & -\sin P \\ \sin P & \cos P \end{bmatrix} \begin{bmatrix} 1 & 0 \\ 0 & 0 \end{bmatrix} \begin{bmatrix} \cos P & \sin P \\ -\sin P & \cos P \end{bmatrix} \begin{bmatrix} 1 \\ 0 \end{bmatrix} E_0$$
$$= \left(r_s \cos A \cos^2 P + r_p \sin A \cos P \sin P \right) E_0 \quad (4)$$

Therefore the light signal received by the detector is

$$I \propto \left|E_f^{\ 2}\right|$$
$$= \eta\left(\cos^2 A\cos^4 P + \frac{1}{4}\sin^2 A\sin^2 2P(\tan\psi)^2 + \frac{1}{2}\sin 2A\sin 2P\cos^2 P\tan\psi\cos\Delta\right) \tag{5}$$

where η is an arbitrary number related to light intensity, ψ and Δ are two ellipsometry parameters. If $A = 2P = \omega_0 t$, then

$$I = I_0 + I_1\cos\omega_0 t + I_2\cos 2\omega_0 t + I_3\cos 3\omega_0 t + I_4\cos 4\omega_0 t \tag{6}$$

where

$$I_0 = \frac{\eta}{4}(7 + 3\tan^2\psi + 2\tan\psi\cos\Delta) + I_b \tag{7a}$$

$$I_1 = \eta(3 + \tan\psi\cos\Delta) \tag{7b}$$

$$I_2 = \eta(2 - \tan^2\psi) \tag{7c}$$

$$I_3 = \eta(1 - \tan\psi\cos\Delta) \tag{7d}$$

$$I_4 = \frac{\eta}{4}(1 + \tan^2\psi - 2\tan\psi\cos\Delta) \tag{7e}$$

In Eq. (7a), I_b is the dc background signal. Hence

$$\rho_0 = \left[\frac{2(I_1 + I_3 - 2I_2)}{I_1 + I_3}\right]^{1/2} \tag{8a}$$

$$\cos\Delta = \frac{I_1 - 3I_3}{\left[2(I_1 + I_3)(I_1 + I_3 - 2I_2)\right]^{1/2}} \tag{8b}$$

or

$$\rho_0 = \left[\frac{9(I_1 + I_3 - 2I_2)}{2(2I_1 + I_2 + 4I_4)}\right]^{1/2} \tag{9a}$$

$$\cos\Delta = \frac{3(I_1 + I_3) - 4(I_2 + 4I_4)}{\left[8(I_1 + I_3)(I_1 + I_3 - 2I_2)\right]^{1/2}} \tag{9b}$$

As for bulk material measured at the incident angle θ in the air, the dielectric constant ε of the sample can be calculated directly from the ellipsometry parameters as follows:

$$\varepsilon = \sin^2\theta + \sin^2\theta\tan^2\theta\left(\frac{1-\rho}{1+\rho}\right)^2 = \varepsilon_1 + i\varepsilon_2 \tag{10}$$

Where $\rho = \tan\Psi \exp(i\Delta)$. ε_1, ε_2 are the real part and imaginary part of the dielectric constant respectively. The refractive index n and absorption coefficient k can be obtained by:

$$n = \frac{1}{\sqrt{2}}\left[\left(\varepsilon_1^2 + \varepsilon_2^2\right)^{1/2} + \varepsilon_1\right]^{1/2} \tag{11a}$$

$$k = \frac{1}{\sqrt{2}}\left[\left(\varepsilon_1^2 + \varepsilon_2^2\right)^{1/2} - \varepsilon_1\right]^{1/2} \tag{11b}$$

The optical system of RPAE is shown schematically in Fig. 4. A Hilger DU560 high-intensity 0.5-m-grating (1200-line/mm) monochromator produced the quasi-mono-chromatic light with a continuum light source provided by a 100 W quartz halogen filament lamp in the 350-800 nm range. The light from the lamp house was conducted by a fused-silica fiber-optic cable to the monochromator. The analyzer and two polarizers were improved types of Glan-Foucault calcite prisms, which were air-gapped in design so that the spectral response could be extended to the UV range if the short-arc xenon lamp were used. The prisms were specially made and strictly tested to have a center transmitted beam deviation of < 1 min.

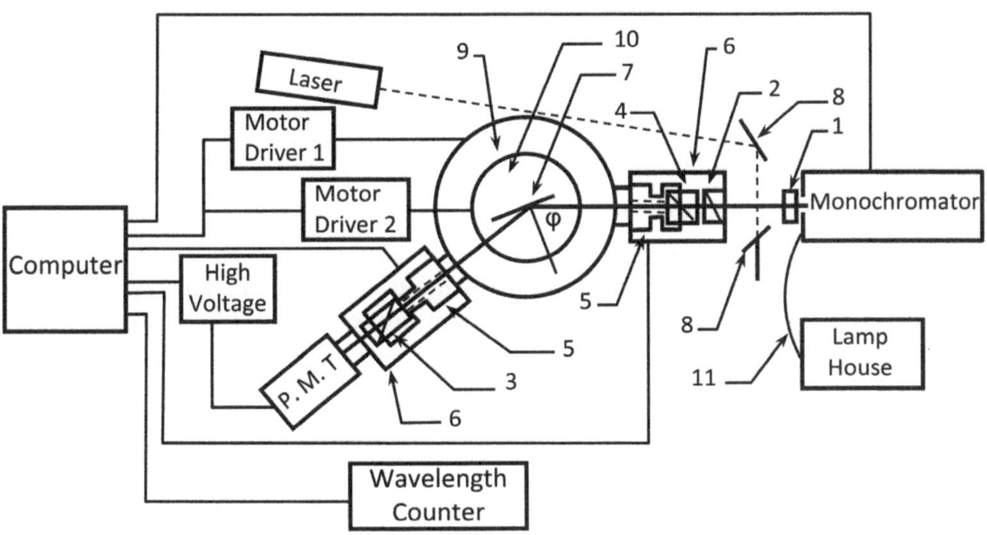

Fig. 4. Schematic diagram of the optical and controlling system of the RPAE: 1, light-collimating lens; 2, the first fixed polarizer P_0; 3, 4, rotating analyzer and polarizer, respectively, directly mounted on the shafts of the stepping motors; 5, stepping motors with hollow shafts; 6, light-shielding boxes; 7, sample; 8, mirrors that guide the laser beam for sample alignment; 9, rotating table connected to the arm that holds the analyzer and photomultiplier (P.M.T.); 10, rotating table connected to the sample-mounting stage; 11, fused-silica fiber-optic cable.

To have fewer parts and to avoid mechanical transmission problems entirely, the rotating analyzer A and polarizer P were directly mounted onto the motor shafts, which were hollow to let light pass through. The outer and inner diameters of the shafts are 9 and 5 mm, respectively. The probe beam size is 3 mm in diameter adjusted by the axis-centered iris diaphragms. In the experiment, the analyzer and polarizer were driven synchronously by two microstepping motors with 10^4 microsteps/revolution, i.e., with a resolution of 0.036 deg/step, instead of with timing belts. Hence the arm, which holds the analyzer and photomultiplier, can move freely according to the incident angle set by the users. As shown in Fig. 4, the arm holding the two polarizers is fixed. The arm, on which the analyzer and photomultiplier with an S20 response are mounted, is connected to the lower rotating table. The sample stage sits on the upper rotating table. The two rotating tables are concentric. In the experiments both rotating tables are driven by two stepping motors to positions exactly coincident with the incident angle with a computer-controlled resolution of 0.001°. The entire optical system is mounted firmly on an optical vibration-isolation table.

3. Data processing for ellipsometry

Optical constants of samples are determined from the change in the polarization state by reflection (or transmission) in ellipsometry. The ellipsometry parameters (ψ, Δ) represent the amplitude ratio and phase difference between p- and s-polarization. However, ellipsometry parameters generally show complicated variations with changes in optical constants and film thicknesses for the investigated samples. Since the ellipsometry parameters can't be related to what we interested directly, some mathematical analysis should be done to get information such as dielectric function, film thickness and so on.

3.1 Effective medium approximation

Ellipsometry is very sensitive to sample surface and interface structures. Hence, to incorporate these structures into an optical model for the investigated sample is necessary in ellipsometry data analysis. The effective medium approximation (EMA) [66] has been applied to calculating the complex refractive indices and dielectric constants of surface roughness and interface layers. In addition, the volume fractions in composite materials can be got from ellipsometry analysis using EMA.

A general equation that describes the EMA model is:

$$\frac{\varepsilon-\varepsilon_h}{\varepsilon+Y\varepsilon_h}=\sum_j f_j\frac{\varepsilon_j-\varepsilon_h}{\varepsilon_j+Y\varepsilon_h} \tag{12}$$

Where, ε_j and f_j are the dielectric function and volume fraction, respectively of jth material. The quantities ε and ε_h are the dielectric functions of the effective medium and the host respectively. Y is a factor related to screening and shape of the inclusion.

According to the above EMA equation structure, the difference of different EMA models depends on the choice of the host material.

For example, the Maxwell-Garnett model just sets the first material which has the largest volume fraction to be the host material (ε_h=ε_1) and then the EMA equation reduces to:

$$\frac{\varepsilon-\varepsilon_1}{\varepsilon+Y\varepsilon_1}=\sum_j f_j\frac{\varepsilon_j-\varepsilon_1}{\varepsilon_j+Y\varepsilon_1} \tag{13}$$

The Bruggeman model assumes that the host material is just the EMA dielectric function ($\varepsilon_h=\varepsilon$) which is self-consistent. For the Bruggeman model the EMA equation reduces to:

$$\sum_j f_j\frac{\varepsilon_j-\varepsilon_1}{\varepsilon_j+Y\varepsilon_1}=0 \tag{14}$$

The Lorentz-Lorenz model is obtained when the host material is chosen as air or vacuum ($\varepsilon_h=1$). The EMA equation reduces to:

$$\frac{\varepsilon-1}{\varepsilon+Y}=\sum_j f_j\frac{\varepsilon_j-1}{\varepsilon_j+Y} \tag{15}$$

3.2 Dielectric function models

Optical constants are closely related to the wavelength of incident light, and it is known as the dispersion relation. In the analysis of ellipsometry data, it is an important step to select appropriate dispersion model for the investigated samples if the dielectric function is unknown.

The Lorentz oscillator model is a classical model which is usually used to describe the dispersion relation of semiconductor and crystalline materials. [6, 7, 67, 68] On the basis of this model, the dielectric function is usually expressed as

$$\varepsilon(E)=1+\sum_j\frac{A_j}{E_{0j}^2-E^2-i\Gamma_jE} \tag{16}$$

It assumes that this material is a sum of the j oscillators. In above expression, the A_j is the strength of jth oscillator in unit of eV^2, Γ is the damping coefficient in unit of eV, E_{0j} in unit of eV is the jth oscillator resonance energy and E is the energy of the light.

Sellmeier model is another model first proposed by W. Sellmeier in 1871. It is best suited to transparent materials such as SiO_2, MgF_2, TiO_2, BK7 optical glass and so on. It can also be used for germanium, silicon, gallium arsenide, *etc.* in infrared spectra region. In the Sellmeier model, the dielectric function can be regarded as a special region of Lorentz model whose ε_2 is 0. It can be expressed by equations:

$$\varepsilon_1=n^2=A+\sum_j\frac{B_j\lambda^2}{\lambda^2-\lambda_{0j}^2} \tag{17a}$$

$$\varepsilon_2=0 \tag{17b}$$

Cauchy model is regarded as an approximate function of Sellmeier model. It was an empirical model first proposed by A. L. Cauchy. The equation of the model is expressed as:

$$n(\lambda) = A + \frac{B}{\lambda^2} + \frac{C}{\lambda^4} + \dots,\ \text{k=0'} \tag{18}$$

The Drude model is best suited to metallic materials such as Al and Au. It tries to explain the properties of metals with the idea that electrons move free between positively charged ionic cores. The dielectric function can be expressed as:

$$\varepsilon(E) = 1 - \sum_j \frac{B_j}{E}\left(\frac{1}{E - i\Gamma_j}\right). \tag{19}$$

Usually one single term is sufficient for accuracy and the following equation is used:

$$\varepsilon = \varepsilon_\infty \left[1 - \frac{E_p^2}{E(E - iv)}\right] \tag{20}$$

Where, ε_∞ is the high-frequency lattice dielectric constant, $E_p=\hbar\omega_p$, ω_p is the plasma angular frequency and v is the electron scattering frequency.

The Forouhi-Bloomer (F-B) theory is first proposed by A. R. Forouhi and I. Bloomer in 1986 [67]. It is applied to parameterize the optical functions of amorphous materials. In their work, they obtained the extinction coefficient k starting from the single electron model with limited excited lifetime. The extinction coefficient k is given by:

$$k(E) = \sum_{i=1}^{q} \frac{A_i (E - E_g)^2}{E^2 - B_i E + C_i} \tag{21}$$

Where A_i, B_i, C_i and band gap E_g are fitting parameters. Then the refractive index n determined by Kramers-Kronig integration is as follows:

$$n(E) = n(\infty) + \sum_{i=1}^{q} \frac{B_{0_i} E + C_{0_i}}{E^2 - B_i E + C_i} \tag{22}$$

$$B_{0_i} = \frac{A_i}{Q_i}\left(-\frac{B_i^2}{2} + E_g B_i - E_g^2 + C_i\right), \tag{23}$$

$$C_{0_i} = \frac{A_i}{Q_i}\left[(E_g^2 + C_i)\frac{B_i}{2} - 2E_g C_i\right], \tag{24}$$

$$Q_i = \frac{1}{2}(4C_i - B_i^2)^{1/2}. \tag{25}$$

The difference between F-B model and classical models is that when E goes large, the $k(E)$ approaches constant and the refractive index n is larger than 1. It is unphysical.

Another model for optical functions of amorphous materials is Tauc-Lorentz (T-L) model which is first proposed by G. E. Jellison *et al.* in 1996 [68]. The parameterization is obtained as a combination of Tauc expression and Lorentz oscillator model for ε_2 of a collection of non-interacting atoms, and $\varepsilon_2(E)$ is given by

$$\varepsilon_2(E) = \frac{AE_0\Gamma(E-E_g)^2}{E[(E^2-E_0^2)^2+\Gamma^2E^2]}\Theta(E-E_g), \tag{26}$$

Where, E_0 is the peak transition energy and Γ is the broadening parameter, Eg is the band gap and A is the prefactor. Θ is the Heaviside Theta function, where $\Theta(E<0)=0$ and $\Theta(E\geq0)=1$.

Although the T-L expression is empirical, it is consistent with Kramers-Kronig relations and the known physical phenomena, within the limitation of the model.

3.3 Data analysis procedure

The ellipsometry data analysis procedure consists of the following steps [6,7]:

1. Constructing an optical model. In the data analysis procedure in ellipsometry, an optical model corresponding to the investigated sample structures must be constructed firstly. An optical model is represented by the complex refractive index and layer thickness of each layer, normally, it consists of an air/thin film/substrate structure. It should be decided if any layer is anisotropic at this stage, and whether or not interface layers are to be modeled as a single effective medium approximation, or is a more complicated graded interface to be used for the sample.
2. Selecting the dielectric functions model for each layer. In some cases, the existing data sets or some kind of parameterization are used. However, the dielectric functions of layers are normally not known, and the dielectric function models described in section 3.2 are employed.
3. Fitting the measured (ψ, Δ) spectra and evaluating the fitting error. When the fitting error σ is large, the optical model or dielectric functions are optimized. Finally, from the optical model and dielectric functions that minimize σ, the optical constants and film thicknesses of the sample are obtained.

Since ellipsometry is a model-based technique, therefore, the resulting best-fit model must be evaluated for fit error and physical meaningfulness. Before the data analysis in ellipsometry, it's necessary and important to make clear the sample structures. In other words, when the optical model or film structures of the investigated sample are not known well, the resulting ellipsometry data analysis must be justified by other characterization methods. There are several techniques, such as scanning electron microscope (SEM), transmission electron microscope (TEM), and atomic force microscope (AFM) have been used to confirm sample structures established in ellipsometry data analysis procedure. This is the disadvantage of the ellipsometry technique. Fortunately, the ellipsometry analysis has a very high-precision, as soon as an analytical method for an investigated sample is established. Moreover, much more reliable ellipsometry results can be obtained, if the ellipsometry data analysis using a data set measured from different incident angles or wavelengths.

4. Application of ellipsometry in stoichiometry

4.1 Surface and interfaces

Ellipsometry is an optical monolayer-sensitive interface analysis method for the investigation of various aspects of surface and interfaces, including surface stoichiometry, surface roughness, adsorption, desorption, and surface strain, etc. In combination with the availability of microcomputers, SE has been developed and widely used for studies of thin films in the ambient, where parameters such as film thickness, composition, and interface roughness are determined. SE is now a popular tool for *in situ* growth control and diagnostics of interfaces.

Here, we will see the ellipsometry application in detecting the surface or interface stoichiometry. Aspnes and Theeten applied SE to study the *in situ* optical properties of the interface between Si and its thermally grown oxide [69]. They found there is an atomically mixed layer of Si and O of average stoichiometry $SiO_{0.4\pm0.2}$, and the thickness of the interlayer is (7±2) Å. Fig. 5 shows the average stoichiometry and thickness of the Si-SiO_2 interface for various substrate orientations and overlayer thicknesses. The refractive-index at λ=5461 Å for $SiO_{0.4\pm0.2}$ has value of 3.2± 0.5. Similar research work by F. Giustino *et al.* also showed that there exists an off-stoichiometric $SiO_x(0<x<2)$ layer of ~ 5-7 Å between the SiO_2 layer and Si substrate [70], and the suboxide region (*i.e.* the region containing the partially oxidized Si atoms) is of higher refractive index in comparison with that of SiO_2 [71]. The results given by F. Giustino *et al.* agree well with those by D. E. Aspnes. The off-stoichiometry interlayer cannot be ignored, especially for ultra-thin film deposited on a substrate. Q. Y. Cai *et al.* studied the evolution of optical constants of silicon dioxide on silicon from ultrathin films to thick films, and found the refractive indices of ultrathin SiO_2 on silicon substrate monotonically increase with decreasing thickness below 60 nm, as shown in Fig. 6 [72]. The increasing of refractive index of ultrathin SiO_2 can be understood as the contribution of the off-stoichiometry silicon oxide layer which has larger refractive index.

Another important factor we are focusing on is the surface or interface roughness, which should be considered carefully to determine the measured dielectric properties of "real" materials, particularly as determined by ellipsornetry. The surface roughness layer is an equivalent mixture layer composed of substrate and ambient materials. The volume fraction of the ambient (usually voids) preset within the surface roughness layer is denoted as f_{void}. Fig.7 shows the (Ψ, Δ) spectra of crystalline silicon (c -Si) obtained from calculation by applying EMA with considering different thicknesses of the surface roughness layer (d_s) [73]. In this calculation, d_s was varied from 0 Å to 50 Å with a step of 10 Å. The incident angle used for the calculation was θ_0= 70°, and f_{void} =0.5. The results show that the surface roughness can be detected effectively by measuring the ellipsometry parameters. A more common way to characterize the surface roughness is to employ both ellipsometry and AFM as reported by several authors [74]. Figure 8 shows the relationship between ellipsometry parameters and rms roughness obtained by calculation from AFM images. Compared with AFM, ellipsometry is regarded as a fast and nondestructive method, and can be applied to in-line measurement.

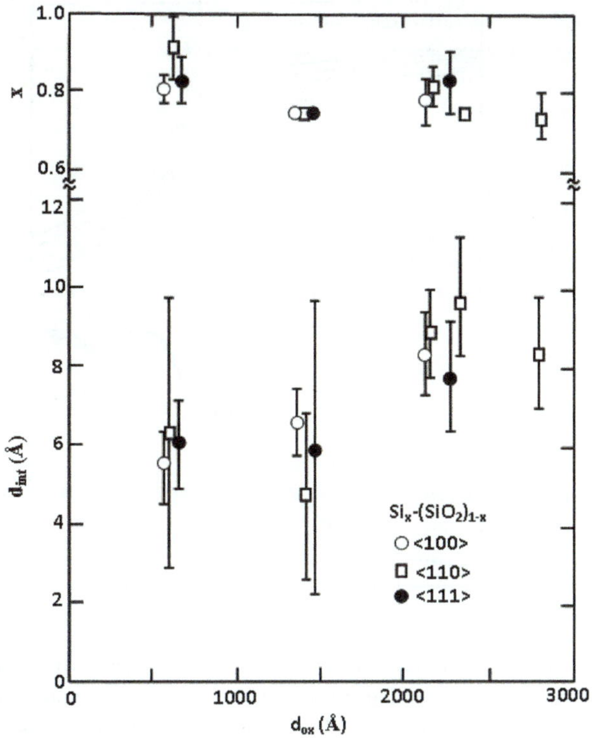

Fig. 5. Average stoichiometry (top) and thickness (bottom) of the Si-SiO_2 interface for various substrate orientations and overlayer thicknesses. After Ref. 69. Reprinted with permission from the American Physical Society, Copyright 1979.

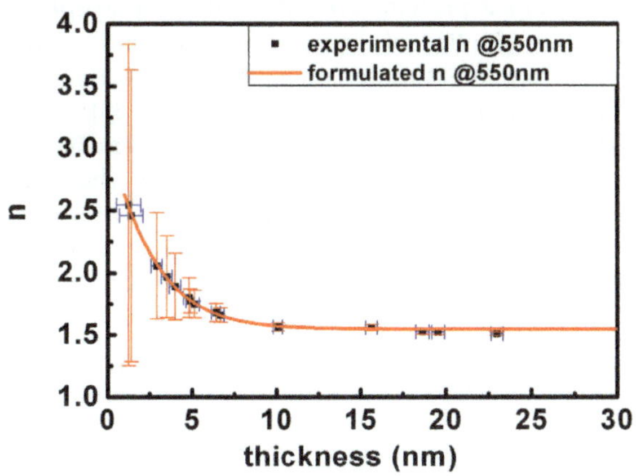

Fig. 6. Experimental and formulated refractive indices at 550 nm for the SiO_2 films with different thicknesses L_{oxide}. After Ref. 72. Reprinted with permission from IOP Publishing, Copyright 2010.

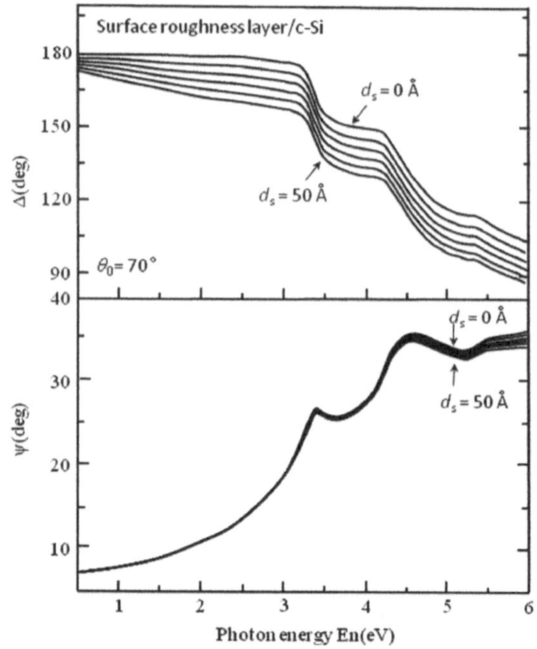

Fig. 7. (Ψ, Δ) spectra of c-Si obtained from different thicknesses of the surface roughness layer (d_s). In this calculation, d_s was varied from 0 Å to 50 Å with a step of 10 Å. The incident angle used for the calculation is θ_0= 70°. After Ref. 7. Reprinted with permission from John Wiley & Sons, Ltd., Copyright 2007.

Recently, SE has been extensively used for the characterization of organic materials and biomaterials [75, 76]. With the development of ellipsometry, it is now possible to obtain monolayer spectroscopy, *e.g.* of a protein layer at a solid/liquid interface, also to get the detailed information on the kinetics of layer formation, and monitor the adsorption processes of biomolecules in aqueous solution on a monolayer scale. With respect to organic materials, polymer thin films [77-81], self-assembled layers [82, 83], Langmuir–Blodgett (LB) films [84, 84], and liquid crystals [86, 87] have been studied intensively using spectroscopic ellipsometry. Here, we will briefly overview biomaterial characterization using the ellipsometry technique, as shown in Fig. 9 [88]. The ellipsometric spectra measured before and after protein adsorption in a porous silicon layer are given in Fig. 9a. The information both about the amount of protein adsorbed and about the adsorption depth was contained in the difference between the two spectra. Four-layer model of a porous silicon layer with human serum albumin (HAS) adsorbed in the pores, which had been achieved by analysis using multilayer models and EMA modeling, was shown in Fig.9b.

4.2 Alloys

An alloy is a homogeneous mixture, either in solution or compound, of two or more elements. Unlike chemical compound, which is formed from a chemical reaction, an alloy is formed from a physical mixture of two or more substance. For example, GaAs is obtained from chemical reaction and it is a compound consisting of Ga atoms bonded to As atoms. So GaAs is not an alloy. However, $Al_xGa_{1-x}As$ is an alloy compound consisting of AlAs and GaAs with a mole ratio of x:(1-x). Si_xGe_{1-x}($0 \leq x \leq 1.0$) is an alloy semiconductor, but SiC is not an physical mixture of Si and C atoms and it is a compound. An alloy with two components is called a binary alloy, and those with three, four, five components are called ternary alloy, quaternary alloy and pentanary alloy, respectively. It has different properties from those of the component elements and usually has better performances than those of materials with pure element. Semiconductor alloys provide a natural means of tuning the magnitude of the forbidden gap and other material parameters so as to optimise and widen the application of semiconductor devices. The optical properties of alloys are important parameters for their applications. SE is an excellent technique with which to investigate the optical response of semiconductors and, in particular, to measure the spectral dependence of the dielectric function. The optical properties of alloys vary with their composition, and alloy composition is an important factor used for achieving maximum tunability. The composition-dependence properties of alloys have been extensively studied using SE method. Here, we will give some examples.

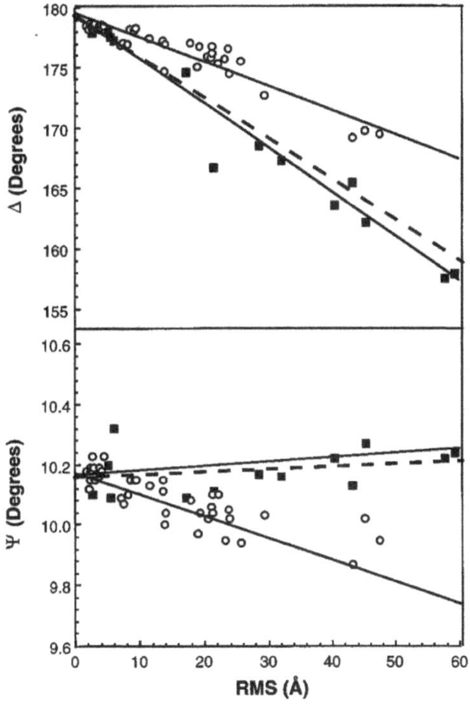

Fig. 8. Correlations between ellipsometric parameters and rms roughness; data for the wet etched sample are shown as the solid squares whereas that from the thermally processed samples is shown as the open circles. Δvs rms roughness is shown in (a) and Ψ vs rms roughness is shown in (b). Fits to the EMA are shown as lines. After Ref. 74. Reprinted with permission from American Institute of Physics, Copyright 1996.

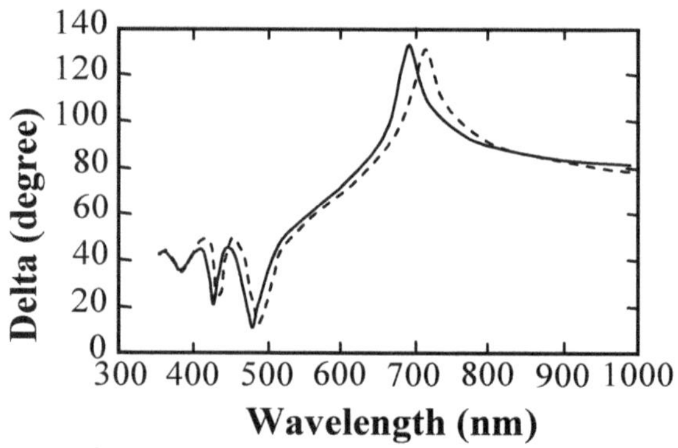

(a)

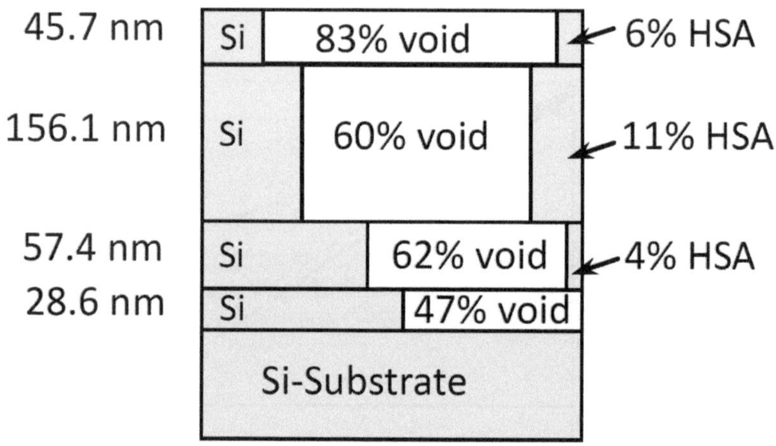

(b)

Fig. 9. (a) *Δ* vs. wavelength before (solid curve) and after (dashed curve) adsorption ofhuman serum albumin in a 288-nm thick porous silicon layer. The measurements were done in a citrate-HCl buffer (pH 4) at a protein concentration of 1 mg/ml. The angle of incidence was 68°. (b) Four-layer model of a porous silicon layer with human serum albumin (HAS) adsorbed in the pores. The voids are filled with buffer solution. The numbers to the left are the thicknesses of the sublayers. After Ref. 88. Reprinted with permission from Elsevier, Copyright 2000.

The $Si_{1-x}Ge_x$ binary alloy system has been widely used in high speed heterojunction bipolar transistors (HBT) and has potential use in optoelectronic devices incorporating light sources and/or detectors [89-91]. It is very important to understand the electrical and optical properties of $Si_{1-x}Ge_x$ binary alloy for using it well. The energy band gaps and optical properties of $Si_{1-x}Ge_x$ binary alloy vary with its composition. C. Pickering *et. al.* applied SE to study thick, relaxed and thin, strained epilayers of $Si_{1-x}Ge_x$ on Si in the range $0 < x < 0.25$ which is the range of importance for the SiGe HBT[92-94]. In Fig. 10, one may find the composition-dependence of the dielectric spectra for $Si_{1-x}Ge_x$ binary alloy. The ε_1 and ε_2 peaks in the dielectric spectra can be seen to be consistent with a monotonic decrease between Si and x=0.2.

R. D'Costa *et al.* reported the composition-dependence of optical transition energies and other critical point parameters as measured with infrared and visible spectroscopic ellipsometry and with photoreflectance for $Ge_{1-y}Sn_y$ alloys [95]. The results were shown in Fig.11. In the figure, the solid squares indicate band gaps obtained from ellipsometry; the empty square represents a photoreflectance measurement. The top dashed line represents a virtual crystal approximation (VCA) pseudopotential calculation [96]. The solid line is a linear interpolation between Ge and α-Sn. The dashed-dotted line is a VCA calculation within a tight-binding formalism [97]. The dotted line is a fit with a quadratic polynomial [95], and the dashed/double-dotted line is obtained using the composition-dependent bowing proposed by Y. Chibane *et al.*[98].

III—V semiconductor alloys are of great importance and now extensively applied in the field of optoelectronic devices like light emitters, extensively applied in the field of optoelectronic devices like light emitters, optical detectors, high speed electronic devices, thermophotovoltaic (TPV) devices, *etc.* III—V semiconductor alloys have a large family, including ternary alloys, such as AlGaP, GalnAs, A1GaSb, GaNP, *etc.*, and quaternary alloys, such as GaInNP, GaInNAs, GaInPAs, GaInAsSb, etc. The parameters of both lattice and the energy band structures of semiconductor alloys are tuned by the composition. An accurate knowledge of band structure enables the identification of alloy compositions where maximum carrier mobility and lifetime are likely to be found for specific bandgap and lattice parameter ranges [99].

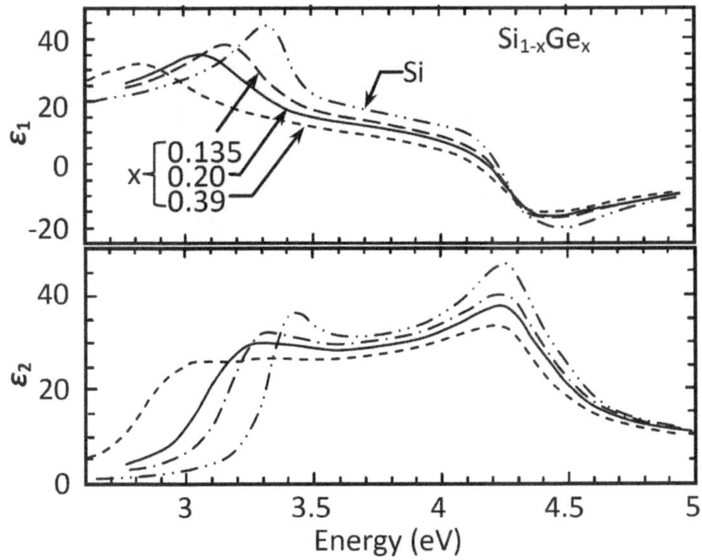

Fig. 10. The dielectric spectra of $Si_{1-x}Ge_x$ (x=0, 0.135,0.20,0.39) binary alloy. After Ref. 92. Reprinted with permission from American Institute of Physics, Copyright 1993.

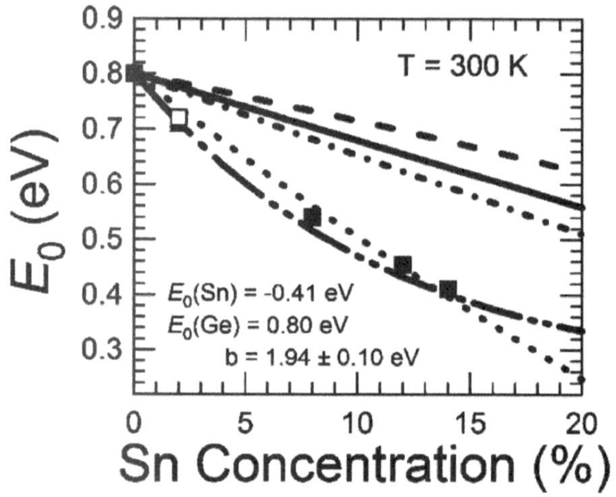

Fig. 11. Composition-dependence of the direct gap E_0 in $Ge_{1-y}Sn_y$ alloys at room temperature. After Ref. 95. Reprinted with permission from American Physical Society, Copyright 2006.

The incorporation of nitrogen into GaAs forms GaAsN, which is promising as active materials for optical-fiber communication networks. M. Schubert *et al.* invested the nitrogen dependence on the band structure of GaAsN alloy using SE and reported the dependence of the critical points (CPs) energies E_1 and $E_1+\Delta_1$ of $GaAs_{1-y}N_y$ ($0.1\%<y<3.7\%$) on the nitrogen composition y [100]. The results are shown in Fig. 12. The dotted lines correspond to the strain-only shift of E_1 and $E_1+\Delta_1$. The dashed line is the alloy-only shift of E_1. The dash-dotted line is the sum of strain and alloying for E_1. Solid lines are linear approximations to the experimental data. With increasing nitrogen composition y, the CP energy E_0 showed red shift, and the CPs E_1 and $E_1+\Delta_1$ blue shift. For nitrogen compositions of $0<y<1.65\%$, the blue shift of the E_1 energy was attributed to the sum of the effects of biaxial (001) strain and alloying.

J. Wagner *et al.* applied SE together with resonant Raman scattering to study the $GaAs_{1-x}N_x$ ($0<x<0.033$) interband transitions involving localized and extended states. The pseudo-dielectric function spectra of $GaAs_{1-x}N_x$($0<x<0.033$) obtained by SE showed a high-energy shift and broadening of the E_1 and $E_1+\Delta_1$ interband transitions with increasing N content[101]. Figure 13 shows the composition-dependence of the E_1 and $E_1+\Delta_1$ band-gap energies, which were determined by the minima in $d^2\langle\varepsilon_2\rangle/dE^2$. Ignoring strain effects, the experimental data can be fitted, using $E_1^{GaN}=7.0$ eV, by $E_1^{GaN}=2.901+2.363\times10^{-2}\,x+1.74\times10^{-4}\,x^2$ (eV) with the N concentration given in %. Similar results have also been reported by other authors [102, 103].

As for III–V–N alloys, taking $GaAs_{1-x}N_x$ for an example, the incorporation of N to GaAs will induce the contraction of the lattice parameter and further reduce the bandgap. While the addition of In or Sb to GaN_xAs_{1-x} will compensate these effects. So it is possible to grow $Ga_{1-y}In_yN_xAs_{1-x}$ or $GaN_xSb_yAs_{1-x-y}$ quaternaries that are lattice-matched to GaAs or InP and the findings are important for using these alloys for a variety of optoelectronic applications [104, 105]. The real and imaginary parts of the bulk dielectric function ε of GaAs and of three

GaInNAs samples are shown in Fig. 14 [106]. It is seen that, with increasing N content, the E_1 and E_1+Δ_1 structures around 2.9–3 eV are smeared due to a disorder in the quaternary compounds.

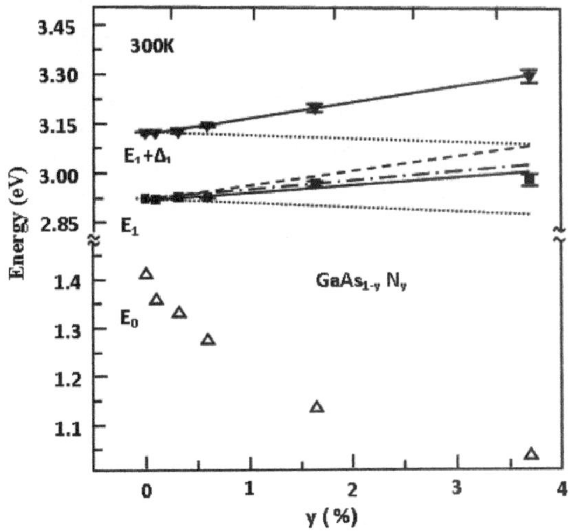

Fig. 12. Room-temperature E_0 and E_1 and E_1+Δ_1 values of $GaAs_{1-y}N_y$. After Ref. 100. Reprinted with permission from American Institute of Physics, Copyright 2000

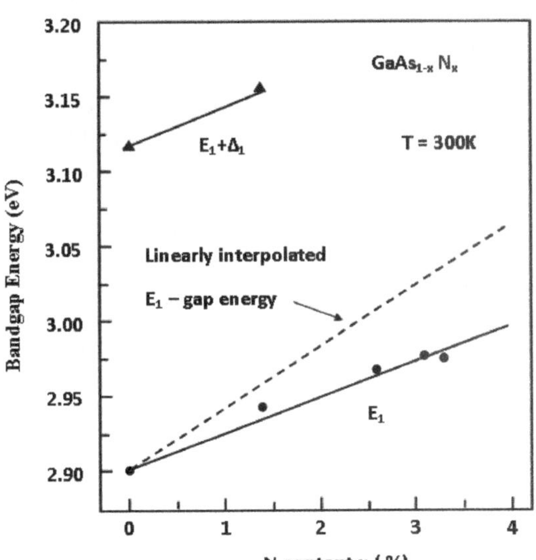

Fig. 13. E_1 and E_1+Δ_1 band gap energy of $GaAs_{1-x}N_x$ at room temperature vs N content x as derived from SE data. Expected linear dependence of E_1 gap energy on x and fitted quadratic dependence are indicated by dashed line and full curve, respectively. After Ref. 101. Reprinted with permission from American Institute of Physics, Copyright 2000.

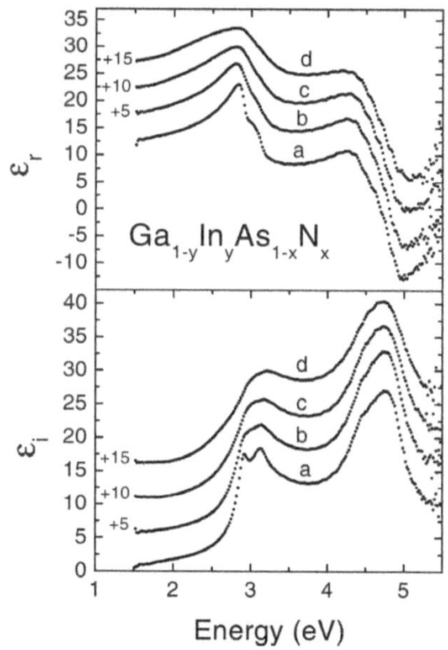

Fig. 14. Real (ε_r) and imaginary (ε_i) parts of the dielectric function in GaInNAs alloys after numerical removal of the influence from the oxide layer: (*a*) GaAs; (*b*) $x = 0.004$, $y = 0.07$; (*c*) $x = 0.0164$, $y = 0.05$; (*d*) $x = 0.0247$, $y = 0.08$. Curves b, c and d are shifted by 5, 10 and 15, respectively. After Ref. 106. Reprinted with permission from IOP Publishing, Copyright 2002.

The AlGaN system can be used in light emitting diodes (LED), metal-oxide-semiconductor field-effect transistor (MOSFET), solar blind photo detectors, surface acoustic wave devices, etc. It has been reasonably well studied. M. Stutzmann *et al.* [107] reported on a systematic study of AlGaN alloys deposited by molecular beam epitaxy (MBE) with the aim to obtain a consistent set of data for the structural, optical and electronic properties of these alloys, which covered the entire compositional range from GaN to AlN. A UV ellipsometer at the Berlin synchrotron facility BESSY I was applied to characterize the samples, and the pseudodielectric functions of AlGaN over a wide energy range (3–17 eV) are shown in Fig. 15. In the figure, one may find that E_1 and E_2 transitions shift to higher energies with increasing Al content, however less pronounced than the fundamental edge transition E_0. The spectra of all samples also show a yet unidentified transition at photon energy of about 13 eV. Antoine-Vincent *et al.* reported optical constants in the 300–600 nm range for several Al compositions of $Al_xGa_{1-x}N$ alloy layers grown on (111) Si substrates [108]. An accurate knowledge of the refractive indices of AlGaN alloys will be helpful for effective design of devices.

Metals and their alloys have important application in information storage devices, sensors, medical devices, *etc.* SE is an excellent tool to measure the dielectric response functions, which contain information on the optical transitions between the occupied and unoccupied electronic band states. K. J. Kim *et al.* reported ellipsometric study of optical transitions in $Ag_{1-x}In_x$ alloys in the 1.2-5.5 eV energy range, and the results are shown in Fig. 16 [109]. As In is added, one can clearly see the splitting of the two transition edges, the strong $L_3 \rightarrow L'_2(E_F)$ (*d* band to Fermi surface near point *L*) around 4.1 eV the weak $L'_2(E_F) \rightarrow L_1$ (Fermi surface near point *L* to a higher conduction band) around 3.8 eV, which were not resolved in pure Ag. With increasing of In concentration, the magnitude of ε_2 at low photon energies increases. This can be interpreted in terms of the increased intraband scattering rate of the conduction electrons below the interband absorption edge due to the added impurities.

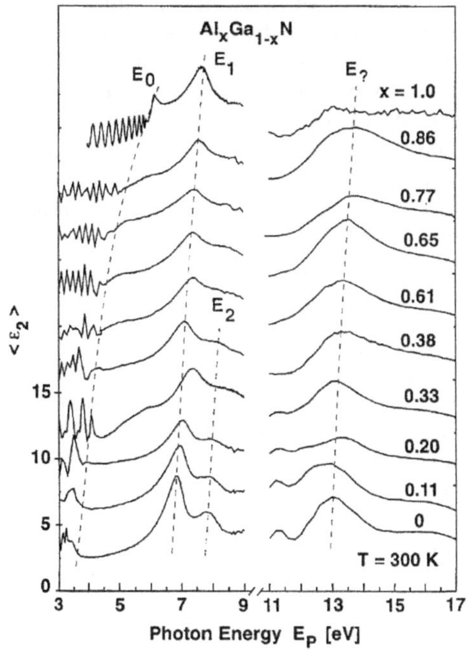

Fig. 15. Imaginary part of the pseudo-dielectric function measured by UV ellipsometry. After Ref. 107. Reprinted with permission from Elsevier, Copyright 1997.

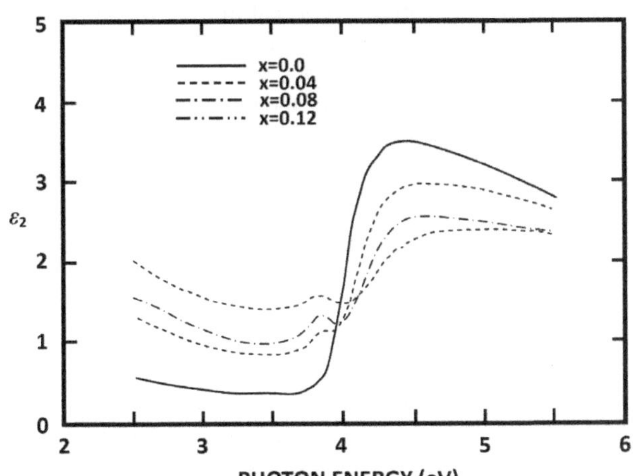

Fig. 16. The ε_2 spectra of various $Ag_{1-x}In_x$ alloys. After Ref. 109. Reprinted with permission from American Physical Society, Copyright 1988.

4.3 Composites

SE with an EMA model is now extensively used to determine the effective refractive index, thickness, and volume fraction of thin nanocomposite films [110-112]. R. Serna *et al.* investigated the optical properties of Bi:Ge composite thin films using SE [113]. The films, grown on chemically cleaned Si (100) substrates at room temperature and produced by pulse laser deposition (PLD) in vacuum, were formed by Bi nanocrystals embedded in amorphous Ge matrices. The Bi:Ge films were grown with 12, 25, 50, 100, or 200 pulses on the Bi target and constant number of pulses on the Ge target (100 pulses). The SE measurement was taken with a step of 10 nm in the 300–800 nm wavelength range. The measured ellipsometry parameters of the composite films were simulated using the Bruggeman and Maxwell-Garnet EMA models in three dimensional isotropic systems considering spherical particles and a standard regression method. For comparison, the Bi content in the films was also measured by Rutherford backscattering spectrometry (RBS). Then the optical constants could be calculated from the ellipsometry parameters supposing an absorbing film lay on a crystalline silicon substrate with the thickness of thinner films (20-40 nm) determined by RBS.

Figure 17 shows the real (n) and imaginary (k) parts of the complex refractive index as a function of the wavelength for the films grown with 25 and 200 pulses on Bi per layer. Only the values of n and k for the films with 25 and 200 pulses on Bi per layer have been included for clarity. For comparison, the values of n and k for pure PLD *a*-Ge [114] and Bi [115] films are included in the figure. One may observe that the curves with 200 and 25 pulses on Bi per layer are close to those of pure Bi and pure a-Ge, respectively. The reason may be attributed to the increase of the crystal size [116].

Dielectric composites have potential applications in the design of optical devices or microelectronic devices [117-119]. MgO is a promising low refractive index oxide with n_L of about 1.62–1.74 (at λ=550 nm) [117], while TiO_2 has a relative high refractive index with n_H of about 2.27–2.48 (at λ=550 nm) [118], and thus it is expected to have a tunable refractive index in the MgO–TiO_2 composite films by controlling the Mg content. So the performance of the optical devices can be improved with the higher refractive index (n_H-n_L) and thinner thickness of the two materials [119]. It has been reported that MgO can tune the band gap of ZnO from 3.4 to 7.8 eV depending on Mg content in MgO–ZnO composite films [120]. An accurate knowledge about the properties of the dielectric composite will broaden their better application. SE is proved to be a powerful tool in the study of the optical properties of those materials.

C. Ye *et al.* reported the optical properties of MgO–TiO_2 amorphous composite films using SE [121]. The MgO–TiO_2 composite thin films were sputtered on *n*-type (100) Si and glass substrates at room temperature by radio frequency magnetron co-sputtering. The growth rate of the composite amorphous film increases with increasing Mg content (atomic ratio of MgO/(MgO+TiO_2)). A dispersion model depending on the interband absorption of thin film is essential for data analysis. The Tauc-Lorentz model especially suitable for amorphous material [68, 122] was used to characterize dielectric function ($\varepsilon=\varepsilon_r+i\varepsilon_i$) of the amorphous MgO–$TiO_2$ composite film. A three-layer optical model was constructed on Si (100) substrate, i.e., SiO_2 interface layer (L_1), MgO–TiO_2 composite layer (L_2), and a surface rough layer (L_3) supposing 50% void and 50% MgO–TiO_2 material in the surface rough layer and using Bruggeman EMA [123]. The best fitting results of layer thickness, E_g, and the value of refractive index (at λ=550 nm) were summarized in Table 1. The data in the table show that with the increase of Mg content the thickness of the surface rough layer (L_3) and the composite layer (L_2) increases, which indicate the increase of growth rate of thin film.

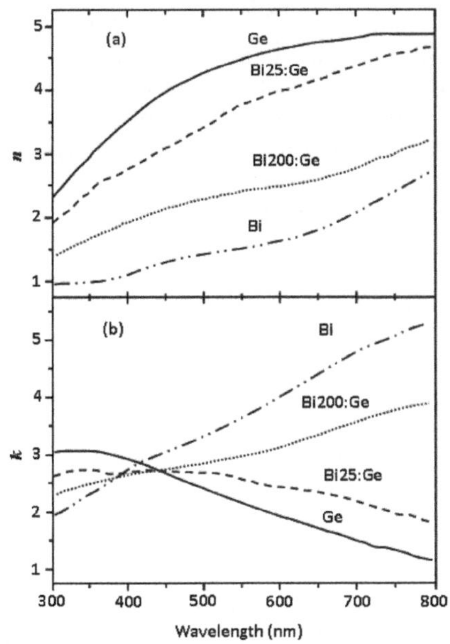

Fig. 17. (a) Real, n and (b) imaginary, k, parts of the complex refractive index as a function of wavelength, for Bi:Ge films with 25 and 200 pulses on the Bi target per layer, and for films of the pure elements (a-Ge and Bi)(see Refs. 114 and 115) grown by PLD under the same conditions for comparison. After Ref. 113. Reprinted with permission from American Institute of Physics, Copyright 1998.

Mg (%)	L_1(nm)	L_2(nm)	L_3(nm)	$L=(L_2+L_3)$ (nm)	n@550nm	E_g(eV) SE	E_g(eV) Abs
20.2%	1.59	50.72	2.36	53.08	2.21	3.36	3.50
41.8%	2.88	54.03	3.32	57.35	2.11	3.42	3.63
51.2%	1.98	57.95	3.91	61.86	2.02	3.48	3.68
62.4%	1.55	95.22	4.61	99.83	1.96	3.80	3.73
77.4%	3.95	135.55	5.44	140.99	1.89	4.05	3.79
85.7%	6.89	197.08	6.42	203.50	1.82	4.08	3.84

Table 1. Best fitting results extracted from the simulation of SE data of MgO-TiO_2 composite films with different Mg contents. [After Ref. 121.]

Figure 18a shows the refractive index in the 190–830 nm wavelength range of the composite films with different Mg contents. In the figure the refractive index for all the films first increases with the wavelength until the maximum value and then decreases. The maximum shifts to shorter wavelength with the increase of Mg content (blue shift). This same behavior has also been observed in different materials [124-126]. The inset in Fig. 18 (a) shows the linear relation between the refractive index (at λ=550 nm) and Mg content in the MgO-TiO_2

composite films. The similar results derived from SiO2-TiO2 amorphous thin film, deposited by double electron-beam co-evaporation, show that the refractive index also decreases linearly as the content of SiO2 in the composite increases [127]. Nevertheless, in other composite thin films, such as $ZnO-In_2O_3$ and $Al_2O_3-TiO_2$, there is no simple linear dependence of the refractive index on its component [128, 129]. The refractive index will be influenced by the packing density and the crystalline quality of the film. But at present, the essential reason of linear change for the refractive index is not very clear, and the possible reason might be the amorphous structure of the composite films with atom cluster mixed together to form $MgO-TiO_2$ composite. Figure 18b shows the extinction coefficient (k) of composite films as a function of the wavelength with different Mg contents. The extinction coefficient firstly increases with the wavelength until the maximum and then decreases rapidly to zero when Mg content lower than 61.4%. While for the composite films with Mg content of 61.4% and higher, the extinction coefficient monotonously decreases until zero. Figure 18b also shows that the extinction coefficient is nearly zero at visible light region and decreases with the increase of Mg content except for the wavelength shorter than 260 nm. The wavelength at which the extinction coefficient is zero moves to short wavelength (blue shift) with the increase of Mg content. The behavior is in accord with the shift of the absorption edge which will be described below in Fig. 19. The consistency lies in the direct proportional relationship between the extinction coefficient and absorption coefficient, $\alpha=4\pi k/\lambda$.

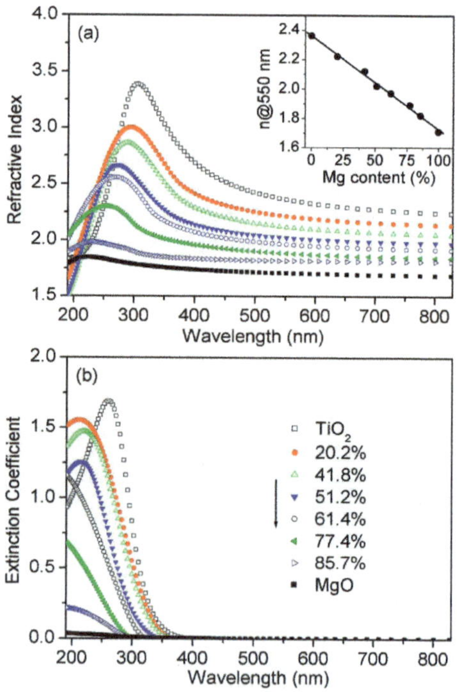

Fig. 18. Refractive index (a) and extinction coefficient (b) of $MgO-TiO_2$ composite films with different Mg contents. The inset shows the dependence of the refractive index at 550 nm on Mg content. After Ref.121. Reprinted with permission from American Institute of Physics, Copyright 2007.

Figure 19 shows the absorption spectra of the MgO-TiO_2 composite films on glass substrate. It is observed that the absorption coefficient decreases and the absorption edge occurs blue shift with the increase of Mg content. The blue shift might derive from the quantum size effect. The optical band gap (E_g) can be obtained from a plot of $(\alpha h\nu)^{1/2}$ versus photon energy ($h\nu$), as shown in the inset in Fig. 19. Extrapolation of the linear portion $(\alpha h\nu)^{1/2}$ to zero gives value of E_g. The results are also listed in Table 1, which shows that the optical band gap of the composite film increases with the increase of Mg content. Compared with E_g extracted from T-L model in SE simulation (shown in Table 1), there are only a slight difference (less than 0.25 eV) between them. In conclusion, a wide energy band oxide MgO as an additive in TiO_2 films can enlarge the optical band gap of the thin films.

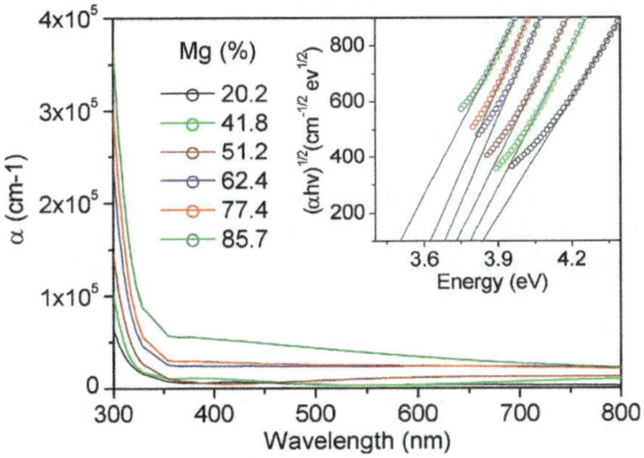

Fig. 19. Absorption spectra of MgO-TiO_2 composite films with different Mg contents. The inset shows the plot of $(\alpha h\nu)^{1/2}$ vs hv for α larger than 10^4 cm^{-1}. After Ref. 121. Reprinted with permission from American Institute of Physics, Copyright 2007.

5. Conclusions

This chapter introduces the principles, measurement techniques, data analysis procedures for ellipsometry, and provides the related applications of ellipsometry, especially in the field of stoichiometry. As examples, we give an overview of the various ellipsometry applications in stoichiometry for surface and interfaces, alloys and composites, etc. It's shown that ellipsometry, either alone or in combination with other techniques, is now a mature technique which has been successfully applied to large variety applications. There will be a bright future for ellipsometry as its combine accuracy, speed, and proven reliability with the huge advantage of nondestructive characterization.

6. Acknowledgements

The authors would like to thank the National Science Foundation (NSF) of China with the contract numbers 60778028, 60938004, 60908005, 11174058, the Science and Technology Commission of Shanghai Municipality (STCSM) project of China with the Grant No. 08DZ1204600 and the No.2 National Science and Technology Major Project of China under Contract No. 2011ZX02109-004, for financially support. D. X. Zhang, W. Lin, X. Yu, Z. J. Xu and F. Zhang are gratefully acknowledged for technical support and fruitful discussion.

7. References

[1] D. E. Aspnes, Spectroscopic ellipsometry of solids, chapter 15 in *Optical Properties of Solids : New Developments*, edited by B. O. Seraphin, North-Holland publishing company, Amsterdam(1976).

[2] R. M. A. Azzam and N. M. Bashara, *Ellipsometry and Polarized Light*, North-Holland, Amsterdam (1977).

[3] R. M. A. Azzam, *Ellipsometry*, chapter 27 in *Handbook of Optics*, Vol. 2, McGraw-Hill, New York (1995).

[4] H. G. Tompkins and W. A. McGahan, *Spectroscopic Ellipsometry and Reflectometry:A User's Guide*, John Wiley & Sons, Inc., New York (1999).

[5] M. Schubert, *Infrared Ellipsometry on Semiconductor Layer Structures: Phonons,Plasmons, and Polaritons*, Springer, Heidelberg (2004).

[6] H. G. Tompkins and E. A. Irene, Eds, *Handbook of Ellipsometry*, William Andrew, New York (2005).

[7] H. Fujiwara, *Spectroscopic Ellipsometry: Principles and applications*, John Wiley & Sons, Ltd (2007).

[8] P. Drude. Über die Gesetze der Reflexion und Brechung des Lichtes an der Grenze absorbierender Kristalle. *Ann. Phys.*, 32 (1887) 584-625.

[9] A. Rothen, The ellipsometer, an apparatus to measure thicknesses of thin surface films, *Rev. Sci. Instrum.*, 16 (1945) 26–30.

[10] W. Paik and J. O'M. Bockris, Exact ellipsometric measurement of thickness and optical properties of a thin light-absorbing film without auxiliary measurements, *Surf. Sci.*, 28 (1971) 61–68.

[11] H. Takasaki. Automatic Ellipsometer. Automatic Polarimetry by Means of an ADP Polarization Modulator III. *Appl. Opt.*, 5(1966)759-764.

[12] J. L. Ord, and B. L. Wills, A Computer-Operated Following Ellipsometer, *Appl. Opt.*, 6(1967)1673-1677.

[13] H. Takasaki, M. Isobe, T. Masaki, A. Konda, T. Agatsuma, and Y. Watanabe, An automatic retardation meter for automatic polarimetry by means of an ADP polarization modulator, *Appl. Opt.*, 3(1964) 345-350.

[14] H. Takasaki, N. Okazaki, and K. Kida,. An automatic polarimeter. II. Automatic polarimetry by means of an ADP polarization modulator, *Appl. Opt.*, 3(1964)833-837.

[15] A. B. Winterbottom, in: E. Passaglia, R. R. Stromberg, J. Kruger (eds.), *Ellipsometry in the Measurement of Surfaces and Thin Films*, Washington, D. C.: National Bureau of Standard Miscellaneous Publication, 256(1964) 97.

[16] H. J. Mathieu, D. E. McClure, R. H. Muller, Fast self-compensating ellipsometer, *Rev. Sci. Instrum.*, 45(1974) 798-802.

[17] K. Postava, A. Maziewski, T. Yamaguchi, R. Ossikovski, Š. Višňovský, and J. Pištora, Null ellipsometer with phase modulation, *Opt. Express*, 12(2004) 6040-6045.

[18] C. V. Kent, and J. Lawson, A Photoelectric Method for the Determination of the Parameters of Elliptically Polarized Light, *J. Opt. Soc. Am.*, 27(1937) 117-119.

[19] P. S. Hauge, F. H. Dill, Design and Operation of ETA, an Automated Ellipsometer, *IBM J. Res. Develop.*, 17(1973) 472-489.

[20] D. E. Aspnes, Fourier transform detection system for rotating-analyzer ellipsometers, *Opt. Commun.*, 8(1973) 222-225.

[21] S. N. Japerson, and S. E. Schnatterly, An Improved Method for High Reflectivity Ellipsometry Based on a New Polarization Modulation Technique, *Rev. Sci. Instrum.*, 40(1969) 761-767.

[22] V. M. Bermudez, and V. H. Ritz, Wavelength-scanning polarization-modulation ellipsometry: some practical considerations, *Appl. Opt.*, 17(1978) 542-552.

[23] W. Budde, Photoelectric Analysis of Polarized Light, *Appl. Opt.*, 1(1962) 201-205.

[24] B. D. Cahan, and R. F. Spanier, A high-speed automatic ellipsometer, *Surf. Sci.*, 16(1969)166-176.

[25] L. Y. Chen, and D. W. Lynch, Scanning ellipsometer by rotating polarizer and analyzer, *Appl. Opt.*, 26(1987) 5221-5228.

[26] L. Y. Chen, X. W. Feng, Y. Su, H. Z. Ma, and Y. H. Qian, Design of a scanning ellipsometer by synchronous rotation of the polarizer and analyzer, *Appl. Opt.*, 33(1994) 1299-1305.

[27] P. S. Hauge, F. H. Dill, A Rotating-Compensator Fourier Ellipsometer, *Opt. Commun.*, 14(1975) 431-437.

[28] P. S. Hauge, Mueller matrix ellipsometry with imperfect compensators, *J. Opt. Soc. Am.*, 68(1978) 1519-1528.

[29] T. M. El-Agez, A. A. El Tayyan and S. A. Taya, Rotating polarizer-analyzer scanning ellipsometer, *Thin Solid Films*, 518(2010) 5610-5614.

[30] D. E. Aspnes, A. A. Studna. High Precision Scanning Ellipsometer, *Appl. Opt.*, 14(1975) 220-228.

[31] R. H. Muller, and J. C. Farmer, Fast, self-compensating spectral-scanning ellipsometer, *Rev. Sci. Instrum.*, 55(1984) 371-374.

[32] Y. T. Kim, R. W. Collins, and K. Vedam, Fast scanning spectroelectrochemical ellipsometry: In-situ characterization of gold oxide, *Surf. Sci.*, 233(1990) 341-350.

[33] C. Chen, I. An, and R. W. Collins, Multichannel Mueller Matrix Ellipsometry for Simultaneous Real-Time Measurement of Bulk Isotropic and Surface Anisotropic Complex Dielectric Functions of Semiconductors, *Phys. Rev. Lett.*, 90(2003) 217402-1~217402-4.

[34] Y. R. Chen, B. Sun, T. Han, C. H. Xu, P. Zhou, X. F. Li, S. Y. Wang, Y. X. Zheng, and L. Y. Chen, Densely folded spectral images of a CCD spectrometer working in the full 200–1000 nm wavelength range with high resolution, *Opt. Express*, 13(2005) 10049-10054.

[35] M. H. Liu, S. X. Pan, Y. R. Chen, Y. F. Wu, Q. Y. Cai, P. H. Mao, Y. X. Zheng, and L. Y. Chen, Path-folded infrared spectrometer consisting of 10 sub-gratings and a two-dimensional InGaAs detector, *Opt. Express*, 17(2009) 14956-14966.

[36] H. Y. You. J. H. Jia, J. K. Chen, T. Han, W. M. Ni, S. Y. Wang, J. Li, R. J. Zhang, Y. M. Yang, L. Y. Chen, and D. W. Lynch, New design of a spectroscopic ellipsometer by using a spectrometer with multiple gratings and a two-dimensional CCD array detector, *Thin Solid Films*, 455-456(2004)84-89.

[37] P. H. Mao, Y. X. Zheng, Y. R. Chen, Q. Y. Cai, R. J. Zhang, and L. Y. Chen, Study of the new ellipsometric measurement method using integrated analyzer in parallel mode, *Opt. Express*, 17 (2009) 8641-8650.

[38] G. Jin, R. Jansson, and H. Arwin, Imaging ellipsometry revisited: Developments for visualization of thin transparent layers on silicon substrates, *Rev. Sci. Instrum.*, 67 (1996) 2930–2936.

[39] D. Tanooka, E. Adachi, and K. Nagayama, Color-imaging ellipsometer: high-speed characterization of in-plane distribution of film thickness at nano-scale, *Jpn. J. Appl. Phys.*, 40 (2001) 877–880.

[40] Q. Zhan and J. R. Leger, High-resolution imaging ellipsometer, *Appl. Opt.*, 41 (2002) 4443–4450.

[41] R. W. Stobie, B. Rao, and M. J. Dignam, Automatic ellipsometer with high sensitivity and special advantages for infrared spectroscopy of adsorbed species, *Appl. Opt.*, 14 (1975) 999–1003.
[42] A. S. Siddiqui and D. M. Treherne, Optical properties of some transition metals at infrared frequencies, *Infrared Physics*, 17 (1977) 33–42.
[43] A. Roseler, Spectroscopic ellipsometry in the infrared, *Infrared Physics*, 21 (1981) 349-355.
[44] A. Roseler and W. Molgedey, Improvement in accuracy of spectroscopic IR ellipsometry by the use of IR retarders, *Infrared Physics*, 24 (1984) 1-5.
[45] A. Roseler, IR spectroscopic ellipsometry: instrumentation and results, *Thin Solid Films*, 234 (1993) 307-313.
[46] F. Ferrieu, Infrared spectroscopic ellipsometry using a Fourier transform infrared spectrometer: some applications in thin-film characterization, *Rev. Sci. Instrum.*, 60 (1989) 3212-3216.
[47] T. E. Tiwald, D. W. Thompson, J. A. Woollam, S. V. Pepper, Determination of the mid-IR optical constants of water and lubricants using IR ellipsometry combined with an ATR cell, *Thin Solid Films*, 313–314 (1998) 718–721.
[48] C. Defranoux, T. Emeraud, S. Bourtault, J. Venturini, P. Boher, M. Hernandez, C. Laviron, T. Noguchi, Infrared spectroscopic ellipsometry applied to the characterization of ultra shallow junction on silicon and SOI, *Thin Solid Films*, 455-456 (2004) 150–156.
[49] R. T. Graf, F. Eng, J. L. Koenig and H. Ishida, Polarization modulation Fourier transform infrared ellipsometry of thin polymer films, *Appl. Spectrosc.*, 40 (1986) 498–503.
[50] A. Canillas, E. Pascual and B. Drevillon, Phase-modulated ellipsometer using a Fourier transform infrared spectrometer for real time applications, *Rev. Sci. Instrum.*, 64 (1993) 2153–2159.
[51] P. S. Hauge, Automated Mueller matrix ellipsometry, *Opt. Commun.*, 17 (1976) 74–76.
[52] P. S. Hauge, Mueller matrix ellipsometry with imperfect compensators, *J. Opt. Soc. Am.*, 68 (1978) 1519–1528.
[53] R. M. A. Azzam, Photopolarimetric measurement of the Mueller matrix by Fourier analysis of a single detected signal, *Opt. Lett.*, 2 (1978) 148–150.
[54] R. M. A. Azzam, K. A. Giardina, and A. G. Lopez, Conventional and generalized Mueller-matrix ellipsometry using the four-detector photopolarimeter, *Opt. Eng.*, 30 (1991) 1583–1589.
[55] D. H. Goldstein, Mueller matrix dual-rotating retarder polarimeter, *Appl. Opt.*, 31 (1992) 6676–6683.
[56] D. A. Ramsey and K. C. Ludema, The influences of roughness on film thickness measurements by Mueller matrix ellipsometry, *Rev. Sci. Instrum.*, 65 (1994) 2874-2881.
[57] G. E. Jellison, Jr, and F. A. Modine, Two-modulator generalized ellipsometry: experiment and calibration, *Appl. Opt.*, 36 (1997) 8184–8189; G. E. Jellison, Jr, and F. A. Modine, Two-modulator generalized ellipsometry: theory, *Appl. Opt.*, 36 (1997) 8190–8198.
[58] E. Compain, B. Drevillon, J. Huc, J. Y. Parey, J. E. Bouree, Complete Mueller matrix measurement with a single high frequency modulation, *Thin Solid Films*, 313–314 (1998) 47–52.
[59] J. Lee, J. Koh, and R. W. Collins, Multichannel Mueller matrix ellipsometer for real time spectroscopy of anisotropic surfaces and films, *Opt. Lett.*, 25 (2000) 1573–1575.

[60] For a review, see C. Chen, I. An, G. M. Ferreira, N. J. Podraza, J. A. Zapien, R. W. Collins, Multichannel Mueller matrix ellipsometer based on the dual rotating compensator principle, *Thin Solid Films*, 455-456 (2004) 14-23.

[61] A. De Martino, E. Garcia-Caurel, B. Laude, B. Drevillon, General methods for optimized design and calibration of Mueller polarimeters, *Thin Solid Films*, 455-456 (2004) 112-119.

[62] J. N. Hilfiker, B. Johs, C. M. Herzinger, J. F. Elman, E. Montback, D. Bryant, and P. J. Bos, Generalized spectroscopic ellipsometry and Mueller-matrix study of twisted nematic and super twisted nematic liquid crystals, *Thin Solid Films*, 455-456 (2004) 596-600.

[63] A. Laskarakis, S. Logothetidis, E. Pavlopoulou, and M. Gioti, Mueller matrix spectroscopic ellipsometry: formulation and application, *Thin Solid Films*, 455-456 (2004) 43-49.

[64] R. M. A. Azzam, A simple Fourier photopolarimeter with rotating polarizer and analyzer for measuring Jones and Mueller matrices," *Opt. Commun.* 25, (1978) 137-140.

[65] L. Y. Chen and D. W. Lynch, Scanning ellipsometer by rotating polarizer and analyzer, *Appl. Opt.* 26, (1987) 5221-5228.

[66] D. E. Aspnes, Optical properties of thin films, Thin Solid Films, 89, (1982) 249-262.

[67] A. R. Forouhi and I. Bloomer, Optical dispersion relations for amorphous semiconductors and amorphous dielectrics, *Phys. Rev. B*, 34 (1986) 7018-7026.

[68] G. E. Jellison, Jr. and F. A. Modine, Parameterization of the optical functions of amorphous materials in the interband region, *Appl. Phys. Lett.* 69 (3), (1996) 371-373.

[69] D. E. Aspnes and J. B. Theeten, Optical Properties of the Interface between Si and Its Thermally Grown Oxide, *Phys. Rev. Lett.* 43 (1979) 1046.

[70] F. Giustino, A. Bongiorno and A. Pasquarello, Atomistic models of the Si (100)-SiO2 interface: structural, electronic and dielectric properties, *J. Phys.:Condens. Matter,* 17 (2005) S2065.

[71] J. A. Moreno, B. Garrido, P. Pellegrino, C. Garcia, J. Arbiol, J. R. Morante, P. Marie, F. Gourbilleau and R. Rizk, Size dependence of refractive index of Si nanoclusters embedded in SiO2,*J. Appl. Phys.*, 98 (2005) 013523.

[72] Q. Y. Cai, Y. X. Zheng, P. H. Mao, R. J. Zhang, D. X. Zhang, M. H. Liu, and L. Y. Chen, Evolution of optical constants of silicon dioxide on silicon from ultrathin films to thick films, *J. Phys. D: Appl. Phys.*, 43(2010)445302.

[73] C. M. Herzinger, B. Johs, W. A. McGahan, J. A. Woollam, and W. Paulson, Ellipsometric determination of optical constants for silicon and thermally grown silicon dioxide via a multi-sample, multi-wavelength, multi-angle investigation, *J. Appl. Phys.*, 83 (1998) 3323-3336.

[74] S. J. Fang, W. Chen, T. Yamanaka,a) and C. R. Helms, Comparison of Si surface roughness measured by atomic force microscopy and ellipsometry, Appl. Phys. Lett. 68, (1996) 2837-2839.

[75] U. Jonsson, M. Malmqvist, and I. Ronnberg, Adsorption of immunoglobulin G, protein A, and fibronectin in the submonolayer region evaluated by a combined study of ellipsometry and radiotracer techniques, *J. Colloid. Interface Sci.*, 103 (1985) 360-372.

[76] F. Tiberg and M. Landgren, Characterization of thin nonionic surfactant films at the silica/wafer interface by means of ellipsometry, *Langmuir*, 9 (1993) 927-932.

[77] Y.-T. Kim, D. L. Allara, R. W. Collins, K. Vedam, Real-time spectroscopic ellipsometry study of the electrochemical deposition of polypyrrole thin films, *Thin Solid Films*, 193/194 (1990) 350-360.

[78] C. L. Bungay, T. E. Tiwald, D. W. Thompson, M. J. DeVries, J. A. Woollam, and J. F. Elman, IR ellipsometry studies of polymers and oxygen plasma-treated polymers, *Thin Solid Films*, 313–314 (1998) 713–717.

[79] L. A. A. Pettersson, F. Carlsson, O. Inganas and H. Arwin, Spectroscopic ellipsometry studies of the optical properties of doped poly(3,4-ethylenedioxythiophene): an anisotropic metal, *Thin Solid Films*, 313–314 (1998) 356–361.

[80] C. M. Ramsdale and N. C. Greenham, Ellipsometric determination of anisotropic optical constants in electroluminescent conjugated polymers, *Adv. Mater.*, 14 (2002) 212–215.

[81] M. Schubert, C. Bundesmann, G. Jakopic, H. Maresch, H. Arwin, N. -C. Persson, F. Zhang, and O. Inganas, Infrared ellipsometry characterization of conducting thin organic films, *Thin Solid Films*, 455–456 (2004) 295–300.

[82] J. P. Folkers, P. E. Laibinis and G. M. Whitesides, Self-assembled monolayers of alkanethiols on gold: comparisons of monolayers containing mixtures of short- and long-chain constituents with CH3 and CH2OH terminal groups, *Langmuir*, 8 (1992) 1330–1341.

[83] A. C. Zeppenfeld, S. L. Fiddler, W. K. Ham, B. J. Klopfenstein, and C. J. Page, Variation of layer spacing in self-assembled hafnium-1,10-decanediylbis(phosphonate) multilayers as determined by ellipsometry and grazing angle X-ray diffraction, *J. Am. Chem. Soc.*, 116 (1994) 9158–9165.

[84] A. Y. Tronin and A. F. Konstantinova, Ellipsometric study of the optical anisotropy of lead arachidate langumuire films, *Thin Solid Films*, 177 (1989) 305–314.

[85] B. Lecourt, D. Blaudez, and J. -M. Turlet, Specific approach of generalized ellipsometry for the determination of weak in-plane anisotropy: application to Langmuir-Blodgett ultrathin films, *J. Opt. Soc. Am. A*, 15 (1998) 2769–2782.

[86] Ch. Bahr and D. Fliegner, Behavior of a first-order smectic-A–smectic-C transition in free-standing liquid-crystal films, *Phys. Rev. A*, 46 (1992) 7657–7663.

[87] M. Schubert, B. Rheinlander, C. Cramer, H. Schmiedel, J. A. Woollam, C. M. Herzinger and B. Johs, Generalized transmission ellipsometry for twisted biaxial dielectric media: application to chiral liquid crystals, *J. Opt. Soc. Am. A*, 13 (1996) 1930–1940.

[88] H. Arwin, "Ellipsometry on thin organic layers of biological interest: characterization and applications", Thin Solid Films 377-378(2000)48-56.

[89] S. S. Iyer, G. L. Patton, J. M. C. Stork, B. S. Meyerson, and D. L. Harame, Heterojunction bipolar transistor using Si-Ge alloys, *IEEE Trans. ED* 36, (1989) 2043-2064.

[90] D. J. Robbins, P. Calcott, and W. Y. Leong, Electroluminescence from a pseudomorphic $Si_{0.8}Ge_{0.2}$ alloy, *Appl. Phys. Lett.* 59, (1991) 1350-1352.

[91] R. P. G. Karunasiri, J. S. Park, and K. L. Wang, $Si_{1-x}Ge_x$/Si multiple quantum well infrared detector, *Appl. Phys. Lett.* 59, (1991) 2588-2590.

[92] C. Pickering, R. T. Carline, D. J. Robbins, W. Y. Leong, S. J. Barnett, A. D. Pitt, and A. G. Cullis, Spectroscopic eliipsometry characterization of strained and relaxed Si1−xGex epitaxial layers, *J.Appl.Phys.*73 (1993)239-250.

[93] C. Pickering and R. T. Carline, Dielectric function spectra of strained and relaxed $Si_{1-x}Ge_x$ alloys (x = 0–0.25), *J. Appl. Phys.*, 75 (1994) 4642–4647.

[94] R. T. Carline, C. Pickering, D. J. Robbins, W. Y. Leong, A. D. Pitt, and A. G. Cullis, Spectroscopic ellipsometry of $Si_{1-x}Ge_x$ epilayers of arbitrary composition $0 \leq x \leq 0.255$, *Appl. Phys. Lett.*, 64 (1994) 1114–1116.

[95] Vijay R. D'Costa, Candi S. Cook, A. G. Birdwell, Chris L. Littler, Michael Canonico, Stefan Zollner, John Kouvetakis, and José Menéndez , Optical critical points of thin-film $Ge_{1-y}Sn_y$ alloys: A comparative $Ge_{1-y}Sn_y/Ge_{1-x}Si_x$ study, Physical Review B 73, (2006) 125207.

[96] R. Ragan and H. A. Atwater, Measurement of the direct energy gap of coherently strained Sn_xGe_{1-x}/Ge(001) heterostructures, *Appl. Phys. Lett.* 77, (2000) 3418-3420.
[97] G. He and H. A. Atwater, Interband transitions in Sn_xGe_{1-x} alloys, *Phys. Rev. Lett.* 79, (1997) 1937-1940.
[98] K. Brunner, K. Eberl, and W. Winter, Near-Band-Edge Photoluminescence from Pseudomorphic $Si_{1-y}C_y$/Si Quantum Well Structures, *Phys. Rev. Lett.* 76, (1996) 303-306.
[99] C. B. Geller, W. Wolf, S. Picozzi, A. Continenza, R. Asahi, W. Mannstadt, A. J. Freeman, E. Wimmer, Computational band structure engineering of III–V semiconductor alloys, *Appl. Phys. Lett.*, 79 (2001) 368-370.
[100] G. Leibiger,V. Gottschalch, B. Rheinla¨nder, J. Sik and M. Schubert, Nitrogen dependence of the GaAsN interband critical points E_1 and $E_1+\Delta_1$ determined by spectroscopic ellipsometry, Appl. Phys. Lett. 77, (2000) 1650-1652.
[101] J. Wagner, K. Kohler, P. Ganser, and N. Herres, GaAsN interband transitions involving localized and extended states probed by resonant Raman scattering and spectroscopic ellipsometry, *Appl. Phys. Lett.*, Vol. 77 (2000) 3592-3594.
[102] J. Sik, M. Schubert,G. Leibiger and V. Gottschalch,G. Wagner, B and-gap energies, free carrier effects, and phonon modes in strained GaNAs/GaAs and GaNAs/InAs/GaAs superlattice heterostructures measured by spectroscopic ellipsometry, J. Appl. Phys., Vol. 89, (2001) 294-305.
[103] G. Leibiger, V. Gottschalch, B. Rheinla¨nder,J. Sik, M. Schubert, Model dielectric function spectra of GaAsN for far-infrared and near-infrared to ultraviolet wavelengths, J. Appl. Phys. 89, (2001) 4927-4938.
[104] S. R. Kurtz, A. A. Allerman, E. D. Jones, J. M. Gee, J. J. Banas and B. E. Hammons, InGaAsN solar cells with 1.0 eV band gap, lattice matched to GaAs, *Appl. Phys. Lett.*, 74 (1999) 729-731.
[105] A. Wagner, C. Ellmers, F. Höhnsdorf, J. Koch, C. Agert, S. Leu, M. Hofmann, W. Stolz, and W. W. Rühle, (GaIn) (NAs)/GaAs vertical-cavity surface-emitting laser with ultrabroad temperature operation range*Appl. Phys. Lett.* 76 (2000)271-272.
[106] C. Skierbiszewski, Experimental studies of the conduction-band structure of GaInNAs alloys, *Semicond. Sci. Technol.*, 17 (2002) 803–814.
[107] M. Stutzmann, O. Ambacher, A. Cros, M.S. Brandt, H. Angerer, R. Dimitrov, N. Reinacher, T. Metzger, R. Höpler, D. Brunner, F. Freudenberg, R. Handschuh, Ch. Deger, Properties and applications of MBE grown AlGaN, *Mat. Sci. Eng.*, B 50 (1997) 212–218.
[108] N. Antoine-Vincent, F. Natali, M. Mihailovic, A. Vasson, J. Leymarie, and P. Disseix,D. Byrne, F. Semond, and J. Massies, Determination of the refractive indices of AlN, GaN, and $Al_xGa_{1-x}N$ grown on (111) Si substrates, *J. Appl. Phys.* 93, (2003) 5222-5226.
[109] K. J. Kim, L. Y. Chen, and D. W. Lynch, Ellipsometric study of optical transitions in Ag1-xInx alloys, *Phys. Rev.* B, 38 (1988) 13107-13112.
[110] D. E. Aspnes, in *Handbook of Optical Constants of Solids*, edited by E. D. Palik, Academic Press, Orlando, FL, Chap. 5, (1985) 89–112.
[111] F. Vega, J. C. G. de Sande, C. N. Afonso, C. Ortega, and J. Siejka, Optical properties of GeO_x films obtained by laser deposition and dc sputtering in a reactive atmosphere. *Appl. Opt.*, 33 (1994) 1203-1208.
[112] J. C. de Sande, F. Vega, C. N. Afonso, C. Ortega, and J. Siejka, Optical properties of Sb and SbO_x films, *Thin Solid Films*, 249 (1994) 195-200.

[113] R.Serna, J. C. G. de Sande, J. M. Ballesteros and C. N. Afonso, Spectroscopic ellipsometry of composite thin films with embedded Bi nanocrystals, *J. Appl. Phys.*, 84(1998) 4509-4516.

[114] J. C. G. de Sande, C. N. Afonso, J. L. Escudero, R. Serna, F. Catalina, and E. Bernabeu, Optical properties of laser-deposited a-Ge films: a comparison with sputtered and e-beam-deposited films, *Appl. Opt.*, 31 (1992) 6133-6138.

[115] J. C. G. de Sande, T. Missana, and C. N. Afonso, Optical properties of pulsed laser deposited bismuth films, *J. Appl. Phys.*, 80 (1996) 7023-7027.

[116] R. Serna, T. Missana, C. N. Afonso, J. M. Ballesteros, A. K. Petford- Long, and R. C. Doole, Bi nanocrystals embedded in an amorphous Ge matrix grown by pulsed laser deposition, *Appl. Phys. A*, 66 (1998) 43-47.

[117] K. H. Nam and J. G. Han, Microstructure and optical properties of MgO films synthesized by closed-field unbalanced magnetron sputtering with additional electron emission, *Surf. Coat. Technol.* 171 (2003) 51-58.

[118] H. Poelman, D. Poelman, D. Depla, H. Tomaszewski, L. Fiermans, and R.De Gryse, Electronic and optical characterization of TiO_2 films deposited from ceramic targets, *Surf. Sci.*, 482–485 (2001) 940-945.

[119] L. Martinu and D. Poitras, Plasma deposition of optical films and coatings: A review, *J. J. Vac. Sci. Technol. A*, 18 (2000) 2619-2645.

[120] J. S. Chen, S. Chao, J. S. Kao, H. Niu, and C. H. Chen, Mixed films of TiO_2-SiO_2 deposited by double electron-beam coevaporation, *Appl. Opt.*, 35 (1996) 90-96.

[121] C. Ye, S. S. Pan, X. M. Teng, and G. H. Li. Optical properties of MgO-TiO_2 amorphous composite films. *J. Appl. Phys.*, 102 (2007) 013520.

[122] R. W. Collins, J. Koh, H. Fujiwara, P. I. Rovira, A. S. Ferlauto, J. A. Zapien, C. R. Wronski, and R. Messier, Recent progress in thin film growth analysis by multichannel spectroscopic ellipsometry, *Appl. Surf. Sci.*, 154-155(2000)217-228.

[123] D. A. G. Bruggemann, Berechnung verschiedener physikalischer Konstanten von heterogenen Substanzen. I. Dielektrizitätskonstanten und Leitfähigkeiten der Mischkörper aus isotropen Substanzen, *Ann. Phys.*, 416 (1935) 636-664.

[124] D. Franta, I. Ohlídal, and D. Petrýdes, Optical characterization of TiO_2 thin films by the combined method of spectroscopic ellipsometry and spectroscopic photometry, *Vacuum*, 80 (2005) 159-162.

[125] M. Serenyi, T. Lohner, P. Petrik, and C. Frigeri, Comparative analysis of amorphous silicon and silicon nitride multilayer by spectroscopic ellipsometry and transmission electron microscopy. *Thin Solid Films*, 515 (2007) 3559-3562.

[126] X. H. Wang, J. L. Shi, Y. Yang, G. J. You, Y. Liu, J. Mi, and S. X. Qian, Synthesis and optical properties of PbS/ZnO composite films, *J. Appl. Phys.*, 95 (2004) 4791-4794.

[127] J. S. Chen, S. Chao, J. S. Kao, H. Niu, and C. H. Chen, Mixed films of TiO_2-SiO_2 deposited by double electron-beam coevaporation, *Appl. Opt.*, 35 (1996) 90-96.

[128] D. H. Kuo and K. H. Tzeng, Characterization and properties of r.f.-sputtered thin films of the alumina–titania system, *Thin Solid Films*, 460 (2004) 327-334.

[129] H. EI Rhaleb, A. E. Naciri, R. Dounia, L. Johann, A. Hakam, and M. Addou, Spectroscopic ellipsometry characterization of ZnO-In_2O_3 systems. *Thin Solid Films*, 455–456 (2004) 384-387.

Part 3

The Influence of Stoichiometry on Intermetallic Compounds Features

6

Stoichiometry in Inter-Metallic Compounds for Hydrogen Storage Applications

Kwo Young
Ovonic Battery Company, Rochester Hills, Michigan, USA

1. Introduction

This chapter is devoted to the discussion of the influence of stoichiometry deviation on the structure, gaseous phase hydrogen storage, and electrochemical properties of some important inter-metallic compounds (IMC). First, the hydrogen storage characteristics and other important features of the IMC will be reviewed and then there will be a discussion of the AB_5, AB_2, A_2B_7, and AB metal hydride (MH) alloy systems with variations in stoichiometry. The experiments that provide the basis for this chapter's discussion are described in detail (experimental set-up and chosen parameters) in earlier publications (Young, et al., 2008, 2009a, 2010a).

The formation of an IMC between two metallic elements with a fixed stoichiometry depends on the heat of formation of the alloy (ΔH_a). With a very positive value of ΔH_a, the two metallic elements in the alloy are immiscible in the liquid and will stay separate during cooling. As the ΔH_a value decreases but is still positive, a eutectic reaction occurs during cooling, and mixtures of each element with limited solubility of the other is formed. As ΔH_a further decreases to a negative value, a solid solution with a broad range of composition between the two elements is expected. Finally, when ΔH_a becomes very negative, an IMC with a fixed stoichiometry is obtained. Based on the experimental data, ΔH_a of an IMC can be predicted from the electron density at the boundary of the Wigner-Seitz atomic cell and the chemical potential of the conducting electrons (De Boer, et al.,1988). Therefore, the stability of an IMC with a stoichiometry of A_mB_n can be calculated theoretically. Among the choices of crystal structures available to certain stoichiometries of an IMC, the one that minimizes the total electron energy dominates. Parameters such as average electron density, atomic radius ratio, stoichiometry, and difference in electronegativity all contribute to the determination of crystal structure.

In contrast to solid solution, the constituent elements for binary IMC usually have large differences in both size and electronegativity. When the electron transfer takes place, the size of the atoms will change toward the ideal ratio. In the case of IMC, the electron being transferred from the element with the lower electronegativity (usually the larger one) to the other element causes the atomic volume to shrink (Pauling, 1987). This charge transfer changes the local electron density and is very important for the storage of hydrogen.

IMCs can have either a limited or a broad range in stoichiometry (defined as the ratio between constituent elements). For example, all the IMCs in the La-Ni binary phase diagram (Fig. 1) exist at vertical lines, or in other words, there is a unique value of the stoichiometry (ratio between Ni and La) for each IMC; in contrast, the IMCs $ZrNi_5$, $ZrNi_3$, and Zr_7Ni_{10} in the Zr-Ni binary phase diagram (Fig. 2) exist at ranges of composition. The line width (also called solubility for a particular IMC) can be traced to the formation of defects and lattice-adjusted contraction (Thoma & Perepezko, 1995). When there are more A atoms present in the IMC, part of the excessive A atoms either move to B-site (anti-site defect) or create vacancy in B-site (vacancy defect) while still maintaining the same crystal structure. Although the radii ratio between A atom and B atom (R_A/R_B) might not be optimum for the structure, the A-B distance will be adjusted by the defect formation and result in a close-to-ideal lattice. For example, the Laves phase IMC $TiCr_2$ with a less than the ideal ratio of 1.225 (ideal is $\sqrt{1.5}$) has a larger solubility on the A-rich side of the AB_2 stoichiometry (Chen, et al., 1994).

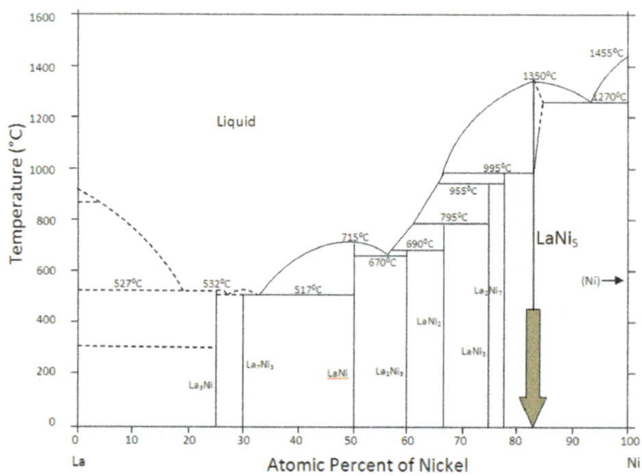

Fig. 1. La-Ni binary phase diagram (data from (Massalski, 1990)).

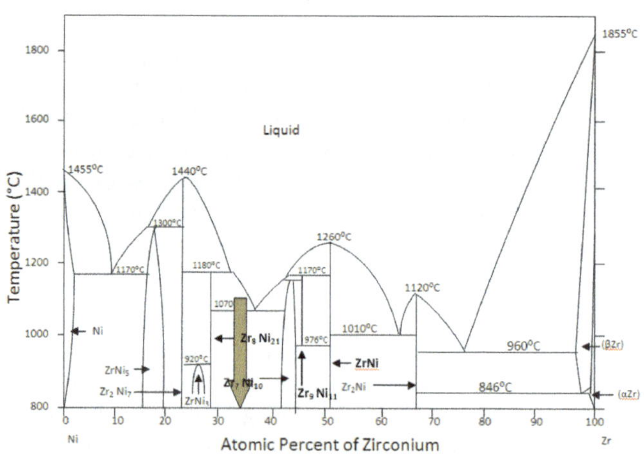

Fig. 2. Ni-Zr binary phase diagram (data from (Massalski, 1990)).

2. Metal hydride alloys

Three different types of chemical bonds can be formed between hydrogen and other elements: the ionic bond, the metallic bond, and the covalent bond for elements with small (Li, Na, for example), medium (Ti, V), and large (N, O) electronegativities, respectively. While the strengths of both ionic and covalent bonds are too large for them to be appropriate for reversible hydrogen storage, the strength of metallic bond can be adjusted for the purpose of hydrogen storage depending on the operation temperature of the application. One way to quantify the metal-hydrogen (M-H) bond strength is by evaluating the heat of hydride formation (ΔH). An application for room temperature, the rechargeable battery for example, would require a ΔH value of between −35 to −30 kJ mol^{-1} H_2. Only expensive metals, such as Pd and V, have the suitable M-H bond strength for room temperature applications. Therefore, the search for hydrogen storage metals was expanded firstly to binary IMCs, such as Mg_2Ni, $LaNi_5$, and TiNi, and then to more complicated and multi-element IMCs.

When hydrogen enters into the bulk of the metal, it contributes its only electron to the conduction band. The remaining proton then goes into the interstitial site among four or six host elements, causing an expansion in the crystal lattice. The metal hydride formed this way, shown in Fig. 3, has the same or very similar crystal structure as the host metal. Driven by a concentration gradient, the proton jumps to the neighboring storage sites through quantum mechanical tunneling. The moving path of the proton in AB_2 C14 MH alloy is along the vertical (*c*-) axis (Fig. 3a), which is different from its movement on the horizontal (*a-b*) plane in AB_5 MH alloy (Fig. 3b). Therefore, a flattened and an elongated unit cell along the *c*-axis are preferable for the ease of proton movement in AB_2 and AB_5 metal hydride (MH) alloys, respectively (Osumi, et al., 1983, Young, et al., 2009b).

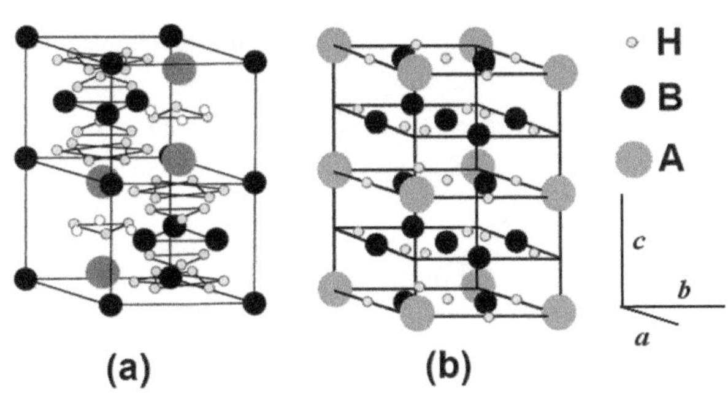

Fig. 3. Crystal structures of C14-AB_2H_4 (a) and AB_5H_6 (b) MH alloys.

The process of hydride formation can be described by the pressure-concentration-temperature (PCT) curves shown in Fig. 4, where the logarithm of the equilibrium hydrogen pressure is plotted against the hydrogen concentration in metal. At a fixed temperature, when the concentration of hydrogen in the bulk is low, the protons are randomly distributed throughout the host metal (α-phase). As the applied hydrogen pressure increases, more hydrogen atoms enter into the metal and start to form MH clusters, similarly to the beginning of ice formation in the case of freezing water. The co-existence of metal (α-phase) and MH (β-phase) eliminates the freedom in pressure and thus contributes to the flat region (plateau) of the PCT curve. With further increase in hydrogen pressure, the transformation into *β*-phase is complete, marks the end of the pressure plateau, and isotherm becomes more vertical. The The PCT curves can be measured with stepwise increases (absorption) or decreases (desorption) in applied hydrogen pressure. These two curves do not align with each other, and the difference in the plateau pressures of the two curves is called the hysteresis of PCT curve. The hysteresis represents the resistance of the metal against the pushing in or pulling out of additional hydrogen atoms. The correlation between the hysteresis and the MH pulverization rate through hydride/dehydride cycles is well established: MH alloys with small hysteresis are less prone to break into small pieces during cycling. Each set of the PCT curves (absorption and desorption) is obtained at one particular temperature. As the temperature increases, $T_3 > T_2 > T_1$ in Fig. 4, the following occur: the regions of *α*- and *β*-phases extend into the plateau region due to the higher entropy in the system, the plateau region shortens, the plateau pressure increases, and hysteresis, reversible and maximum hydrogen storage capacities reduce. By connecting all the boundary points between α-phase, $\alpha + \beta$ mixed phase, and *β*-phase, a dome-shaped curve is obtained. As the temperature rises above a certain critical value (T_3 in Fig. 4), the plateau region of the PCT curve completely disappears. This critical temperature is very material-dependent. In the case of $LaNi_5$ based MH alloys, the systems are very ordered, and the critical temperatures are very high. Therefore, most of the PCT curves observed around room temperature are very flat. On the other hand, the multi-phase $(Ti, Zr)Ni_2$ based MH alloys have lower critical temperatures due to their high degrees of disorder inherent with the alloys and produce PCT curves without recognizable plateau region. Slope factor, defined as the percentage of storage capacity within a pressure range (usually from 0.01 to 0.5 MPa), is used to quantify the slope of the PCT isotherm.

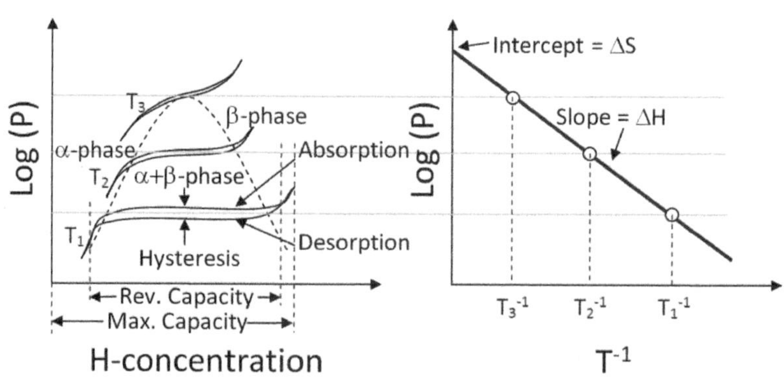

Fig. 4. Schematics of PCT isotherm (left) and the corresponding Van't Hoff diagram (right).

The plateau pressures in the PCT curves can be used to calculate the changes in enthalpy (ΔH) and entropy (ΔS) by the equation

$$\Delta G = \Delta H - T\Delta S = RT \ln P \tag{1}$$

where R is the ideal gas constant and T is the absolute temperature. When the logarithm of the pressure is plotted against the reciprocal of the absolute temperature as shown in Fig. 4,

ΔH and ΔS can be calculated from the slope and the *y*-axis intercept of the line, respectively. The value of ΔH can be used to quantify the M-H bond strength. MH with a lower ΔH (more negative) will release more heat during the formation of a more stable hydride with a lower equilibrium pressure, indicating both a stronger M-H bond and a larger interstitial site for hydrogen. For an ordered MH, ΔS is around the standard entropy of hydrogen gas (−130 J K^{-1} mol^{-1} H_2). ΔS will be higher for a less ordered MH.

Besides the application of solid hydrogen storage, MH alloys also play a very important role in nickel/metal hydride (Ni/MH) rechargeable battery as the negative electrode. The chemical reaction during charge/discharge is:

$$M + Ni(OH)_2 \underset{\text{discgarge}}{\overset{\text{charge}}{\rightleftarrows}} MH + NiOOH \quad (2)$$

During charge, the water molecule near the surface of the MH electrode splits into a proton (H^+) and a hydroxide ion (OH^-) by the applied voltage. While OH^- moves into the electrolyte, H^+ enters into the bulk of the MH and is neutralized by the electron coming from the current collector. Charging/discharging in the electrochemical environment is very similar to hydriding/dehydriding in the gaseous phase; these two reactions can be linked by the Nernst Equation:

$$\text{Open circuit voltage vs. Hg/HgO} = -0.934 - 0.0291 \log P_{H2} \quad (3)$$

where Hg/HgO is the reference electrode commonly used in alkaline solution. This correlation explains the similarity between the PCT curves (Fig. 4) and the electrochemical charge/discharge curves illustrated in Fig. 5. The origin for the voltage difference between charge and discharge curves is quite complicated and contains at least the ohmic resistance in the connector and substrate, resistance from surface reaction, limited hydrogen diffusion, and electrolyte concentration difference. At a higher rate (gray curves in Fig. 5), the charge voltage is higher while the discharge voltage and the obtained capacity are lower, mostly due to the increase in current (i) as

$$V = OCV + iR \text{ (during charge) and} \quad (4)$$

$$V = OCV - iR \text{ (during discharge)} \quad (5)$$

where OCV is the open circuit voltage and R is the internal resistance.

Currently, there are four main MH alloy families used or proposed to be used as the negative electrode for Ni/MH rechargeable battery: the AB_5, AB_2, A_2B_7 and AB alloy families. Rare earth element-based AB_5 alloy family is the most commonly used, with the misch-metal (Mm) (mixture of light rare earth elements, such as La, Ce, Nd, and Pr) as the A-elements and Ni, Mn, Al, and Co as the B-elements. Ni is an indispensably key element in alkaline electrochemistry due to its corrosion resistance and catalytic nature for water splitting/recombination. Other modifiers all have different contributions. The main advantages of AB_5 MH alloy are its high power capability as a result of the high Ni-content and superior cycle life. The transition metal-based AB_2 alloy (with Zr and Ti as the A-elements and B-elements similar to those of the AB_5 alloy) offers a reduction in raw material cost due to the elimination of rare earth elements and also has the advantage in storage capacity. In A_2B_7 MH alloy, a small amount of Mg is added to the mixture of A-elements (which is composed of rare earth elements) to stabilize the crystal structure. The special super-lattice structure of A_2B_7 MH alloy provides a compromise in performance between AB_2 and AB_5. The last alloy family for Ni/MH application is the Mg-Ni based AB alloy. The much weaker M-H bond of Mg when compared to other A-elements, such as La, Ce, Zr, and Ti, reduces the desirable stoichiometry down to 1:1. The abundance of the raw material and high storage capacity of AB alloy make it a potential candidate to further reduce cost and improve capacity. The impact of stoichiometry on the structures, gaseous phase hydrogen storage properties, and electrochemical properties for these four alloy families will be presented in the remaining sections.

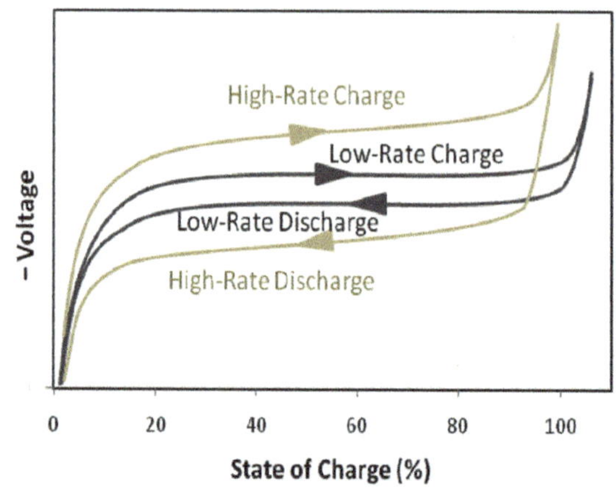

Fig. 5. Schematics of charge and discharge curves under a high (gray) and a low (black) charge densities

3. Stoichiometry in $LaNi_5$-based metal hydride alloys

According to the phase diagram (Fig. 1), eight IMCs exist in the La-Ni binary system. As the Ni-content increases, the ΔH of the hydride increases and forms a straight line in Fig. 6. Since the operational conditions for Ni/MH battery are room temperature and 1 atm, the suitable ΔH value for the MH alloy is around −40 kJ mol^{-1} H_2 (by assigning ΔS to the average value of −130 J K^{-1} mol^{-1} H_2 in formula (1)). In the IMCs of La and Ni, only ΔH of $LaNi_5$ falls into the range suitable for room temperature Ni/MH application. As a result, $LaNi_5$-based MH alloy was chosen as the negative electrode in Ni/MH battery from early on and still dominates today's market.

The following is a description of the sample preparation for the experimental work backing this publication. Five Mm-based AB_5 alloys were prepared by arc melting. The composition of ingot prepared by arc melting is not as uniform as the commercial product produced by induction melting. The designed compositions of these alloys are listed in Table 1 with increasing B/A stoichiometric ratios from 4.8 (hypo) to 5.2 (hyper). The base alloy, AB5.0, is commonly used in today's high-power Ni/MH rechargeable battery and achieves a good balance among different performances. The ingots were annealed in vacuum (1×10^{-8} torr) at 960°C for 8 hours before being mechanically crushed into −200 mesh powder

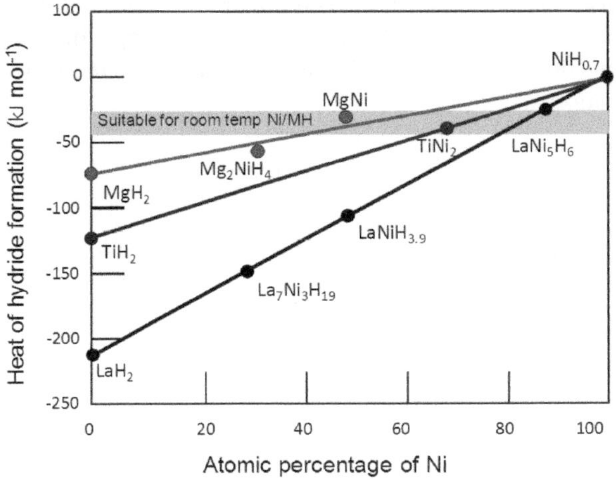

Fig. 6. Relationship between heat of hydride formation and atomic percentage of Ni in the alloy for AB (MgNi), AB_2 ($TiNi_2$), and AB_5 ($LaNi_5$) inter-metallic compounds.

Alloy	La	Ce	Pr	Nd	Ni	Co	Mn	Al	B/A
AB4.8	10.9	4.5	0.5	1.5	59.6	12.6	5.9	4.7	4.8
AB4.9	10.7	4.4	0.5	1.4	59.8	12.7	5.9	4.7	4.9
AB5.0	10.5	4.3	0.5	1.4	60.0	12.7	5.9	4.7	5.0
AB5.1	10.3	4.2	0.5	1.4	60.2	12.7	5.9	4.7	5.1
AB5.2	10.2	4.2	0.5	1.4	60.4	12.8	5.9	4.7	5.2

Table 1. Designed compositions in at. % and B/A stoichiometry ratios of AB_5 alloys.

3.1 Influences on microstructure

In Fig. 1, $LaNi_5$ is represented by a straight line in the phase diagram and shows no solubility at room temperature. If the liquid is quenched very quickly, some hyper-stoichiometry is possible, as can be seen from the dash line near 1270°C; however, annealing and slow cooling will eliminate this hyper-stoichiometry. After annealing, a La_2Ni_7 and a Ni phase are expected in the cases of hypo- and hyper-stoichiometries, respectively. In a commercial AB_5 alloy suitable for general performance requirements, A is mainly Mm (combination of La, Ce, Pr, and Nd) while B contains Ni, Co, Mn, and Al. The multi-element nature of the alloy system will increase the solubility of particular phases since anti-site or vacancy defects become easier to form in multi-element systems.

The phases of the five AB_5 alloys were first studied by x-ray diffractometer (XRD); the results are presented in Fig. 7. Almost all peaks can be attributed to a $CaCu_5$ structure. Ni peak starts to appear when the stoichiometry becomes ≥ 5.0. The lattice parameters and ratio, unit cell volume, and phase abundances obtained from XRD for each alloy are listed in Table 2. As the B/A stoichiometric ratio increases from 4.8 to 5.2, both *a* and *c* lattice constants first increase and then decrease. The maximums occur at a B/A stoichiometric ratio of 4.9, where the c/a ratio reaches the minimum, suggesting a higher pulverization rate during hydride/dehydride cycling. The Ni phase abundance increases with the increase in B/A stoichiometric ratio.

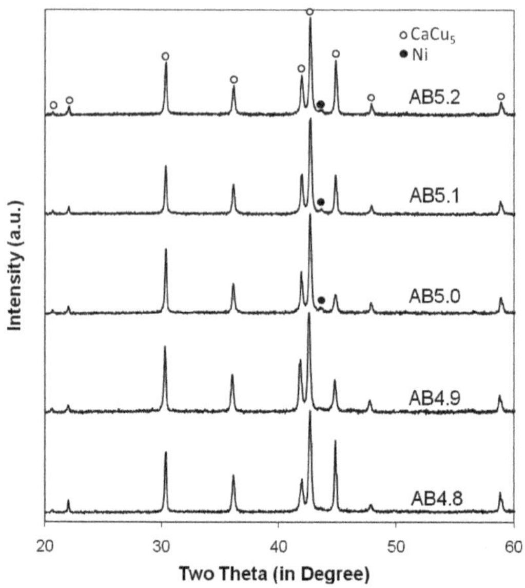

Fig. 7. XRD patterns of AB_5 alloys using Cu-K_α as the radiation source.

Alloy	a (Å)	c (Å)	c/a	Unit cell volume (Å^3)	$CaCu_5$ %	Ni %
AB4.8	4.984	4.052	0.813	87.17	100.0	ND
AB4.9	5.002	4.055	0.811	87.86	100.0	ND
AB5.0	4.986	4.052	0.813	87.23	99.6	0.4
AB5.1	4.984	4.051	0.813	87.14	99.4	0.6
AB5.2	4.985	4.051	0.813	87.18	99.2	0.8

Table 2. Lattice constants and ratios, unit cell volumes, and phase abundances of $CaCu_5$ and Ni of AB_5 alloys from XRD analysis.

The microstructures were further studied by scanning electron microscope (SEM); the resulting backscattering electron images (BEI) are shown in Fig. 8. These micrographs were chosen due to the presence of multi-phase features and do not represent the general phase distribution of the alloys. Four distinctive phases, *A*, *B*, *C*, and *D*, can be identified from these micrographs. Phase *A*, with the brightest contrast due to its high average atomic weight, is the misch-metal with the composition very close to the raw material as indicated by x-ray energy dispersive spectroscopy (EDS) analysis. Phase *B* has a composition close to AB, which is the phase on the left-hand side of the eutectic point in Fig. 1. The AB phase abundance is too small to be detected by XRD analysis. Phase *C* is the main AB_5 phase, where the B/A stoichiometric ratio increases from 5.17, 5.29, 5.49, 5.48, finally to 5.47 with the increase in designed B/A values. In other words, the solubility of the AB_5 phase extends to about 5.5 in this case as the alloy's B/A stoichiometric ratio increases. Further increase of the B-content in the alloy promotes the formation of Ni-phase as seen from Fig.1 and Table 2. Phase *D* with the darkest contrast is a mixture of transition metals and has an average composition of $Al_9Mn_{21}Co_{20}Ni_{50}$, which is high in Al and Mn and low in Ni when compared to the raw material.

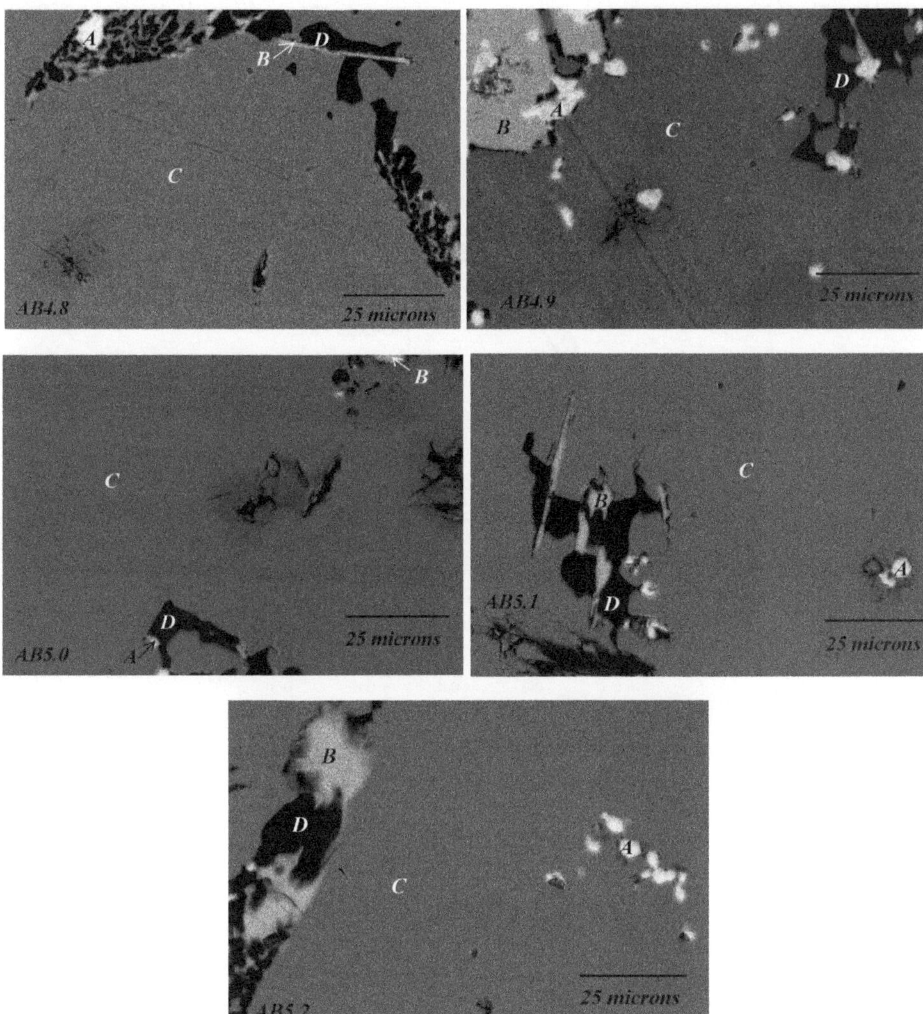

Fig. 8. SEM backscattering images for AB5 alloys. Letters A, B, C, and D indicate metallic rare earth elements, LaNi, $LaNi_5$, and transition metals, respectively.

3.2 Influences on gaseous hydrogen storage

The gaseous phase hydrogen storage characteristics were studied by PCT measurements performed at 30 and 60°C. The 30°C PCT isotherms from the five AB_5 alloys are plotted in Fig. 9 and summarized in Table 3. As the designed B/A stoichiometric ratio increases, both the maximum and reversible hydrogen storage capacities measured at 30°C first increase and then decrease. This trend is the same as that in unit cell volume. The larger unit cell contributes to a more stable hydride with lower equilibrium pressure and higher storage capacity. The isotherm from AB4.9 is flatter than those from other alloys, indicating a higher uniformity in the main phase. The PCT hysteresis from AB4.9 is the highest and corresponds to its lowest *c*/*a* ratio among all alloys, suggesting a more difficult tunneling path for protons to hop within the structure. ΔH and ΔS were calculated from the plateau pressures measured at 30 and 60°C by formula (1) and are listed in Table 3. ΔH and ΔS values are about the same.

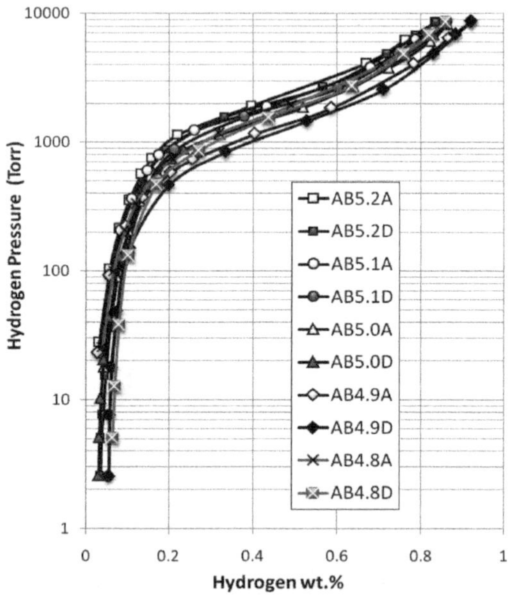

Fig. 9. PCT isotherms measured at 30°C for AB_5 alloys. Open and solid symbols represent data taken during absorption and desorption, respectively.

Alloy	Max. H-storageat 30°C (wt. %)	Reversible H-storageat 30°C wt. %)	Plateau pressure at 30°C Des. (torr)	30°C slope factor	30°C hysteresis	$-\Delta H$ (kJ mol^{-1} H_2)	$-\Delta S$ (J K^{-1} mol^{-1} H_2)
AB4.8	0.86	0.80	1683	0.76	0.06	30	104
AB4.9	0.92	0.86	1479	0.87	0.10	30	105
AB5.0	0.86	0.83	1555	0.80	0.01	30	105
AB5.1	0.86	0.80	1944	0.75	0.06	29	105
AB5.2	0.84	0.80	2068	0.72	0.08	30	106

Table 3. Summary of gaseous phase hydrogen storage properties of AB_5 alloys.

3.3 Influences on electrochemical properties

The discharge capacities for the five AB_5 alloys were measured at two discharge current densities (100 and 8 mA g^{-1}), and the results are listed in Table 4. The full capacity, defined as the capacity measured at the slowest discharge rate (8 mA g^{-1}), increases first and then decreases with the increase in B/A stoichiometric ratio. The stoichiometric AB5.0 alloy shows the highest capacity. The high-rate dischargeability (HRD) values (ratios between the capacities measured at 100 and 8 mA/g discharge currents) measured at the third cycle are listed in Table 4. The maximum HRD value was obtained from a slightly hyper-stoichiometric AB5.1 alloy.

Alloy	Full capacity (mAh g^{-1})	High-rate capacity (mAh g^{-1})	Ratio (High-rate dischargeability)	OCV at 50% DOD (V)
AB4.8	216	203	0.94	−0.921
AB4.9	246	233	0.95	−0.914
AB5.0	249	237	0.95	−0.922
AB5.1	214	205	0.96	−0.927
AB5.2	215	200	0.93	−0.931

Table 4. Summary of electrochemical properties of AB_5 alloys.

OCV measured at 50% depth-of-discharge (DOD) of the MH electrode against the Hg/HgO reference electrode for each alloy is listed in Table 4. In the AB_5 alloy series, the changes in OCV are small due to their similar plateau pressures. In general, OCV decreases with the increase in B/A stoichiometric ratio, which correlates well with the increase in plateau pressure by formula (3). Lower OCV value is a desirable trait because it will contribute to higher cell voltage and better power in the sealed battery configuration.

4. Stoichiometry in Laves phase-based AB_2 metal hydride alloys

Compared to La, Ti has a lower M-H bond strength and thus requires a smaller amount of Ni to adjust the heat of hydride formation to fall within the desired range. According to the graph that plots ΔH vs. atomic percentage of Ni (Fig. 6), a stoichiometry of approximately AB_2 would be suitable for room temperature Ni/MH application. However, the IMC of $TiNi_2$ is not available according to the Ti-Ni binary phase diagram. As seen from the Zr-Ni phase diagram in Fig. 2, Zr, a common element used to partially replace Ti in order to adjust the hydrogen storage, does not have an IMC at $ZrNi_2$ either. Since $TiCr_2$, $ZrCr_2$, $TiMn_2$, and $ZrMn_2$ are all available based on the phase diagrams (Fig. 10, for example), the design of AB_2 alloy starts from the formulation of $(Ti, Zr)(Ni, Cr, Mn)_2$. There are three AB_2 Laves phases, namely, C14, C15, and C36. From the Zr-Cr phase diagram in Fig. 10, the order of phase formation during cooling is C14 first, followed by C36, and finally the C15 phase. Due to the existence of defects, $ZrCr_2$ has a broad range of solubility, allowing the designed composition to be off-stoichiometric.

Five transition element-based AB_2 alloys were prepared by arc melting, and the designed compositions are listed in Table 5. The designed B/A stoichiometric ratio varies from 1.8

(hypo) to 2.2 (hyper). The average electron density (e/a) of each alloy, based on the numbers of outer-shell electrons of the constituent elements, is listed in Table 5 and can be used to predict the dominant phase (C14 or C15) (Shaltiel, et al., 1977). The base alloy, AB2.0, is rich in C14 phase and a candidate for replacing AB_5 alloy in Ni/MH batteries. Its microstructure was studied extensively with a combination of SEM and transmission electron microscope previously (Boettinger, et al., 2010, Bendersky, et al., 2010). In order to preserve the secondary phases, which serve as an electrochemical catalyst, these AB_2 alloys were not annealed (Young, et al., 2010b).

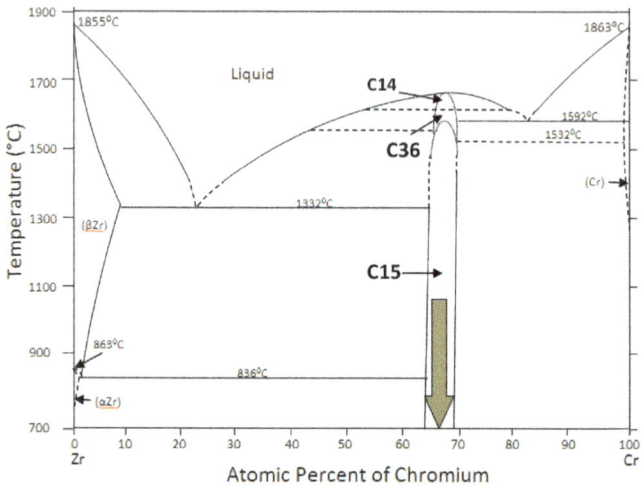

Fig. 10. Zr-Cr binary phase diagram (data from (Massalski, 1990)).

Alloy	Zr	Ti	V	Cr	Mn	Co	Ni	Al	Sn	B/A	e/a
AB1.8	22.4	13.3	9.7	8.2	5.4	1.5	38.9	0.4	0.3	1.8	6.85
AB1.9	21.7	12.9	9.8	8.4	5.5	1.5	39.5	0.4	0.3	1.9	6.90
AB2.0	21.0	12.5	10.0	8.5	5.6	1.5	40.2	0.4	0.3	2.0	6.95
AB2.1	20.2	12.0	10.2	8.7	5.7	1.5	41.0	0.4	0.3	2.1	7.01
AB2.2	19.6	11.7	10.3	8.8	5.8	1.6	41.6	0.4	0.3	2.2	7.05

Table 5. Designed compositions in at. %, B/A stoichiometry ratios, and average electron densities (e/a) of AB_2 alloys.

4.1 Influences on microstructure

Fig. 11 compares the XRD patterns from the five AB_2 alloys. As the e/a value increases, the main phase shifts from C14 to C15. Besides the main Laves phases, TiNi phase, which is associated with the solid-state transformation of non-Laves secondary phases during cooling, is also present. The C14 lattice constants a and c, c/a aspect ratio, unit cell volume, and phase abundances of each alloy are listed in Table 6. In this series of AB_2 alloys, a and c

decrease monotonically with the increase in stoichiometry. In the case of hypo-stoichiometry, the excessive A-site atoms move to the B-site and slightly enlarge the unit cell. On the contrary, the larger B-site elements (however, still smaller than the A-site elements) will occupy the A-site and shrink the unit cell in the case of hyper-stoichiometry (Kodama, et al., 1995). The stoichiometric AB2.0 alloy has the highest or close to the highest *c/a* aspect ratio, so theoretically it should be the least stable alloy during hydride/dehydride cycling. The C14 unit cell volume decreases with the increase in B/A stoichiometric ratio. The smaller unit cell will contribute to an unstable hydride with higher plateau pressure and lower storage capacity. As the B/A stoichiometric ratio increases, more of the main C14 phase was replaced by the C15 phase due to the increase of the *e/a* value. The phase abundance of TiNi also reduces with the increase in B/A stoichiometric ratio. The TiNi phase has a higher A-atom content than the Laves phase does and therefore is not favorable in the case of B-rich hyper-stoichiometry.

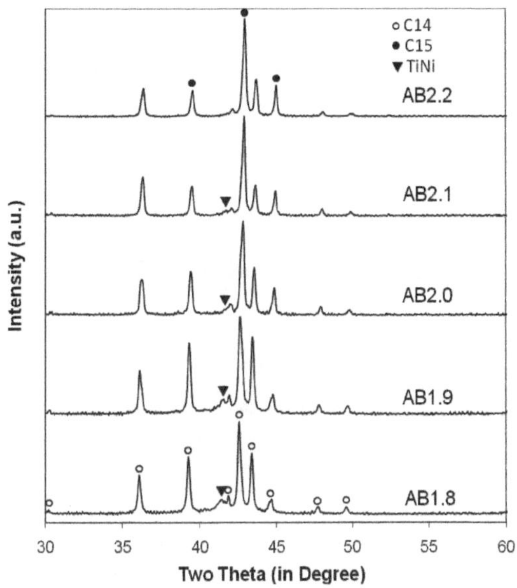

Fig. 11. XRD patterns of AB_2 alloys using Cu-K_α as the radiation source.

Alloy	*a* (Å)	*c*(Å)	*c/a*	Unit cell volume (Å^3)	C14 %	C15 %	TiNi %
AB1.8	4.9932	8.1471	1.632	175.9	74.0	11.1	14.9
AB1.9	4.9831	8.1381	1.633	175.0	74.8	17.0	8.2
AB2.0	4.9771	8.1288	1.633	174.4	70.7	25.7	3.6
AB2.1	4.9634	8.1005	1.632	172.8	62.3	36.7	1.0
AB2.2	4.9564	8.0882	1.631	172.1	60.0	40.0	0.0

Table 6. C14 lattice constants and ratios, unit cell volumes, and phase abundances of C14, C15 and TiNi of AB_2 alloys from XRD analysis.

Four regions with different contrasts can be observed in most of the SEM BEI micrographs shown in Fig. 12. Region *A* with the brightest contrast is the Zr_7Ni_{10} phase, which was not observed in XRD due to its small abundance. The light gray region *B* has a composition close to AB and can be related to the TiNi phase found in XRD analysis. Both regions *A* and *B*, with nominal compositions of A_7B_{10} and AB, become smaller and even diminish as the B/A stoichiometric ratio increases. Region *C* is the main AB_2 phase. As the designed B/A stoichiometric ratio increases, the B/A stoichiometric ratio in the main AB_2 phase evolves from 1.85 to 1.97, 2.05, 2.01 and 1.99. This result demonstrates that the AB_2 phase has a wider solubility on the hypo-side than on the hyper-side in the C14-predominaing case. The last feature of the micrographs is the dark spot with very high Zr-content, which is identified as ZrO_2 inclusion (region *D*). This oxide is formed through oxidation of Zr segregated out from the melt; it is not the residual oxide in the raw material as reported before in a study of ZrO_2 inclusions in the Al-containing AB_2 alloys [Young, et al., 2010c]. As the B/A stoichiometric ratio varies, the size and amount of ZrO_2 inclusions do not change.

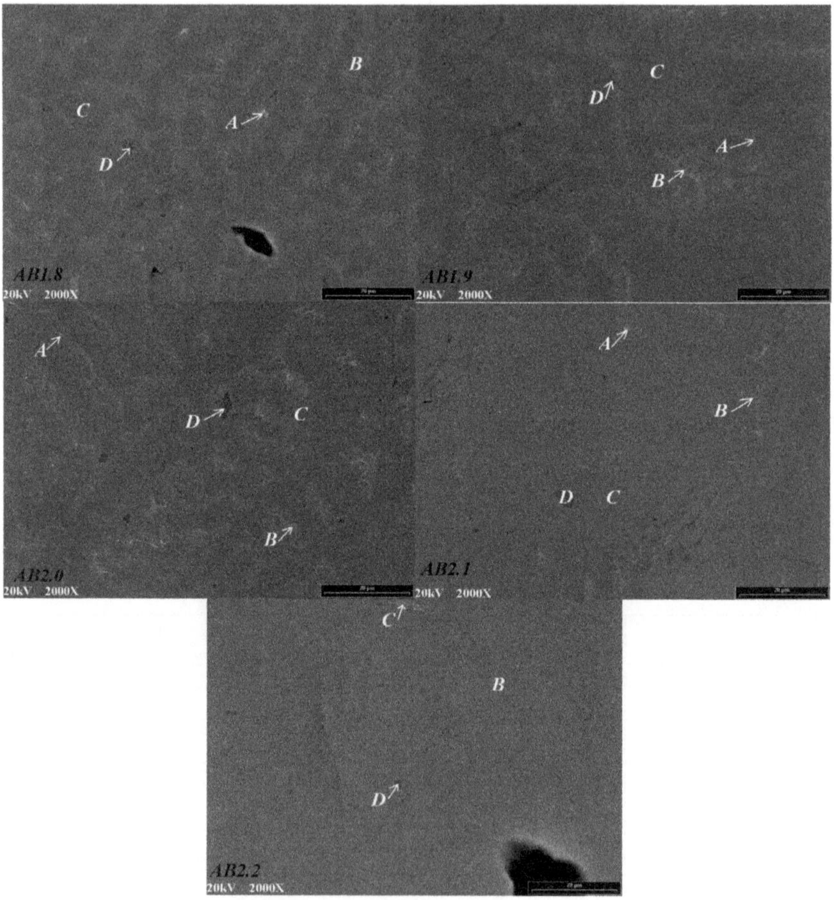

Fig. 12. SEM backscattering images for AB_2 alloys. Letters *A*, *B*, *C*, and *D* indicate Zr_7Ni_{10}, TiNi, AB_2, and ZrO_2, respectively.

4.2 Influences on gaseous hydrogen storage

The PCT absorption and desorption isotherms measured at 30°C for the five AB_2 alloys are plotted in Fig. 13 and summarized in Table 7. As the B/A stoichiometric ratio increases, the maximum hydrogen storage capacity first increases and then decreases. The highest maximum capacity was obtained from AB1.9 alloy. For the alloys with B/A stoichiometric ratio ≥ 2.0, the maximum capacities decrease as a result of the shrinking lattice unit cell. In alloy AB1.8, the phase abundance of C14, which has less storage capacity compared to C15 (Young, et al. 2010b), is the highest and thus decreases the maximum storage capacity. The relationship between B/A stoichiometric ratio and reversible gas phase storage is similar to what is observed in maximum storage capacity: as the B/A stoichiometric ratio first increases, the reversible storage increases due to the higher plateau pressure and the higher C15 phase abundance with higher gaseous reversibility (Young, et al. 2010b). With further increase in B/A stoichiometric ratio, the reversible storage capacity decreases for the same reason as the maximum storage capacity. The maximum reversible capacity happens at AB2.0. Due to the high degree of disorder in AB_2 MH alloys, the critical temperature (top of the dome expressed by a dashed line in Fig. 4) is much lower than it is in AB_5 MH alloys. Therefore, all the PCT isotherms lack a well-defined plateau pressure, or in other words, a metal (α)-hydride (β) co-existing region. In order to compare the hydrogen equilibrium pressures among the alloys, the mid-point pressure, defined as the pressure where half of the desorption is complete, is calculated and listed in Table 7. The mid-point pressure increases as a result of shrinking lattice unit cell as the B/A stoichiometric ratio goes up. With the increase in B/A stoichiometric ratio, the PCT isotherm becomes flatter except the last AB2.2 alloy. Slightly hyper-stoichiometry (2.1) shows the most ordered structure.

Alloy	Max. H-storage at 30°C (wt. %)	Reversible H-storage at 30°C (wt. %)	Mid-point pressure at 30°C Des. (torr)	30°C slope factor	30°C hysteresis	$-\Delta H$(kJ mol^{-1} H_2)	$-\Delta S$ (J K^{-1} mol^{-1} H_2)
AB1.8	1.30	0.78	150	0.59	0.69	29	77
AB1.9	1.44	1.09	200	0.66	0.59	28	81
AB2.0	1.40	1.16	500	0.79	0.09	24	77
AB2.1	1.31	1.13	700	0.81	0.13	23	76
AB2.2	1.11	0.91	1100	0.78	0.25	21	71

Table 7. Summary of gaseous phase hydrogen storage properties of AB_2 alloys.

For AB2.0-stoichiomtery alloys, the hysteresis is related to the average electron density (Young, et al. 2009b). The PCT hysteresis first decreases and then increases with the increase in B/A stoichiometric ratio. The hysteresis minimum happens at a B/A stoichiometric ratio of 2.0. Therefore, from a cycle stability point of view, the stoichiometric AB2.0 alloy is preferable and contradicts from the c/a ratio prediction in the XRD analysis. ΔH and ΔS values, calculated from the mid-point pressures of 30 and 60°C desorption isotherms, are listed in Table 7. The value of $-\Delta H$ decreases as the B/A stoichiometric ratio decreases, which implies the weakening of the M-H bond strength and correlates well with the shrinking lattice. The measured ΔS values from these AB_2 alloys are similar and higher than those in AB_5 alloys (around −105 J K^{-1} mol^{-1} H_2), which suggest higher degrees of disorder in AB_2 alloys and are consistent with the steeper PCT isotherms.

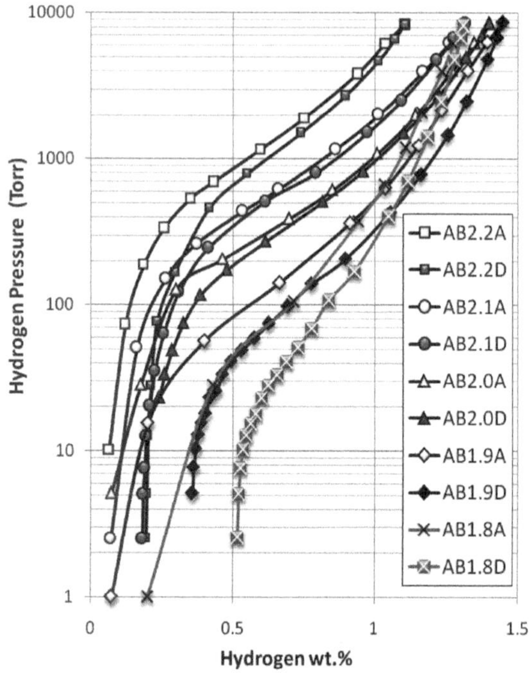

Fig. 13. PCT isotherms measured at 30°C for AB_2 alloys. Open and solid symbols represent data taken during absorption and desorption, respectively.

4.3 Influences on electrochemical properties

The discharge capacities for the five AB_2 alloys were measured, and the results are listed in Table 8. The full capacity increases first and then decreases with the increase in B/A stoichiometric ratio. The stoichiometric AB2.0 alloy shows the highest capacity. The HRD values measured at the tenth cycle are listed in Table 8. The maximum value occurs at AB2.1. Similar to the case of AB_5 alloys, slightly hyper-stoichiometry increases the HRD.

OCV measured at 50% DOD of the MH electrode against Hg/HgO reference electrode for each alloy is listed in Table 8. In the AB_2 alloy series, OCV decreases with the increase in B/A stoichiometric ratio, which correlates well with the increase in mid-point pressure by formula (3). This observation is consistent with that from the AB_5 alloy series. Similar influences of stoichiometry on various properties of other two series of AB2 alloys predominated by C15 and C14/C15 phases were also reported previously (Young, et al. 2011b & 2011c).

Alloy	Full capacity ($mAhg^{-1}$)	High-rate capacity ($mAh\ g^{-1}$)	HRD	OCV at 50% DOD (V)
AB1.8	369	144	0.39	−0.859
AB1.9	369	323	0.88	−0.882
AB2.0	401	357	0.89	−0.890
AB2.1	333	308	0.92	−0.902
AB2.2	283	210	0.74	−0.915

Table 8. Summary of electrochemical properties of AB_2 alloys

5. Stoichiometry in NdMgNi-based A_2B_7 metal hydride alloys

La_2Ni_7 is an IMC with slightly lower Ni-content than $LaNi_5$ as shown in the La-Ni phase diagram (Fig. 1). ΔH of La_2Ni_7 is below the suitable range for room temperature operation as shown in Fig. 6. In order to increase ΔH, a small amount of Mg is added to replace La and brings ΔH into the suitable range for room temperature operation. So far, three MH alloys of the A_2B_7 family have been used as the negative electrode in Ni/MH rechargeable battery. Kohno et al. first introduced $La_5Mg_2Ni_{23}$ with a potential capacity of over 400 mAh g^{-1} (Kohno, et al., 2000) (Mm, Mg)(Ni, Co, Al)$_{3.3}$ alloys were used in high-capacity 2700 mAh AA-size battery by Sanyo (Yasuoka, et al., 2006) where Mm is the misch metal. Finally, (Nd, Mg)$_2$(Ni, Al)$_7$ alloys were used for the low self-discharge AA battery also introduced by Sanyo.

Five Nd_2Ni_7-based alloys with different B/A stoichiometric ratios were prepared by induction melting. The target compositions of these alloys together with the designed B/A stoichiometric ratios ranging from 3.3 to 3.7 are listed in Table 9. These alloys were annealed in 1 atm of argon at 900°C for 5 hours.

Alloy	Nd	Zr	Mg	Ni	Co	Al	B/A
AB3.3	20.3	0.2	2.7	73.1	0.1	3.5	3.3
AB3.4	19.9	0.2	2.6	73.7	0.1	3.5	3.4
AB3.5	19.4	0.2	2.6	74.1	0.1	3.5	3.5
AB3.6	19.0	0.2	2.5	74.6	0.1	3.5	3.6
AB3.7	18.6	0.2	2.5	75.0	0.1	3.6	3.7

Table 9. Designed compositions in at. % and B/A stoichiometry ratios of A_2B_7 alloys.

5.1 Influences on microstructure

The Nd-Ni phase diagram is very similar to that of La-Ni shown in Fig. 1. Under the equilibrium condition, the Nd_2Ni_7 phase has a very limited solubility. The influence of off-stoichiometry to microstructure was studied by XRD analysis, and the results are plotted in Fig. 14. Besides the reflection peaks from the Nd_2Ni_7 main phase, secondary phases of Nd, NdNi, $NdNi_2$, and $NdNi_5$ can be identified from the patterns. The lattice constants of the Nd_2Ni_7 phase were calculated and are listed in Table 10 together with the phase abundances of all observed phases. In general, as the B/A stoichiometric ratio increases, both c/a ratio and the unit cell volume decrease as seen from the shifting in the peak into higher angle illustrated by the line shown in Fig. 14; the phase abundance of Ng_2Ni_7 remains about the same; the phase abundance of NdNi decreases; and the phase abundances of $NdNi_2$ and $NdNi_5$ increase except the last AB3.7 alloy.

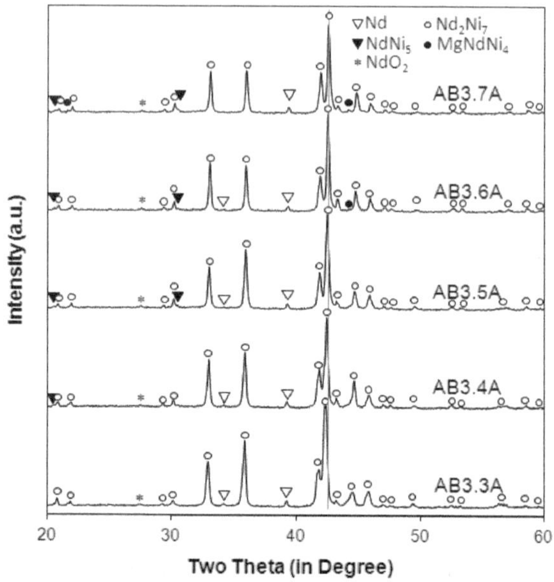

Fig. 14. XRD patterns of A_2B_7 alloys using Cu-K_α as the radiation source.

Alloy	*a* (Å)	*c* (Å)	*c*/*a*	Unit cell volume (Å³)	Nd_2Ni_7 %	Nd %	NdNi %	$NdNi_2$ %	$NdNi_5$ %
AB3.3	5.0036	24.4325	4.883	529.7	97.6	0.9	0.5	0.4	ND
AB3.4	5.0056	24.3597	4.866	528.6	97.2	1.0	0.3	0.5	0.3
AB3.5	5.0012	24.3716	4.873	527.9	96.7	1.4	0.2	0.4	0.5
AB3.6	4.9986	24.3145	4.864	526.1	95.6	1.0	ND	1.2	1.3
AB3.7	4.9968	24.2844	4.860	525.1	95.7	1.6	ND	1.3	0.6

Table 10. Nd_2Ni_7 lattice constants and ratios, unit cell volumes, and phase abundances of Nd_2Ni_7, Nd, NdNi, $NdNi_2$, $NdNi_5$ of A_2B_7 alloys from XRD analysis.

The microstructures were further studied by SEM-EDS, and the resulting BEI micrographs are shown in Fig. 15. Five regions (identified by letters *A* to *E*) can be distinguished by backscattering contrast. The brightest spot *A* is the Nd metal. The size and abundance of Nd does not change significantly with the change in B/A stoichiometric ratio. According to EDS, region *A* has the second highest Zr-content (~ 0.3 %) next to region *E*, which is entirely composed of ZrO_2. The second brightest region, denoted by *B*, has a composition close to NdNi. The Mg-content in this region is small when compared to that of the main phase.

Region *C* is the main Nd_2Ni_7 phase and the B/A stoichiometric ratio changes from 2.91 to 3.27, 3.33, 3.33, and 3.32 as the designed B/A stoichiometric ratio increases from 3.3 to 3.4, 3.5, 3.6, and 3.7. With the incorporation of Mg in the Nd-site, the stable main phase stoichiometry is between 2.91 and 3.33. Since the extra B elements (Ni and Al) are likely to be pushed to $NdNi_5$, which is rich in B, the B/A stoichiometric ratio in the main phase is lower compared to the designed ratio. Furthermore, from the XRD analysis, the $NdNi_5$ phase can be observed in most alloys, and the amount of $NdNi_5$ phase increases with the increase in B/A stoichiometric ratio. Only the alloy with a designed B/A stoichiometry of 3.3 has too low of $NdNi_5$-content to be detected by XRD. The Mg-content in the main phase is about 3.5%, which is higher than the average Mg-content of the designed compositions. The fourth phase, region *D*, has a composition close to $NdNi_2$ with the highest Mg-content of about 12 to 15%. The SEM micrographs show appreciable amount of region *D* in all alloys. However, the XRD analysis only shows a small amount of $NdNi_2$ in the last two alloys, AB3.6 and AB3.7. The areas where the SEM micrographs were taken are usually those with the most numbers of phases and should not be used to quantify the phase abundances. Region *E*, with the darkest contrast, is from ZrO_2.

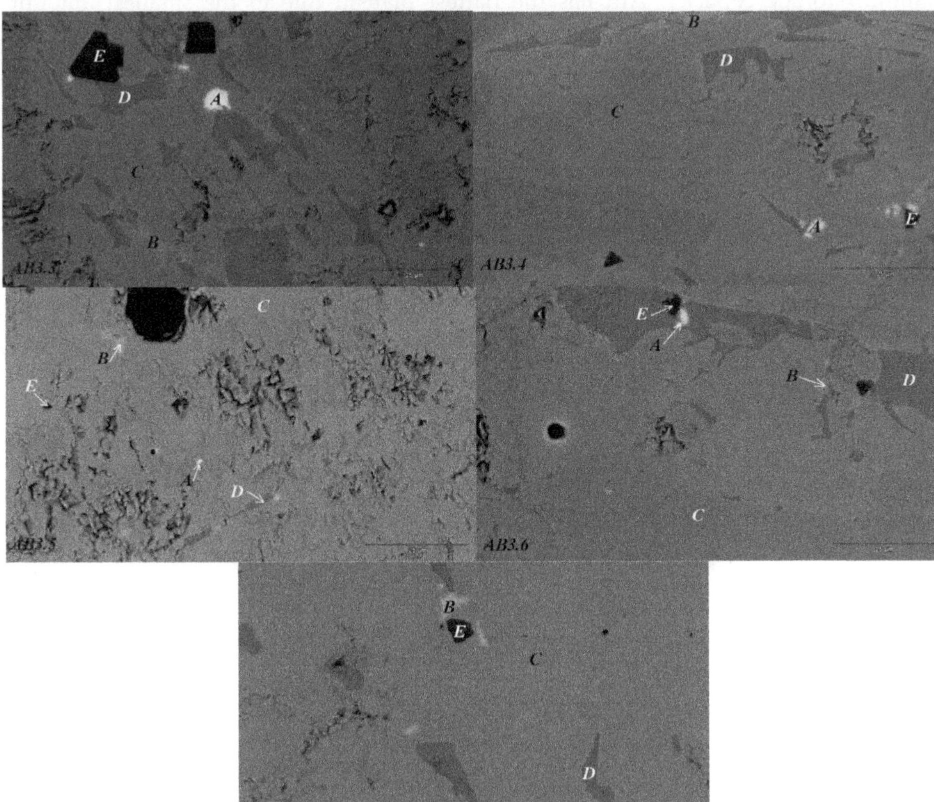

Fig. 15. SEM backscattering images for A_2B_7 alloys. Letters *A*, *B*, *C*, *D*, and *E* indicate Nd metal, NdNi, Nd_2Ni_7, $NdNi_2$, and ZrO_2, respectively.

5.2 Influences on gaseous hydrogen storage

The PCT isotherms measured at 30°C for three five A_2B_7 alloys are shown in Fig. 16 and summarized in Table 11. As the B/A stoichiometric ratio increases, the maximum hydrogen storage capacity increases in the beginning and stabilizes when the designed stoichiometry becomes ≥ 3.5; the reversible hydrogen storage capacity increases; the mid-point pressure increases by a small amount; the isotherm becomes flatter; and the hysteresis, ΔH, and ΔS remain about the same. The changes in maximum hydrogen storage capacity are similar to those in B/A stoichiometric ratio in the main Nd_2Ni_7 phase. The composition in the main phase is optimized for maximum storage capacity at B/A = 3.33. Although both the reversible capacity and the plateau pressure increase with the increase in B/A stoichiometric ratio, the increase in plateau pressure is too small to account for the large increase in reversible storage capacity. The secondary phases such as $NdNi_2$ and $NdNi_5$ are most likely to contribute to the increase in reversible hydrogen storage as the case of the ZrNi and Zr_7Ni_{10} phases in AB_2 alloy (Young, et al., 2011a). Comparing the PCT results among the AB_5, AB_2, and A_2B_7 families yields the following observations: AB_2 has the highest maximum storage capacity, AB_5 has the flattest plateau due to the smallest amount of secondary phases, and A_2B_7 has the most ordered MH structure as seen from its lowest ΔS value.

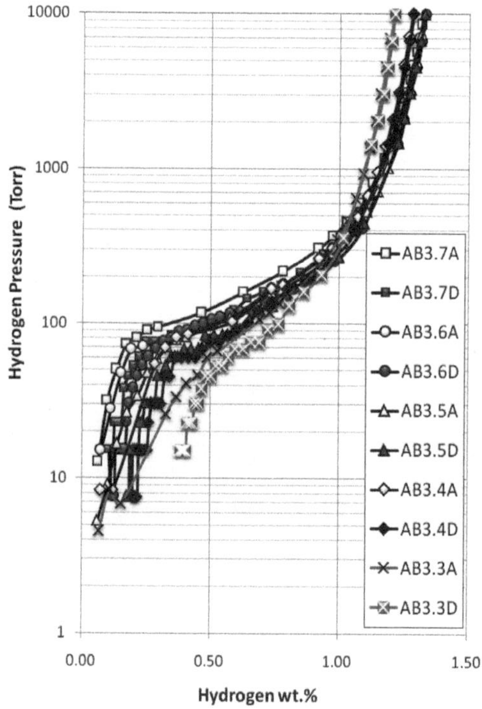

Fig. 16. PCT isotherms measured at 30°C for A_2B_7 alloys. Open and solid symbols represent data taken during absorption and desorption, respectively.

Alloy	Max. H-storage at 30°C (wt. %)	Reversible H-storage at 30°C (wt. %)	Mid-point pressure at 30°C Des. (torr)	30°C slope factor	30°C hysteresis	-ΔH (kJ mol^{-1} H_2)	-ΔS (J K^{-1} mol^{-1} H_2)
AB3.3	1.22	0.84	123	0.58	0.22	40	115
AB3.4	1.30	1.10	140	0.65	0.19	40	118
AB3.5	1.34	1.16	139	0.71	0.18	41	121
AB3.6	1.34	1.22	148	0.76	0.18	40	119
AB3.7	1.34	1.24	164	0.77	0.19	40	118

Table 11. Summary of gaseous phase hydrogen storage properties of A_2B_7 alloys.

5.3 Influences on electrochemical properties

Similar to the case of AB_5 alloys, these A_2B_7 alloys were easy to activate. All alloys reached the highest capacities within the first three cycles without any pre-treatment. The full and high-rate discharge capacities at the third cycle for each alloy are listed in Table 12. Identical to the maximum storage in gaseous phase, both the full and high-rate capacities increase and stabilize when the designed B/A becomes ≥ 3.5 due to the similar B/A stoichiometric ratio in the main Nd_2Ni_7 phase. The maximum HRD happens at alloy AB3.6. Further increase in B/A stoichiometric ratio reduces the HRD due to the reduction in $NdNi_5$ phase as indicated by the XRD analysis. OCV, measured at 50% DOD of the MH electrode against $Ni(OH)_2$/NiOOH counter electrode, shows a decrease with the increase in B/A stoichiometric ratio. According to formula (3), a higher PCT plateau pressure should generate a higher cell voltage. An increase from 123 (AB3.3) to 164 torr (AB3.7) should correspond to an increase of 0.004 volts. The voltage drop with the increase in B/A stoichiometric ratio in A_2B_7 alloys requires additional investigation.

Alloy	Full capacity (mAh g^{-1})	High-rate capacity (mAh g^{-1})	HRD	OCV at 50% DOD (V)
AB3.3	316	299	0.95	1.238
AB3.4	331	317	0.96	1.238
AB3.5	341	332	0.97	1.234
AB3.6	340	332	0.98	1.227
AB3.7	340	331	0.97	1.225

Table 12. Summary of electrochemical properties of A_2B_7 alloys.

6. Stoichiometry in MgNi-based AB metal hydride alloys

MgNi-based MH alloys have attracted much of researchers' attention due to the abundance of Mg and Ni resources. Mg metal forms a stable hydride of MgH_2 which only desorbs hydrogen at temperature above 400°C. According to the schematic of ΔH vs. Ni % in Fig. 6, a stoichiometry of Mg : Ni close to 1:1 is suitable for room temperature operation. However, according to the Mg-Ni binary phase diagram illustrated in Fig. 17, only two IMCs are allowed at room temperature under the equilibrium condition, namely, Mg_2Ni and $MgNi_2$. The former is a MH alloy but only desorbs hydrogen at 250°C; the latter will not form any stable hydride at 1 atm. The desirable MgNi single phase (indicated by the gray arrow in Fig. 17) is not allowed. Therefore, quick quench methods under the non-equilibrium conditions, such as RF sputtering, laser ablation, and combination of melt-spin and mechanical alloying, were used to prepare the MgNi alloys for hydrogen storage and Ni/MH rechargeable battery applications.

Five thin films were prepared with the RF-sputtering technique simultaneously with a compositional gradient target, 20 mtorr Ar gas background, 200 watt of rf power in a MRC 8″ multi-target system. The substrate was a Ni tab taped down to a copper plate maintained at 0°C by circulating water through the imbedded pipeline. The composition varies from Mg : Ni = 48:52 to 52:48, 57:43, 61:39, and 65:35. The deposition time was 2 hours, and the average film thickness was about 1.5 microns.

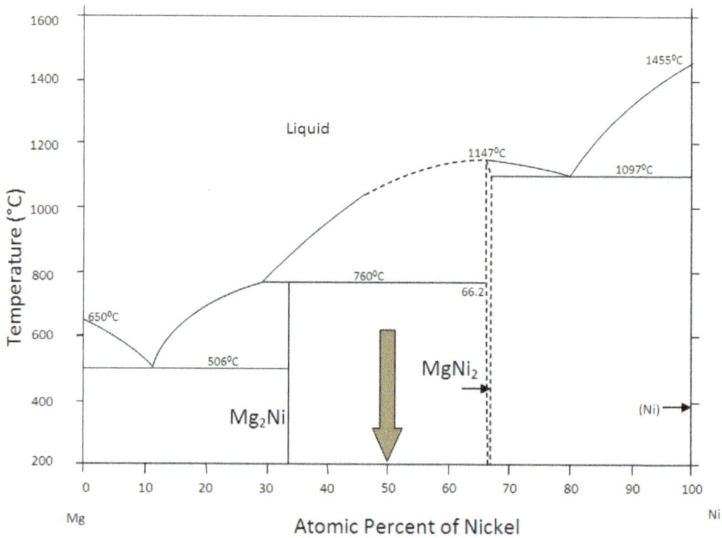

Fig. 17. Mg-Ni binary phase diagram (data from (Massalski, 1990)).

6.1 Influences on microstructure

The XRD patterns obtained from these five thin films on Ni substrate are shown in Fig. 18. The two strongest defection peaks are from the Ni-substrate. The microcrystalline phase in the film has two peaks at around 20 and 40° from the broadening of $MgNi_2$ and Mg_2Ni peaks. The positions of microcrystalline peak will not change with the film composition. While the first broad peak of microcrystalline can be seen at 20°, the second broad peak of microcrystalline (around 40°C) is unable to be discerned since it overlaps with another broad peak from the amorphous phase in the film. The center of the amorphous peak can be related to the average distance between two neighboring atoms by the Ehrenfest formula (Guinier, 1963):

$$2r\sin\theta = 1.2\lambda \quad (6)$$

where r is the average distance between two neighboring atoms and λ is the wavelength of the x-ray (in this case, 1.5418Å). The shift of the amorphous peak to the lower angle is due to the increase in the inter-atomic distance with the higher Mg (larger than Ni)-content in the thin film. As the Mg content in the alloy increases above 52%, deflection peaks from the crystalline Mg_2Ni start to appear. Film with a composition close to 1:1 ($Mg_{48}Ni_{52}$ and $Mg_{52}Ni_{48}$) shows mainly amorphous phase with smaller amounts of microcrystalline phase. With the increase in Mg-content, the Mg_2Ni crystalline phase starts to form.

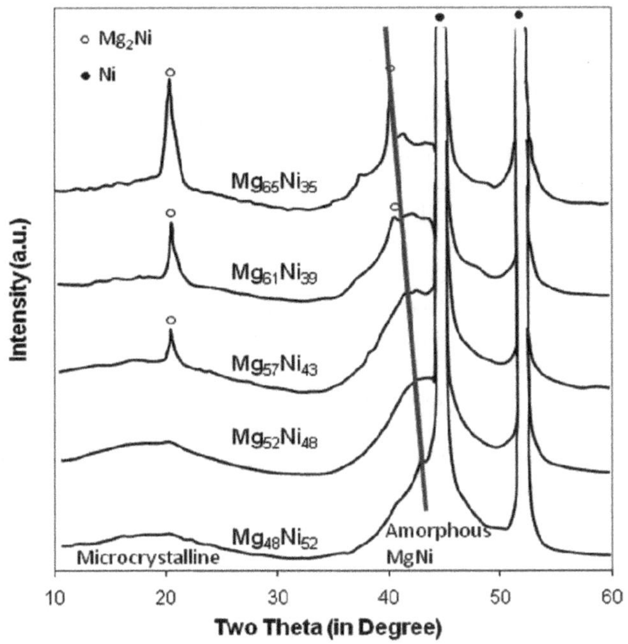

Fig. 18. XRD patterns of MgNi thin films deposited on Ni-substrate using Cu-K_α as the radiation source. The straight line illustrates the shift of amorphous peak into lower angle as the Mg-content in the film increases.

6.2 Influences on electrochemical properties

The film thickness and capacity measured at a discharged rate of 20 mA g^{-1} of each film are listed in Table 13. As the Mg-content increases, the film becomes thinner, and the discharge capacity first increases and then decreases. The maximum capacity happens at a composition of $Mg_{52}Ni_{48}$, which is mainly amorphous in nature. Both the microstructure and electrochemical capacity are strongly affected by the film composition, or in other words, stoichiometry.

Alloy composition	Film thickness (μm)	Discharge capacity ($mAh\ g^{-1}$)
$Mg_{48}Ni_{52}$	2.02	302
$Mg_{52}Ni_{48}$	1.82	327
$Mg_{57}Ni_{43}$	1.64	260
$Mg_{61}Ni_{39}$	1.45	75
$Mg_{65}Ni_{35}$	1.32	20

Table 13. Summary of electrochemical properties of AB alloys.

7. Conclusion

The influences of stoichiometry on the structures, gaseous phase hydrogen storage properties, and electrochemical properties of four families of MH alloys were studied. In the cases of AB_5, AB_2, and A_2B_7, as the B/A stoichiometric ratio increases, the sizes of the unit cell of main phases decrease, resulting in less stable hydrides with higher hydrogen equilibrium pressures, the stoichiometry in the main phase increases and then becomes stabilized at the stoichiometric alloy. Further increases in the average stoichiometry increases the abundance of certain secondary phases but do not change the stoichiometry in the main phase. This suggests that the alloys under the current study have a wider composition range on the hypo-stoichiometric side (A-rich) than they do on the hyper-stoichiometric side (B-rich). The stoichiometric alloys usually have the highest gaseous phase storage capacities (both maximum and reversible). Slightly over-stoichiometric alloys are more suitable for high-power applications due to their higher HRD values. In the case of AB alloys, both the microstructure and electrochemical properties are strongly related to the film composition (stoichiometry). The optimal composition is $Mg_{52}Ni_{48}$ considering its high capacity and amorphous content.

8. References

Bendersky, L.A., Wang, K., Boettinger, W.J., Newbury, D.E., Young, K. & Chao, B. (2010). Examination of Multiphase (Zr, Ti)(V, Cr, Mn, Ni)$_2$ Ni-MH Electrode Alloys: Part II. Solid-State Transformation of the Interdendritic B2 Phase. *Metallurgical and Materials Transactions A*, Vol.41, pp.1891-1906, ISSN 1073-5623

Boettinger, W.J., Newbury, D.E., Wang, K., Bendersky, L.A., Chiu, C., Kattner, U.R., Young, K. & Chao, B. (2010). Examination of Multiphase (Zr, Ti)(V, Cr, Mn, Ni)$_2$ Ni-MH Electrode Alloys: Part I. Dendritic Solidification Structure. *Metallurgical and Materials Transactions A*, Vol.41, pp.2033-2047, ISSN 1073-5623

Chen, K.C., Allen, S.M. & Livingston, J.D. (1994). Stoichiometry and Alloying Effects on the Phase Stability and Mechanical Properties of $TiCr_2$-Base laves Phase Alloys. *High-temperature ordered intermetallic alloys VI.*, pp. 1401, MRS Symposium Proceeding Vol. 364, Material Research Society, ISBN 1-558-99265-0, Pittsburgh, Pennsylvania

De Boer, F.R., Boom, R., Mattens, W.C.M., Miedema, A.R. & Niessen, A.K. (1988). *Cohesion in Metal*, pp. 4, North-Holland, ISBN 0-444-87098-9, Amsterdam

Guinier, A. (1963). *X-ray Diffraction in Crystals, Imperfect Crystals and Amorphous Bodies*, pp.73, Freeman, ISBN 0-486-68011-8, San Francisco

Kodama, T., Anada, H. & Kaminaka, H. (1995). The Site Occupancies for the Excess Manganese Atoms and the Third Element Niobium in the Intermetallic Compound $ZrMn_2$. *Journal of Alloys and Compounds*, Vol. 224, pp.70-75, ISSN 0925-8388

Kohno, T., Yoshida, H., Kawashima, F., Inaba, T., Sakai, I., Yamamoto, M. & Kanda, M. (2000). Hydrogen Storage Properties of New Ternary System Alloys: La_2MgNi_9, $La_5Mg_2Ni_{23}$, La_3MgNi_{14}. *Journal of Alloys and Compounds*, Vol.311, pp.L5-L7 , ISSN 0925-8388

Massalski, T.B. (1990). *Binary Alloy Phase Diagram*, ASM International, ISBN 0-87170-403-X, Ohio, USA.

Osumi, Y., Suzuki, H., Kato, A., Oguro, K., Kawai, S. & Kaneko, M. (1983). Hydrogen Absorption-desorption Charateristics of Mm-Ni-Al-M and Mm-Ni-Mn-M Alloys (Mm = misch metal). *Journal of Less-common Metals*, Vol.89, pp.287-292 , ISSN 0022-5088

Pauling, L. (1987). Factors Determining the Average Atomic Volumes in Intermetallic Compounds. *Proceedings of the National Academy of Science of the United State of America*, Vol.84, pp.4754-4756, ISSN 002708424

Shaltiel, D., Jacob, I. & Davidov, D. (1977). Hydrogen Absorption and Desorption Properties of AB_2 Laves-phase Pseudobinary Compounds. *Journal of Less-common Metals*, Vol.53, pp.117-131, ISSN 0022-5088

Thoma, D.J. & Perepezko, J.H. (1995). A Geometric Analysis of Solubility Ranges in Laves Phases. *Journal of Alloys and Compounds*, Vol.224, pp.330-341, ISSN 0925-8388

Yasuoka, S., Magari, Y., Murata, T., Tanaka, T., Ishida, J., Nakamura, H., Nohma, T., Kahara, M., Baba, Y. & Teraoka, H. (2006). Development of High-capacity Nickel-metal Hydride Batteries Using Superlattice Hydrogen-absorbing Alloys. *Journal of Power Sources*, Vol.156, pp.662-666, ISSN 0378-7753

Young, K., Fetcenko, M.A., Ouchi, T., Li, F., & Koch, J. (2008). Structural, Thermodynamic, and electrochemical Properties of $Ti_xZr_{1-x}(VNiCrMnCoAl)_2$ C14 Laves Phase Alloys. *Journal of Alloys and Compounds*, Vol.464, pp.238-247, ISSN 0925-8388

Young, K., Ouchi, T., Koch, J. & Fetcenko, M.A. (2009a). The Role of Mn in C14 Laves Phase Multi-component Alloys for NiMH Battery Application. *Journal of Alloys and Compounds*, Vol. 477, pp.749-758 , ISSN 0925-8388

Young, Y., Ouchi, T. & Fetcenko, M.A. (2009b). Pressure-composition-temperature Hysteresis in C14 Laves Alloys: Part 1. Simple Ternary Alloys. *Journal of Alloys and Compounds*, Vol.480, pp.428-433, ISSN 0925-8388

Young, K., Huang, B., Regmi, R.K., Lawes, G. & Liu, Y. (2010a). Comparisons of Metallic Clusters Imbedded in the Surface Oxide of AB_2, AB_5, and A_2B_7 Alloys. *Journal of Alloys and Compounds*, Vol.506, pp.831-840, ISSN 0925-8388

Young, K., Ouchi, T., Huang, B., Chao, B., Fetcenko, M.A., Bendersky, L.A., Wang, K. & Chiu, C. (2010b). The Correlation of C14/C15 Phase Abundance and Electrochemical Properties in the AB_2 Alloys. *Journal of Alloys and Compounds*, Vol.506, pp.841-848, ISSN 0925-8388

Young, K., Regmi, R., Lawes, G., Ouchi, T., Reichman, B., Fetcenko, M.A. & Wu, A. (2010c). Effects of Aluminum Substitution in C14-rich Multi-component Alloys for NiMH Battery Application. *Journal of Alloys and Compounds*, Vol.490, pp.282-292, ISSN 0925-8388

Young, K., Nei, J., Ouchi, T. & Fetcenko, M.A. (2011a). Phase Abundances in AB_2 Metal Hydride Alloys and Their Correlations to Various Properties. *Journal of Alloys and Compounds*, Vol.509, pp.2277-2284 , ISSN 0925-8388

Young, K., Ouchi, T., Yang, J. & Fetcenko, M.A. (2011b) Studies of off-Stoichiometric AB_2 Metal Hydride Alloy: Part 1. Structural Characteristics. *International Journal of Hydrogen Energy*, Vol.36, pp.11137-11145, ISSN 0360-3199

Young, K., Ouchi, T., Yang, J. & Fetcenko, M.A. (2011c) Studies of off-Stoichiometric AB_2 Metal Hydride Alloy: Part 2. Hydrogen Storage and Electrochemical Properties. *International Journal of Hydrogen Energy*, Vol.36, pp.11146-11154, ISSN 0360-3199

Part 4

A Stoichiometric Approach to the Analysis of Metal Oxides Properties

7

Determination of Thermodynamic and Transport Properties of Non-Stoichiometric Oxides

Mauvy Fabrice[1] and Fouletier Jacques[2]
[1]*ICMC Bordeaux, CNRS, Université de Bordeaux, UPR 9048, 33608 Pessac, Cedex,*
[2]*LEPMI, UMR 5279, CNRS, Grenoble INP, Université de Savoie, Université Joseph Fourier, BP75. 38402 Saint Martin d'Hères, France*

1. Introduction

The purpose of this chapter is to describe methods for determining the oxygen stoichiometry (bulk and surface oxygen activities) and transport properties in non-stoichiometric oxides using solid electrolyte cells. These oxides are generally referred to as "*mixed ionic-electronic conductors*" (*MIEC*). It is customary to consider that in a MIEC, the ionic or electronic transport number is higher than 0.01 and that the total electrical conductivity is sufficiently high, i.e., higher than 10^{-5} S.cm^{-1}. Emphasis will be given on sources of error. Devices for overcoming these sources of error will be described. As the aim of this chapter is limited and cannot be an exhaustive review, the readers are referred to excellent general handbooks dealing with solid state electrochemistry, with chapters devoted to electrochemistry of non-stoichiometric oxides [Gellings & Bouwmeester, 1997; Kharton, 2011; Rickert, 1982; Sorensen, 1981] or review papers [Heyne, 1982; Weppner & Huggins, 1978].

2. Control and measurement of oxygen activity

Precise control and monitoring of the oxygen pressure in the experimental chamber is required for the determination of thermodynamic and transport properties in MIECs. Electrochemical devices have been developed since more than thirty years, allowing the control of the oxygen pressure in the 1 - 10^{-27} bar range in various gas mixtures or under partial vacuum. Solid electrolyte microprobes have also been proposed for the local determination of the oxygen activity on the surface of a non-stoichiometric oxide.

2.1 Measurement of oxygen pressure in a gas phase and under partial vacuum

The conventional oxygen sensor, shown Figure 1 a, is based on a closed yttria-stabilized zirconia tube [Kleitz et al., 1992]. The electrodes are made of platinum paste. The outer electrode, in contact with air, forms the reference electrode. The cell *emf* obeys the Nernst law:

$$E_{th} = \frac{RT}{4F} \ln \frac{P_{O_2}}{P_{(O_2)_{air}}} = \frac{RT}{4F} \ln \frac{P_{O_2}}{0.2} \quad (1)$$

A miniaturized sensor (Figure 1 b) with enclosed metal-metal oxide reference system (Pd-PdO, Co-CoO) has been developed and commercialized[1] [Fouletier & Vitter, 1980]. The main advantage of this device concerns the possibility of in situ measurements without circulation of the analyzed gas.

It has been shown that under partial vacuum or for high temperature operation, the sensor could become inaccurate due to the disturbing effect of the oxygen semipermeability flux through the zirconia tube, reaching the measuring electrode. To overcome the problem, it has been proposed to use the "zirconia point electrode" shown in Figure 1 c. The platinum measuring electrode is replaced by a conical piece of zirconia (see also Figure 4 a) pressed in contact with the inner surface of the zirconia tube, which is not platinized. The actual measuring electrode is a platinum coating deposited on the top of the zirconia probe. In this design, due to the shape of the probe, the oxygen semipermeability flux does not reach the measuring electrode since the current follows the path of lowest resistance. Consequently, the measuring electrode remains in equilibrium with the gas phase. As developed in Section 2.2.1, this zirconia point electrode can be used for the measurement of oxygen activity on the surface of a non-stoichiometric oxide.

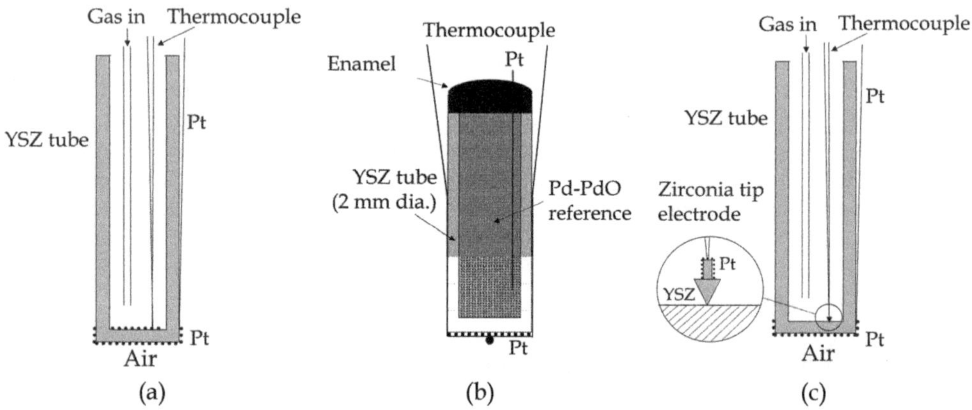

Fig. 1. Schematic drawing of oxygen sensors: (a) conventional type (air reference electrode), (b) minisensor with Pd-PdO reference system, (c) sensor with zirconia tip measuring electrode, [Fouletier, 1982/83].

2.1.1 Pump-sensor device

This device is widely used for the control and monitoring of oxygen content in gas mixtures flowing in the experimental chamber. The gas circuit is schematized in Figure 2 a. When nominally pure gases (Ar, N_2, He, etc.) are used, according to the Faraday law, the oxygen mole fraction X in the flowing gas obeys the following equation:

$$X = X^{\circ} + 0.209\, I/D \tag{2}$$

[1] SETNAG, Marseille

where X° is the oxygen mole fraction in the gas supplied to the pump, D is the gas flow rate in $L.h^{-1}$ NTP and I is the current intensity in A. As shown in Figure 2 b, the theoretical equation (2) is verified in the 10^{-7} - 1 mole fraction range: the oxygen mole fraction is determined by the oxygen sensor (using Nernst equation (1)) for various current intensities I passing through the pump [Fouletier et al., 1975].

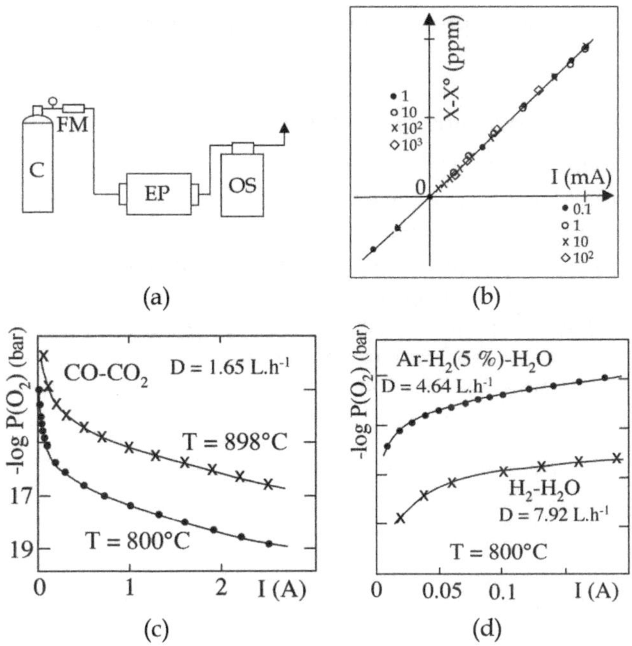

Fig. 2. (a): Gas circuit, C: gas container, FM: flowmeter, EP: electrochemical pump, OS: oxygen sensor; (b): Faraday's law test ($T_{sensor} = 900°C$, $X° = 9.1 \times 10^{-7}$, $D = 11.1$ $L.h^{-1}$, experimental slope: 1.96×10^{-2} A^{-1}, theoretical slope: 1.88×10^{-2} A^{-1}); (c): Reduction of flowing carbon dioxide for two temperatures of the sensor, (d): Oxidation of Ar – H_2 (5 %) mixture and pure hydrogen. The full lines are the theoretical curves, according to equations (3) and (4) [Caneiro et al., 1981; Fouletier et al., 1984].

The same device can be used for monitoring the composition of CO_2-CO and Ar-H_2O-H_2 mixtures. In that case, pure (H_2, CO_2) or premixed gases, such as Ar – 5 % H_2 are used. According to the Faraday's law, the equilibrium oxygen pressure vs. the current intensity passing through the pump obeys the following equations:

CO-CO_2 mixture

$$P_{O_2} = \left(2.392\frac{D}{I} - 1\right)^2 \exp\left(21.05 - \frac{68150}{T}\right) \text{(in bar)} \quad (3)$$

Ar-H_2-H_2O mixture

$$P_{O_2} = \left(2.392q\frac{D}{I} - 1\right)^{-2} \exp\left(13.278 - \frac{59571}{T}\right) \text{(in bar)} \quad (4)$$

where q is the hydrogen mole fraction in the feed gas, and T is the absolute temperature in the experimental chamber.

The association of an electrochemical oxygen pump and an oxygen sensor allows the monitoring of oxygen partial pressure in a flowing gas in the range 1 bar - 10^{-27} bar with an accuracy of 2 %.

2.1.2 Oxygen pressure domains of ideal response

It should be pointed out that the accuracy of the equilibrium oxygen pressure control depends on the *buffer capacity* δ of the gas. The buffer capacity can be defined as the number of moles of oxygen required for changing the chemical potential of 1 kJ/mole of gas. The variation of δ with oxygen pressure, at various temperatures, for Ar-O_2 and CO_2/CO mixtures is given in Fig. 3. It has been shown previously that the buffer capacity of the gas has to be higher than 10^{-6} mole, at 800°C. Consequently, at this temperature, the pressure domains in which the oxygen pressure is accurately controlled are: 1 - 10^{-6} bar in Ar-O_2 mixtures and 10^{-10} - 10^{-27} bar in CO-CO_2 mixtures. Obviously, these pressure domains depend on temperature. The same type of curve is obtained with H_2/H_2O mixtures.

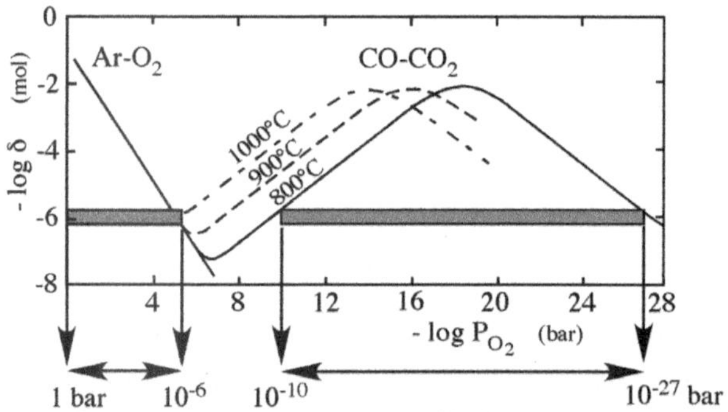

Fig. 3. Variation of the buffer capacity of inert gas-O_2 and CO-CO_2 mixtures, with indication of the oxygen pressure domains of correct utilization of the oxygen sensor at 800°C [Fouletier, 1982/83].

2.2 Control and measurement on an oxide surface

Basically, the oxide ion activity in a MIEC can be determined using the electrochemical cell:

Pt - Reference system / oxide electrolyte / MIEC - Pt. The measurement of the emf of the electrochemical chain allows the determination of the oxygen activity in the MIEC. The following sections are devoted to the description of devices allowing the determination of the oxide activity on the surface of a MIEC, and of the deviation from equilibrium of the oxide surface.

2.2.1 Metallic and solid electrolyte probes

It has been shown that in case of oxygen transfer through the studied sample, the activity of oxygen $a^*_{O_2}$ on the surface of the pellet can be noticeably different from that in the bulk. The oxygen activity $a^*_{O_2}$ can by defined as an equivalent oxygen pressure $P^*_{O_2}$. This surface activity can be measured using a ceramic point electrode (either a zirconia- or a ceria-based microprobe, Figure 4 a), as depicted in Section 2.1. The schematic drawing of the cell is given, Figure 4 b. The oxygen activity on surface of the sample (MO) results from the balance of the oxygen adsorption and desorption fluxes and the oxygen flux through the sample. Consequently, the oxygen activity on the surface of the sample can be very different from the oxygen partial pressure in the gas ($a^*_{O_2} = P^*_{O_2} \neq P_{O_2}$) and it can be written:

$$\frac{1}{2}O_{2(\text{surface})} + 2e^-_{MO} = O^{2-}_{MO} \text{ leadingto}: \frac{1}{2}\mu^*_{O_2} + 2\mu_e^{MO} = \mu^{MO}_{O^{2-}} \quad (5)$$

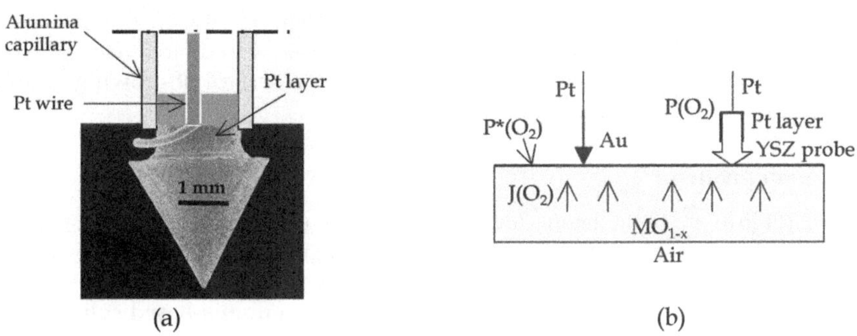

Fig. 4. (a): Ceramic point electrode; (b): Schematic drawing of the cell allowing the measurement of the oxygen activity on the surface of a MIEC.

The potential change in the electrochemical chain can be summarized as ($E = \phi^{Pt\text{-}I} - \phi^{Pt\text{-}II}$):

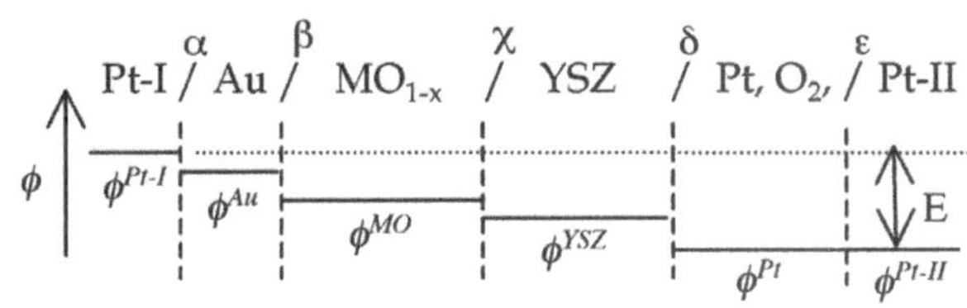

Fig. 5. Qualitative variation of the potential within the chain.

Gold is used as metallic point electrode due to its poor catalytic activity for oxygen electrode reaction; consequently, gold can be considered as an electronic probe on the MIEC and: $\tilde{\mu}_e^{Au} = \tilde{\mu}_e^{MO}$. It is also assumed that there is no oxygen flux at the interface between the ceramic point electrode and the pellet ($\tilde{\mu}^{MO}_{O^{2-}} = \tilde{\mu}^{YSZ}_{O^{2-}}$). The MIEC surface is considered as an equipotential. The emf E of the chain can be easily obtained from the following assumptions:

- Electronic equilibrium at interfaces α and β: $\tilde{\mu}_e^{Pt-I} = \tilde{\mu}_e^{Au} = \tilde{\mu}_e^{MO}$

- Ionic equilibrium at the interface χ: $\tilde{\mu}_{O^{2-}}^{MO} = \tilde{\mu}_{O^{2-}}^{YSZ}$
- Electrochemical equilibrium at the interface δ:

$$\frac{1}{2}O_{2(gas)} + 2e^{-}_{Pt} = O^{2-}_{YSZ} \quad \text{leading to}: \frac{1}{2}\mu_{O_2} + 2\tilde{\mu}_e^{Pt} = \tilde{\mu}_{O^{2-}}^{YSZ}.$$

- Electronic equilibrium at the interface ε: $\tilde{\mu}_e^{Pt-II} = \tilde{\mu}_e^{Pt}$

Considering that $\mu_{O_2} = RT\ln P_{O_2}$ and $\mu^*_{O_2} = RT\ln P^*_{O_2}$ we obtain easily the emf E of the chain:

$$E = \phi^{Pt-I} - \phi^{Pt-I1} = \frac{RT}{4F}\ln\left[\frac{P^*_{O_2}}{P_{O_2}}\right] \tag{6}$$

Obviously, if the surface of the sample is in equilibrium with the gas phase, the emf is nil. Provided the oxygen pressure in the gas phase near the surface is known (as an example using a microsensor described in Section 2.1), the oxygen activity on the surface can be determined (see Section 4.1). Similar designs using calcia-stabilized zirconia cells have been tested for continuous monitoring of oxygen activity on the surface of growing scale during high temperature oxidation of metals [Akida et al., 2008].

2.2.2 Oxygen minisensor

Mini-probes (Figure 6) have been developed for the determination of thermodynamic properties of MIEC as a function of composition. Zirconia- or thoria-based tubes, a few mm in diameter, have been used. A metal-metal oxide system serves as a reference. Such cells have been used for the measurement of oxygen potential in urania-based solid solutions or for continuous control of oxygen redistribution in UO_{2+x} under a thermal gradient [Ducroux et al., 1980; Une & Oguma, 1982].

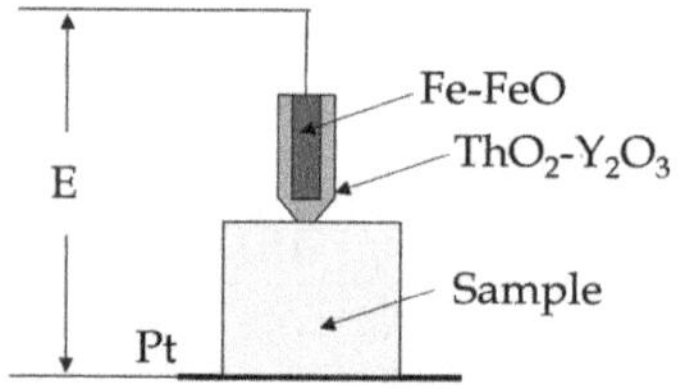

Fig. 6. Solid electrolyte mini-probe.

3. Determination of nonstoichiometry in oxides

Various methods based on the measurement of different physico-chemical characteristics have been developed. Most of the methods measure changes of the non-stoichiometry ratio x; however, with the help of a defect model, the absolute value of x can often be obtained [Riess & Tannhauser, 1982]. Thermogravimetry under controlled oxygen pressure is the most widely used method [Caneiro et al., 1982, 2011; Kharton et al.]. Other methods are based on the coulometric titrations. The oxygen activity in MIECs can also be determined using potentiometric cells involving a solid electrolyte. In the following sections, only the techniques involving solid electrolytes designs will be described.

3.1 Thermogravimetric investigation under controlled oxygen partial pressure

Thermal stability and oxygen content versus temperature and oxygen partial pressure are very important parameters in the case of non-stoichiometric compounds. Oxygen non-stoichiometry (δ) is generally measured using electronic microbalances under controlled atmospheres [Kharton et al., 2008; Tsipis et al., 2008]. About 1 g of powder sample is placed in a silica basket suspended by platinum wires from the beam of the microbalance. The oxygen partial pressure is controlled by the introduction of gas mixtures (Ar-O_2, Ar-H_2, dry or wet) into the sample chamber [Nakamura et al., 2009a], [Kiselev et al., 2011; Nakamura et al., 2009b]. First of all, the equilibrium between the sample and the surrounding gas phase is checked by controlling that both the weight of the sample and the oxygen partial pressure (recorded by a zirconia sensor introduced into the sample chamber) reached constant value. Figure 7a shows a typical weight change curve of $Nd_{2-x}NiO_{4+\delta}$ in Ar-H_2 mixed atmosphere to determine the starting oxygen content of the sample [Zhao et al., 2008].

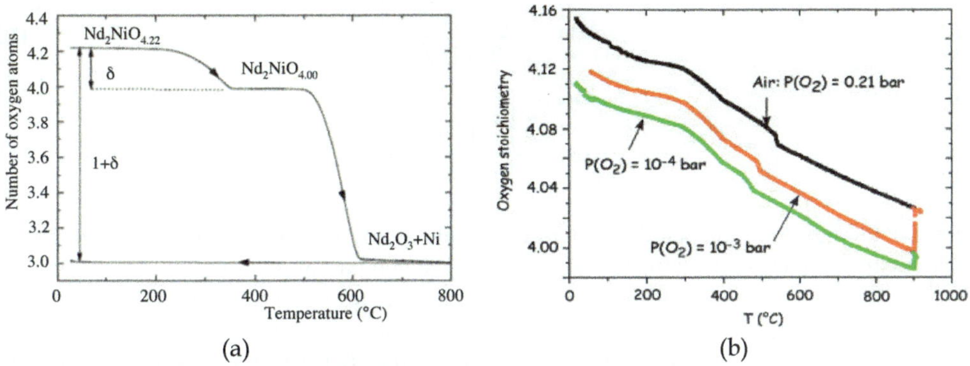

Fig. 7. Thermogravimetric analysis (TGA) (a) weight loss measurements during reduction of $Nd_2NiO_{4.22}$ in H_2 (5%) / Ar atmosphere ; (b) oxygen stoichiometry versus temperature of $Nd_{1.95}NiO_{4+\delta}$, under three oxygen pressures.

The change in oxygen content is determined from variation of the weight of the sample Δw_S according to the following relation:

$$\Delta\delta = (M_S/M_O).\ (\Delta w_S/w_S) \qquad (7)$$

where $\Delta\delta$, M_S, M_O, w_S are the variation of oxygen nonstoichiometry, the molar formula weight of the sample and oxygen atom and the weight of the specimen, respectively. The experimental error due to the buoyancy is negligibly small compared to the weight variation of the sample due to release or incorporation of oxygen.

In the case of figure 7a, two weight changes are observed. The first one occurring at 350°C corresponds to the loss of interstitial oxygen and the reduction of Ni^{3+} to Ni^{2+}. The second weight loss is assigned to the total reduction of $Nd_2NiO_{4+\delta}$ to Nd_2O_3 and Ni metal. The absolute value of the oxygen content is determined from the weight change of the sample during the decomposition in H_2 atmosphere. The decomposition reaction can be expressed by:

$$Nd_{2-x}NiO_{4+\delta} \rightarrow (2-x)/2\ Nd_2O_3 + Ni + (2+2\delta+3x)/2\ O \qquad (8)$$

From the weight change, the δ value is deduced and the Ni^{3+} concentration can be estimated. This mixed valency parameter can be linked to the electronic conductivity of the compound.

The change of molar Gibbs energy $\Delta G°$ for reaction (8) represents the partial molar Gibbs energy of oxygen atoms per 1 mol and can be expressed as follows [Caneiro et al., 2011; Patrakeev et al 1995]:

$$\Delta G° = -\ 0.5\ RT\ Ln\ PO_2 \qquad (9)$$

where R, T and PO_2 are the gas constant, the temperature and the oxygen partial pressure, respectively. Taking into account the Gibbs-Helmholtz equation, and equation (9), the changes of standard partial molar enthalpy $\Delta H°$ and entropy $\Delta S°$ of oxygen release process per 1 mol of oxygen atoms can be deduced from equations (10) and (11) respectively:

$$\Delta H° = -\frac{R}{2}\left[\frac{\partial \ln P_{O_2}}{\partial(1/T)}\right]_\delta \qquad (10)$$

$$\Delta S° = \frac{R}{2}\left[\frac{\partial(T.\ln P_{O_2})}{\partial T}\right]_\delta \qquad (11)$$

In order to determine the standard thermodynamic quantities, experimental dependences of the non-stoichiometry (δ) versus temperature and oxygen partial pressure are determined (see Fig. 7b) [Mauvy et al., 2009]. If the plots into the coordinates R/2 Ln PO_2 vs. $(1/T)_\delta$ and RT/2 Ln PO_2 vs. $(T)_\delta$ give linear relationship within the values of δ studied, it allows to access to standard partial molar enthalpy $\Delta H°$ and entropy $\Delta S°$ (Fig. 8a). This linear behaviour indicates that the values of standard thermodynamic quantities are essentially independent of the temperature in the working temperature range. These values can be estimated from linear regression coefficients at given oxygen content. Figure 8b shows the calculated values of the standard enthalpy $\Delta H°$ for oxygen release process for $Nd_{1.95}NiO_{4+\delta}$ compound.

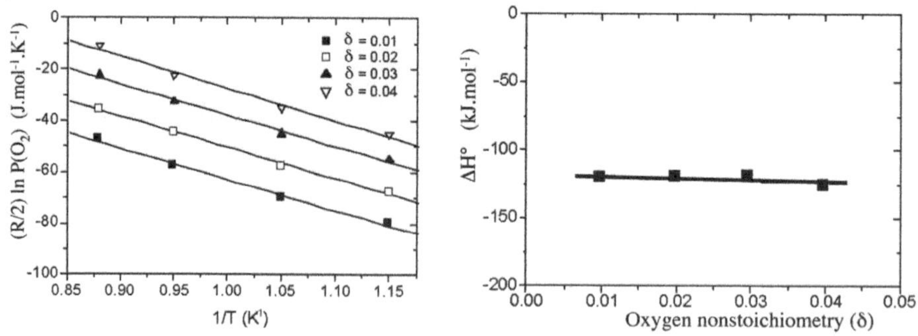

(a) (b)

Fig. 8. (a) R/2 Ln PO_2 vs. (1/T)□ plots of $Nd_{1.95}NiO_{4+\delta}$; (b) partial molar enthalpy of $Nd_{1.95}NiO_{4+\delta}$.

It can be noticed that a linear behaviour of both thermodynamic parameters ΔH° and ΔS° versus δ, can be related to a random distribution of non-interacting point defects as components of ideal solution. Generally, when the nonstoichiometry increases, noticeable deviation from linearity can be observed. In the case of $Nd_{1.95}NiO_{4+\delta}$ oxide, ideal-solution-like state means that the interaction among defect species is nearly constant regardless of the defect concentration.

3.2 Pump-sensor device

The pump-sensor device described in section 2.1.1 can be used for the determination of the stoichiometry ratio in oxides using the device shown Fig. 9 a [Meas et al., 1978].

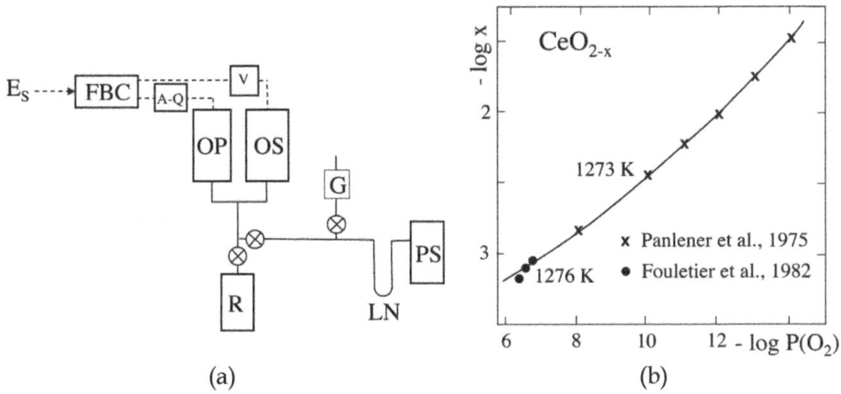

Fig. 9. (a): Experimental device for the study of gas-solid reaction under constant oxygen pressure using a pump-sensor device, G: vacuum gauge, LN: liquid nitrogen trap; (b): Variation of the stoichiometry ratio of CeO_{2-x} at 1273 K.

The experimental reactor R is connected to an oxygen sensor OS (using a zirconia probe as measuring electrode), an oxygen pump OP and a mechanical pumping system PS. A feed-back controller FBC compares the sensor emf E given by the sensor to a set value E_S and adjust the oxygen pressure in the experimental vessel by passing an appropriate current through the electrochemical pump. As a preliminary treatment, the system was initially outgassed under a pressure lower than 10^{-8} bar and then equilibrated with a well-controlled low oxygen pressure, typically 10^{-7} bar. Starting from this equilibrium state, a typical experiment is simply performed by changing the set voltage E_S and recording the resulting pumping current passing through the oxygen pump and its integral which is proportional to the oxygen amount exchanged between the solid and the gas. Following an identical procedure, measurements were first carried out without sample in the reaction vessel. The quantities of oxygen involved in the gas-solid equilibrium are calculated by difference. This device has been used for oxygen adsorption studies on zeolite or on stabilized zirconia [Meas et al. 1978]. As an example, in figure 9 b, is plotted the variation of the stoichiometry ratio of CeO_{2-x} as a function of oxygen partial pressure, at 1273 K [Fouletier et al., 1982].

3.3 Coulometric titration

The coulometric titration method can be defined as charge controlled mass transfer of mobile component between electrodes of electrochemical cell. This method is a very sensitive control of the composition of the nonstoichiometric phase, combined with emf measurements that yield very accurate thermodynamic information [Tretyakov et al., 1997].

As reported in Fig. 10, an yttria stabilized zirconia tube is used as electrolyte for a galvanic cell [Mizusaki et al., 1991; Nakamura et al., 2009a]. Platinum paste is painted on the outside of the tube to stick the Pt mesh whereas Pt mesh is attached to the sample powder as the inside electrode.

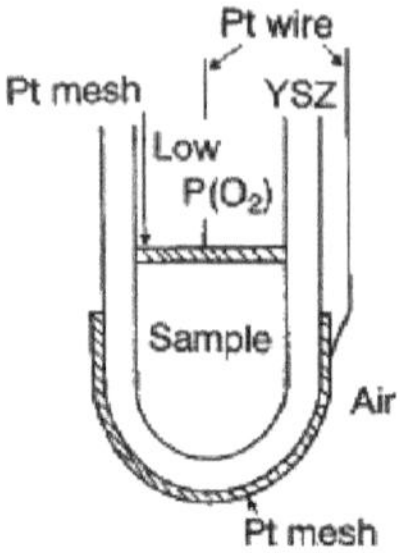

Fig. 10. Schematic diagram of the coulometric titration cell.

After evacuation and fulfilment with Ar gas procedures, the amount of oxygen, which is extracted or incorporated to the sample, is controlled by the electric charge passed through the cell. After specified amount of electric charge is applied, the electromotive force is measured to determine the equilibrium oxygen partial pressure in the tube. By considering that the amount of oxygen inside the tube is negligibly small, the oxygen amount, which migrated between the sample oxide and the gas phase, is small enough to be neglected. Then, $\Delta\delta$ can be calculated according to the relation:

$$\Delta\delta = (I.t) / (2.F.M_S) \quad (12)$$

where I , t, F and M_S are the current, the time, the Faraday constant and the molar weight of the sample, respectively.

Typical examples of coulometric titration curves are reported in Fig. 11 for $La_2NiO_{4+\delta}$ compound at different temperatures [Nakamura et al, 2009b].

It can be noticed that data points near the plateau of δ versus log PO_2 region contain larger uncertainty than other data points because equilibrium potential varies easily by the small variation of oxygen content near the plateau region. According to Wagner theory, the slope of the δ versus log PO_2 curve for nonstoichiometric compounds shows minimum value at the stoichiometric composition [Wagner, 1971].

Investigations with oxygen concentration cells revealed that the major problem is a non-electrochemical transport of oxygen through oxide ion electrolytes. Such leakage introduces uncontrolled and excessive changes in the oxide composition and appears to be the principal cause of instability.

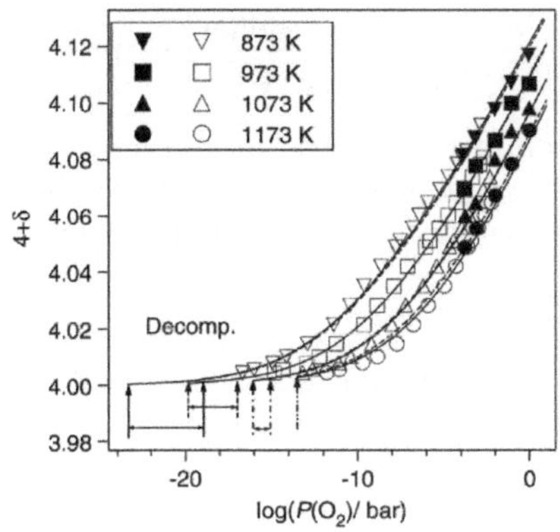

Fig. 11. Oxygen nonstoichiometry of $La_2NiO_{4+\delta}$. Open symbols and closed symbols are measured by coulometric titration and TG analysis respectively [Nakamura et al., 2009b].

3.4 Gas release method

The stoichiometry ratio in a non-stoichiometric oxide can be monitored by passing a current through it [Fouletier & Kleitz, 1978; Fouletier et al, 1982]. The experimental set-up is schematized, Fig. 12a. It consists of an inert gas cylinder, an oxygen pump, the experimental vessel and an oxygen sensor. The sample is a cylindrical pellet with platinum electrodes deposited on its bases. The sample is electrochemically reduced by passing a direct current. The amount of oxygen extracted from the sample, and the corresponding variation of the stoichiometry ratio, is determined by integration of the oxygen content deduced from the downstream oxygen sensor emf (Fig. 12b) [Levy et al., 1988].

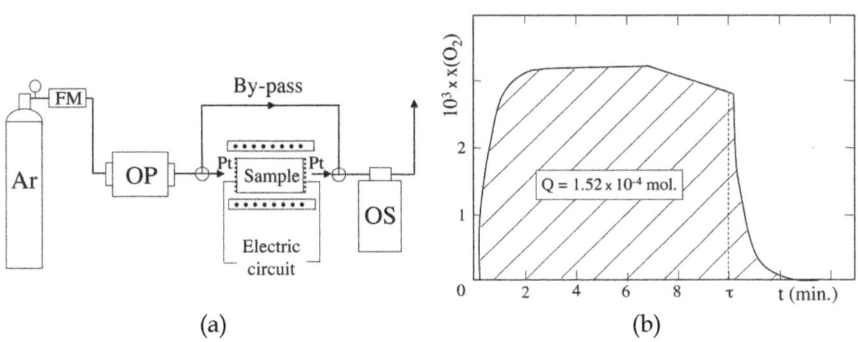

Fig. 12. (a): Experimental set-up for the monitoring of the stoichiometry ratio in a MIEC, FM: flowmeter, OP: oxygen pump, OS: oxygen sensor; (b): Variation of the oxygen mole fraction $x(O_2)$ in the flowing gas during reduction process (ZrO_2 - 12 m/o Y_2O_3, T = 1015 K, $I_{reduction}$ = 0.01 A, gas flow rate : 1.8×10^{-3} mol.s^{-1}).

As an example, for the oxide $Zr_{1-y}Y_{2y}O_{2+y-x}$, (y is the dopant concentration), the stoichiometry ratio x is calculated from the equation:

$$x = 0.4\frac{M_S D}{m(1+y)V_m}\int_0^t\left(\exp\left[\frac{4FE(t)}{RT}\right]-\exp\left[\frac{4FE(0)}{RT}\right]\right)dt \qquad (13)$$

where E(t) represents the emf of the oxygen sensor during the reduction process, E(0) is the sensor emf before electrochemical reduction, D the gas flow rate, V_m the volume of a mole of gas, M_S the sample molecular weight, and m the sample weight.

3.5 EMF method

According to the pioneering works of Kiukkola & Wagner [Kiukkola & Wagner, 1957], the electrochemical chain is the following: Pt - Reference system / oxide electrolyte / MIEC - Pt. As described in Section 2.2.1, the emf of the cell allows the measurement of the oxygen activity in the MIEC, provided that there is no mass transfer through the oxide surface. The reference system is either a gas (pure oxygen, air) or a metal-metal oxide mixture (Fe-FeO, Cu-Cu_2O, Pd-PdO, etc.). Examples of measurements on non-stoichiometric oxides are numerous [Mari et al., 1977; Nakamura & Fujino, 1987; Otobe et al., 2009; Porat & Riess, 1994].

The two main problems concern the reactivity between the MIEC and the solid electrolyte and the appearance of electronic conductivity in the solid electrolyte inducing polarization phenomena. The appropriate choice of the solid electrolyte can noticeably reduce the chemical reactivity: as an example, ceria-based electrolyte or apatite is less reactive than stabilized zirconia [Mauvy et al., 2009]. Double-electrolyte cells have been proposed to extend the oxygen activity range of thermodynamic cell measurements compared to the range of a single-electrolyte cell arrangement [Shores & Rapp, 1971; Tretyakov & Muan, 1969].

We will focus on the use of cone-shaped MIEC. The experimental set-up is schematized, Figure 13a, and a photograph of a nickelate point electrode is shown, Figure 13b. Air is used as a reference electrode, and the cone-shaped MIEC is gently pressed in contact with the

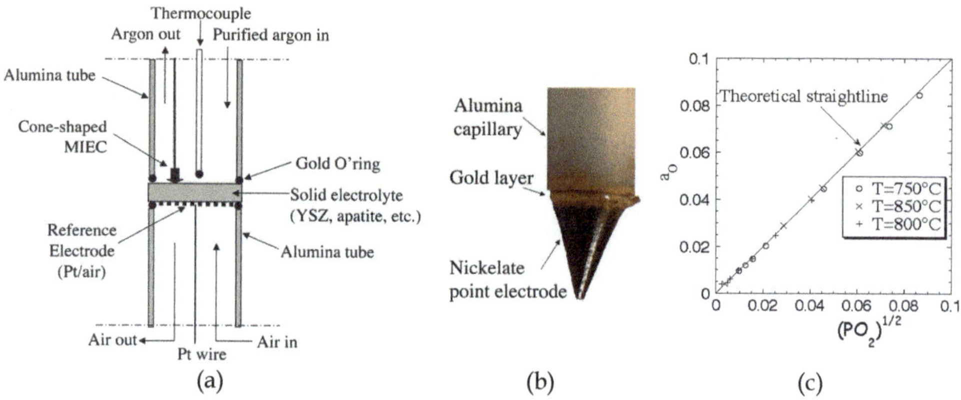

Fig. 13. (a): Experimental set-up for emf measurement; (b): Photograph of a nickelate point electrode ($Ni_{1.95}NiO_{4+\delta}$); (c): Variation of the oxygen activity coefficient in the nickelate as a function of temperature for various oxygen activity values in the gas phase [Mauvy et al., 2009].

solid electrolyte (zirconia-based pellet, apatite, etc.). If available, a solid electrolyte closed-end tube can be used. The main advantages concern the possibility of simultaneous measurements with various MIEC materials, the ease to handle, the reactivity between the MIEC and the solid electrolyte is minimized, a reduced response time to reach thermodynamic equilibrium.

As an example, in Figure 13c is plotted the variation of the oxygen activity (a_O) in the nickelate phase $Ni_{1.95}NiO_{4+\square}$ as a function of the activity of oxygen in the gas ($P_{O_2}^{1/2}$). The equilibrium between the nickelate and the gas phase is observed in the whole investigated temperature range.

4. Transport properties investigations in MIEC oxides

Because of the charge equilibrium within the oxide, the electronic partial conductivity (electrons or holes) σ_{el} and the ionic partial conductivity σ_i are linked to the oxygen non-stoichiometry and, consequently, are two key parameters characterizing a MIEC. The methods developed for deconvoluting the ionic and electronic contributions to the total conductivity of a MIEC allow either the measurement of the partial conductivities, the diffusion coefficient of mobile defects or the transport numbers, either t_{el} or t_{ion}. Only the methods involving a solid electrolyte cell are briefly described. Reviews papers can be referred [Heyne, 1982; Rickert, 1982; Riess, 1997; Weppner & Huggins, 1978].

4.1 Oxygen permeation

Assuming that there is no oxygen gradient in the gas phase, the overall oxygen permeation rate through a non-stoichiometric oxide may be schematically decomposed into three elementary steps: ionic defect bulk diffusion (step 1), surface exchange between oxygen and oxygen vacancies (step 2 (high pressure side, $P_{O_2}^{rich}$) and step 3 (low pressure side, $P_{O_2}^{lean}$)). As previously proposed [Bouwmeester et al., 1992] the membrane may be divided into three zones, as schematically shown in Fig. 14a. $P_{O_2}^{rich}$ and $P_{O_2}^{lean}$ are the oxygen pressure in the gas phase, $P_{O_2}^{*rich}$ and $P_{O_2}^{*lean}$ are the actual corresponding oxygen activities on both surfaces of the membrane.

The experimental cell for oxygen permeation measurements, including metallic and ceramic point electrodes is shown Figure 14 b.

The chemically driven oxygen flux through a mixed-conducting oxide was first modelled using Wagner's theory, assuming that both oxide surfaces are in equilibrium with the imposed gas atmospheres ($P_{O_2}^{rich} = P_{O_2}^{*rich}$ and $P_{O_2}^{lean} = P_{O_2}^{*lean}$) [Heyne, 1977; Wagner, 1957].

According to this theory, the steady oxygen permeation flux density j_{O_2} (in mol.cm^{-2}.s^{-1}) controlled by bulk diffusion in a mixed conductor is given by:

$$j_{O_2} = \frac{RT}{16F^2L} \int_{P_{O_2}^{lean}}^{P_{O_2}^{rich}} \frac{\sigma_i \sigma_e}{(\sigma_i + \sigma_e)} d\ln P_{O_2} \tag{14}$$

where all the symbols have their original meaning and L is the membrane thickness.

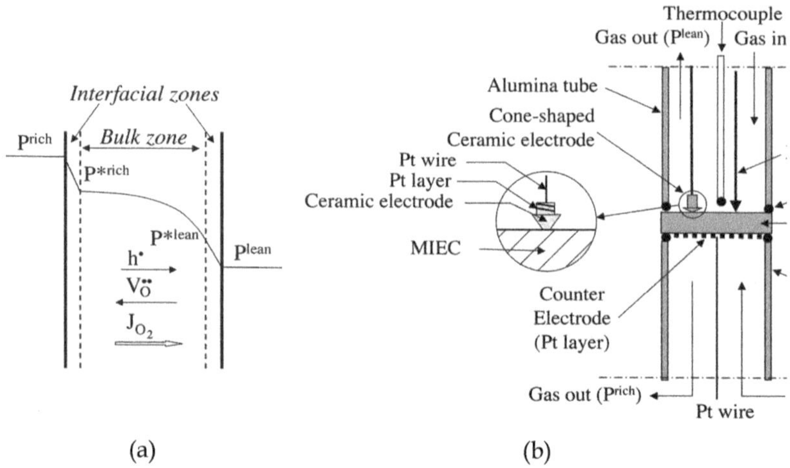

Fig. 14. (a) Schematic drawing of the oxygen pressure gradients within a permeation membrane, (b) Permeation set-up, with magnified drawing of the ceramic point electrode.

In the case of mixed conductors in which the electronic conduction predominates, i.e., $\sigma_e >> \sigma_i$, the integral in the Wagner relation involves only σ_i over the oxygen partial pressure gradient:

$$j_{O_2} = \frac{RT}{16F^2L} \int_{P_{O_2}^{lean}}^{P_{O_2}^{rich}} \sigma_i \, d\ln P_{O_2} \tag{15}$$

Two cases can be considered:

- Taking an average value for σ_i or assuming σ_i to be constant simplifies the equation so that j_{O_2} is directly proportional to σ_i and $\ln\left(P_{O_2}^{rich} / P_{O_2}^{lean}\right)$:

$$j_{O_2} = \frac{RT\sigma_i}{16F^2L} \ln \frac{P_{O_2}^{rich}}{P_{O_2}^{lean}} = \frac{J_{O_2}}{L} \ln \frac{P_{O_2}^{rich}}{P_{O_2}^{lean}} \tag{16}$$

According to Möbius [Möbius, 1986], j_{O_2} is the oxygen permeation flux density and J_{O_2} is the specific oxygen permeability (in $mol.cm^{-1}.s^{-1}$).

- In case of simple defect model, neglecting the formation of defect association, the oxygen nonstoichiometry δ and the conductivity σ_i are proportional to $P_{O_2}^{1/n}$. Substitution into Eq. 15 with subsequent integration leads to a simplified expression for oxygen permeation:

$$J_{O_2} = \frac{\alpha}{L}\left[P_{rich}^{1/n} - P_{lean}^{1/n}\right] \text{ with } \alpha = \frac{RT\sigma_i^{\circ}}{16F^2} \tag{17}$$

σ_i° is the value of the ionic conductivity at unit oxygen pressure.

As indicated by Bouwmeester [Bouwmeester et al., 1994], for high values of n, Eq. (16) is obtained by expanding Eq. (17) as a power series and truncation after the first term.

Kleitz et al. [Kleitz et al., 1973; Fouletier et al., 1975] were the first to demonstrate that the existence of a nonvanishing semipermeability flux through a solid electrolyte induces a deviation from equilibrium on both sides of the membrane. They expanded the Wagner theory to account for partial control of surface reactions on the transport kinetics through stabilized zirconia. This approach has been applied to mixed ionic-electronic oxides [Bouwmeester et al., 1994; Chen et al., 1997; Geffroy et al., 2011; Xu & Thomson, 1999].

According to equation 16, in case of negligible effect of the surface exchange rate, the oxygen permeation flux density j_{O_2} is proportional to $[\ln(P^{rich}/P^{lean})]$, and the specific oxygen permeability J_{O_2} is independent of the membrane thickness [Kharton et al., 1999].

In case of limiting effect of the surface exchange rate, the specific oxygen permeability increases with the membrane thickness due to a decreasing role of the exchange kinetics (see Figure 15).

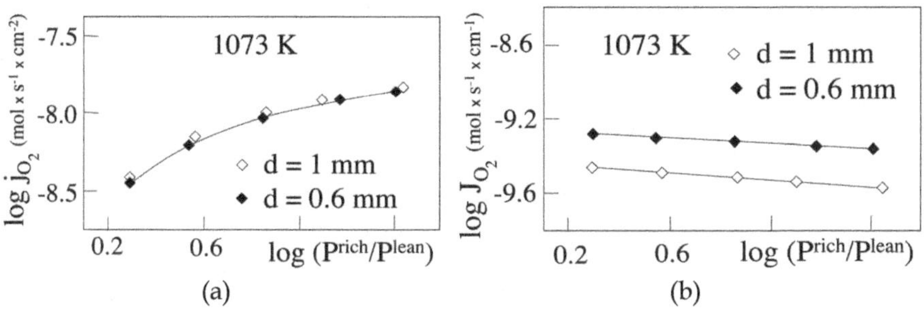

Fig. 15. (a): Variation of the oxygen permeation flux density (a) and of the specific oxygen permeability (b) with the oxygen pressure gradient, from [Kharton et al., 1999].

Another approach was to define the ratio h = k/D* where k is the surface exchange coefficient and D*, the tracer diffusion coefficient [Carter et al., 1992; Steele, 1992] or a critical thickness L_d (which is the reciprocal of the parameter h), at which the oxygen flux is half of that expected for a diffusion controlled process [Bouwmeester et al., 1994)]. The higher the L_d characteristic thickness (or the smaller the h value), the more the surface exchange contributes to the oxygen permeation flux. Examples of L_d values are given in [Gellings & Bouwmeester, 1997, p. 505]: L_d varies from 0.2 μm for $La_{0.5}Sr_{0.5}MnO_{3-\delta}$ to 0.03 cm for $La_{0.6}Sr_{0.4}Co_{0.4}Ni_{0.6}O_{3-\delta}$ at 700°C.

Oxygen permeation flux data for various MIECs have been recently compiled [Kharton, 1999; Sunarso et al., 2008].

The ionic conductivity can be deduced from the measured oxygen semipermeability flux using the following equation, in which the oxygen activities P* are taken into account instead of the oxygen pressures in the gas:

$$J_{O_2} = \frac{RT\sigma_i^{\circ}}{16F^2L}\left[P_{rich}^{*1/n} - P_{lean}^{*1/n}\right] \tag{18}$$

The oxygen activity on both sides of the pellet, i.e., $P_{O_2}^{*rich}$ and $P_{O_2}^{*lean}$, can be measured using ceramic point electrodes as described in section 2.2.1 [Geffroy et al., 2011; Mauvy et al., 2007].

Another approach has been to correlate the oxygen permeation flux to oxygen partial pressures [Xu & Thomson, 1999]. Assuming that the electronic conductivity is high and constant within the MIEC membrane, the oxygen flux can be written as;

$$J_{O_2} = \frac{D_V}{2L}\left([V_O^{\bullet\bullet}]^{lean} - [V_O^{\bullet\bullet}]^{rich}\right) \tag{19}$$

D_V is the diffusion coefficient of oxygen vacancies.

In case of mixed control of the oxygen flux (bulk diffusion and surface exchange), the concentration of oxygen vacancies at both faces of the membrane is governed by the reaction:

$$\frac{1}{2}O_2 + V_O^{\bullet\bullet} \xleftrightarrow{k_f, k_r} O_O^{\times} + 2h^{\bullet} \tag{20}$$

where k_f and k_r are the forward and reverse reaction rate constants.

Under stationary conditions, the oxygen permeation can be expressed as:

$$J_{O_2} = k_f\left(P^{rich}\right)^{1/2}\left[V_O^{\bullet\bullet}\right]^{rich} - k_r = k_r - k_f\left(P^{lean}\right)^{1/2}\left[V_O^{\bullet\bullet}\right]^{lean} \tag{21}$$

leading to:

$$J_{O_2} = \frac{D_V k_r\left[\left(P^{rich}\right)^{1/2} - \left(P^{lean}\right)^{1/2}\right]}{2Lk_f\left(P^{rich} \times P^{lean}\right)^{1/2} + D_V\left[\left(P^{rich}\right)^{1/2} + \left(P^{lean}\right)^{1/2}\right]} \tag{22}$$

The three parameters D_V, k_f and k_r can be estimated by fitting the experimental results as function the oxygen pressures on both sides of the membrane.

4.2 EMF – Faradic efficiency methods

The following cell has been used for the measurement of the ionic transport in a MIEC:

Pt, O_2 (P') / MIEC / Pt, O_2 (P")

According to Wagner theory, the emf of the cell obeys the equation:

$$E = \frac{RT}{4F}\int_{P©}^{P"} t_{ion}\, d\ln P_{O_2} \tag{23}$$

where t_{ion} is the oxide ions transport number. In case of a pure ionic conductor, equation (23) becomes the Nernst law E_{th} (see equation (1)). Equation (23) can often be simplified as:

$$E = \overline{t}_{ion} E_{th} \tag{24}$$

where $\overline{t}_{ion}$ is the average ionic transference number.

This method has been applied to oxide electrolytes. However, it should be pointed out that the ionic transport number has to be not too small. In case of a non-stoichiometric oxide the resulting oxygen permeation flux will polarize one or both electrodes, and this method is not recommended.

The oxygen ionic transport number t_O can also be measured using the Faradic Efficiency method, i.e., t_O is the ratio between the oxygen ionic current and the total current driven through the sample by an applied electrical field. However, in case of noticeable electrode polarization, the measured transport number can differ from the actual value.

Kharton and co-workers [Kharton & Marques, 2001; Kharton et al., 2001; Kharton et al., 2007] have proposed modified EMF (EMF) and Faradic Efficiency (FE) methods and combination of these techniques (EMF-FE), taking into account the electrode polarization. The set-up is schematized in figure 16a. The cell includes an oxygen pump, an oxygen sensor and the studied membrane, all parts being separated by insulating layers. Figure 16b compares the results obtained with these techniques on $La_{0.9}Sr_{0.1}Ga_{0.8}Mg_{0.2}O_{3-\delta}$: due to polarization phenomena, the classical methods give an overestimated electronic transport number.

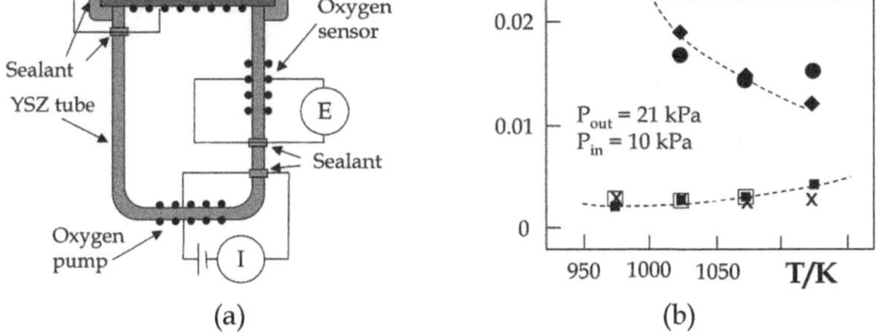

Fig. 16. (a): Electrochemical cells used for EMF and Faradic Efficiency measurements; (b) variation of the average electronic transport numbre of $La_{0.9}Sr_{0.1}Ga_{0.8}Mg_{0.2}O_{3-\delta}$ determined by different techniques, from [Kharton et al., 2007].

4.3 Patterson diagrams

This method was initially proposed by Patterson [Patterson, 1971]. It is based on the P_{O_2} dependence of the electrical conductivity. In Figure 17a is plotted the variation of log σ as functions of log (P_{O_2}) and of the reciprocal temperature. As shown in Figure 17b, at a given temperature, three oxygen pressure domains can often be defined. In the medium oxygen pressure range, the conductivity is constant and $\sigma = \sigma_i$; at low P_{O_2}, the conductivity increase is ascribed to an additional n-type electronic conductivity ($\sigma = \sigma_i + \sigma_n$) and at high P_{O_2} the conductivity increase is due to an additional p-type conductivity ($\sigma = \sigma_i + \sigma_p$). Assuming that the ionic conductivity is constant over the whole oxygen pressure range, the partial electronic conductivity can be easily calculated.

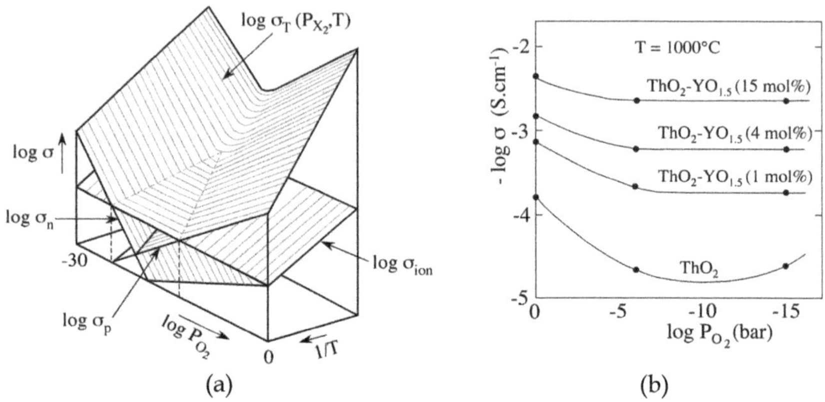

Fig. 17. (a): Schematic representation of log σ surfaces over log PO_2, 1/T space for $\sigma = \sigma_{ion}$, σ_n, σ_p and σ_{total}, from [Patterson, 1971]; (b): Log σ vs. log PO_2 for the system ThO_2-Y_2O_3.

This method can be considered as a fast and simple screening tool. It should be pointed out that the technique is convenient for electronic transport numbers higher than 1 % and when there is an oxygen partial pressure domain in which the conductivity is purely ionic.

4.4 Hebb-Wagner method

The Hebb-Wagner polarization technique has been developed either for the determination of electron and hole conductivity in ionic conductors [Hebb, 1952; Joshi & Wagner, 1975; Wagner, 1957] or for the measurement of ionic conductivity in MIECs [Riess, 1996; Wiemhöfer et al., 2002]. Basically, the method consists in using a reversible electrode and blocking electrodes to suppress the predominant charge carrier and thus enable measurement of the minority species. The main limitations of the method have been reviewed [Riess, 1996] and new experimental set-ups have been proposed.

The initially proposed cell is schematized in Figure 18a. Assuming that the MIEC is cationic conductor (mobile defects: $M_i^{\bullet}$), the interface M/MIEC (1) is a reversible electrode, and the interface C/MIEC (4) is an ion blocking electrode. The polarization of the voltage is chosen so that the mobile species tend to be depleted from the blocking electrode. Under steady-state condition, the ionic current is eliminated ($I = I_{el}$) and the theoretical I-V relation, according to the Wagner's theory is the following:

$$I = I_{el} = -\frac{RTS}{FL}\left[\sigma_e(0)\left(1-e^{-\frac{FV}{RT}}\right)+\sigma_h(0)\left(e^{-\frac{FV}{RT}}-1\right)\right] \tag{24}$$

$\sigma_e(0)$ and $\sigma_h(0)$ are the partial electron and hole conductivities in the MIEC at the interface (1), L and S are, respectively, the length and cross-sectional area of the MIEC.

By measuring the current through the MIEC sample and the applied voltage V, the electronic conductivity is determined.

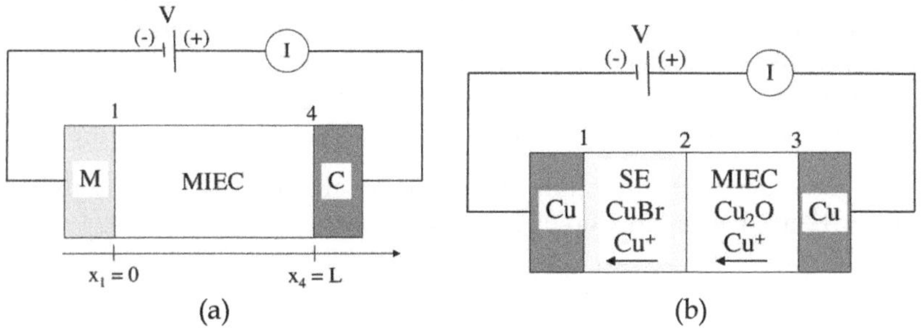

Fig. 18. (a): Two-point Hebb-Wagner polarization cell, C: ion-blocking electrodes, M: reversible electrode, MIEC: MIEC containing positive mobile ionic defects; (b): Two-point Hebb-Wagner polarization cell for the measurement of σ_i in Cu_2O from [Riess, 1992b].

An analogous method has been developed for determining the ionic conductivity in a MIEC. The blocking electrode for electronic defects is a solid electrolyte (SE), that conducts the same ions as the MIEC ($M_i^{\bullet}$ defects, is the experiment described, see Figure 18b):

$$(-)\ M/\text{Solid Electrolyte}\ (M_i^{\bullet})/\text{MIEC}\ (M_i^{\bullet}, e', h^{\bullet})/M\ (+)$$

Multi-electrode set-ups have been proposed by Riess, eliminating experimental problems such as errors due to overpotential at both electrodes or allowing the simultaneous measurement of σ_{el} and σ_i in a MIEC [Riess, 1992a, 1992b, 1996].

4.5 Short-circuit method ("zero driving force" method)

The ionic conductivity in a MIEC can be determined using the "short-circuiting" method [Riess, 1991]. Both electrodes are reversible systems with different compositions. The electronic current is brought to zero by short-circuiting the MIEC on a low impedance amperometer (V = 0). Consequently, it can be demonstrated that the ionic resistance R_i obeys the following equation [Riess, 1991]:

$$R_i = \frac{E_{th}}{I} \tag{25}$$

where E_{th} is the Nernst voltage determined by the compositions of the reversible electrodes, and I is the short-circuit current.

The main advantage of the method is that the contribution of electronic defects to the conductivity is eliminated without requirement of a blocking electrode. The sources of errors have been discussed [Riess, 1991, 1992b, 1997] and the method was applied to mixed conducting pyrochlores [Riess et al., 1992].

4.6 Conductivity vs. non-stoichiometry ratio

The electrical conductivity of electrochemically-reduced oxides is measured by impedance spectroscopy. The gas circuit is schematized in Figure 19a. During the first stage (indexes 1) the sample is reduced electrochemically (according to the procedure described in Section 3.2.3) and the stoichiometry ratio is determined using the oxygen sensor. Then the gas flowing in the experimental cell is purified using an oxygen getter (indexes 2). It has been checked that the sample composition remains constant over several days. The electrical conductivity of the reduced sample is measured as a function of temperature.

Figure 19b gives the Arrhenius plot of the electrical conductivity of stabilized zirconia (ZrO_2 - Y_2O_3, 12 m/o) as a function of the stoichiometry ratio.

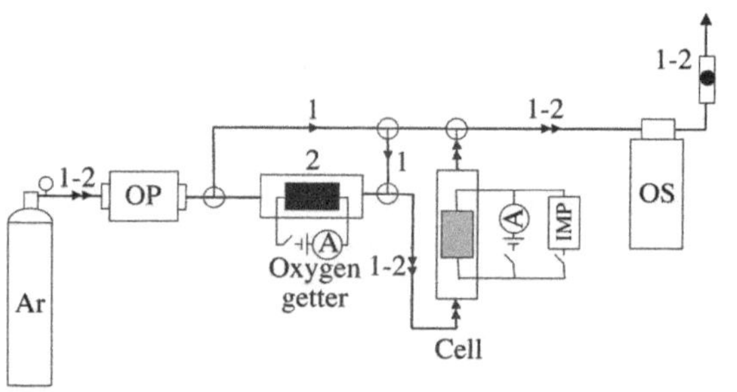

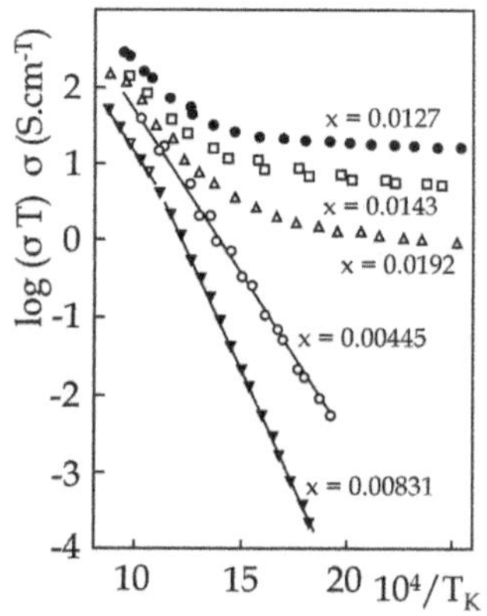

Fig. 19. (a) Gas circuit, OP: oxygen pump, OS: oxygen sensor, IMP: impedancemeter, (1): reduction process, (2): conductivity measurement; (b) Arrhenius plot of the electrical conductivity for various non-stoichiometry ratio x (from Levy et al., 1988).

4.7 Relaxation methods

In chemical relaxation experiments, an abrupt change of chemical potential of one of the constituent elements, usually PO_2 for oxide materials is imposed on a sample under constant temperature. Physical properties such as weight and volume of the sample are recorded versus time until a new thermodynamically equilibrium state is reached. Because electrical conductivity is much more sensitive to change of oxygen chemical potential in the atmosphere than are the other properties as weight, considerable changes in conductivity can be observed even when the oxygen partial pressure change of the corresponding nonstoichiometry is very small. This makes the conductivity relaxation method more easily applicable to a wider variety of materials than other methods such as thermogravimetry analysis [Lane et al., 2000; Ma et al. 1997; ten Elshof et al., 1997]. The transient behaviour in the re-equilibration process is recorded and analyzed by fitting the relaxation data to the solution of Fick's second law with appropriate boundary conditions [Crank, 1975].

A classical experimental setup, used for this type of study, is schematically drawn in Fig. 20a. Electrical conductivity relaxation experiments are performed using the four-probe method. Four platinum wires are connected to the sample using platinum paste (see Fig. 20b). Various atmospheres surrounding the sample are obtained by flowing a mixture of oxygen and nitrogen into the sample chamber, using two mass flow controllers or by introducing air. In both cases, the same flow rate is used. During the change of atmosphere, the flow rate is high enough to stabilize PO_2 inside the cell within a short duration compared with the relaxation time. Oxygen partial pressure steps of small amplitude are required to assign the relaxation process to a defined oxygen partial pressure. Moreover, this amplitude has to be small enough to assume constant the mobility of the defects during the relaxation process.

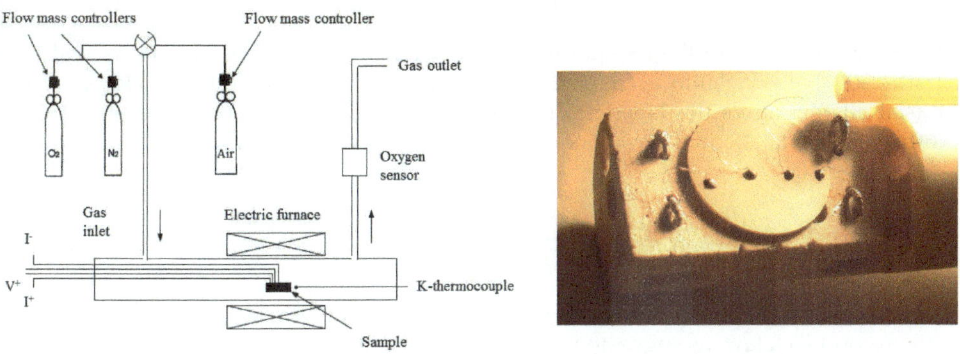

(a) (b)

Fig. 20. Schematic drawing of the experimental setup (a) and picture (b) of the 4-electrode sample holder.

In the case of $La_2Cu_{0.5}Ni_{0.5}O_{4+\delta}$ compound, oxygen is incorporated into the crystal lattice during the oxidation process or released from the lattice during the reduction step, as it can be observed on Fig. 21 when the sample is subjected to a sudden change of PO_2 [Mauvy et al., 2004].

During the relaxation process, nonstoichiometry spreads through the sample by lattice diffusion, which is driven by concentration gradients of defects. The ionic conductivity of $La_2Cu_{0.5}Ni_{0.5}O_{4+\delta}$ can only be contributed by oxygen vacancies ($V_O^{\bullet\bullet}$) and interstitial oxygen ions (O_i''). Theoretically, the oxygen vacancy concentration should decrease with increase in oxygen partial pressure. So, if the oxygen vacancy is the controlling defect, then one should expect the ionic conductivity to decrease with increase PO_2. However, experimental observations (see Fig. 21), indicate that ionic conductivity increases. This suggests that O_i'' is the predominant defect in this working conditions (high PO_2). Because the mobility of cationic defects is much lower than that of oxide defects and holes, chemical diffusion can be considered as the process of diffusion of interstitial oxide ions and counterdiffusion of holes.

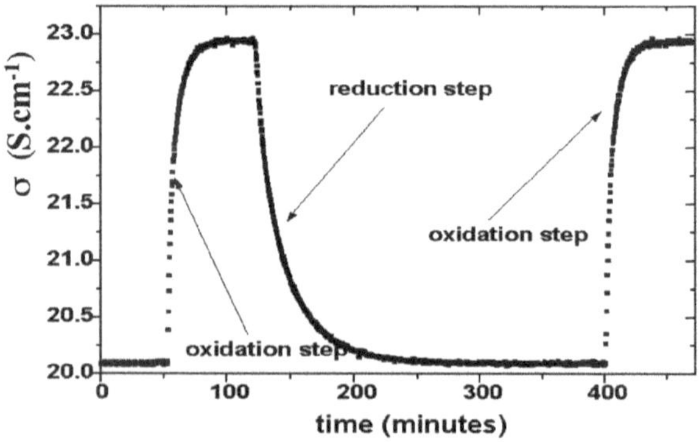

Fig. 21. Conductivity relaxation plots for $La_2Cu_{0.5}Ni_{0.5}O_{4+\delta}$ at 700°C for oxidation and reduction steps change in PO_2 between 0.21atm and 0.01atm.

The transient behaviour in the re-equilibration process can be described by the second Fick law [Sitte, 2001]. Taking into account the aspect ratio of the sample, the thickness of the pellet (slab like) controls the kinetic of the oxygen re-equilibration process. Consequently, the diffusion can be treated as a one-dimensional mechanism leading to the following equation:

$$\frac{dC}{dt} = \tilde{D}\frac{d^2C}{dx^2} \tag{26}$$

If the chemical diffusion is the rate-determining step, the following equation can be derived from the second Fick law, as suggested by Crank [Crank, 1975]:

$$\frac{M_t}{M_\infty} = 1 - \sum_{n=0}^{\infty} \frac{8}{(2n+1)^2\pi^2} \times \exp\left[-\frac{(2n+1)^2\pi^2\tilde{D}t}{4L^2}\right] \tag{27}$$

where L is the diffusion length and t the time. M_t / M_∞ represent the ratio of the mass of diffused oxide ions at time t, to that obtained for an infinite time.

Assuming constant the number of charges and the mobility of the charge carriers, the apparent conductivity is given by integrating the local conductivity all over the sample. Finally, the conductivity ratio can be directly related to the mass ratio:

$$\frac{M_t}{M_\infty} = \frac{\sigma_{app}(t) - \sigma_{app}(0)}{\sigma_{app}(\infty) - \sigma_{app}(0)} \tag{28}$$

In this formula σ(0), σ (t) and σ (∞) denote the apparent conductivity at t = 0 (initial), at time t (in the course of relaxation) and for t → ∞ (after reaching a new equilibrium state), respectively. The experimental data and the fitting curve in the form of fractional conductivity change as function of time are shown in Fig. 22. The chemical diffusion coefficient $\tilde{D}$ is then obtained by least-square fitting of the relaxation data.

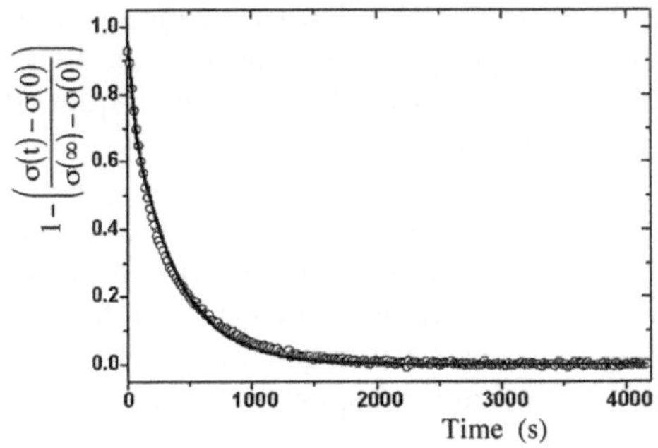

Fig. 22. Normalized and fitted conductivity data and fitting curve vs. time (at 700°C - PO_2 step from 0.01 atm to 0.21 atm).

Fig. 23 shows the Arrhenius plots of $\tilde{D}$ measured on two different samples of the same composition $La_2Cu_{0.5}Ni_{0.5}O_{4+\delta}$, but with two different thicknesses. Oxygen chemical diffusion coefficients, obtained for both samples, are equal over a wide temperature range. This observation confirms that the relaxation process is not controlled by the surface exchange, but by the bulk diffusion (i.e., this latter process is the limiting step).

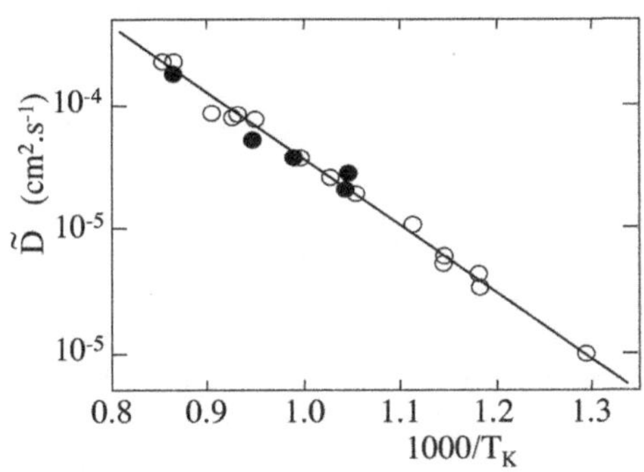

Fig. 23. Arrhenius plots of the oxygen chemical diffusion coefficient ($\tilde{D}$) measured on two samples with two different diffusion lengths: L_1 = 1.3mm (○) and L_2 = 0.7mm (●)

5. Conclusion

Determination of nonstoichiometry in oxides is a key point in the search for new materials for electrochemical applications. In recent decades, owing to their current and potential applications (electrodes in fuel cells, insertion electrodes, membranes of oxygen separation, gas sensors, catalytic materials, etc.), various methods of precise characterization of MIECs have been proposed, either the measurement of the defect concentrations and the stoichiometric ratio as functions of the oxide composition, of the surrounding oxygen pressure and of temperature, or the transport properties. There are different methods to determine the electrical properties of MIECs and, more specifically, the ionic and electronic contributions. The most appropriate method depends on different parameters, i.e., the total electrical conductivity of the studied oxides, the ionic and electronic transport numbers, the temperature domain, etc. As examples, the characterization of solid electrolytes requires the determination of electronic transport numbers as low as 10^{-4}; oxygen leakage currents through interconnect materials in solid oxide fuel cells require the measurement of ionic conductivity in a practically pure electronic conductor; in ceramic membranes devoted to oxygen permeation, one of the most important criteria concerns the high ionic and electronic conductivities of the oxide. In the case of MIEC oxides used as oxygen electrode, the electro-catalytic activity is related to the properties of mixed conduction and, therefore, of non-stoichiometry. The determination of the level of ionic conductivity is fundamental for the material selection but this characterization is generally difficult to carry out because of the high electronic contribution to the total transport properties.

In this chapter all the proposed methods have not been described. The objective was to focus on the sources of errors, which may render the results useless. Deviation from equilibrium of the MIEC surface due to oxygen semipermeability flux can lead to erroneous measurements. Experimental set-ups, which allow overcoming these experimental difficulties, were described.

6. References

Akida, K.; Ueda, M.; Kawamura, K. & Maruyama, T. (2008). Continuous Monitoring of Oxygen Chemical Potential at the Surface of Growing Oxide Scales during High Temperature Oxidation of Metals, *Mater. Trans.* , Vol. 49, No 3, pp. 629-636

Bouwmeester, H.J.M.; Kruidhof, H.; Burggraaf, A.J. & Gellings, P.J. (1992). Oxygen semipermeability of erbia-stabilized bismuth oxide, *Solid State Ionics*, Vol. 53-56, pp. 460-468

Bouwmeester, H.J.M.; Kruidhof, H. & Burggraaf, A.J. (1994). Importance of the surface exchange kinetics as rate limiting step in oxygen permeation through mixed-conducting oxides, *Solid State Ionics*, Vol. 72, pp. 185-194

Caneiro, A.; Bonnat, M. & Fouletier, J. (1981). Measurement and regulation of oxygen content in gases using solid electrolyte cell, IV - Accurate preparation of CO_2-CO and H_2O-H_2 mixtures, *J. Applied Electrochem.*, Vol. 11, pp. 83-90

Caneiro, A.; Bavdaz, P.; Fouletier, J. & Abriata, J.P. (1982). Adaptation of an electrochemical system for measurement and regulation of oxygen partial pressure to a symmetrical thermogravimetric analysis system developed using a Cahn 1000 electrobalance, *Rev. Sci. Instrum.*, Vol. 53, pp. 1072-1075

Caneiro, A.; Mogni, L.; Grunbaum, N. & Prado, F. (2011). Physicochemical properties of non-stoichiometric oxides - Mixed conductors, Part I & II, *J. Therm. Anal. Calorim.*, Vol. 103, pp. 597-606 & Vol. 104, pp. 781-788

Carter, S.; Selcuk, A.; Chater, R.J.; Kajida, J.; Kilner, J.A. & Steele, B.C.H. (1992). Oxygen transport in selected nonstoichiometric perovskite-structure oxides, *Solid State Ionics*, Vol. 53-56, pp. 597-605

Chen, C.H.; Bouwmeester, H.J.M., van Doorn, R.H.E., Kruidhof, H. & Burggraaf, A.J. (1997). Oxygen permeation of $La_{0.3}Sr_{0.7}CoO_{3-d}$, *Solid State Ionics*, Vol. 98, pp. 7-13

Crank, C. (1975). *The mathematics of diffusion*, 2nd Edn., Clarendon Press, Oxford, pp.44-68

Ducroux, R.; Fromont, M.; Jean Baptiste, Ph. & Pattoret, A. (1980). Mesures en continu de la redistribution de l'oxygène sous gradient thermique dans UO_{2+x}, *J. Nucl. Mat.*, Vol. 92, pp. 325-333

Ducroux, R. & Jean Baptiste, Ph. (1981). Mesure du potentiel d'oxygène dans le système $U_{0.7}Ce_{0.3}O_{2+x}$ à l'aide d'une minisonde à électrolyte solide, *J. Nucl. Mat.*, Vol. 97, pp. 333-336

Fouletier, J.; Fabry, P. & Kleitz, M. (1976). Electrochemical semipermeability and the electrode microsystem in solid oxide electrolyte cells, *J. Electrochem. Soc.*, Vol. 123, pp. 204-213

Fouletier, J. & Kleitz, M. (1978). Direct determination of the electrical conductivity-non stoichiometry relationship in ionically-conducting metallic oxides, *J. Electrochem. Soc.*, Vol. 125, pp. 751-755

Fouletier, J. & Vitter, G. (1980). Gaseous oxygen gauges : characteristics and applications, In: *Applications of Solid Electrolytes*, T. Takahashi, A. Kozawa (Eds.), pp. 108-113, JEC Press Inc., Cleveland

Fouletier, J.; Vitter, G. & Kleitz, M. (1981). Measurement and regulation of oxygen content in gases using solid electrolyte cells, III - Oxygen pump-gauge, *J. Applied Electrochem.*, Vol. 5, pp. 111-120

Fouletier, J.; Meas, Y.; Fouletier, M. & Kleitz, M. (1982). Oxidation studies at constant oxygen pressure or constant oxidation rate, In: *Reactivity of Solids*, Vol. 1, K. Dyrek, J. Haber, J. Nowotny (Eds.) , p. 138-144, Elsevier, ISBN 83-01-03585-4, Amsterdam

Fouletier, J. (1982/83). Gas analysis with potentiometric gas sensors, *Sensors and Actuators*, Vol. 3, pp. 295- 314

Fouletier, J.; Siebert, E. & Caneiro, A. (1984). Accurate monitoring of low oxygen activity in gases with conventional oxygen gauges and pumps, In: *Science and Technology of Zirconia II*, Adv. In Ceram., Vol. 12, N. Claussen et al. (Eds.), pp. 618-626, Am. Ceram. Soc., Columbus

Geffroy, P.M.; Vivet, A.; Fouletier, J.; Richet, N.; Del Gallo, P. & Chartier, T. (2011). Influence of Oxygen Surface Exchange on Oxygen Semi-Permeation through $La_{1-x}Sr_xFe_{1-y}Ga_yO_{3-\delta}$ Dense Membrane, *J. Electrochem. Soc.*, Vol. 158, No 8, pp. B971-B979

Gellings, P.J. & Bouwmeester, H.J.M. (Eds). (1997). *The CRC Handbook of Solid State Chemistry*, CRC Press, ISBN 0-8493-8956-9, London

Hebb, M.C. (1952). Electrical Conductivity of Silver Sulfide, *J. Chem. Phys.*, Vol. 20, pp. 185-190

Heyne, L. (1977). Electrochemistry of Mixed Ionic-Electronic Conductors, In: *Solid Electrolytes*, Topics in Applied Physics, S. Geller (Ed.), p. 169-221, Springer, Berlin

Joshi, A.V.; Wagner Jr., J.B. (1975). Electrochemical Studies on Single Crystalline CuCI Solid Electrolyte, *J. Electrochem. Soc.*, Vol. 122, pp. 1071-1080

Kharton, V.V.; Yaremchenko, A.A.; Kovalevsky, A.V.; Viskup, A.P.; Naumovich, E.N. & Kerko, P.F. (1999). Perovskite-type oxides for high-temperature oxygen separation membrane, *J. Memb. Sci.*, Vol. 163, pp. 307-317

Kharton, V.V. &, Marques, F.M.B. (2001). Interfacial effects in electrochemical cells for oxygen ionic conduction measurements - I. The e.m.f. method, *Solid State Ionics*, Vol. 140, pp. 381-394

Kharton, V.V.; Viskup, A.P.; Figueiredo, F.M.; Naumovich, E.N. Yaremchenko, A.A. & Marques, F.M.B. (2001). Electron-hole conduction in Pr-doped $Ce(Gd)O_{2-\delta}$ by faradaic efficiency and emf measurements, *Electrochim. Acta.*, Vol. 46, pp. 2879-2889

Kharton, V.V.; Shaula, A.L. & Marques, M.F.B. (2007). Oxygen ion transport number: assessment of combined measurement methods, *Ionics*, Vol. 13, pp. 163-171

Kharton, V.V. ; Tsipis, E.V. ; Naumovich, E.N. ; Thursfield, A. ; Patrakeev, M.V. ; Kolotygin, V.A. ; Waerenborgh, J.C. & Metcalfe, I.S. (2008). Mixed conductivity, oxygen permeability and redox behavior of K_2NiF_4-type $La_2Ni_{0.9}Fe_{0.1}O_{4+\delta}$, *J. Solid State Chem.*, Vol. 181, pp. 1425-1433

Kiukkola, K.; Wagner, C. (1957). Measurements on galvanic Cells Involving Solid Electrolytes, *J. Electrocheml. Soc.*, Vol. 104, No 6, pp. 379-387

Kleitz, M.; Siebert, E.; Fabry, P. & Fouletier, J. (1992). Solid State Electrochemical Sensors, In: *Sensors - A Comprehensive Survey*, W. Göpel, J. Hesse, J.N. Zemel (Eds), Chemical and Biochemical Sensors, Part I, Vol. 2, pp. 341-428, VCH, ISBN 3-527-26768-9, Weinheim

Lane, J.A. & Kilner, J.A. (2000). Measuring oxygen diffusion and oxygen surface exchange by conductivity, *Solid State Ionics*, Vol. 136-137, pp. 997-1001

Lankhorst, M.H.R. & Bouwmeester, H.J.M. (1997). Determination of Oxygen Nonstoichiometry and Diffusivity in Mixed Conducting Oxides by Oxygen Coulometric Titration, *J. Electrochem. Soc.*, Vol. 144, No 4, pp. 1261-1267

Levy, M.; Fouletier, J. & Kleitz, M. (1988). Model for the electrical conductivity of reduced stabilized zirconia, *J. Electrochem. Soc.*, Vol. 135, pp. 1584-1589

Mari, C.M.; Pizzini, S.; Manes, L. & Toci, F. (1977). A Novel Approach to the Oxygen Activity Microdetermination of Oxides by EMF Measurements, *J. Electrochem. Soc.*, Vol. 124, No 12, pp. 1831-1836

Mauvy, F.; Bassat, J.M.; Boehm, E.; Dordor, P.; Grenier, J.C. & Loup, J.P. (2004). Chemical oxygen diffusion coefficient measurement by conductivity relaxation - Correlation between tracer diffusion coefficient and chemical diffusion coefficient, *J. Eur. Ceram. Soc.*, Vol. 24, pp. 1265-1269

Mauvy, F.; Boehm, E.; Bassat, J.M.; Grenier, J.C. & Fouletier, J. (2007). Oxygen permeation fluxes through $La_2Cu_{0.5}Ni_{0.5}O_{4+\delta}$ dense ceramics: comparison with oxygen diffusion coefficients, *Solid State Ionics*, Vol. 178, pp. 1200-1204

Mauvy, F.; Lalanne, C.; Bassat, J.M.; Grenier, J.C.; Brisse, A., Sauvet, A.L.; Barthet, C. & Fouletier, J. (2009). Electrochemical study of the $Nd_{1.95}NiO_{4+\delta}$ oxide electrolyte interface, *Solid State Ionics*, Vol. 180, pp. 1183-1189

Meas, Y.; Fouletier, J.; Passelaigue, D. & Kleitz, M. (1978). Nouveau montage d'étude des réactions d'oxydation à pression d'oxygène constante, *J. Chim. Phys.*, Vol. 75, pp. 826-834

Mizusaki, J, Tagawa, H, Naraya, K & Sasamoto, T. (1991). Nonstoichiometry and thermochemical stability of the perovskite-type $La_{1-x}Sr_xMnO_{3-\delta}$, *Solid State Ionics*, Vol. 49, pp. 111-118

Möbius, H.H. (1986). Oxygen Current Density Coefficient of Oxidic Materials as a Parameter for Selection and Development of Electrodes with Solid Electrolytes, Extended Abstracts of the *37th Meeting ISE, Int. Soc. of Electrochem.*, Vilnius, USSR, Vol. 1, p. 136-139

Nakamura, A. & Fujino, T. (1987). Thermodynamic study of UO_{2+x} by solid state emf technique, *J. Nucl. Mat.*, Vol. 149, No 1, pp. 80-100

Nakamura, T., Yashiro, K., Sato, K & Mizusaki, J. (2009a). Oxygen nonstoichiometry and chemical stability of $Nd_{2-x}Sr_xO_{4+\delta}$, *J. Solid State Chem.*, Vol. 182, pp. 1533-1537

Nakamura, T., Yashiro, K., Sato, K & Mizusaki, J. (2009b). Oxygen nonstoichiometry and defect equilibrium in $La_{2-x}Sr_xO_{4+\delta}$, *Solid State Ionics*, Vol. 180, pp. 368-376

Otobe, H.; Akabori, M. & Arai, Y. (2009). Oxygen potential measurements of $Am_{0.5}Pu_{0.5}O_{2-x}$ by EMF method, *J. Nucl. Mat.*, Vol. 389, pp. 68-71

Panlener, R.J.; Blumenthal, R.N., Garnier, J.E. (1975). A Thermodynamic Study of Nonstoichiometric Cerium Oxide, *J. Phys. Chem. Solids*, Vol. 36, No 11, pp. 1213-1222

Porat, O.; Riess, I. (1994) Defect chemistry of $Cu_{2-y}O$ at elevated temperatures. Part. I: Non-stoichiometry, phase width and dominant point defects, *Solid State Ionics*, Vol. 74, pp. 229-238

Rickert, H. (1982). *Electrochemistry of Solids - An introduction*, Springer-Verlag, ISBN 0-387-11116-6, Berlin

Riess, I. & D.S. Tannhauser, D.S. (1982). Control and measurement of stoichiometry in oxides, In: *Reactivity of Solids*, Vol. 1, K. Dyrek, J. Haber, J. Nowotny (Eds.), Elsevier, pp. 503-517

Riess, I. (1991). Measurement of electronic and ionic partial conductivities in mixed conductors without the use of blocking electrodes, *Solid State Ionics*, Vol. 44, pp. 207-214

Riess, I. (1992). Four point Hebb-Wagner polarization method for determining the electronic conductivity in mixed ionic-electronic conductors, *Solid State Ionics*, Vol. 51, No 3-4, pp. 219-229

Riess, I. (1992b). Recent investigations into the properties of mixed ionic electronic conductors, In: *Solid State Ionics*, Balkanski, M.; Takahashi, T. & Tuller, H.L. (Eds.), pp. 475-485, Elsevier, ISBN 0-444-89354-7, Amsterdam

Riess, I.; Kramer, S. & Tuller, H.L. (1992). Measurement of ionic conductivity in mixed conducting pyrochlores by short circuit method, In: *Solid State Ionics*, Balkanski, M.; Takahashi, T. & Tuller, H.L. (Eds.), pp. 499-505, Elsevier, ISBN 0-444-89354-7, Amsterdam

Riess, I.; Safadi, R. & Tuller, H.L. (1994). Problems with Hebb-Wagner polarization measurements due to overpotentials and decomposition of the sample, *Solid State Ionics*, Vol. 72, pp. 3-6

Riess, I. (1996). Review of the limitation of the Hebb-Wagner polarization method for measuring partial conductivities in mixed ionic electronic conductors, *Solid State Ionics*, Vol. 91, pp. 221-232

Riess, I. (1997). Electrochemistry of mixed ionic-electronic conductors, In: *The CRC Handbook of Solid State Chemistry*, Gellings, P.J. & Bouwmeester, H.J.M. (Eds), pp. 223-268, CRC Press, ISBN 0-8493-8956-9, London

Shores, D.A.; Rapp, R.A. (1971). Solid Bielectrolyte Cells for Thermodynamic Measurements, *J. Electrochemical Soc.*, Vol. 118, No 7, pp.1107-1111

Sitte, W., Bucher, E.,Benisik A; Preis W., (2001) Oxygen nonstoichiometry and ionic transport properties of $La_{0.4}Sr_{0.6}CoO_{3-\delta}$,. *Spectrochimica Acta*, 2001, Part A57, 2071-2076

Sorensen T.O. (Ed.). (1981). *Nonstoichiometric oxides*, Academic Press, ISBN 0-12-655280-0, London

Sunarso, J.; Baumann, S.; Serra, J.M.; Meulenberg, W.A.; Liu, S.; Lin, Y.S. & Diniz da Costa, J.C. (2008). Mixed ionic-electronic conducting (MIEC) ceramic-based membranes for oxygen separation, *J. Memb. Sci.*, Vol. 320, pp. 13-41 ten Elshof, J.E.; Lankkhorst, M.H.R. & Bouwmeester, H.J.M. (1997). Chemical diffusion and oxygen exchange of $La_{0.6}Sr_{0.4}Co_{0.6}Fe_{0.4}O_{3-\delta}$, *Solid State Ionics*, Vol. 99, pp. 15-22

Tretyakov, J.D. & Muan, A. (1969). A New Cell for Electrochemical Studies at Elevated Temperatures Design and Properties of a Cell Involving a Combination of Thorium Oxide-Yttrium Oxide and Zirconium Oxide-Calcium Oxide Electrolytes, *J. Electrochem. Soc.*, Vol. 116, No 3, pp. 331-334

Tretyakov, J.D., Maiorova, A.F. & Berezovskaya, Y.M. (1997). Electrochemical Titration in Study of Nonstoichiometric Compounds, *Key Engineering Materials*, Vol. 125-126, pp. 283-316

Tsipis, E.V.; Naumovich, E.N. ; Shaula, A.L. ; Patrakev, M.V. ; Waerenborgh, J.C. & Kharton, V.V. (2008). Oxygen nonstoichiometry and ionic transport in $La_2Ni_{0.9}Fe_{0.1}O_{4+\delta}$, *Solid State Ionics*, Vol. 179, pp. 57-60

Patrakeev, M.V.; Leonidov, I.A. ; Lakhtin, A.A. & Kozhevnikov, V.L. (1995), Oxygen Nonstoichiometry of $(Nd_{2/3}Ce_{1/3})_4(Ba_{2/3}Nd_{1/3})_4Cu_6O_{16+x}$, *J. Solid State Chem.*, Vol. 120, pp. 146-150

Une, K. & Oguma, M. (1982). Thermodynamic properties of nonstoichiometric urania-gadolinia solid solutions in the temperature range 700-1100°C, *J. Nucl. Mat.*, Vol. 110, pp. 215-222.

Wagner, C. (1957). Galvanic Cells with Solid Electrolytes Involving Ionic and Electronic Conduction, In: *Proceedings of the International Committee Electrochemical Thermodynamics and Kinetics*, Chap. 8.5, 7th C.I.T.C.E. Meeting, 1955, Butterworths, London, pp. 361-377

Wagner, C. (1971). The determination of small deviations from the ideal stoichiometric composition of ionic crystals and other binary compounds, *Progress in Solid State Chemistry*, vol.6, pp. 1-15

Weppner, W.; Huggins, R.A. (1978). Electrochemical methods for determining kinetic properties of solids, *Ann. Rev. Mater. Sci.*, Vol. 8, pp. 269-311

Wiemhöfer, H.D.; Bremes, H.G.; Nigge, U. & Zipprich, W. (2002). Studies of ionic transport and oxygen exchange on oxide materials for electrochemical gas sensors, *Solid State Ionics*, Vol. 150, pp. 63-77

Xu, S.J. & Thomson, W.J. (1999). Oxygen permeation rates through ion-conducting perovskite membranes, *Chem. Eng. Sci.*, Vol. 54, pp. 3839-3850

Zhao, H.; Mauvy, F.; Bassat, J.M.; Fourcade, S.; Lalanne, C.; Boehm, E. & Grenier, J.C. (2008). New cathode materials for IT-SOFC: phase stability, oxygen exchange and cathode properties of $La_{2-x}NiO_{4+\delta}$, *Solid State Ionics*, Vol. 179, pp. 2000-2005

8

Molar Volume, Ionic Radii in Stoichiometric and Nonstoichiometric Metal Oxides

Andrzej Stokłosa
Cracow University of Technology,
Faculty of Chemical Engineering and Technology, Cracow,
Poland

1. Introduction

Metal oxides are a group of compounds that are now intensively studied because of their wide use in modern ceramic materials, composites, nano-composites, etc., with new special properties. Many oxides are used as electronic materials for their electrical, magnetic and optical properties. The mechanisms of corrosion of metals and their alloys, as well as the mechanism of reduction in metallurgical processes depends on metal oxides formed in these processes. In contrast to the main group metal oxides, generally stoichiometric, the transition metal oxides feature bigger diversity in their properties. Their electrical properties vary from metallic to semiconductor. Many oxides show significant deviations from stoichiometry. The transition metals form numerous oxide phases with ions in different oxidation states, depending on the oxygen pressure. They are a group of compounds in which structure and properties depend strongly on the metal type because of a change in interactions of ionic-covalent character. An especially important issue becomes the relation between the structure and ionic radii sizes on the one hand and properties on the other. One of the important issues is a change in spheres occupied by ions (ionic radii) dependent on the interaction character, coordination number, oxidation degree (formal and real charge), metal type in the case of hetero-metallic oxides, dopants, etc. A particularly complex problem occurs in the case of oxides. In compact structures, in which close packing of anions is assumed, not all voids are occupied by cations. Besides, ionic defects occur in significant concentrations in many oxides. In these oxides, in spite of the conservation of the coordination number of cations, often having different oxidation states, distortions of coordination polyhedra occur which results in various bond lengths in the crystal. Next problem is the size of the sphere occupied by ions in polymorphic forms differing by lattice parameters and often having deformed structures. For the cases mentioned above, the problem of the ion radius has not been solved practically.

For many years attempts have been made to find relatively simple parameters which could allow to investigate the changes in the properties of crystalline compounds. One of first parameters was ionic radius. A method of its calculation was proposed as early as in 1926 by Goldschmidt (Goldschmidt, 1926) and many other methods were described later (Pauling, 1927; Zachariansen, 1931; Ahrens, 1952, 1953). To differentiate the ability of ions to interact, Cartledge (Cartledge, 1928, 1930) and Ahrens (Ahrens, 1952, 1953) introduced a parameter being the ratio of ion charge or its ionisation energy to ion radius. The above parameters have been used in mineralogy for geochemical characteristics of ions, their weathering resistance or to determine the sequence of crystallisation of different phases in an alloy. A widely used parameter differentiating the properties of atoms of the elements is the scale of electronegativity introduced by Pauling (Pauling, 1932, 1960). Also other experimental scales, developed later, are in common use (Mulliken, 1934, 1935; Sanderson, 1976; Allen, 1998; Allred & Rochow, 1958; Görlich, 1989a). For the purpose of characterisation of compounds or ions Parr (Donnelly & Parr, 1978; Parr et al., 1983) introduced absolute electronegativity. It is possible to calculate it theoretically, basing on the density functional theory (Hohenberg & Kohn, 1964, Kohn, et al., 1996). The properties of ions or molecules can be described also by values derived from electronegativity, as acidity and basicity (Pearson, 1997; Komorowski, 1993) or effective nuclear charge of the ion (Görlich, 1989b, 1990; Stokłosa et al., 2004). For the purpose of classification of the bonds in solid state, Pauling (Pauling, 1929) introduced bond ionicity and bond strength parameters. The above concepts were developed in a bond-valence model proposed by Browns (Brown, 1992, 2002), relating bond energy to its length. Most of the above parameters are evaluated basing on the quantities determined for isolated ions and they can be used in the analysis of ionic compounds. However, they are less useful in the analysis of the properties of oxides. This is due to the existence of ionic-covalent bond in oxides. Its character varies significantly with the position in the table of elements.

The development of the solid state theory has created strong theoretical basis for the description of interactions between atoms in a crystal. The electronic structure model, resulting from these interactions, allows identifying electrical, magnetic and optical properties of a compound. Today, the development of computational methods and the numerical capabilities of big computers allow calculating the interactions in quite big crystalline clusters, and thus to determine their properties. Nevertheless there is no data allowing detailed analysis of the properties of the oxides to be performed.

Hence, in spite of the development of chemistry and physics of solids, there are still open problems: Why some transition metal oxides show significant deviations from stoichiometric compositions (e.g. oxides of Ti, V, Nb, Fe)? Why many metals form few or a dozen of oxide phases? What is the reason for high melting temperatures of some phases, while others have low melting temperatures, so in what way the bond energy varies and why, etc.

The present work proposes to introduce an effective (mean) ionic radius (ionic sphere) based on molar volume obtained from X-ray studies for comparison of the properties of metal oxides depending on the ion type and size. It would allow distinguishing groups of oxides, not merely simple oxides but also complex ones, containing various cations (dopes) and following or predicting their properties, which was practically impossible when based on tabulated values of ionic radii. The analysis of the relationship between molar volume of metal oxides and their ion radii was discussed. A linear dependence was found for ionic radii versus the cube root of molecular volume standardised per oxygen atom ($M_{a/b}O$) of MO oxides of alkali earth metals, M_2O oxides of alkali metals and MO_2 oxides of f-electron metals, and parameters of the linear equation were calculated. This equation makes it possible to predict mean ionic radii of metals in compact structures of polymorphic forms of oxides, mixed-valence oxides and oxides showing deviation from stoichiometry, otherwise not accessible.

Presented also is the problem of changes in distances between oxygen atoms (ions), depending on the metal type, metal charge, polarization capabilities, composition and crystallographic structure of metal oxides. Changes in ion packing density in oxides have also been presented. It has been shown that these changes depend not only on ionic radii, but also on the interaction character (interatomic bonds).

2. Molar volume and composition of metal oxide

One of parameters characterizing ions/atoms in a crystal, and therefore characterizing the properties of the compound is molar volume being a result of ion radii sizes and their interactions. It can be easily calculated using lattice parameters. However, there are difficulties in its use because individual types of oxides have different compositions. It is therefore possible to compare molar volumes of oxides of different metals, but of the same type (e.g. MO, M_2O_3) and preferably having the same structure. It is nevertheless difficult to draw conclusions from changes of molar volumes of the oxides of the same metal differing by composition and structure type. That is why analyses of structures and properties of various oxides, which can be found in monographs on crystal chemistry and solid state chemistry (Adams, 1974; O'Keeffe & Navrotsky, 1981; Wells, 1990; Rao & Raveau, 1995) are performed usually for a given type of oxides.

There are few oxides that - considering a given type - have ideal geometric structure. Besides, many oxide phases show deviation from the stoichiometry, or metal ions are in different oxidation states, e.g. in ordered phases (Magnéli phases). A distortion of coordination polyhedra occurs in these oxides and as a result M-O bond lengths are differentiated. Hence, the volume of an unit cell is well defined, however, it is much more difficult to determine "real" ion positions and distances between them. Besides, unit cells of oxides differing by structure and composition contain different numbers of molecules, making the comparison of their volumes impossible.

2.1 Effective molar volumes of oxides

A parameter which could be used to compare molar volumes of different oxides is the volume of the oxide of composition $M_{a/b}O$ (containing one mole of oxygen atoms). Therefore, using the structural data: structure type, lattice parameters or X-ray density, and number of molecules in an unit cell it is possible to calculate its volume. Dividing the volume of an unit cell ($V_{M_aO_b}$) by the number of molecules (Z) in an unit cell and by the number of oxygen atoms in the M_aO_b molecule and multiplying it by the Avogadro number (N_A) one can obtain a volume further called *effective molar volume* of an oxide, containing one mole of oxygen atoms (V_{eff}) (Stokłosa & Laskowska, 2008a):

$$((V_{M_aO_b}/Z)/b)N_A = V_{eff} \tag{1}$$

The dependence of the effective molar volume (V_{eff}) of transition metal oxides on their composition (on the a/b ratio) is presented in Fig. 1. In the case of non-stoichiometric compounds of $M_{1-y}O_{1-x}$ type (oxides of V, Ti, Nb etc.) the real contents of metal and oxygen in an unit cell of these oxides were taken into account and then recalculated to the volume containing one mole of oxygen atoms. The above relations have been obtained for oxides undergoing mutual transformations.

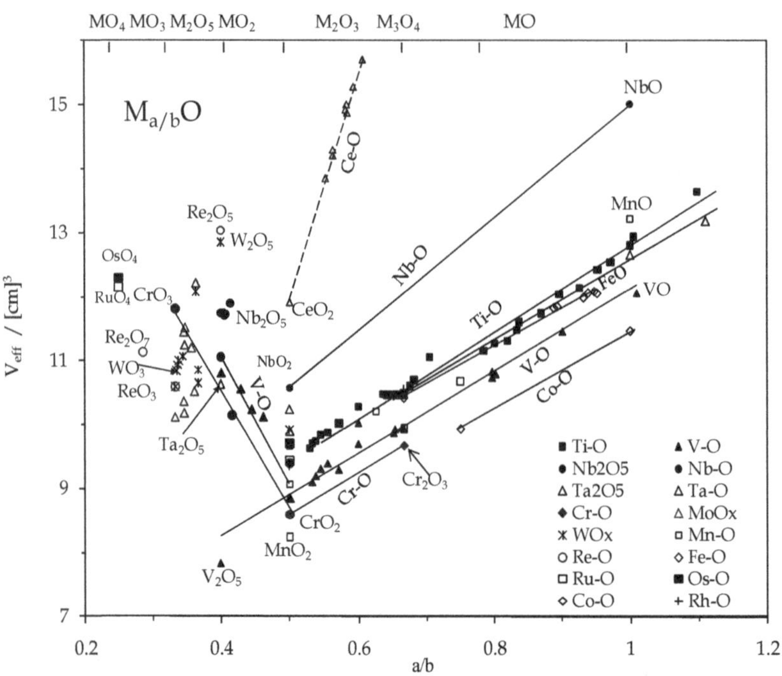

Fig. 1. The dependence of the effective molar volume (V_{eff}) of the oxides $M_{a/b}O$ on the composition (a/b) for the transition metal oxides and Ce oxides.

As it can be seen in Fig.1, the dependence of the effective molar volume of oxides on the a/b ratio shows linear character in the composition range $0.5<a/b<1$. This character is conserved not only for the oxides having ions with the same charge (2+, 3+) but also in the case of oxides showing deviation from the stoichiometry, for many oxide phases (Magnéli phases) with ions in different oxidation states and, as in the case of $Fe_{1-y}O$ or M_3O_4 oxides with spinel structure, when metal ions have different coordination number, 4 or 6 (tetrahedral or octahedral).

As it can be seen in Fig. 1, in the case of the oxides with cations in higher oxidation state $M^{z+}>M^{4+}$ ($a/b<0.5$) the effective molar volume (V_{eff}) increases. In the case of oxides of Cr, V and Re the lines obtained have negative slope, while linear correlations are also rather good. The increase of molar volumes but lack of correlations occur in the case of many Magnéli phases or other ordered oxides of W, Mo, Nb with compositions in the range of $0.25<a/b<0.5$. In the case of V_2O_5 two significantly different volumes were determined. The V_2O_5 oxide with monoclinic structure has small volume, correlating with volumes of the oxides VO → VO_2, while V_2O_5 with rhombohedral structure, having bigger volume, correlates with the oxides having composition $a/b<0.5$.

On the other hand, the dependence of the molar volume (V_{eff}) on the composition for f-electron metal oxides is presented in Fig. 2.

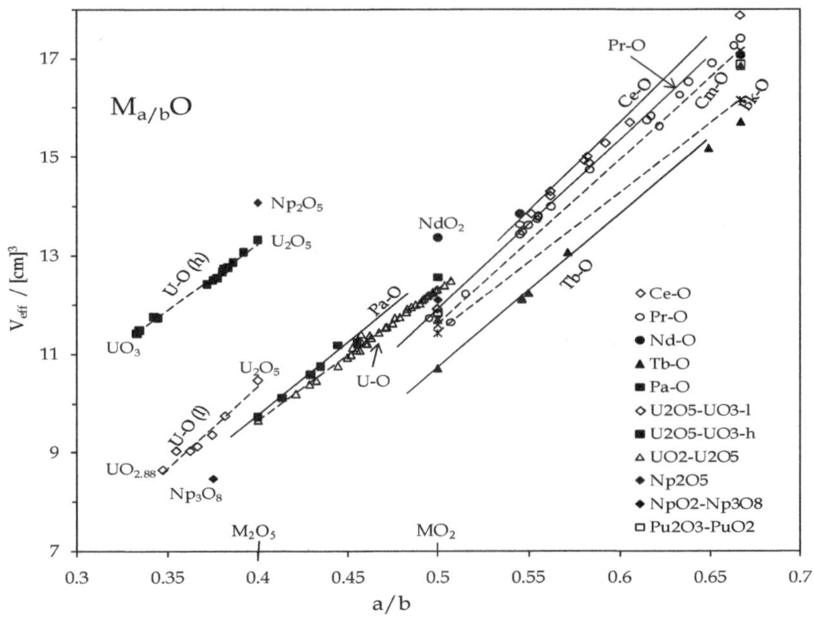

Fig. 2. The dependence of the effective molar volume (V_{eff}) of the oxides $M_{a/b}O$ on the composition (*a/b*) for *f*-electron metal oxides, (—) - lanthanides, (- -) - actinides.

As it can be seen in Fig. 2, for the oxides with compositions in the range of 0.666>*a/b*>0.4, for systems containing many phases (oxides of Ce, Pr, Tb, Pa, U) as well as for systems containing only oxides M_2O_3 and MO_2 (with exception of the Nd-O system), the slopes of the obtained lines are similar. In the case of uranium oxide phases three groups of crystallographic data were found. Apart from oxide phases with compositions varying in the range of 0.4<*a/b*<0.5, phases of the same or close composition significantly differing by volume were found in the composition range of 0.33<*a/b*<0.4. The obtained dependencies have practically the same slopes.

Furthermore, the dependence of the molar volume on the composition for oxides of alkali metals and other metals forming M_2O oxides and also for alkaline earth metal oxides. is presented in Fig. 3 and for group 11-15 oxides is depicted in Fig. 4.

A bigger number of phases in this group of oxides is observed only in the case of row 6 metal oxides (Tl, Pb and Bi), and the obtained correlations are worse in comparison to other oxides. The slope for the oxides of Pb, Bi is three times bigger than that for the transition metal oxides. For comparison, the dependence of V_{eff} on the composition for the Ce-O system oxides was marked in Fig. 1 and 4. As it can be seen, either values and slopes for the above lines for oxide phases of *f*-electron metals significantly deviate from the relationship obtained for the oxides of *d*-electron metals. On the other hand, the obtained linear function parameters are close to the parameters for the main group metal oxides (Bi, Pb).

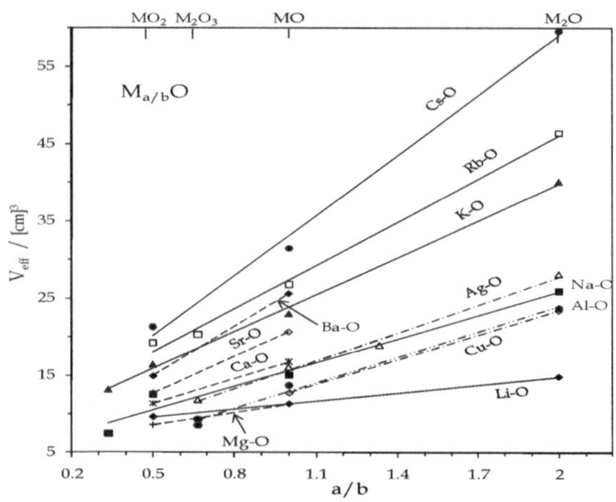

Fig. 3. The dependence of the effective molar volume (V_{eff}) of the oxides $M_{a/b}O$ on the composition (*a/b*) for the alkali metal oxides, the oxides of Al, Ag, Cu and the alkaline earth metal oxides.

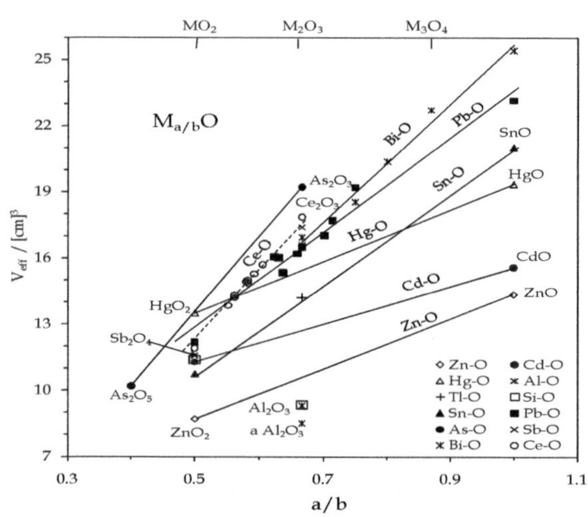

Fig. 4. The dependence of the effective molar volume (V_{eff}) of the oxides $M_{a/b}O$ on the composition (*a/b*) for the group 11-15 metal oxides and Ce oxides.

2.2 Compact structures

The obtained linear character of the dependence of the effective molar volume (V_{eff}) on the composition (*a/b*) in the range of composition changes of $2>a/b>0.5$ (in the case of transition metal oxides $1>a/b>0.5$, and $0.666>a/b>0.35$ in the case of *f*-electron metal oxides), should be considered as a general relationship for the phases undergoing mutual transformations. The observed decrease of the molar volume of the oxides (V_{eff}) is a result of smaller quantity of metal ions per one mole of oxygen, and of smaller radii of ions in higher oxidation states (3+, 4+). The character of the change of molar volume results from the tendency of the individual oxides to form compact structures, typical for compounds with non-oriented ionic bond or ionic-covalent one. The compact structures are structures where anions form the close-packaging lattice and cations are located in its voids. The term close-packed structures is explicitly defined, in the case of metals it equals the closest possible packaging of spheres. However, there is no precise definition of compact structure for ionic compounds (Von Meerssche & Dupont, 1976; Wells, 1990; West, 1996)). Doubts emerge especially if the systems have low symmetry, when significant number of voids in anion sublattice is not occupied by cations or/and the anionic sites are not occupied (there are anion vacancies). Besides, the distances between oxygen ions are much bigger than their radii.

Considering the results obtained one can state that compact structures are adopted by:

- $M_{a/b}O$ oxides with composition in the range of $2>a/b>0.5$ (in the case of actinide oxides up to 0.33) having effective molar volumes of oxides ($M_{a/b}O$) of the same metal decreasing linearly when the metal content decreases.
- Polymorphic forms of these oxides will also have compact structures, because slight changes of their molar volumes result from the change of interactions between ions/atoms caused by the change of temperature or pressure.
- Compact structures will be adopted also by the oxides of other metals, having the same compositions and structures as the oxides classified as compact structures.

Typical compact structures adopted by the individual oxides and oxide phases are NaCl-type structures (MO oxides), Al_2O_3-type hexagonal structures and rutile-type (TiO_2) structures with which most MO_2 oxides crystallize. In the case of *f*-electron metal oxides these structures are: cubic Mn_2O_3-type structures, and CaF_2-type structures in MO_2 oxides. With the decrease of the metal content relative to oxygen in an oxide, the number of voids in the anionic sublattice not occupied by cations increases. As a result one can observe the transition from the close-packed structures of anions to a structure formed by coordination polyhedra connected by edges and/or corners despite the fact that the anionic sublattice practically does not change (the structures still have the character of compact structures). In many cases also other types of structures may appear, with lower symmetry but with similar volume, e.g. as in the case of homologue series V_nO_{2n-1}. The decrease of system symmetry occurs as a result of distortion of the coordination polyhedra or as a result of ordering of point defects. The above perturbations cause the deviations from the ideal geometric structure.

Starting from the M_2O or MO oxides, having biggest chemical potential of metal, as a result of the increase of oxygen potential in gas atmosphere, the transformation into an oxide with lower metal content is forced (the M/O ratio decreases), where ions must be in higher oxidation states (to compensate the oxygen charge) and atoms tend to the closest possible packing. The effective molar volumes of the individual oxides are successively smaller and this change is continuous in the case of formation of point defects, or stepwise when the ordered structures are formed but the same linear character of the dependence of V_{eff} on the a/b ratio is practically conserved.

The formation of compact structures is fully consistent with the fact that the arrangement of atoms in the space result from their sizes and the optimal interactions between atoms/ions, being superior relative to the ideal structure with highest possible symmetry.

2.3 Framework structures

The increase of the effective molar volume with the decrease of the metal content, observed for most oxides, in the range of 0.25<*a/b*<0.5, is also justified. The increase of the amount of oxygen atoms relatively to metal together with the presence of ions with high charges (M^{5+} and/or M^{6+}) cause the increase of the repulsive forces that lead to the increase of the effective molar volume of an oxide comparing to that of a MO_2 type oxide. The presence of ions in high oxidation states, as well as a significant contribution of the covalent bond (*spd* hybrid orbital's) causes that these oxides have significant bond energies, high enough to maintain the stability of the crystal. The observed linear increase of the volume V_{eff} with the decrease of the *a/b* ratio for the oxides of Cr, Re, V metals, and possibly also for other systems, comparing to the MO_2 oxide, will be probably present in the case of the reversible transformation of these oxides (thermodynamically stable forms). The lack of correlation of the molar volume and the composition observed for many oxides of W, Mo, Nb etc. can indicate that these phases could be thermodynamically unstable. Most of these phases form homologues of type M_nO_{2n-m}, which usually are created in the reduction process, when the oxygen is removed from the lattice sites (the mosaic structures or shear structures are created) (Rao & Raveau, 1995). The phases formed in this way can accept the oxygen, but they are not created by the oxidation of MO_2 oxide or metal. The formation of phases of this type does not result in the significant changes in the structure, despite the changes of system (crystal symmetry). There occurs ordering or agglomeration of point defects, the shear structures are formed, etc. The change of the volume of an unit cell may be small (or its volume can be a multiple of the unit cell), but the volume of the oxide containing one mole of oxygen atoms will significantly increase.

All these oxide phases can be included in the oxides with *framework structure*. These are oxides ($M_{a/b}O$) with composition a/*b*<0.5 and effective molar volumes bigger than those for the oxides of compact structures (MO_2) and the smallest metal content. In these oxides the coordination polyhedra can be identified, connected by corners, edges, etc. Most of these structures can be derived from the structure of the oxide ReO_3 where the MO_6 octahedrons form a regular lattice. Other cations can easily be included in the structure of this type of oxides. This leads to the stabilization of the structure, and as a result, one can obtain the perovskite-type structures or their variations.

3. Molar volume of oxides and their ionic radii

A parameter characterizing ions in a crystal, beside the ion formal charge, is the ion radius. Even if, according to quantum mechanics, ion radius has no theoretical meaning, it has become a useful parameter, defining space (a sphere) occupied by an ion in a crystal. Goldschmidt (Goldschmidt, 1926, 1931) assuming the additivity of interionic distance variations in compounds with a common ion and using the radii of F^- and O^{2-} calculated by Wasastjerna (Wasastjerna, 1923) gave the first scale of ionic radii. Other methods of ionic radii calculation were proposed by Pauling (Pauling, 1927), Zachariansen (Zachariansen, 1931), Ahrens (Ahrens, 1952, 1953). The scale of effective ionic radii presently used was designed and verified by Shannon and Prewitt (Shannon & Prewitt, 1969, 1970, 1976) and Jia (Jia, 1991). It is based on the radii of F^- = 0.068 nm or O^{2-} = 0.140 nm. As it is known, the ion radius is dependent on the coordination number and on the ionic-covalent character of interionic interactions. The values of ionic radii calculated by various methods generally differ, due to the assumptions made and depending on the group of compounds used for the scale construction. A particularly complex problem occurs in the case of oxides. In compact structures, in which close packing of anions is assumed, not all voids are occupied by cations. Besides, ionic defects occur in significant concentrations in many oxides. In these oxides, in spite of the conservation of the coordination number of cations, often having different oxidation states, distortions of coordination polyhedra occur and, as a result, there are various bond lengths in the crystal (Wells, 1990; West, 1996; Brown, 1992 2002; Luehen et al., 1987; Kunz & Brown, 1995; Nespolo et al., 2001; Lalik, 2005). Next problem is the size of the sphere occupied by ions in polymorphic forms differing by lattice parameters and often having deformed structures. For the cases mentioned above, the problem of ion radius has not been solved practically, it was, however, discussed in many works (Paulin, 1929; Fumi & Tosi, 1964; Phillips & Williams, 1965; Shannon & Prewitt, 1969, 1970, 1976; Brown, 2002).

3.1 Ionic radii and molecular volume of oxides

Shannon and Prewitt (Shannon & Prewitt, 1969, 1970, 1976) performed a verification of ionic radii in isotypic compounds, basing on the linear relation between the unit cell volume and the ionic radius (r^3). Next, Brisse and Knop (Brisse & Knop, 1968) have demonstrated a linear relation between the radius and the cube root of the unit cell volume $(V_{M_2O_3})^{1/3}$ for the M_2O_3 lanthanide oxides. A close relation between the radius and the cube root of the unit cell volume (V_{MO}) exists in the case of oxides with cubic NaCl-type structure, because for the MO oxides $\sqrt[3]{V_{MO}}$ is a sum of the diameters of metal ion ($2r_M^c$) and oxygen ion ($2r_O^c$):

$$\sqrt[3]{V_{MO}} = 2(r_M^c + r_O^c) = \sqrt[3]{4\,V_{eff}^m} \quad \equiv \quad \sqrt[3]{V_{eff}^m} = \sqrt[3]{2}(r_M^c + r_O^c) \tag{2}$$

Similarly, the cube root of the volume occupied by the MO molecule $(V_{eff}^m)^{1/3}$ is a linear function of the bond length. A question emerges - whether the cube root of the volume containing one molecule $(V_{eff}^m)^{1/3}$ of any oxide ($M_{a/b}O$) is a simple function of ionic radii, analogously as in the case of oxides with NaCl-type structure (Eq.2).

The relation between the cube root of the effective molecular volume of oxides $(V_{eff}^m)^{1/3}$ and the radius of respective cations (r_M) is plotted in Fig. 5 for a series of oxides. The volume of the molecule of $M_{a/b}O$ oxide was calculated dividing the unit cell volume by the number of molecules (Z) and by the number of oxygen atoms (b) in the M_aO_b oxide. It will be further called *the effective molecular volume* ($(V_{M_aO_b}/Z)/b = V_{eff}^m$) (Stokłosa & Laskowska, 2008b). As it can be seen in Fig.5, points from quite a big group of oxides are located on three approximately parallel lines. The highest line is formed by the alkali metal oxides with anti-CaF_2 structure and with metal ion coordination number 4 (CN=4). The points for the ZnO, BeO, SiO_2 and GeO_2 oxides with hexagonal structure (CN=4) lie close to this line. The points for PtO and PdO oxides, with tetragonal structure, in which metal ions have square planar coordination lie further away. Relatively close are located points for the PbO oxide with orthorhombic structure and HgO with tetragonal structure. The CuO and AgO oxides with monoclinic structure (CN=4) show bigger deviation. Significant molar volumes are found for the Ag_2O and Cu_2O oxides with rather untypical structure, where metal ions have coordination number of 2. On the other hand, very good correlation is found for a series of oxides having coordination number 6. These are MO alkaline earth metal oxides and *d*-, *f*-electron metal oxides, with NaCl-type cubic structure, M_2O_3 with α-Al_2O_3-type hexagonal structure, MO_2 oxides with rutile-type tetragonal structure and few main group metal oxides: Al_2O_3, In_2O_3, Tl_2O_3, with α-Al_2O_3-type structure.

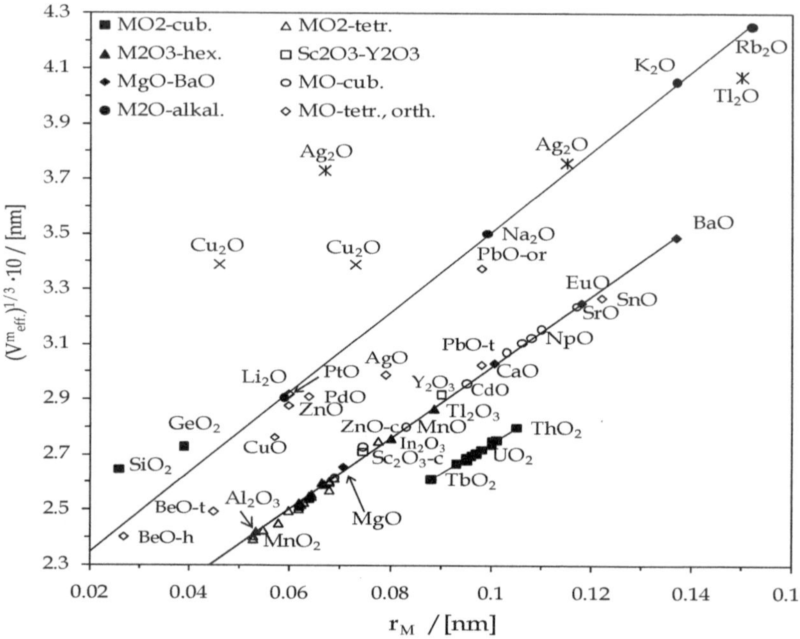

Fig. 5. The relation between the cube root of the molecular volume of oxides ($\sqrt[3]{V_{\text{eff}}^{\text{m}}}$) and the radius of the respective metal ions.

The lowest line is formed by the MO_2 oxides of *f*-electron metals with CaF_2-type cubic structure with metal ion coordination number 8. The parameters α and β of the linear functions:

$$\sqrt[3]{V_{\text{eff}}^{\text{m}}} = \alpha\, r_{\text{M}} + \beta \qquad (3)$$

formed by individual groups of oxides are collected in Table 1 of the work (Stokłosa & Laskowska, 2008b). The lines were obtained for the oxides of the same type, of the same metal and for the oxides of metals from the same group of the periodic table. In the case of *d*-electron oxides from the same group all the oxides with various oxidation states were considered. The correlation coefficients for most lines are high, $R^2>0.98$, and the α parameter (slope) for individual groups of oxides varies only slightly.

3.2 Relative radii of metal ions

The obtained correlations between $(V_{eff}^{m})^{1/3}$ and radii of metal ions with a given coordination number indicate that they are characteristic of compact structures (see Chapter 2). Assuming therefore, as previously (Shannon & Prewitt, 1969, 1970), constant oxygen ion radius and the hypothesis of compact packing of anions in the oxide and using experimental relations between $(V_{eff}^{m})^{1/3}$ and the radii of metal ions with a given coordination number, for the oxides with "ideal" geometric structure a method of calculation of the relative radius of the sphere occupied by metal ion in a given oxide can be proposed. According to the obtained results, parameters of such equations can be evaluated basing on M_2O alkali metal oxides for ions with coordination number 4, MO alkaline earth metal oxides for ions with coordination number 6 and MO_2 f-electron metal oxides for metal ions with coordination number 8 and. These oxides do not show deviation from stoichiometry (all sites in the structure are occupied) and considering the character of interionic interactions it can be assumed that the oxygen ion radius is constant. Thus, these equations are:

for oxides with coordination number 4:

$$r_M = 0.6868\sqrt[3]{V_{eff}^{m}} - 0.1731 \text{ [nm]} \tag{4}$$

for oxides with coordination number 6:

$$r_M = 0.7622\sqrt[3]{V_{eff}^{m}} - 0.1304 \text{ [nm]} \tag{5}$$

for oxides with coordination number 8:

$$r_M = 0.91\sqrt[3]{V_{eff}^{m}} - 0.1494 \text{ [nm]} \tag{6}$$

The parameters of these equations have been evaluated using molecular volumes of oxides calculated basing on lattice parameters and the table values of ionic radii by Shannon and Prewitt (Shannon & Prewitt, 1969, 1970, 1976) (see Table 1, Stokłosa & Laskowska, 2008b). The values of calculated relative radii of metal ions (relative to oxygen ion, r_O=0.140 nm) in oxides forming compact structures, using parameters of Eq.(4)-(6) and experimental values of molecular volumes in a series of selected oxides are listed in Table 2 of the work (Stokłosa & Laskowska, 2008b). The trend lines marked in Fig. 5 and 6 and 7 have been evaluated for oxides used in the evaluation of the parameters of Eq.(4)-(6).

In Fig. 6 the points corresponding to table radii of M^{4+} ions (points ◆) and the calculated values (points Δ) for r_O = 0.14 nm were marked for the MO_2 oxides of *f*-electron metals.

As it can be seen, the difference between these radii is about 0.002 nm, which indicates that the interionic interactions lead to a shortening of bond lengths and if the oxygen ion radius is assumed to be 0.14 nm, the metal ion radii should be smaller by about 0.002 nm. On the other hand, if we assume oxygen ion radius for coordination number 4 as 0.138 nm, i.e., the value proposed by Shannon and Prewitt (Shannon & Prewitt, 1969, 1970), then metal ion radii do not differ much from the table values.

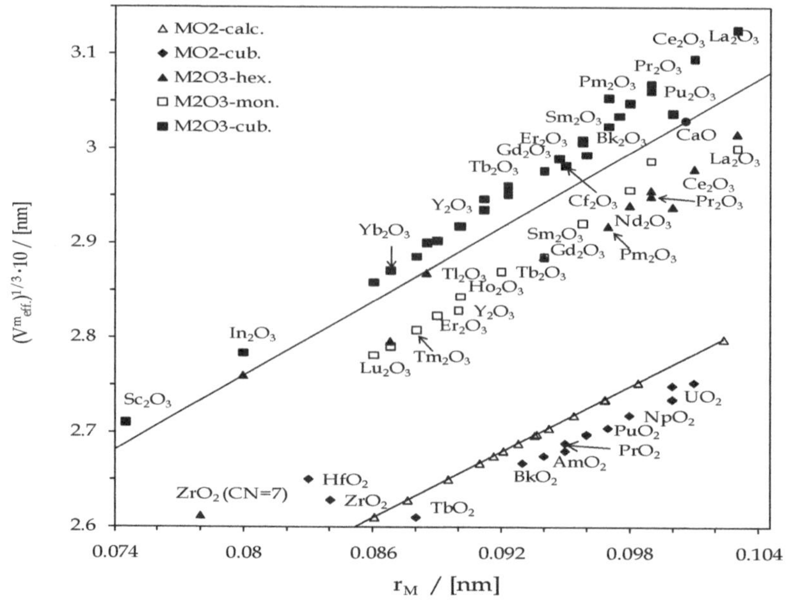

Fig. 6. The relation between the cube root of the molecular volume of *f*-electron metal oxides $(V^{m}_{eff})^{1/3}$ and the radius of cations with coordination number 6 and 8 (the trend lines were evaluated for alkaline earth metal oxides and for MO_2 *f*-electron metal oxides (r_M calculated for r_O=0.140 nm)).

3.3 Relative ionic radii in polymorphic forms of oxides

The introduced concept of relative radius of metal ion (relative radius of the sphere occupied by ion) in compact structures opens up the possibility for evaluation of the relative radius (sphere) of metal ions also in polymorphic forms of a given oxide. Using the experimental values of the effective molecular volume of a given polymorphic oxide form and respective equations (Eq.(4)-(6)) it is possible to calculate the effective radius of the sphere occupied by metal ion in individual polymorphic forms. The calculated values of ion radii in a series of polymorphic oxide forms are given in Table 2 of the work (Stokłosa & Laskowska, 2008b). For illustration, the points corresponding to the $(V^{m}_{eff})^{1/3}$ values for polymorphic forms of many oxides and to the metal ion radii (Shannon & Prewitt, 1969, 1970, 1976) have been plotted in Fig. 6 and Fig. 7.

As can be seen from Fig. 6, M_2O_3 oxides of *f*-electron metals with La_2O_3-type hexagonal structure and Sm_2O_3-type monoclinic structure show practically the same relationship, which indicates that in the monoclinic systems a shift of atoms occurs, but real deformation of structure is small. These oxides have smaller volumes than it would result from the relations for other oxides of this type, with ions having coordination number 6, which may be considered anomalous. They have high cohesive enthalpies and high melting temperatures, being a result of strong interactions between the M^{3+} ions and the oxygen.

Therefore, as a result of interionic interactions a shortening of bond lengths and also of oxygen ion radius occurs. It is big enough that in the M_2O_3 oxides of *f*-electron metals with hexagonal and monoclinic structure the effective radii of metal ions should be smaller by about 0.040 nm (with $r_O = 0.14$ nm). Bigger molecular volumes are found for M_2O_3 oxides of *f*-electron metals with cubic structure being formed at higher temperatures, e.g., by the reduction of MO_2 oxide.

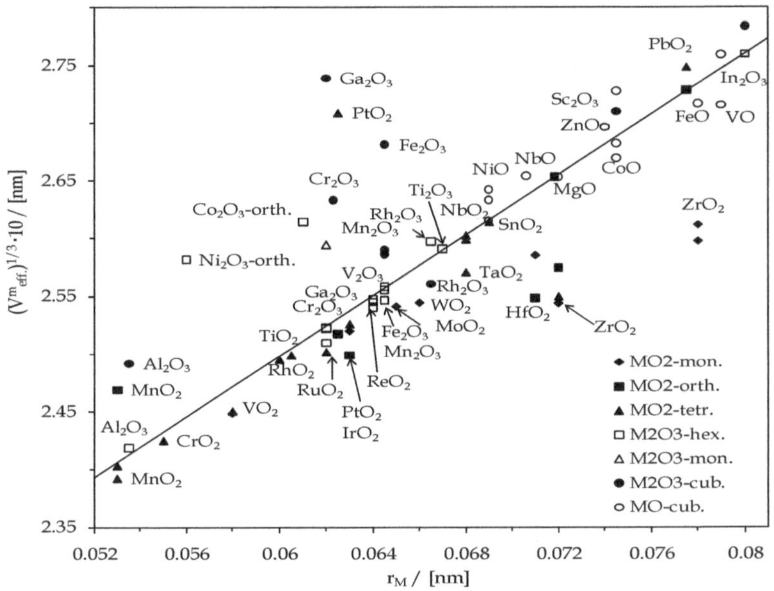

Fig. 7. The relation between the cube root of the molecular volume of oxides and their polymorphic forms $(V_{\text{eff}}^{\text{m}})^{1/3}$ and the radius of cations with coordination number 6 (the trend lines were evaluated for the alkaline earth metal oxides)

As can be seen from Fig. 7 the polymorphic forms of oxides: MnO_2 with orthorhombic structure, Al_2O_3, Cr_2O_3 with cubic structure, Mn_2O_3, Fe_2O_3, and PbO_2 with tetragonal structure, Rh_2O_3 with hexagonal structure, In_2O_3 with cubic structure have molecular volumes bigger than those resulting from their radii (tabulated values). This fact indicates that at higher temperatures, in which these forms are stable, the sphere occupied by ions in this oxide became larger and in such a structure larger radii should be attributed to metal ions (assuming constant oxygen ion radius). On the other hand, the oxides: ReO_2, PtO_2 with tetragonal structures and Rh_2O_3 with an orthorhombic structure have smaller molecular volumes than it would result from the size of their ions. The ions of these metals strongly interact with oxygen ions, which leads to the shortening of bond lengths (and also oxygen ion radius) and as a result, also the molecular volume decreases. Besides, a distortion of coordination polyhedra occurs in these oxides, which results in different M-O bond lengths, and thus also "radii" of ions are different. Most of these forms were created under high pressure.

3.4 Mean relative ionic radii in non-stoichiometric oxides and mixed-valence oxides

In the case of *d*-electron oxides the increase in oxygen pressure leads to a change in the concentration of ionic defects or at a given oxygen pressure a transformation occurs, often into a mixed-valence oxide. In this type phases metal ions are in different oxidation states and their ratio, in the case of defect concentration change, undergoes continuous changes or assumes a defined value, as in spinel structures or in ordered phases, e.g., Magnéli phases. As it was shown in the chapter 2, in spite of the fact that in the structure of the above oxide phases (M_aO_b) there are cation and/or anion vacancies or cations occupy octa- or/and tetrahedral voids (as in spinel structures), their molar volumes are linear functions of the composition (in the composition range of $0.5<a/b<1$). These phases are thermodynamically stable, they undergo mutual transformations and they have compact structures. The above facts indicate that for this type of oxide phases a *mean effective metal ion radius* could be introduced. Such radius would indicate trends in the changes in the interactions of metal ions depending on their content (oxygen composition), ion charge, but also variable ratio of ions in different oxidation states. Their positions in the case of non-stoichiometric compounds are practically impossible to determine (due to electron delocalization). Thus, using Eq.(5) and the values of the effective molecular volume for oxide phases with metal ions with coordination number 6, one can calculate the mean effective radius of metal ions in a given oxide (relative to constant oxygen radius). The mean relative ionic radii in Ti and V oxides showing deviations from stoichiometric compositions and in mixed-valence oxides is given in Table 3 of the work (Stokłosa & Laskowska, 2008b).

3.5 Alternative method for mean ionic radii calculation

The linear relations between the effective molar volume of the oxides of the same metal ($M_{a/b}O$) and their composition, discussed in the chapter 2 and the dependence of $(V_{\text{eff}}^{\text{m}})^{1/3}$ on ionic radii indicate that there should be also a simple correlation between $(V_{\text{eff}}^{\text{m}})^{1/3}$ and the a/b ratio. The relation between $(V_{\text{eff}}^{\text{m}})^{1/3}$ and the a/b ratio for the oxides of Ti and V, and for the oxides of Ce, Pr and Tb is shown in Fig. 8. As can be seen, linear relationships have been obtained, with a high correlation coefficient, $R^2>0.98$. Table 4 in the work (Stokłosa & Laskowska, 2008b) contains parameters of the obtained linear functions:

$$\sqrt[3]{V_{\text{eff}}^{\text{m}}} = A\, a/b + B \tag{7}$$

Comparing the dependence of $(V_{\text{eff}}^{\text{m}})^{1/3}$ on the ion radii (r_M) and on the a/b ratio (Eq.(3) and Eq.(7)) for the oxides of the same metal exhibiting compact structures, one can obtain the equation allowing the mean effective radius of ions in oxides ($M_{a/b}O$) with a given composition (a/b ratio) to be predicted:

$$\overline{r_M} = (A\, a/b + B - \beta)/\alpha = \gamma\, a/b + \delta \tag{8}$$

Thus, this is an alternative method for the estimation of the mean ion radius in mixed-valence oxides or in non-stoichiometric oxides, using values of metal ion radii in various oxides of the same metal. The above method allows for the prediction of a mean ionic radius in oxides having so much deformed structure that it is difficult to determine the coordination number of metal ions or this number changes with the oxidation state.

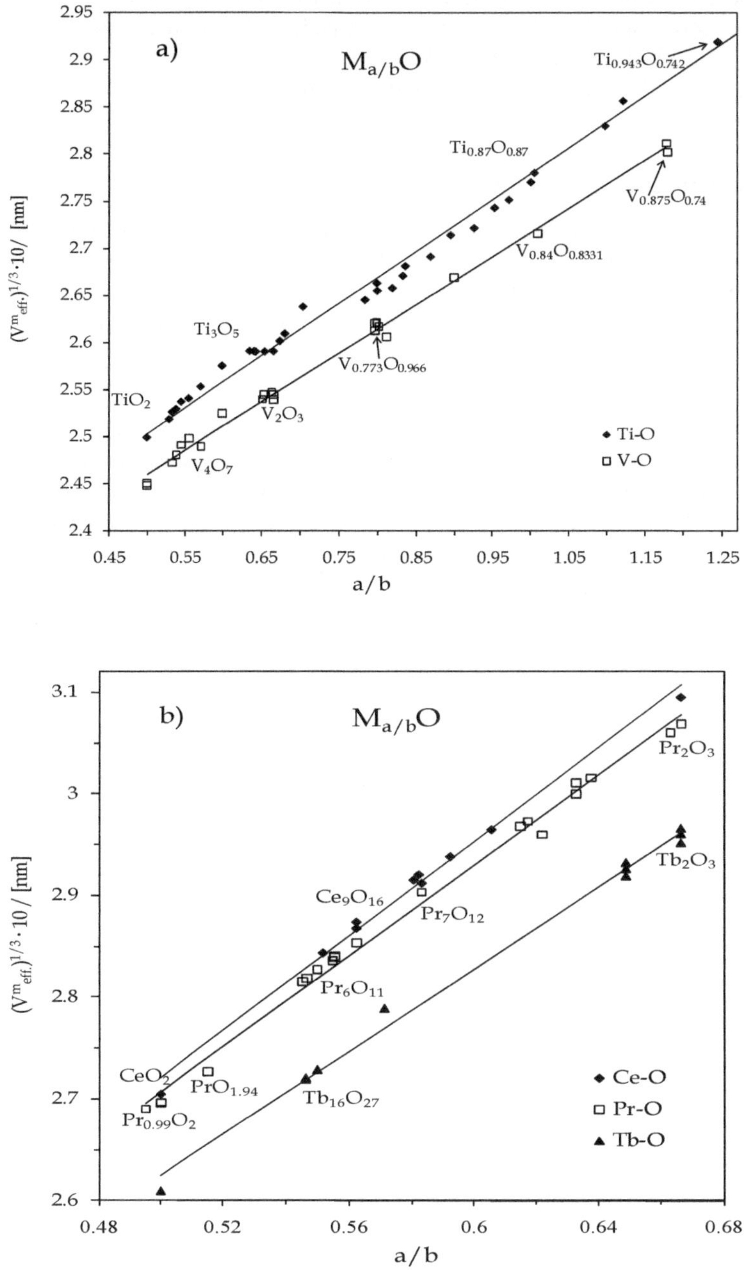

Fig. 8. The relation between the cube root of the molecular volume $(V^m_{eff})^{1/3}$ of oxide phases $M_{a/b}O$ and the a/b ratio for the metal oxides: a) Ti and V, b) Ce, Pr and Tb.

Assuming the table values of the radii of M^{4+} and M^{3+} ions the values of mean radii of Ti and V ions in a series of oxide phases of these metals have been calculated. They are given in in Table 3 of the work (Stokłosa & Laskowska, 2008b). Similarly, ion radii in the oxide phases of Ce, Pr and Tb have been calculated, which was impossible with the previous method.

3.6 Distances between oxygen ions in oxides

In most metal oxides the compact structure the oxygen anions sublattice has the symmetry of a close-packed structure with cations in its voids. Due to the size of cation radii and repulsive interactions, mainly between oxygen ions, distances between oxygen nodes are normally higher than the sum of their radii. The parameter indicating the difference in space-filling by ions and the change of bond character in oxide (M-O) is the distance between oxygen atoms.

The interionic distances can be easily calculated for non-deformed structures (ideal geometric structures), thus in the oxides with NaCl- or CaF_2 -type cubic structures. In oxides with Al_2O_3 -type hexagonal structure or TiO_2 -type tetragonal structure there are different bond lengths and different distances between oxygen atoms. For example, in TiO_2 (rutile) the shortest distance between oxygen atoms is d = 0.252 nm (Vos, 1977; Meogher & Lager, 1979), whereas in SiO_2 (quartz) d= 0.260 nm, and in its high-pressure form d= 0.216 nm (Shannon & Prewitt, 1969, 1976). Furthermore, in Ti_2O_3 d= 0.279 nm (Rice & Robinson, 1977), V_2O_3 d = 0.279 nm (Robinson, 1975), Fe_2O_3 d = 0.262 nm (Newnham &Haan, 1962), and in α-Al_2O_3 d= 0.252 nm (Shannon & Prewitt, 1969). As it can be seen, in many oxides the shortest O-O distance is smaller than 0.28 nm (twice the table value of oxygen ion radius, r_O = 0.14 nm). For the analysis of distances between oxygen ions in oxides with typical structures, the relation between the difference between spheres occupied by oxygen atoms (Δ=d - 0.28) and the effective radius of metal ion has been presented in Fig. 9 (Stokłosa & Laskowska, 2011). The reference distance is equal to a double oxygen ion radius, $2r_O$=0.28 nm. The distances d between oxygen atoms were calculated using lattice parameters, assuming ideal structure of an oxide of a given type. As it can be seen in Fig. 9 many linear correlations were obtained. A correlation is found for the M_2O alkali metal oxides. In the Li_2O oxide the distance between spheres occupied by oxygen atoms is 0.04 nm, in the case of K_2O 0.175 nm, and in Rb_2O as much as 0.198 nm. Similarly good linear relation was found in the case of MO oxides with NaCl-type cubic structure (the trend lines were evaluated for the alkaline earth metal oxides).

The smallest M^{2+} ion is V^{2+} and also in VO oxide the distance between the spheres occupied by the oxygen ion is the shortest and equals only 0.013 nm. It should be noted that VO and TiO show significant point defects concentrations in cation sublattice and oxygen sublattice and in NbO oxide 25% lattice nodes is unoccupied. With the increase of metal ion radius also the distance between oxygen ions increases and in the case of BaO it equals 0.179 nm.

In the case of MO_2 oxides with TiO_2-type tetragonal structure with deformed MO_6 octahedrons, for the analysis of the changes in differences Δ the distances between oxygen atoms laying in planes (000) and (001) were compared. This distance equals c parameter (d = c in tetragonal lattice). As it can be seen in Fig. 9, the correlations were obtained separately for two oxide groups, which indicates that the structures of these oxides are deformed in different ways.

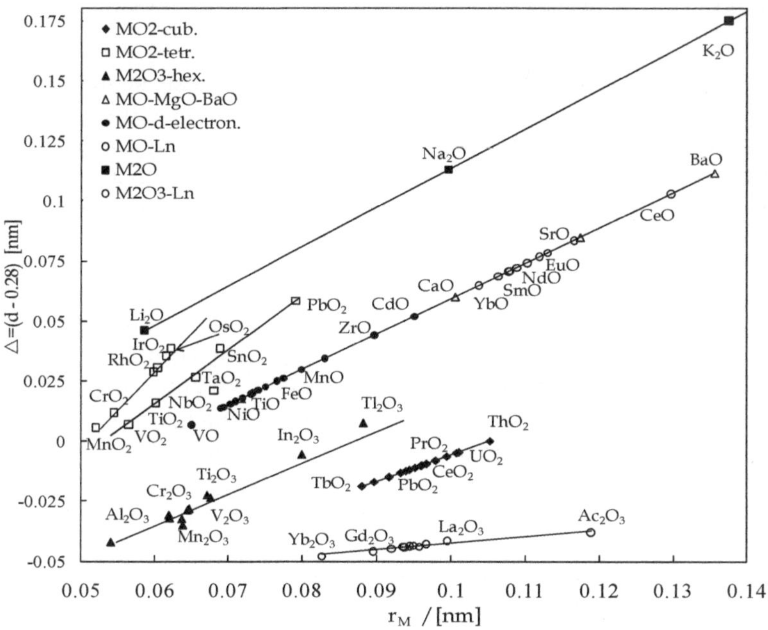

Fig. 9. The dependence of the distances between the spheres occupied by oxygen atoms ($\Delta=d-2r_O$) where $r_O=0{,}28$nm) and the ionic radii of metals in M_2O oxides with anti-CaF_2 structure, M_2O_3 oxides with α-Al_2O_3 structure, and MO_2 oxides with CaF_2 and TiO_2 structure.

In the case of M_2O_3 oxides of α-Al_2O_3-type, having similarly deformed oxygen octahedrons, the O-O distance equal a half of the lattice parameter was assumed in calculations. As it can be seen in Fig. 9, for M_2O_3, with the exception of Tl_2O_3, negative values of Δ have been obtained, and this function has linear character similarly as functions for other oxides. The smallest radius is found for the aluminium ion, 0.054 nm. It is comparable to the void size in an ideal close-packed structure of oxygen ions, the void radius being 0.058 nm (assuming the oxygen radius equals 0.14 nm). Negative values of distances between the spheres occupied by oxygen ions (Δ) indicate that these spheres in the M_2O_3 with α-Al_2O_3 structure undergo deformations. This is related to the fact that only 2/3 of octahedral voids in hexagonal close-packing structure of oxygen ions is occupied by metal atoms, whereas optimal interactions span through the entire volume of the crystal. This leads to shifts of metal ions and also of oxygen ions comparing to the ideal close-packed structure. As a result, most *d* distances between oxygen atoms are bigger than 0.28 nm. Small distances between spheres occupied by oxygen atoms in these oxides indicate that in these oxides strongly polarized M-O bond must be present. A confirmation of this fact is the highest cohesive enthalpy of α-Al_2O_3 and also its melting temperature; similarly high melting temperatures are observed for remaining M_2O_3 oxides with hexagonal structures. If the deformation of oxygen atoms arrangement had not occurred, and if the spheres occupied by oxygen ions had been in contact, then, as it can be seen from Fig. 9, the oxygen ion radius would have been smaller by 0.021 nm. In this case the radius of octahedral void (inter) is 0.049 nm and the radius of the sphere occupied by the aluminium ion could not be bigger than this value.

As it can be seen in Fig. 9, the MO_2 oxides of *f*-electron metals similarly have negative values of parameter Δ (the distance between the spheres occupied by oxygen ions). In the structure of these oxides, contrasting with the previous ones, metal ions have an arrangement typical for close-packing lattices. If we assume, that in TbO_2 oxide the oxygen ion spheres come in touch, then the oxygen ion radius should be $r_O = 0.131$ nm. It is therefore smaller than that proposed by Shannon and Prewitt ($r_O = 0.138$ nm for CN = 4 (Shannon & Prewitt 1969)). If we assume that this value of oxygen ion radius in other oxides remains constant, then, as it can be seen in Fig. 9, the distances between the spheres occupied by oxygen ions increase with the increase of metal ion radius in the next MO_2 oxides and for the ThO_2 oxide this distance will equal 0.019 nm. In the case of MO_2 lanthanide and actinide oxides, their bond energies and melting temperatures increase with the increase of the distances between oxygen ions (highest melting temperature is observed for ThO_2) (Lide, 2004; Stokłosa & Laskowska, 2007a). The increase of the bond energies can be an effect of the increase in the M-O interaction related to the electronic configuration of M^{4+} ions and their polarizing interaction ability. Considering the properties of a M^{4+} ion, the increase of bond energy can be a result of the repulsive contribution to the interaction following the increase of the distances between oxygen ions when the metal ion radius increases. Therefore, in the case of MO_2 oxides of *f*-electron metals, small distances between the spheres occupied by oxygen ions can lead to the increase of the repulsive interactions, which will imply the weakening of bond energies in these oxides, on the contrary to the situation for the *d*-electron oxides.

The present analysis of the distances between the spheres occupied by the oxygen ion indicates that its radius in many oxides must change, especially in the case of small M^{3+} and M^{4+} ions and strong polarizing interaction. Shannon and Prewitt (Shannon & Prewitt, 1969, 1976) have widely discussed the change of metal ion radius depending on the coordination number. Ziółkowski and Dziembaj (Ziółkowski & Dziembaj, 1985a, 1985b), developing this problem, has proposed a method for the evaluation of the ion radius for coordination number 0 (radius of free ion, r_0), regardless of the environment. He also proposed an electrostatic hover model of crystal structure, permitting the change of both the size and form of the sphere occupied by the metal ion with a minimal radius r_0.

3.7 Packing density

The size of the ionic radii and the character of their interactions, dependent on the metal type, is reflected in the space packing density. Knowing the values of ionic radii one can calculate the packing density of ions in individual oxides. The relation between the packing fraction of ions and the metal ion radius is presented in Fig.10. The relative metal ion radii were used and the oxygen ion radius was assumed to be 0.14 nm. The packing fraction was calculated as the ratio of the volume occupied by ions (V_{at}) to the volume of the unit cell calculated from lattice parameters (V_X).

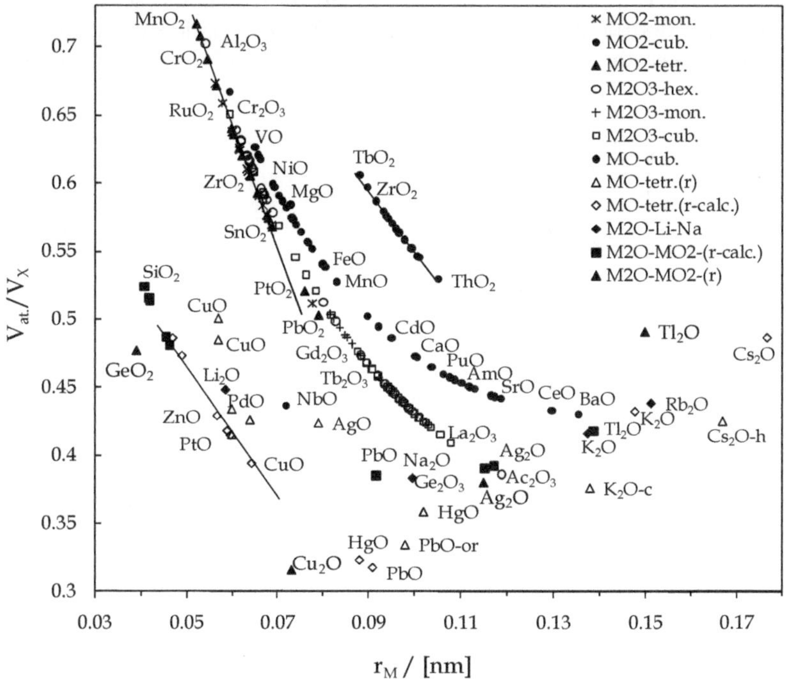

Fig. 10. The dependence of the packing density of a series of metal oxides with compact structures on the ionic radii calculated using relative ionic radii.

As it can be seen in Fig. 10, a monotonic relationship between the packing fraction and metal ion radii was obtained for the oxides of the same type forming compact structures. It is worth noticing that the packing fractions of oxides of the same type, of different metal differ rather significantly. The MO_2 oxides, except for PtO_2 and PbO_2, lay on a straight line, whereas the curve determined by M_2O_3 and MO_2 oxides deviates with the increase of the ionic radius. On the other hand, the MO_2 *f*-electron metal oxides, with relatively big radii, have a significant packing fraction, close to the MO *d*-electron metal oxides, having metal ions of about 30% smaller. A different relationship character is observed for alkali metal oxides that, as it was mentioned, have an anti-CaF_2 structure, and one should expect the relationship character similar to this for oxides with a CaF_2 structure. (In the structure of these oxides metal ions have an arrangement typical for close-packing lattice with oxygen ions occupying its voids). The points for Cu_2O, Ag_2O and Tl_2O lay close to the curve for the alkali metal oxides. That indicates that the composition of these oxides (M_2O) approaches them considering packing to the alkali metal oxides. On the other hand, SiO_2 and GeO_2 oxides and MO oxides with hexagonal structure form a separate straight line. Close to it is the point for Li_2O. Despite the small ionic radii of these metals, these oxides have a small packing fraction, related to a characteristic structure of framework bonds typical for covalently bonded compounds (diamond-type). For comparison, the points calculated for ions with CN = 4 (Eq.(4)) were plotted for the oxides of this group. The calculated packing fraction does not change much.

An important issue is a change in packing density in oxides of the same metal, thus in oxides with ordered structures and showing significant deviation from the stoichiometry. The relations between the packing fraction and the *a/b* ratio for the oxides of Ti, V and Ce, Pr, Tb are shown in Fig. 11.

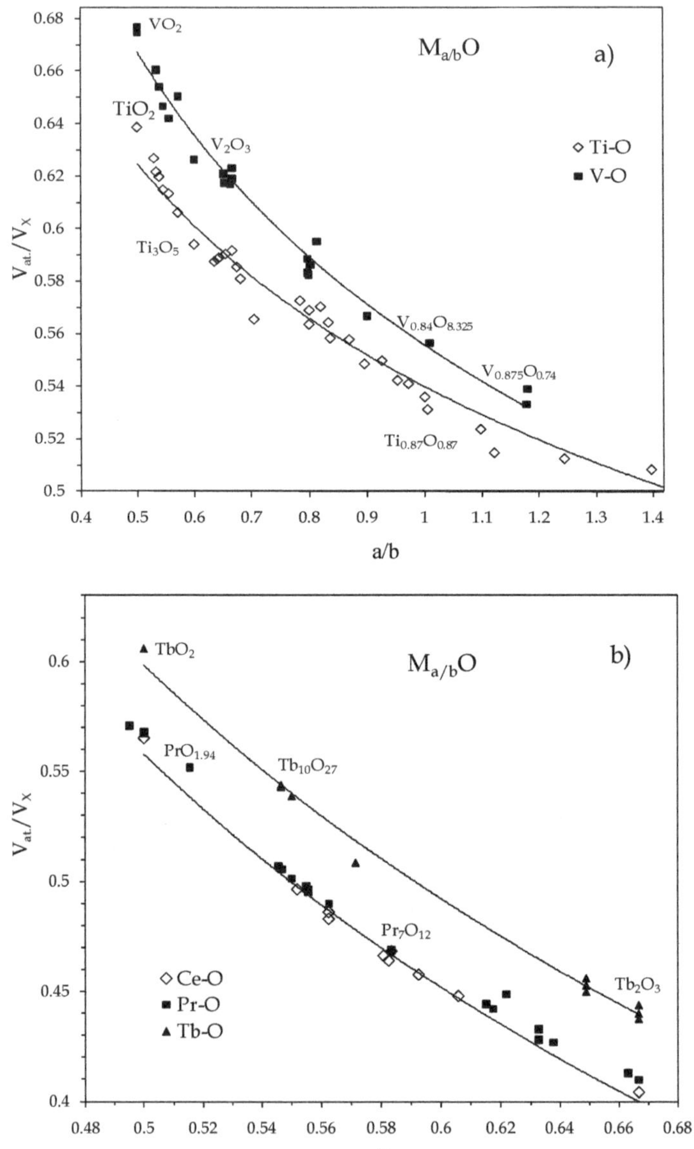

Fig. 11. The dependence of the packing density in the oxide phases (M_aO_b) of metals: a) Ti and V, b) Ce, Pr and Tb on the *a/b* ratio, calculated using mean ionic radii.

Average ionic radii were used in the calculations. This fraction, as it can be seen, decreases monotonically with the increase of metal content in the oxide (increase of *a/b* ratio). It should be noted that the determined ionic radii or interionic distances reflect only ion positions in a frozen structure (in which the lattice parameters are determined). At high temperatures and in the equilibrium state the displacements of ions/atoms occur, so their positions can differ from those in frozen structures. Before all, the optimization of interactions and also of interatomic distances occur. Nevertheless information about the possibilities for modification of the sphere occupied by ions, and of packing fraction, is important for many crystal properties at high temperatures, e.g. diffusion of ions. This results from the fact that moving ions cause local deformation in their close environment.

4. Conclusions

Basing on the presented analysis of the molar volume of oxides the following conclusions can be formulated:

- The effective molar volume of the oxide $V_{eff}= V_{M_{a/b}O}$ (per one mole of oxygen atoms) is a parameter allowing to compare oxides with different compositions, structures and with ions in different oxidation states.
- In the composition range $1>a/b>0.5$ (and in the case of actinide oxides up to $a/b<0.35$) the oxides undergoing mutual transformations have compact structures, and the effective molar volumes of the oxides of the same metal decrease with the decrease of its content (with decreasing *M/O* ratio). This type of relationship can be an additional criterion to classify an oxide of a given structure as a compact structure. The polymorphic variations of these oxides will have compact structures, because slight changes of their molar volumes, caused by changes of ion positions are the result of the changes of temperature or pressure.
- In the composition range $0.25<a/b<0.5$ the oxides form usually structures with bigger volume than the MO_2 oxides. The observed increase of the effective volume with the decrease of the metal content is the result of the increase of the effective repulsive forces contribution caused by the increase of the amount of oxygen atoms relatively to metal and the strong repulsive interaction of ions with high charges, M^{5+} and/or M^{6+}. This leads to the formation of structures with loose space packaging (comparing to the MO_2 oxides). The presence of ions in high oxidation states, as well as a significant contribution of the covalent bond (the formation of hybrid *spd* orbital) results in quite high cohesive enthalpies of these oxides. As a result, the formed framework structures are stable.
- The $\sqrt[3]{V_{eff}^{m}}$ parameter (cube root from the molecular volume – the volume of an oxide containing one $M_{a/b}O$ molecule) for the oxides with composition $0.5<a/b<2$ (from M_2O to MO_2) forming compact structures for individual groups of oxides with coordination numbers 4, 6 or 8 shows linear dependence on the ion radius.
- Using experimental relations for alkali metal oxides, alkaline earth metal oxides and MO_2 *f*-electron metal oxides with CaF_2 structure, the effective metal ion radii relative to constant oxygen ion radius equal 0.140 nm may be determined for oxides with compact

structures. This method allows the relative radius (sphere radius) of metal ion in polymorphic form of oxides to be calculated.

- Using the concept of a compact structure and the above method of relative ionic radii calculation, mean ion radii in oxides showing deviations from the stoichiometric composition and in mixed-valence oxides, or generally in ordered phases may be determined. Such radius indicates trends in the changes in interactions of metal ions depending on ion charge, their number (oxide composition), but also on the variable ratio of ions in different oxidation states of the metal.
- Assuming radii values determined in relation to constant oxygen radius (r_O=1.4 nm) a linear dependence of the difference in distances between spheres occupied by the oxygen ion (Δ) on the radius of the metal ion was obtained. For a series of M_2O_3 oxides with α-Al_2O_3 structure and MO_2 lanthanides and actinides negative values were obtained of the difference between the distance between oxygen atoms and their double radius; this indicates that the real sphere occupied by oxygen ions must be smaller than the assumed oxygen radius (0.14 nm) or/and lattice deformations occur. The oxygen radius depends on the metal type and on its polarizing abilities (strength of polarized M-O bond).
- The packing density depends on the ion radius and on the character of interatomic interaction (with ionic-covalent character) which is reflected in the different character of the relation between the packing density and cation radius. These relations are obviously different in the case of MO and M_2O_3 oxides of *d*- and *f*-electron metals and MO_2 oxides of *f*-electron metals compared to alkali metal oxides and metal oxides with framework structure with dominating covalent interaction.
- The work (Stokłosa & Laskowska, 2007a, 2007b) presented the study on the properties of oxides of main group metals and *d*- and *f*-electron metals in relation to the bond energy and the cube root of molecular volume. The cohesive enthalpy per one mole of oxygen atoms was used to compare the bond energy of oxides of various composition. The value of this enthalpy was calculated from thermochemical data. A linear relation between the cohesive enthalpy and the composition was found for oxide phases of the same metal. The changes in electrical properties of the oxides and the concentration of electronic and ionic defects in relation to the above parameters have been discussed. The reasons for the formation of many oxide phases of the same metal, the formation of ordered phases and the formation of phases having significant deviation from the stoichiometric composition were investigated.

5. Acknowledgments

The author are very grateful to Drs S.S. Kurek and B. Laskowska for valuable help and discussion.

6. References

Adams, D.M. (1974). *Inorganic Solids, An Introduction to Concepts in Solid State Structural Chemistry*, Wiley, London, ISBN 0471004707

Ahrens, L.H. (1952). The use of Ionization Potentials. Part 1. Ionic Radii of the Elements *Geochim. et Cosmochim. Acta*, Vol.2, No. 3, pp. 155-169

Ahrens, L.H. (1953). The use of Ionization Potentials. Part 2. Anion Affinity and Geochemistry, *Geochimica Cosmochimca Acta*, Vol.3, No.1, pp 1-29

Allen, L.C. (1998). Electronegativity and the Periodic Table, in: *Encyclopedia of Computational Chemistry*, P. Schleyer, (Ed.), J. Wiley, New York

Allred, A.L. & Rochow, E.G. (1958). A Scale of Electronegativity Based on Electrostatic Force, *J. Inorg. Nucl. Chem.* Vol.5, No.4, pp.264-268

Brisse, F. & Knop, O. (1968). Pyrochlores. III. X-Ray, Neutron, Infrared, and Dielectric Studies of $A_2Sn_2O_7$ Stannates, *Canad. J. Chem.*, Vol.46, pp. 859-873

Brown, J.D. (1992). Chemical and Steric Constraints in Inorganic Solids, *Acta Cryst. B*, Vol.48, pp. 553-572

Brown, J.D. (2002). The Chemical Bond in Inorganic Chemistry. The Bond Valence Model, Oxford Univ. Press, New York, ISBN 0 19 850870 0

Cartledge, G.H. (1928). Studies on the Periodic System. I. The Ionic Potential as a Periodic Function, II. The Ionic Potential and Related Properties, *J. Am. Chem. Soc.*, Vol.50, No.11, pp.2855–2872

Cartledge, G.H. (1930). Studies on the Periodic System. III. The Relation between Ionizing Potentials and Ionic Potentials, *J. Am. Chem. Soc.*, Vol.52, No.8, pp.3076–3083

Donnelly, R.A. & Parr, R.G. (1978). Elementary properties of an energy functional of the first-order reduced density matrix *J. Chem. Phys.* 69, No.10, 4431-4440

Fumi, F.G. & Tosi, M.P. (1964). Ionic Sizes and Born Repulsive Parameters in the NaCl-type Alkali Halides—I: The Huggins-Mayer and Pauling Forms, II. The generalized Huggins-Mayer Form, *J. Phys. Chem. Solids*, Vol.*25*, No.1, pp. 31-52

Gazquez, J.L. & Ortiz, E. J. (1984). Electronegativities and Hardnesses of Open Shell Atoms, *J. Chem. Phys.* Vol.81, No.6, pp.2741-2749

Goldschmidt, V.M. (1926). Geochemische Verteilungsgesetze der Elemente, VII, *Die Gesetze der Kristallchemie*, Skrifter, Norsk Videnkaps Akad. Oslo, I. Math. Naturwid. Kl, Ig..

Goldschmidt, V.M. (1931). *Fortschr. Mineral.*, Vol.15, pp. 73

Görlich, E. (1989a). The Effective Nuclear Charges and their Ralelation to the Paulin's Electronegativity Scale, *Z. Phys. Chem. Leipzig*, Vol.270, No.2, pp.384-388

Görlich, E. (1989b). The Effective Nuclear Charges and a Standardized Ionicity Scale, *Z. Phys. Chemie Leipzig*, Vol.270, No.1, pp.961-967

Görlich, E. (1990). The Effective Nuclear Charges and the Softness of the Atomic Cores, Ralelation to the Paulin's Electronegativity Scale, *Z. Phys. Chem. Leipzig*, Vol.271, pp 169-174

Hohenberg, P. & Kohn, W. (1964). Inhomogeneous Electron Gas, *Phys. Rev. B: Solid State*, Vol.136, No.3B, pp. B864–B871

Jia, Y.Q. (1991). Crystal Radii and Effective Ionic Radii of the Rare Earth Ions, *J. Solid State Chem.* Vol.95, No.1, pp.184-185

Kohn, W.;. Becke, A.D., &. Parr R.G. (1996). Density Functional Theory of Electronic Structure, *J. Phys. Chem.*, Vol.*100, No.*31, pp.12974–12980

Komorowski, L. (1993) Hardness Indices for Free and Bonded Atoms, In: *Structure and Bondind.*, Vol.80. pp. 45-70

Kunz, M. & Brown, I.D. (1995). Out-of-Center Distortions around Octahedrally Coordinated d^0 Transition Metals, *Journal of Solid State Chemistry*, Vol.115, pp.395-406

Lalik, E. (2005). Shannon Information as a Measure of Distortion in Coordination Polyhedral, *J. Appl. Cryst.*, Vol.38, 152-157

Lide D.R. (ed) (2004). *CRC Handbook of Chemistry and Physics*, CRC Press, Boca Raton, 08493-0485-7

Luehen, H.; Elsenhans, U. & Stamm, U. (1987). Low-Symmetric Coordination Polyhedra - Pseudosymmetry and Idealization, *Acta Cryst. A*, Vol.43, pp. 187-194

Meogher, E.P. & Lager, G.A. (1979). Polyhedral Hermale Xpansion in that Ion, Polymorphs: Refinement of the Crystal Structures Rutile and Brookite at High Temperature, *Canadian Mineralogist*, Vol.17, pp. 77-85

Mulliken, R.S. (1934). A New Electroaffinity Scale; Together with Data on Valence States and on Valence Ionization Potentials and Electron Affinities, *J. Chem. Phys.* Vol.2, pp.782-794

Mulliken, R.S. (1935). Electronic Structures of Molecules XI. Electroaffinity, Molecular Orbitals and Dipole Moments, *J. Chem. Phys.* Vol.3, pp. 573-586

Nespolo, M.; Ferraris, G.; Ivaldi, G. & Hoppe, R. (2001). Charge distribution as a tool to investigate structural details. II. Extension to hydrogen bonds, distorted and hetero-ligand polyhedral, *Acta Cryst. B*, Vol.57, pp.652-664,

Newnham, R.E. & de Haan Y.M. (1962). Refinement of the *a* Al_2O_3, Ti_2O_3, V_2O_3 and Cr_2O_3 Structures, *Zeitschrift für Kristallographie*: Vol. 117, No. 2-3, pp. 235-237

O'Keeffe, M.; Navrotsky, A. (1981) (Eds.), *Structure and Bonding in Crystals*, Academic Press. New York, QD911:O4

Parr, R.G., &. Pearson, R.G, (1983). Absolute Hardness: Companion Parameter to Absolute Electronegativity, *J. Am. Chem. Soc.*, Vol.*105, No.*26, pp. 7512–7516,

Pauling, L. (1927). The sizes of ions and the structure of ionic crystals, *J. Am. Chem. Soc.*, Vol.49, No.3, pp. 765-790

Pauling, L. (1929). The Principles Determining the Structure of Complex Ionic Crystals, J. *Am. Chem. Soc.* Vol.51, No.4, pp. 1010-1026

Pauling, L. (1932). The nature of the Chemical Bond. iv. the Energy of Single Bonds and the Relative Electronegativity of Atoms, *J. Am. Chem. Soc.,, 54* No.9, pp. 3570–3582

Pauling, L. (1960). *The Nature of the Chemical Bond*, Cornell Univ. Press, Ithaca, New York,.

Pearson, R.G. (1988). Absolute Electronegativity and Hardness, Application to Inorganic Chemistry, *Inorg.Chem.* Vol.27, pp. 734-740,

Pearson, R.G. (1997). *Chemical Hardness*, J. Wiley-VCH, New York, ISBN 3-427-29482-1

Phillips, C.S.G. & Williams, R.J.P. (1965). *Inorganic Chemistry*, Oxford Univ. Press, New York

Rao, C.N.R. (Ed) (1993). *Chemistry of Advanced Materials*, Blackwell, Oxford, ISBN 0-632-03385-1

Rao, C.N.R. & Raveau, B. (1995). *Transition Metal Oxides,* VCH Publ. New York, ISBN 1-56081-647-3

Rice, C.E. & Robinson WR (1977). High-Temperature Crystal Chemistry of Ti_2O_3: Structural Changes Accompanying the Semiconductor-Metal Transition, *Acta Crystall.* B Vol.33, pp.1342-1348

Robinson, W.R. (1975). High-temperature Crystal Chemistry of V_2O_3 and 1% Chromium-Doped V_2O_3, *Acta Crystall* B Vol.31. pp.1153-116

Sanderson, R.T. (1976). *Chemical Bonds and Bond Energy,* Academic Press. New York

Shannon, R.D & Prewitt, C.T. (1969). Effective Ionic Radii in Oxides and Fluorides, *Acta Cryst. B,* Vol.25, pp.925-946

Shannon, R.D. & Prewitt, C.T. (1970). Revised Values of Effective Ionic Radii, *Acta Cryst. B,* Vol.26, pp.1046-1048

Shannon, R.D. (1976). Revised Effective Ionic Radii and Systematic Studies of Interatomic Distances in Halides and Chalcogenides, *Acta Cryst. A,* Vol.32, pp.751-767

Slater, J.C. (1930). Atomic Shielding Constants, *Phys. Rev.* Vol.36, pp.57–64

Stokłosa, A; Zajęcki, J. & Kurek, S.S. (2004). Effective Nuclear Charge of the Ion, *Materials Science-Poland,* Vol. 22. No.1, 35-45

Stokłosa, A. & Laskowska, B. (2007a). The Bond Energy and the Composition of Metal Oxides, *High Temperature Materials and Processes,* Vol.26, No.2, pp. 93-102

Stokłosa, A. & Laskowska, B. (2007b). Trends in the Changes of Properties of Oxides in Relation to Band Energies and Interatomic Distances. *High Temperature Materials and Processes,* Vol.26, No.5-6, pp. 317-339

Stokłosa A. & Laskowska, B. (2008a) Molar Volume and Composition of Oxides of the Same Metal, *J. Chem. Crystallogr,* Vol.38, pp. 279-284

Stokłosa, A. & Laskowska, B. (2008b). Ionic Radii in Mixed-Valance and Nonstoichiometric Metal Oxides and Their Polymorphic Forms, *J. Chem. Crystallogr,* Vol.38, pp. 913-925

Stokłosa, A. & Laskowska, B. (2011). Influence of Ionic Interactions and Ion Sizes on Their Packing Density in Metal Oxides, *J. Chem. Crystallogr,* Vol.41, pp. (on line)

Von Meerssche, M. & Dupont, J.F. (1976). *Introduction à la cristallographie et à la Chimie structurale,* OYEZ, Luven

Vos, K (1977). Reflectance and Electroreflectance of TiO_2 Single Crystals. II. Assignment to Electronic Energy Levels, *J. Phys. C*, Vol.10, No.19, pp. 3893- 3917

Wasastjerna, J.A. (1923). *Soc. Sci. Fennica, Commentationes Phys. Math.I,* Vol.38, pp.12

Wells, A.K. (1990). *Structural Inorganic Chemistry,* Clarendon Press, Oxford

West, A.R. (1996). *Basis Solid State Chemistry,* J. Wiley, Chichester, New York, ISBN 0-471-98755-7 (hbk.)

Zachariansen, W.H. (1931). *Z. Kristallogr.,* Vol.80, pp.137

Ziółkowski, J. & Dziembaj L. (1985a). Empirical Relationship beteen Individual Cation-Oxygen Bond, *Journal of Solid State Chemistry,* Vol.57, No.3., pp. 291-299

Ziółkowski, J. (1985b). New Ralation between Ionic Radii, Bond Length, and Bond Strength, *Journal of Solid State Chemistry*, Vol.57, No.3. pp. 269-290

9

Oxygen Potentials and Defect Chemistry in Nonstoichiometric $(U,Pu)O_2$

Masato Kato
Japan Atomic Energy Agency
Japan

1. Introduction

Nuclear reactors have been developed to secure stable long-term energy supply. Light water reactors (LWRs) have been operated on a commercial basis in many countries. In LWRs, uranium oxide is used as a fuel. Plutonium which is another element that can be used for generating nuclear energy is produced by a nuclear reaction during burning of uranium oxide fuel and it is extracted from spent fuel when they are reprocessed. Uranium and plutonium mixed oxide s (MOX) fuels containing Pu of less than and more than 15% have been developed for LWRs and fast reactors, respectively. These oxides are fabricated into pellets by powder metallurgy. In the pellet production process, dimensions, density and oxygen-to-metal (O/M) ratio of the pellets are managed as important fuel specifications. The O/M ratio significantly affects various physical properties. Therefore it is essential to understand the O/M ratio dependence on physical properties for analyzing irradiation behavior.

Uranium and plutonium mixed oxide $(U,Pu)O_2$, which is used in nuclear fuels, is a substitutional solid solution compound having the CaF_2 type fluorite structure as shown in Fig.1 (Markin & Street, 1967; Sari et al., 1970; Kato & Konashi, 2009). $(U,Pu)O_2$ is stable in a wide range region of hyper- and hypo-stoichiometric compositions. The O/M ratio in nonstoichiometric $(U,Pu)O_2$ is dominated by the numbers of oxygen vacancies and interstitial oxygen atoms. The stoichiometry of $(U,Pu)O_2$ significantly affects various physical properties such as lattice parameter, melting temperature, and thermal conductivity. (Kato et al., 2011a) Therefore many studies on O/M ratio dependence of oxygen potential have been carried out by various methods. The oxygen potential of $(U,Pu)O_2$ drastically changes with a slight O/M change in the near stoichiometric region. Such drastic change may cause a large uncertainty in oxygen potential measurements, and the measured data may be scattered in a range of more than ±100 kJ/mol.

Oxygen potentials have been measured by various methods including the E.M.F. and the gas equilibrium method using thermo-gravimetric measurements. The E.M.F. method cannot be applied to measurements above 1473 K temperature, and it is difficult to use it to determine slight O/M changes in the near stoichiometric region. In the thermo-gravimetric measurements, various gas mixtures like CO/CO_2 and H_2O/H_2 are employed for controlling the oxygen potential. The oxygen potential region control depends on the kind of gases used for the measurements. So, it is crucial to understand the relationships among the used gas types, oxygen potential and O/M ratio. Furthermore, for the thermo-gravimetric measurements as well, it is very difficult to determine the stoichiometric composition in the high temperature region because of the O/M ratio continuous variation for the near stoichiometric composition.

In previous studies (Kato et al., 2009a, 2011b 2011c), the stoichiometric compositions in $(U,Pu)O_2$ have been determined based on defect chemistry. The relationship between oxygen partial pressure and deviation x from stoichiometric composition has been analyzed in non-stoichiometric oxides. Kosuge (1993) used statistical thermodynamics considerations for description of non-stoichiometric compounds, and Karen (2006) reported a point-defect scheme for them. Recently their methods have been applied for nonstoichiometric $(U,Pu)O_2$, and experimental data, accurately measured in the near stoichiometric region, were analyzed as a function of temperature. In this report the measurement data and the measurement technique were reviewed and analysis results based on defect chemistry were summarized.

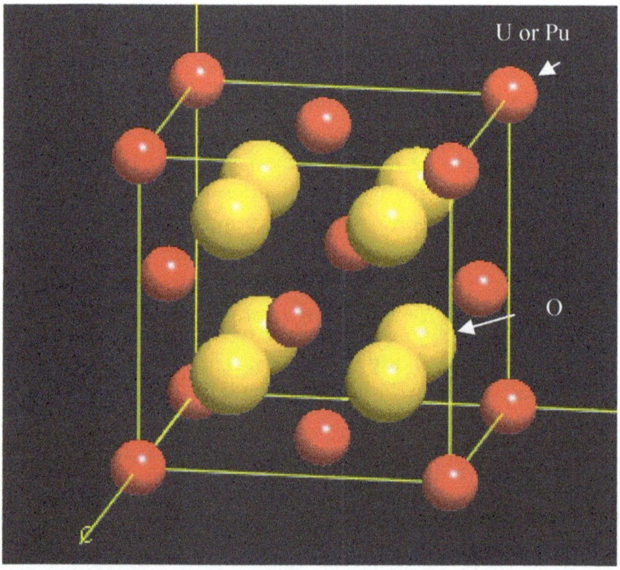

Fig. 1. CaF_2 type fluorite structure

2. Experimental data of oxygen potential in $(U,Pu)O_2$

The measurement data of $(U,Pu)O_2$ are listed in Table 1. The relationships between oxygen potential and O/M ratio were obtained as functions of temperature and Pu content. Almost all data that were measured in the 1960's and 1970's had large uncertainty. Recently Kato et al. (2005, 2009b, 2011b, 2011c) accurately obtained a large number of data in $(U,Pu)O_2$ by in-situ analysis. Fig.2 shows the measurement data as functions of O/M ratio and Pu content. The data were measured in both regions of hyper- and hypo-stoichiometric compositions and they increased with Pu content. Fig.3 shows the oxygen potential $\Delta\bar{G}_{O_2}$ of $(U_{0.8}Pu_{0.2})O_{2\pm x}$. The oxygen potential is calculated by eqs.(1)-(2).

References	Pu (%)	O/M	Temp. (K)	Data points
Woodley, 1981	10	1.96-2.00	1273-1473	17
Sørensen,1976	10	1.98-1.99	1222-1674	12
Markin & McIver, 1965	11	1.95-2.08	1073-1373	45
Kato et al., 2011b	12	1.988-2.0408	1673-1873	76
Chilton & Kirkham, 1976	15	2.002-2.106	1518-1823	24
Javed, 1973	20	1.920-1.989	1273-1973	20
Sørensen,1976	20	1.99-2.00	1359-1674	17
Mari et al., 1977	20	1.96-1.99	757-1276	24
Tetenbaum, 1974	20	1.92-1.96	2150-2550	25
Kato et al., 2009b,2011c	20	1.979-2.0088	1473-1873	105
Vasudeva Rao et al., 2006	21	1.964-2.00	1073-1473	53
Chilton & Edwards, 1980	23	1.980-2.030	1523-1822	114
Woodley &Adamson, 1985	25	1.90-2.00	1222-1674	68
Vasudeva Rao et al., 2006	28	1.975-2.00	1073-1473	19
Markin & McIver, 1965	30	1.86-2.06	1073-1373	39
Kato et al., 2005, 2011c	30	1.964-2.028	1273-1873	153
Chilton & Kirkham, 1976	31	1.99-2.07	1518-1823	52
Woodley, 1981	40	1.90-2.00	1273-1473	23

Table 1. Measurement data of (U,Pu)O_2

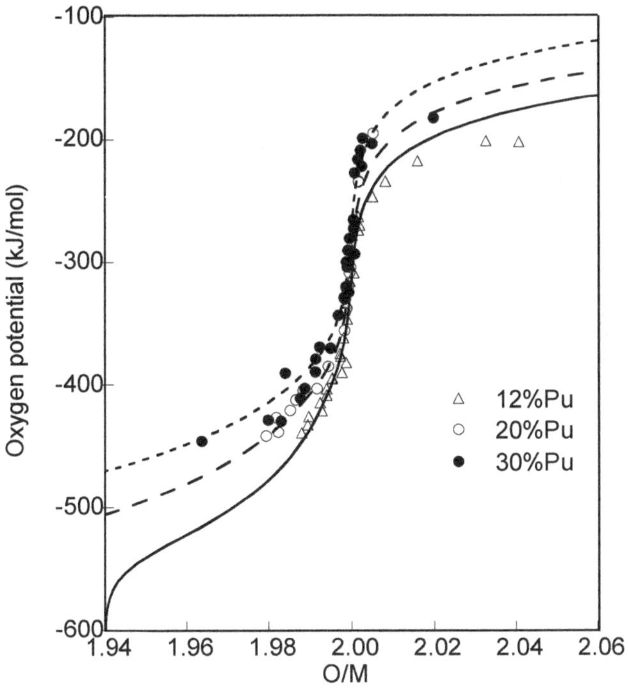

Fig. 2. Oxygen potential at 1873 K as function of Pu content and O/M ratio (Kato et al., 2009b, 2011c)

$$\Delta\overline{G}_{O_2} = RT\ln P_{O_2} \tag{1}$$

$$\Delta\overline{G}_{O_2} = \Delta\overline{H}_{O_2} - T\cdot\Delta\overline{S}_{O_2}\, , \tag{2}$$

Here R is the gas constant (8.3145 J/K/mol), T is absolute temperature $\Delta\overline{H}_{O_2}$ is partial molecular enthalpy of O_2 and $\Delta\overline{S}_{O_2}$ is partial molecular entropy of O_2. Eq.(2) predicts that the oxygen potential varies in a linear manner as a function of temperature. The data, however, have an inflection point at hypo-stoichiometric composition of 2.00>O/M>1.95 as shown in Fig.3. This change is consistently explained based on defect chemistry as mentioned later.

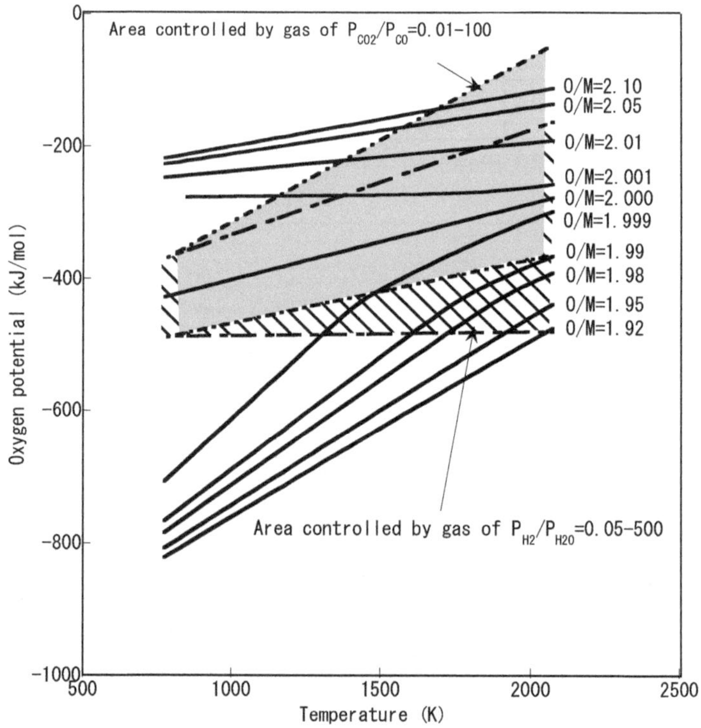

Fig. 3. Oxygen potential of $(U_{0.8}Pu_{0.2})O_{2\pm x}$

3. Measurement techniques

The measurements in previous studies were carried out by the E.M.F. and gas equilibrium methods. The former can accurately measure oxygen potentials, but the measurement data are only limited to a temperature range between 1123 K and 1423 K. Besides, it is difficult to determine the O/M ratio for the near stoichiometric composition. In the gas equilibrium method, a thermo-gravimeter is employed, which can measure a small deviation of the O/M ratio. Here, the measurement technique by the gas equilibrium method is explained. A thermo-gravimetry and differential thermal analysis system (TG-DTA) was put in glove box and used to measure oxygen potential (Kato et al., 2005, 2009b, 2011b, 2011c). The oxygen partial pressure of the flowing gas was monitored at the inlet and outlet of the apparatus using stabilized zirconia oxygen sensors. Gas mixtures of H_2/H_2O and CO/CO_2 were used for controlling the measurement atmosphere. The oxygen partial pressures are determined by the reactions of eqs.(3) and (4) in each gas system.

$$CO + 1/2\, O_2 \leftrightarrow CO_2 \tag{3}$$

$$\Delta\bar{S}_{O_2} \leftrightarrow H_2O \tag{4}$$

The regions controlled by both gas mixtures are shown by shaded and gray areas, respectively, in Fig.3. The H_2/H_2O mixture gives control at lower oxygen potential. Previously, Kato et al. (2011b, 2011c) used the H_2/H_2O mixture to measure hyper- and hypo-stoichiometric (U,Pu)O_2. Considering the reaction of eq.(4), the equilibrium may be expressed by a free energy of formation, ΔG_f, given by eqs.(5) and (6).

$$\Delta G_f = -RT \cdot \ln \frac{P_{H_2O}}{P_{H_2} \cdot P_{O_2}^{1/2}} \tag{5}$$

$$\Delta G_f = -246440 + 54.8 \cdot T \tag{6}$$

The P_{O_2} values in the flowing gas from the measurement apparatus were monitored at 973 K with the stabilized zirconia sensors at the inlet and the outlet. The ratio of P_{H_2}/P_{H_2O} at the outlet sensor can be determined by eqs.(5) and (6), and P_{O_2} at measurement temperatures was calculated under the assumption that the ratio of P_{H_2}/P_{H_2O} at the sensor position was the same at any sampling position. Therefore, the oxygen potential can be determined by controlling the ratio of P_{H_2}/P_{H_2O} in the flowing gas.

4. Statistical thermodynamics of point defects

The relation between x and P_{O_2} in non-stoichiometric compounds was analyzed based on statistical thermodynamics of point defects in a previous report (Kosuge, 1993) and eq.(7) was derived.

$$\frac{P_{O_2}(x)}{P_{O_2}(0)} = 1 + \frac{x^2 + x\sqrt{x^2 + 4C^2}}{2C^2} \tag{7}$$

Here $P_{O_2}(0)$ represents P_{O_2} at the stoichiometric composition and C is the intrinsic fraction of defects. Eq.(7) can be applied in the near stoichiometric region. Data for (U,Pu)O_2 in the near stoichiometric region have been obtained by Kato et al. (2011b, 2011c). Curves of

$P_{o_2}(0)/P_{o_2}(x)$ versus x are shown in Fig.4. Eq. (7) was a least square fit using the experimental x and $P_{O2}(x)$ as a parameter of temperature and the relationship between x and $P_{O2}(x)$ was determined. The $P_{O2}(x)$ and C were evaluated from Fig.4; they are shown in Table 2. Temperature dependence on oxygen potential is given by eq. (2). Therefore, oxygen potential of stoichiometric $(U_{0.8}Pu_{0.2})O_2$ was obtained from the data in Table 1 according to the following equation.

$$\Delta\overline{G}_{O_2} = -517.5\times10^3 + 116.5\cdot T \tag{8}$$

Eq.(8) shows that $\Delta\overline{H}_{O_2}$ and $\Delta\overline{S}_{O_2}$ at stoichiometric composition are -517.5 kJ/mol and 116.5 J/mol/K, respectively. Kato et al. (2011c) summarized their data in MOX containing 12, 20 and 30% Pu. The $\Delta\overline{H}_{O_2}$ values were almost the same and the $\Delta\overline{S}_{O_2}$ values decreased with Pu content. In the present work the oxygen potential of $(U_{1-y}Pu_y)O_{2.00}$ dependence on Pu content was estimated by eq.(9).

$$\Delta\overline{G}_{O_2} = -521.2\times10^3 - (117.8\cdot y + 92.1)\cdot T \tag{9}$$

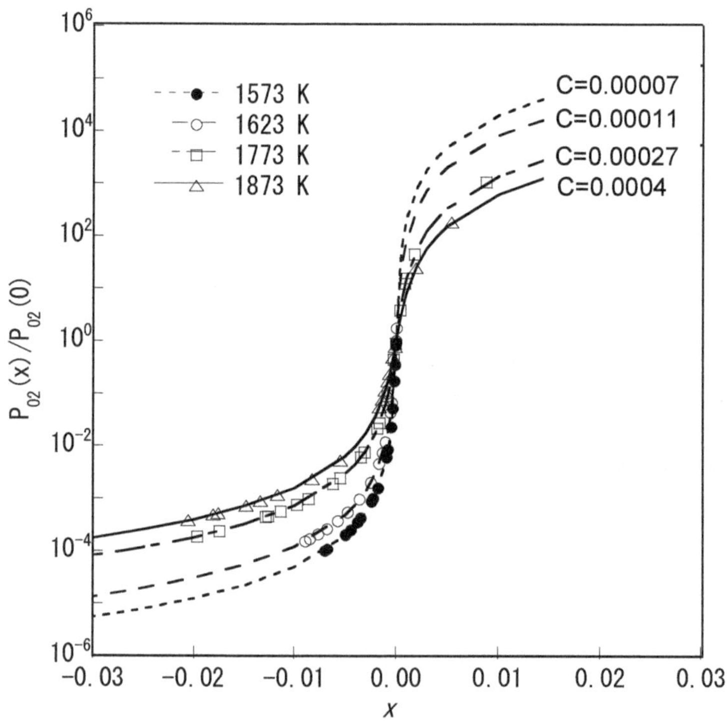

Fig. 4. Curves of $P_{o_2}(0)/P_{o_2}(x)$ versus x in $(U_{0.8}Pu_{0.2})O_{2\pm x}$

Temperature (K)	$P_{o_2}(0)$ (atm)	C
1573	7.93E-12	0.00007
1623	2.68E-11	0.00011
1773	6.88E-10	0.00027
1873	4.48E-09	0.00040

Table 2. $P_{o_2}(0)$ and C in $(U_{0.8}Pu_{0.2})O_2$

The intrinsic fraction of defects C may be described by the following equation:

$$C = A \cdot \exp\left(-\frac{E}{2\kappa T}\right) \tag{10}$$

where E is the formation energy for intrinsic ionization, A is a constant and κ is the Boltzmann constant. E was estimated to be 2.9 eV from the relationship between C and $1/T$. The value of E predicts the band gap energy. The band gap energies of UO_2 andPuO_2 were reported to be 2.7 and 2.5 eV, respectively (Nakamura & Fujino, 1986; Naito et al., 1980). The value of E in this work was slightly higher compared with other data.

5. Defect equilibria — Kröger-Vink diagram

The nonstoichiometry of various compounds has been analyzed using Kröger - Vink diagrams (Kröger & Vink, 1957; Kofstad, 1972; Sørensen, 1981; Karen, 2009). In this section, the notation of Kröger - Vink was used. (Kröger & Vink, 1957) Deviation x in $(U,Pu)O_{2\pm x}$ is proportional to $P_{o_2}^{\pm 1/n}$, where n is a characteristic number identifying the defect (Kofstad, 1972; Sørensen, 1981). Relationships between x and P_{o_2} in $(U_{0.8}Pu_{0.2})O_{2\pm x}$ were analyzed, and are shown in Fig.5. The slopes of n=-4, and -2 were observed in hypo-stoichiometric MOX, and that of n=+2 was observed in hyper-stoichiometric MOX. It was reported that n=-4 changed to n= -3 for $(U_{0.7}Pu_{0.3})O_{2\pm x}$.

Then, defect reactions of eqs. (11)-(16) are considered for defect equilibria in $(U_{0.8}Pu_{0.2})O_{2\pm x}$.

$$O^{\times}_{O} + M^{\times}_{Pu} \rightarrow \left(V_O^{\bullet\bullet} M'_{Pu}\right)^{\bullet} + e' + \frac{1}{2}O_2 \tag{11}$$

$$2O_O^{\times} + O_2 \rightarrow \left(2O_i^a 2O_i^b 2V_O^{\bullet\bullet}\right)' + h^{\bullet} \tag{12}$$

$$Nill \rightarrow e' + h^{\bullet} \tag{13}$$

$$O_O^{\times} \rightarrow V_O^{\bullet\bullet} + O_i'' \tag{14}$$

$$O_O^{\times} \rightarrow V_O^{\bullet\bullet} + 2e' + \frac{1}{2}O_2 \tag{15}$$

$$\frac{1}{2}O_2 \rightarrow O_i'' + 2h^{\bullet} \tag{16}$$

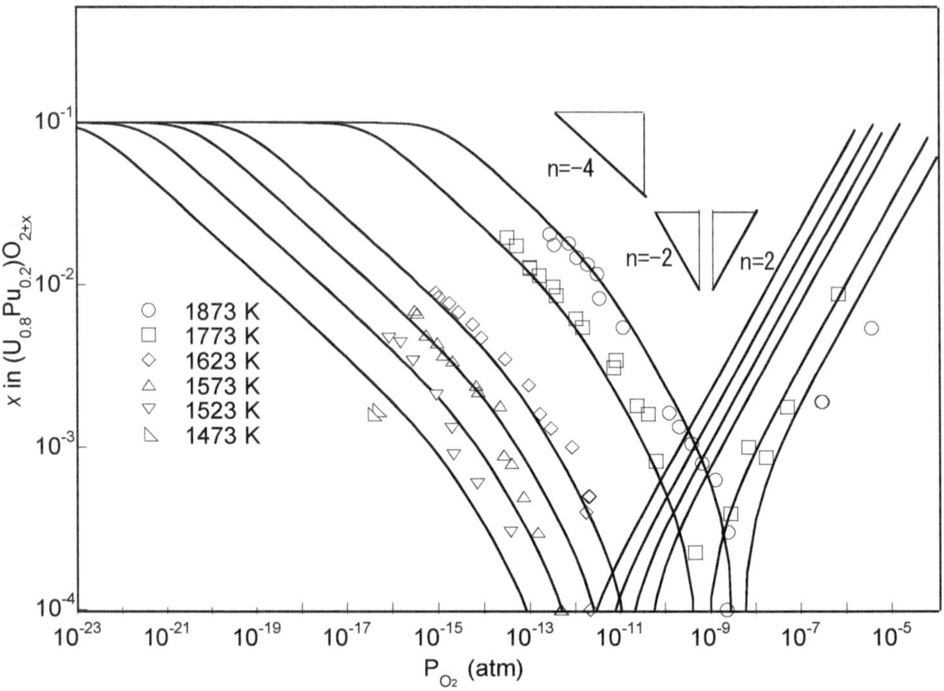

Fig. 5. The P_{o_2} versus x in $(U_{0.8}Pu_{0.2})O_{2\pm x}$

The equibrium constants in the defect reactions (11)-(16) are described as Eqs.(17)-(22), respectively.

$$K_{Re} = \left[(V_O^{\bullet\bullet} Pu')^{\bullet}\right][Pu']P_{O_2}^{1/2} \tag{17}$$

$$K_{Ox} = \left[\left(2O_i^a 2O_i^b 2V_O^{\bullet\bullet}\right)'\right][h^{\bullet}]P_{O_2}^{-1} \tag{18}$$

$$K_i = [h^{\bullet}][e'] \tag{19}$$

$$K_F = [V_O^{\bullet\bullet}][O_i''] \tag{20}$$

$$K_{Vo} = [V_O^{\bullet\bullet}][e']^2 P_{O_2}^{1/2} \tag{21}$$

$$K_{Oi} = [O_i''][h^{\bullet}]^2 P_{O_2}^{-1/2} \tag{22}$$

Two types of diagrams were proposed depending on the dominant defects of the near stoichiometric region; these are Frenkel defects or intrinsic defects.

Region I: Near stoichiometric region

In the case of (U,Pu)O$_{2\pm x}$, it was reported that intrinsic ionization of eq.(13) is dominant in the near stoichiometric region. So, eq.(20) can be written as

$$\left[h^{\bullet}\right]=\left[e'\right]=K_i^{1/2}. \tag{23}$$

When inserting eq.(23) into eqs.(21) and (22), $\left[V_O^{\bullet\bullet}\right]$ and $\left[O_i''\right]$ are given by eqs.(24) and (25).

$$\left[V_O^{\bullet\bullet}\right]=\frac{K_{Vo}}{\left[e'\right]^2}P_{O_2}^{-1/2}=\frac{K_{Vo}}{K_i}P_{O_2}^{-1/2} \tag{24}$$

$$\left[O_i''\right]=\frac{K_{Oi}}{\left[h^{\bullet}\right]^2}P_{O_2}^{1/2}=\frac{K_{Oi}}{K_i}P_{O_2}^{1/2} \tag{25}$$

If the deviation x is given by

$$x=\left[V_O^{\bullet\bullet}\right] \quad or \quad \left[O_i''\right], \tag{26}$$

the value of n is +2 and -2 in the near stoichiometric region, which is consistent with the relationship shown in Fig.5.

Region II: Reducing region

In the defect reaction of eq.(11), the electro-neutrality condition is given by

$$\left[\left(V_O^{\bullet\bullet}Pu'\right)^{\bullet}\right]=\left[e'\right]. \tag{27}$$

When inserting eq.(27) into eq.(17), $\left[V_O^{\bullet\bullet}\right]$ can be written by

$$\left[V_O^{\bullet\bullet}\right]=\left[\left(V_O^{\bullet\bullet}Pu'\right)^{\bullet}\right]=K_{\mathrm{Re}}^{1/2}P_{O_2}^{-1/4}. \tag{28}$$

The value of n is derived as -4.

Region III: Oxidizing region

In the defect reaction of eq.(12), the electro-neutrality condition is given by

$$\left[\left(2O_i^a 2O_i^b 2V_O^{\bullet\bullet}\right)'\right]=\left[h^{\bullet}\right] \tag{29}$$

When inserting eq.(29) into eq.(18), $\left[O_i''\right]$ can be written by

$$\left[O_i''\right]=2\left[\left(2O_i^a 2O_i^b 2V_O^{\bullet\bullet}\right)'\right]=2K_{Oi}^{1/2}P_{O_2}^{1/2} \tag{30}$$

The value of n is derived as +2.

The measured data shown in Fig.5 were fitted by eqs.(24), (25), (29) and (30), and the equilibrium constants were estimated as a function of temperatures. Eqs.(31)-(34) were obtained from the relationship between equilibrium constants and $1/T$.

$$K' = \frac{K_{Vo}}{K_i} = 493.4\exp\left(\frac{-372.0\times10^3}{RT}\right) \tag{31}$$

$$K'' = \frac{K_{Oi}}{K_i} = 5.095\times10^{-5}\exp\left(\frac{175.6\times10^3}{RT}\right) \tag{32}$$

$$K_{Re} = 80420\exp\left(\frac{-515.0\times10^3}{RT}\right) \tag{33}$$

$$K_{Ox} = 4.14\times10^{-8}\exp\left(\frac{291.0\times10^3}{RT}\right) \tag{34}$$

Eqs. (24), (25), (29), (30) and (31)-(34) could represent the Kröger - Vink diagram as shown in Fig.6. The calculation results were consistent with the measured data. In addition, the equilibrium constants, K_i, K_F, K_{Vo} and K_{Oi}, were obtained as eqs.(35)-(38).

$$K_i = 106400\exp\left(\frac{-286.0.0\times10^3}{RT}\right) \tag{35}$$

$$K_F = 0.2006\exp\left(\frac{-226.5\times10^3}{RT}\right) \tag{36}$$

$$K_{Vo} = 5.248\times10^7\exp\left(\frac{-658.0\times10^3}{RT}\right) \tag{37}$$

$$K_{Oi} = 4.41\times10^{-3}\exp\left(\frac{5.0\times10^3}{RT}\right) \tag{38}$$

Eqs.(35)-(38) can predict the transformation energy of defects. The predicted values are shown in Table 3 and they had close agreement with other data.

The O/M ratio of $(U,Pu)O_2$ is an important parameter for development of nuclear fuels. So, it is essential to know the O/M ratio as functions of temperature and oxygen potential. If deviation x is equivalent to $[V_O^{\bullet\bullet}]$ and $[O_i'']$, the O/M ratio can be written by

$$O/M = 2.00 - [V_O^{\bullet\bullet}] + [O_i'']. \tag{39}$$

In the hypo-stoichiometric region, $[V_O^{\bullet\bullet}]$ is described by eqs.(24) and (28). In addition, it is assumed that $[V_O^{\bullet\bullet}]$ is limited by eq.(40) because the valence of Pu is +3.

$$[V_O^{\bullet\bullet}] = 0.01. \tag{40}$$

Therefore, $[V_O^{\bullet\bullet}]$ is estimated by the following equation using eqs.(24), (28) and (40).

$$[V_O^{\bullet\bullet}] = \left[\left(\frac{K_{Vo}}{K_i}P_{O_2}^{-1/2}\right)^{-3} + \left(K_{Re}^{1/2}P_{O_2}^{-1/4}\right)^{-3} + (0.01)^{-3}\right]^{-1/3} \tag{41}$$

$[O_i'']$ is estimated by the following equation using eqs.(25).

$$[O_i''] = \left[\left(\frac{K_{Oi}}{K_i} P_{O_2}^{1/2} \right)^{-2} + \left(2K_{Oi}^{\ 1/2} P_{O_2}^{1/2} \right)^{-2} \right]^{-1/2} \quad (42)$$

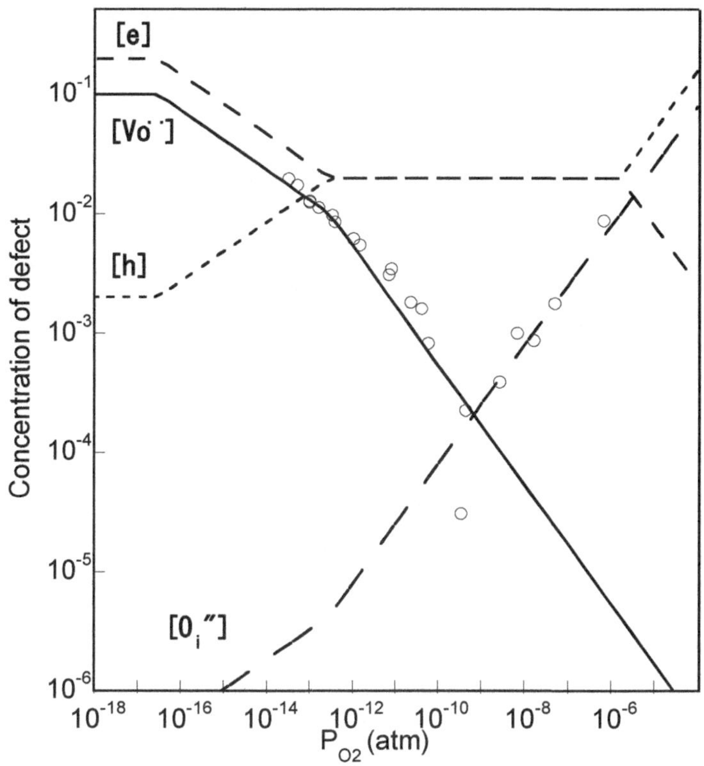

Fig. 6. Calculation results of defect concentration at 1773 K in $(U_{0.8}Pu_{0.2})O_{2\pm x}$

When inserting eqs.(38) and (39) into eq.(36), the O/M ratio can be represented. The calculation results are shown in Fig.5; lines as functions of temperature and oxygen partial pressure were in good agreement with the experimental data. The figure shows that the dominate defect changed with temperature in hypo-stoichiometric oxide with constant O/M ratio. So, two forms of eq.(2) were obtained, and the curve of the temperature dependence of the oxygen potential had a flexion point as shown in Fig 3.

In the reducing region, the relationship of n changed from -4 to -3 with Pu content. The n=-3 was observed in $(U_{0.7}Pu_{0.3})O_{2-x}$. In this region, the defect reaction of (43) was assumed for describing the Kröger - Vink diagram. The defect concentration was given by eq.(44) and the relationship of n=-3 was obtained.

$$2O^{\times}{}_O + 2M^{\times}{}_{Pu} \rightarrow \left(2V_O^{\bullet\bullet}2M'_{Pu}\right)^{\bullet} + 2e' + \frac{1}{2}O_2 \tag{43}$$

$$\left[V_O^{\bullet\bullet}\right] = 2\left[\left(2V_O^{\bullet\bullet}2Pu'\right)^{\bullet\bullet}\right] = (2K_{\mathrm{Re}})^{1/3}P_{O_2}^{-1/3} \tag{44}$$

Defect reaction	Material	Transformation energy of defect (eV)	References
$Nill \rightarrow e' + h^{\bullet}$	MOX	3.0	This work
	UO$_2$	2.1	Nakamura & Fujino, 1986
	PuO$_2$	2.5	Stan & Cristea, 2005
$O_O^{\times} \rightarrow V_O^{\bullet\bullet} + O_i''$	MOX	2.3	This work
	UO$_2$	5.3	Nakamura & Fujino, 1986
$O_O^{\times} \rightarrow V_O^{\bullet\bullet} + 2e' + \frac{1}{2}O_2$	MOX	6.8	This work
	PuO$_2$	4.8	Stan & Cristea, 2005
$\frac{1}{2}O_2 \rightarrow O_i'' + 2h^{\bullet}$	MOX	-0.05	This work
	UO$_2$	-0.2	Nakamura & Fujino, 1986
$O_O^{\times} + Pu^{\times} \rightarrow \left(V_O^{\bullet\bullet}Pu'\right)^{\bullet} + Pu' + \frac{1}{2}O_2$	MOX	5.3	This work
$2O_O^{\times} + O_2 \rightarrow \left(2O_i^a 2O_i^b 2V_O^{\bullet\bullet}\right)' + U^{\bullet}$	MOX	-3.0	This work
	UO$_2$	-2.6	Nakamura & Fujino, 1986

Table 3. Comparison of transformation energy of defects

6. Application to O/M control

The O/M ratio is one of the most important parameters in controlling a reactor using (U,Pu)O_2 nuclear fuels. The fuels are used in the hypo-stoichiometric composition range to control the fuel and cladding chemical interaction. The O/M ratio is adjusted in a sintering process which is carried out in an Ar/H_2 gas mixture atmosphere. In this atmosphere, the oxygen potential is determined by eqs. (1), (5) and (6). According to the equations, the oxygen partial pressure is determined by controlling the P_{H_2} / P_{H_2O} ratio and temperature.

The O/M ratios of (U,Pu)O_2 are represented by the derived eqs.(36), (38) and (39) and are shown in Fig.7 as functions of P_{H_2} / P_{H_2O} and temperatures. In the low temperature region of less than 1000 K, the O/M ratio is almost 2.0, despite P_{H_2} / P_{H_2O} ratio in the atmosphere. In this temperature range, it is almost impossible to adjust to a hypo-stoichiometric composition in the H_2/H_2O gas atmosphere. In the higher temperature range, the O/M ratio decreases with increasing P_{H_2} / P_{H_2O} ratio. The O/M ratio approaches 1.65 and remains at almost 2.00 at 1973 K in the atmosphere of $P_{H_2} / P_{H_2O} = 200$ and $P_{H_2} / P_{H_2O} = 2$, respectively. The (U,Pu)O_2 fuels with various O/M ratios may be produced using the relationships of Fig.7.

However, the O/M ratio increases during the cooling process after the sintering. So, control of cooling rate may be needed to obtain low O/M fuel. Furthermore, the kinetics evaluation of O/M change is essential to get adjustment to an accurate O/M ratio.

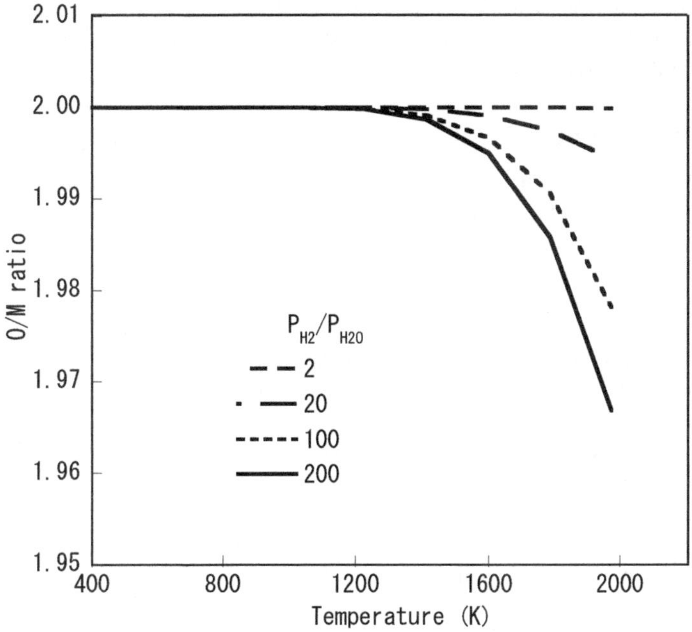

Fig. 7. O/M variation of $(U_{0.8}Pu_{0.2})O_{2\pm x}$ as a function of P_{H_2}/P_{H_2O}

7. Summary

In $(U,Pu)O_2$ the oxygen nonstoichiometry significantly affects various physical properties, so it is essential to know the oxygen potential as functions of O/M ratio and temperatures for development of nuclear fuels. Many studies on oxygen potential have been carried out. However, the data are scattered because of the difficulty of measurement.

Recently, data were accurately measured by the gas equilibrium method using thermogravimetry. The data were analysed based on defect chemistry. The O/M ratio varies continuously with changing oxygen potential for the near stoichiometric composition, so it is very difficult to determine the stoichiometric composition in experiments. The stoichiometric composition was determined by analysis of statistical thermodynamics of point defects. The analysis results predicted the band gap energy of 2.9 eV. Additionally, the data were analyzed by a Kröger - Vink diagram. The defect reactions were assumed from experimental data, concentrations of defects were evaluated as functions of temperature and oxygen partial pressure, and the equation for the O/M ratio was derived. The derived equation can represent the O/M ratio as functions of temperature and P_{H_2}/P_{H_2O} and can be applied for the sintering process to adjust the O/M ratio.

In the development of nuclear fuels, it is also important to understand the oxygen potential dependence on Pu, Am and Np contents. Pu content is known to increase the oxygen potential. Several groups have investigated the effect of Am and Np addition in $(U,Pu)O_2$ (Kato et al., 2009;Nakamichi et al., 2011). They found that the addition of Am caused the oxygen potential to increase, but Np addition did not affect the oxygen potential. The derived equation for O/M ratio will be extended to represent the O/M ratio dependence on Pu and Am contents in future fuels.

8. Acknowledgement

The author is grateful to Dr. K. Konashi and Dr. S. Nagai for his fruitful discussions. The authors are pleased to acknowledge Mr. T. Sunaoshi, Mr. T. Tamura and Mr. H. Uno for his collaboration

9. References

Chilton, G. R. & Edwards, J. (1980). Oxygen Potentials of $U_{0.77}Pu_{0.23}O_{2\pm X}$ in the temperature range 1523 - 1822K, *United Kingdom Atomic Energy Authority Northern Division Report, ND-R-276(W)*

Chilton, G. R. & Kirkham, I. A. (1976).The Determination of Oxygen Potentials of Hyperstoichiometric U-Pu Dioxides in the Temperature Range 1500 - 1800K, *Plutonium 1975*, pp.171-180

Edwards, J.; Wood, R. N. & Chilton, G. R. (1985). Oxygen Potentials of Uranium-Plutonium Oxides in the Near Stoichiometric Region, *Journal of Nuclear Materials*, Vol.130, pp.505-512

Javed, N. A. (1973). Thermodynamic Behavior of (U, Pu) Mixed-Oxide Fuels, *Journal of Nuclear Materials*, Vol.47, pp.336-344

Karen, P. (2006). Nonstoichiometry in oxides and its control, *Journal of Solid State Chemistry*, Vol.179, pp.3167-3183

Kato, M.; Tamura, T.; Konashi, K. & Aono, S. (2005). Oxygen Potentials of Plutonium and Uranium Mixed Oxide, *Journal of Nuclear Materials*, Vol.344, p.p.235-239

Kato, M. & Konashi, K. (2009). Lattice parameters of $(U,Pu,Np,Am)O_{2-x}$, *Journal of Nuclear Materials*, Vol.385, pp. 117–121, 2009

Kato, M. ; Konashi, K. & Nakae, N. (2009a). Analysis of oxygen potential of $(U_{0.7}Pu_{0.3})O_{2\pm x}$ and $(U_{0.8}Pu_{0.2})O_{2\pm x}$ based on point defect chemistry, *Journal of Nuclear Materials*, Vol.389, pp.164-169

Kato, M.; Tamura, T. & Konashi, K. (2009b). Oxygen potentials of Mixed Oxide Fuels for Fast Reactors, *Journal of Nuclear Materials*, Vol.38, pp.419–423

Kato, M. ; Maeda, K.; Ozawa, T.; Kashimura, M.& Kihara, Y. (2011a). Physical Properties and Irradiation Behavior Analysis of Np- and Am-Bearing MOX Fuels, *Journal of Nuclear Science and Technology*, Vol.48, pp.1-8

Kato, M. ; Takeuchi, K. ; Uchida, T. ; Sunaoshi , T. & Konashi, K. (2011b). Oxygen potential of $(U_{0.88}Pu_{0.12})O_{2\pm x}$ and $(U_{0.7}Pu_{0.3})O_{2\pm x}$ at high temperatures of 1673-1873K, *Journal of Nuclear Materials*, Vol. 414, p.p.120–125

Kato, M.; Nakamichi, S.; Takeuchi, K. & Sunaoshi, T. (2011c). Measurement of oxygen potential in $(U_{0.8}Pu_{0.2})O_{2\pm x}$ at high temperatures of 1773, and 1873K, *CALPHAD; Computer Coupling of Phase Diagram and Thermochemistry*, Vol.35, p.p. 623-626

Kofstad, P. (1972). *Nonstoichiometry, Diffusion and Electrical Conductivity in Binary Metal Oxides*, John Wiley and Sons, New York

Kosuge, K. (1993). *Chemistry of Non-Stoichiometric Compounds*, Oxford University, Press, New York, pp. 21–26

Kröger, F.A. & Vink, H. (1957). *Solid State Physics*, Academic Press, New York, 1957

Mari, C.M.; Pizzini, S.; Manes, L. & Toci, F. (1977). A Novel Approach to the Oxygen Activity Microdetermination of Oxides by EMF Measurements, *Journal of Electrochemistry Society* Vol.56, pp.1831-1836

Markin, T.L. (1967). The Plutonium-Oxygen and Uranium-Plutonium-Oxygen Systems: A Thermochemical Assessment, *Technical Report Series No.79, IAEA*

Markin, T. L. & McIver E. J. (1965). Thermodynamic and phase studies for plutonium and uranium -plutonium oxides with application to compatibility calculations, *3rd International Conference on Plutonium 1965*, p.p. 845-857

Markin, T. L. & Street, R. S. (1967). The Uranium-Plutonium-Oxygen Ternary Phase Diagram, *Journal of Inorganic and Nuclear Chemistry*, Vol.29, pp.2265-2280

Naito, K.; Tsuji, T.; Ouchi, K.; Ouchi, K.; Yahata, T.; Yamashita, T. & Tagawa, H. (1980). Electrical conductivity anomaly in near-stoichiometric plutonium dioxide, *Journal of Nuclear Materials*, Vol.95, pp.181–184

Nakamichi, S.; Kato, M. & Tamura, T. (2011). Influences of Am and Np on oxygen potentials of MOX fuels, *CALPHAD ; Computer Coupling of Phase Diagram and Thermochemistry*, Vol.35, p.p. 648-651

Nakamura, A. & Fujino, T. (1986). Thermodynamic analysis on point defects of UO_{2+x} at relatively small deviation from stoichiometry between 600 and 1400°C, *Journal of Nuclear Materials*, Vol.140, pp.113–130

Sari, C.; Benedict, U. & Blank, H. (1970). A Study of the Ternary System UO_2-PuO_2-Pu_2O_3, *Journal of Nuclear Materials*, Vol.35, pp.267-277

Sørensen, O. T. (1976). Thermodynamic Studies at Higher Temperatures of the Phase Relationships of Substoichiometric Plutonium and Uranium/Plutonium Oxides, *Plutonium and Other Actinides 1975*, pp.123-131

Sørensen, O. T. (1981). *Nonstoichimetric Oxides*, Academic Press, New York, p.p.1-59, 1981

Stan, M. & Cristea, P. (2005). Defects and oxygen diffusion in PuO_{2-x}, *Journal of Nuclear Materials*, Vol.344, pp.213–218

Tetenbaum, M. (1974), High Temperature Vapourisation behaviour of Hypostoichiometric U-Pu-O and U-Nd-O Solid Solutions, *Thermodynamics of Nuclear Materials*, IAEA, Vienna, (1974) 305-320

Vasudeva Rao, P. R.; Bagawde, S. V.; Ramakrishna, V. V. & Patil, S. K. (2006). Oxygen potential and thermal conductivity of (U, Pu) mixed oxides, *Journal of Nuclear Materials*, Vol.348, pp.329–334

Woodley, R. E. & Adamson, M. G. (1985), The Oxygen potential of Near- and Non-Stoichiometric urania-25mol% Plutonia Solid Solutions: A Comparison of Thermogravimetric and Galvanic Cell Measurements, *Journal of Nuclear Materials*, Vol.82, pp. 65-755

Woodley, R. E. (1981), Oxygen Potentials of Plutonia and Urania Solid Solutions, *Journal of Nuclear Materials*, Vol.96, pp. 5-14

Part 5

The Importance of Stoichiometry in Electrochemical Applications

10

Synthesis and Stoichiometric Analysis of a Li-Ion Battery Cathode Material

Norlida Kamarulzaman[1] and Mohd Hilmi Jaafar[2]
[1]*Centre For Nanomaterials Research, Institute of Science, Universiti Teknologi MARA, Shah Alam, Selangor,*
[2]*Centre For Foundation Study In Sciences, University of Malaya, Kuala Lumpur, Malaysia*

1. Introduction

Generally, stoichiometry deals with the quantitative relationship of reactants and products in a chemical reaction. Strictly speaking in areas of materials science, solid-state physics and chemistry, stoichiometric compounds are compounds that have simple ratios to each other, that is, A_xB_y, where A, B = chemical element and x,y = natural numbers. There is no limit to the number of elements in the molecule.

Scientists nowadays modify materials such that their ratios are no longer simple ratios to each other but fractions, for example, $Li_{0.1}Mg_{0.9}O$, $Li_{1.1}Mn_2O_4$, $LiMn_{1.7}Fe_{0.3}O_2$, $Al_{1.9}Cr_{0.1}O_3$, and countless others. Strictly speaking, these compounds are called non-stoichiometric compounds. There are various reasons for obtaining non-stoichiometric compounds. It is found that processes such as substitution or doping will change, modify or enhance a material's physical and chemical behaviour. Scientists will then have a way to design materials to suit their applications. This can be seen in the well known semi-conductor, Si, whereby doping it with elements of different oxidation states can produce either a p-type or n-type silicon. This technology is very important because nearly all electrical devices are silicon based.

The most common cathode materials used in commercial Li-ion batteries is $LiCoO_2$ (Amatucci et al. 1996; Antolin et al. 2004; Ding et al. 2005). This is because $LiCoO_2$ gives high energy density as well as good cycling behaviour. However, in terms of economy and availability of elements, Co is considered expensive, not so abundant (Tanaka et al. 2001) and also considered a toxic heavy metal (Shin et al. 2004). Therefore, modification of the material or a complete change of material is warranted in order to have a sustainable technology.

There are other reasons for modifying the stoichiometry of cathode materials. It is found that substitution and doping of cathode materials improve the materials' electrochemical behaviour in terms of specific capacity and energy density (Li et al. 2002; Park et al. 2001; Veluchamy et al. 2001). Secondly, it may stabilize the material and improve cycling performance (Fey et al. 2003). Thirdly, it may lower the cost of production by reducing the usage of certain expensive metals such as cobalt and utilizing less expensive metals such as

Mn or Fe. Last but not least, it may reduce toxicity of the materials and contribute towards a healthier and safer living environment.

2. Substitution and doping

The word doping has been widely used by scientists. It originates from the work done on semiconductors to modify their characteristics. The technique was first introduced in the U.S.A. by John Woodyard of Sperrygyroscope Company (U.S. Patent 1950) but work was also done by Morgan Teal and Morgan Sparks of Bell Labs (U.S. Patent 1953). There are two kinds of doping, that is, substitutional and interstitial doping. As an example, $Li_{1.1}Mn_2O_4$ is obtained by interstitial doping while $Li[Mn_{1.7}Fe_{0.3}]O4$ is obtained by substitutional doping. Observe that the spinel stoichiometry of AB_2O_4 is conserved for the later material but not for the former material. Substitutional doping is therefore taken as the introduction of foreign elements into the host material to take the place of some proportion of the original host chemical or element. Substitutional doping, thus, preserves the crystal structure of the compound while interstitial doping may change or modify the structure. The important point is that, for substitutionally doped compounds, the positions of the atoms are precisely known in the crystal lattice whereas for an interstitially doped compound, the atoms/ions may sit in interstitials whereby the position is not precisely known. Thus, it is quite problematic to analyze stoichiometries of interstitially doped materials in terms of the position of the foreign elements in the crystal structure of the materials.

Obtaining pure, single phase substituted materials is not an easy task. The thermodynamic properties of the elements involved must be taken into account. As an example, if some of the Mn in $LiMn_2O_4$ is to be substituted with Fe, to produce say, $LiMn_{1.7}Fe_{0.3}O_4$, the reaction temperature must be carefully controlled because the enthalpies of the oxides of the metal/transition metals are generally lower than that of the $LiMn_2O_4$ material. The enthalpy of formation is given for the metal-transition metal oxides as follows:

α-Fe_2O3 at 298.15 K $\quad \Delta_f H^o = -824.2$ kJ/mol (Lide 1995)

β-MnO_2 at 298.15 K $\quad \Delta_f H^o = -520.0$ kJ/mol (Lide 1995)

Mn_2O_3 at 298.15 K $\quad \Delta_f H^o = -959.0$ kJ/mol (Lide 1995)

$LiMn_2O_4$ at 876.0 K $\quad \Delta_f H^o = -1404.2$ kJ/mol (Yamaguchi et al. 1995)

As can be seen, iron and manganese oxides are easier to form than $LiMn_2O_4$ or $LiMn_{1.7}Fe_{0.3}O_4$ materials and the probability of their formation is high. Therefore, careful analysis has to be done of a TG/DSC thermal profile of $LiMn_{1.7}Fe_{0.3}O_4$ precursors obtained for the identification of a suitable calcination temperature that will not result in the formation of the metal/transition metal oxide impurities in the calcined sample. Optimally, for this example, the temperature chosen should be high enough to produce pure $LiMn_{1.7}Fe_{0.3}O_4$ material, by-passing the lower temperatures of formation for the metal/transition metal oxides. Normally, for good electrochemical characteristics of the $LiMn_{1.7}Fe_{0.3}O_4$ material, the calcination temperature chosen is above 800 °C and the precursor has to be placed in an already heated furnace of the right temperature.

Choosing the right calcination temperature will suppress the formation of the impurity metal oxides in the sample. The high temperature heating will force the Fe ions into the $LiMn_2O_4$ lattice and produce stoichiometric $LiMn_{1.7}Fe_{0.3}O_4$ material. A well chosen calcination temperature will produce pure, single phase $LiMn_{1.7}Fe_{0.3}O_4$ material with good electrochemical behaviour.

3. Synthesis routes

Cathode materials used in Li-ion batteries consist of quite complex stoichiometries, containing two or more cations and one or more anions. Examples of some common cathode materials are $LiCoO_2$, $LiCo_{1-x}Ni_xO_2$, $LiMn_2O_4$, $LiMn_{1-x}Fe_xO_4$ and $LiFePO_4$. Many synthesis methods can be used in obtaining a particular chemical compound suitable for use as cathodes in Li-ion batteries. The choice of a synthesis method depends on the available starting materials, the apparatus accessible to the scientist and the desired characteristics of the final product. The common synthesis methods used are the solid-state reaction, the sol-gel, the combustion and the hydrothermal methods. There are many other methods available but only the first three will be discussed here.

The solid state reaction method involves starting materials which are normally powders. The materials are firstly mixed together to form a homogeneous mixture and normally a grinding process using a mortar and pestle is used. The mixed powder will be pressed into pellets where they will then undergo a precalcining process. Usually the precalcined temperature is at a lower temperature of around 500 to 650 °C (Ohzuku et al. 1999; Zou et al. 2005; Li et al. 2000; Wu et al. 2001). The precalcined pellets will be ground and pelletized again whence they will be reacted at a high temperature of between 700 to 1200 °C (Ohzuku et al. 1999; Zou et al. 2005; Li et al. 2000; Wu et al. 2001). The pellets will normally be reground again. The advantage of this method is ease and simplicity. The disadvantage is that it normally yields impure final products and large particle size that may not produce the best electrochemical performance.

The sol-gel or Pechini synthesis method is now very popular due to several reasons such as versatility in the method (especially for partial cation substitution synthesis), more homogeneous particle size (compared to the solid-state reaction method), ease in obtaining phase pure products, a lower calcining temperature in some instances and better electrical behaviour (Lu et al. 2001; Huang et al. 2000). The starting materials should be compounds that will easily dissolve in a solvent. Normally hydroxides, acetates or nitrates of the metals will be used. There may or may not be a chelating agent employed. Chelating agents are normally acids. The starting materials will be dissolved in the solvents by stirring for some time and the chelating agent will be added until a gel is formed. Then, a slow drying process is done until a precursor material is acquired. Usually, a thermal profile of the precursor is taken and analyzed in order to identify suitable calcining temperatures. The materials will be calcined using temperatures from 500 to 1200 °C

The combustion synthesis method has several advantages over the other two methods. It is a relatively cheap, easy and rapid way of obtaining the precursors. Normally, the precursor obtained is pure, of very fine particles and they are more homogeneous in terms of size and shape. The particle size is often sub-micron but this depends on the calcination temperatures and the time duration of the heating process. The synthesis method also involves the dissolution of the metal starting materials in a solvent. An oxidant such as glycine, amines or acids is added and the mixture is then slowly heated. A combustion reaction will eventually occur which is over in a few seconds. The precursor material is obtained and later calcined. There are a few considerations to be made with this type of synthesis method. If the reaction is very vigourous, then calculations for 5 g or less of final product should be done and a proper synthesis environment must be considered. These are safety precautions that should be taken to avoid explosions.

4. Techniques for the analysis of material stoichiometries

Cathode materials usually have complex stoichiometries as mentioned in section 1.0 above. Therefore, substantiation of the stoichiometries of the synthesized products is necessary. Often, the stoichiometries of the synthesized materials may not be exactly as calculated from molar ratio considerations. This is because the steps involved in the synthesis methods may introduce errors. For example, the weighing of the starting materials themselves is a crucial step and if not done carefully can be a serious source of error. Calcination procedures may also produce errors in stoichiometries due to certain processes such as Li vapourization at high temperatures of 850 °C and above producing lithium deficient materials (Ohzuku et al. 1990). The resulting final product may therefore be slightly different from the calculated values. That is why it is important to experimentally determine the stoichiometries of the final products.

There are many ways to elucidate the stoichiometries of cathode materials. Methods such as energy dispersive X-Ray spectroscopy (EDS), X-Ray diffraction (XRD), neutron diffraction (ND), X-Ray photoelectron spectroscopy (XPS), X-Ray fluorescence (XRF), inductively coupled plasma (ICP) and time of flight secondary ion-mass spectrometry (tof-SIMS) are some examples of methods that can give a measure of the stoichiometries of materials. Some methods such as EDS, XRD and ND are non-destructive techniques but others such as ICP and tof-SIMS are destructive. This is often due to the sample preparation methods and not the measurement technique itself. An exception is tof-SIMs where the principle of the method itself is destructive of the samples. The methods to measure the stoichiometries of materials have varying accuracies and most have limitations as to the range of atomic number (Z) of the elements in the periodic table that can be detected. Only EDS, XRD, ND and XPS will be discussed here. One important factor that is emphasized is that the samples used must be phase pure except for the diffraction methods. This is because diffraction can qualify and quantify the presence of different phases while the other methods cannot. However, EDS has one advantage, that is, it has high spatial resolution especially in the TEM.

4.1 Energy dispersive X-ray spectroscopy *via* SEM and TEM

EDS is a relatively easy method of elucidating stoichiometries of materials (Echelin et al. 1992; Goldstein et al. 1995). However, several considerations have to be made before a measurement is done. EDS normally has to be used in conjuction with a SEM or TEM. The accelerating voltage of SEM and TEM is not the same whereby it may affect the accuracies of the EDS results obtained. TEM also has a much smaller probe current area than SEM and this affects some measurements involving very small dimensions or areas under consideration.

An EDS detector is not unlike an elemental analyzer. It is normally integrated with a SEM or TEM. Particular advantage here is that scientists can choose the spots or area on the sample of which to be analyzed. With modern EDS detectors, the accuracy of the measurements are acceptably high and another advantage of the technique is that it is appreciably easier than other elemental methods such as mass-spectroscopy. When energetic electrons in a SEM or TEM column hit an atom of a sample, it may eject an inner orbital electron. A hole is created and the atom becomes unstable. Characteristic X-Rays will be produced when electrons in the outer orbital jumps to the inner orbital to occupy the vacancy. These characteristic X-Rays are used to identify an element *via* an EDS detector (normally, Si(Li)). This is the basic principle of elemental detection in EDS. Fig. 1(a) below illustrates the process of the ejection of the inner orbital electron by an incident energetic electron (from a SEM/TEM electron gun) and Fig. 1(b) shows the filling in of the hole by an outer orbital electron and the production of characteristic electromagnetic waves accompanying the process.

There are considerations to be made in using EDS to deduce stoichiometries of materials. Firstly, the detection limit of EDS is from the atomic number $Z > 3$ or $Z > 11$ depending on the type and quality of the EDS detectors used. Therefore, light elements cannot be detected by EDS. Detection limits is typically about 0.1 weight per cent or more of elements in the sample (subject to the quality of the EDX detector). Detection energies range from 1 to 30 keV in a SEM and higher in a TEM due to the greater accelerating voltages possible in a TEM. Another limitation of the EDS technique is the overlapping of peaks of the characteristic X-Rays of the elements. For example, the peaks of Mn K_α , Ti K_α, and Cr K_β are very close together and overlaps. Care has to be taken when samples contain elements with very close K and L lines.

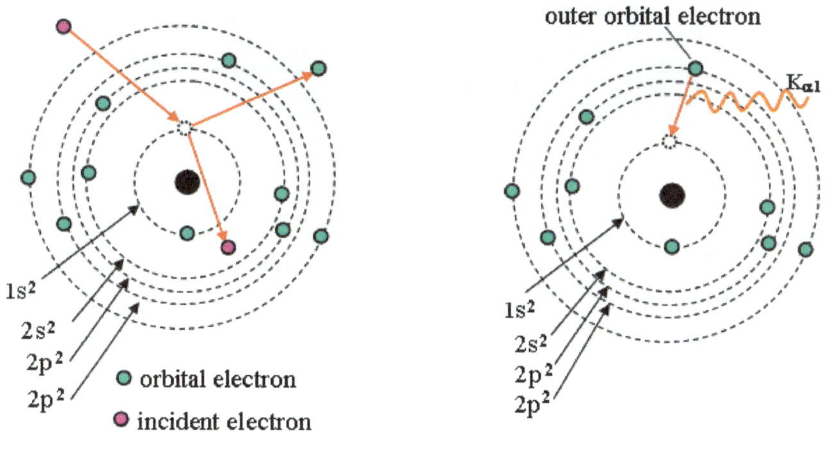

Fig. 1. The process of X-Ray production in EDX measurements of samples (a) showing how an inner orbital electron is ejected out from an electron coming from the electron gun of a SEM/TEM and (b) the production of characteristic X-Rays from the transition of an outer electron orbital to the inner orbital

The interactions of the incident electron beam with the sample can be broadly divided into two, that is, elastic and inelastic. In elastic collisions, both the momentum and kinetic energy is conserved. However, for inelastic collisions, momentum is conserved but not kinetic energy. The dominant scattering event in a SEM or TEM basically depends on the incident electron beam energy and the atomic number (Z) of the elements in the sample. If the dominant scattering event is elastic, the interaction volume will have a wider width whereas if it is inelastic, it will have depth and the electrons will lose energy as it penetrates deeper into the sample. Thus, the interaction volume in a sample is dependent on the electron beam energy and the Z of the elements in the sample. The interaction volume of EDS is generally about a few microns into the sample and the depth for K and L lines are not exactly the same (Joy 1995; Oxfor EDS Manual). This is illustrated in Fig. 2. EDS is therefore, not considered a surface analysis technique and is more representative of the bulk. However, it is different in the EDX TEM if the measurement is done on a very small particle or very thin section.

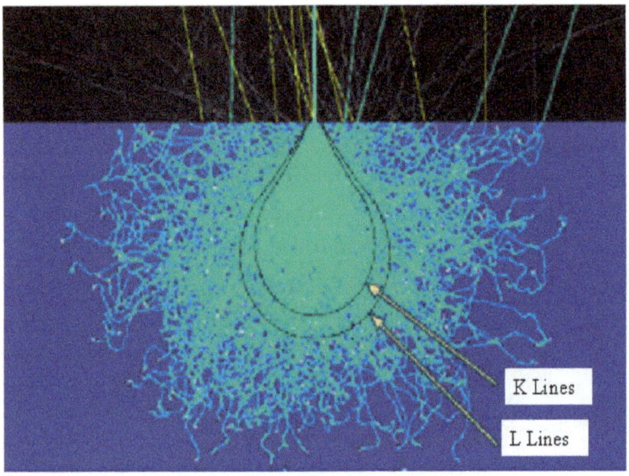

Fig. 2. The interaction of incident electrons with a sample (Oxford EDS Manual). The interaction volume for the K and L lines are not the same.

The EDS measurements shown in Fig. 3 are taken using optimized optical conditions such as distance of sample to electron beam pole piece, accelerating voltage, position of sample to EDS detector, detector settings, etc. as described in the respective manuals. The electron accelerating voltage for SEM is 15 kV while that for TEM is 200 kV. The EDS equipment models used in the measurements here is the Oxford INCA X-Max 51-XMX 0021 integrated with the SEM (JEOL JSM-7600F) and INCAx-sight 6498 integrated with the TEM (JEOL JEM-2100F). The EDS results obtained *via* SEM and TEM of the cathode material $LiCo_{0.3}Ni_{0.3}Mn_{0.3}Cr_{0.1}O_2$ shown in Fig. 3 (a) and (b) respectively clearly show the presence of the transition metals, Co, Ni, Mn and Cr. The results are summarized in Table 1 and the comparison is also made with the calculated synthesized values. It is observed that the EDS results from both SEM EDS and TEM EDS show transition metal content quite close to the calculated values. The deviation is about 1-4% for SEM EDS and 1.6-9% for TEM EDS respectively for Co, Ni and Mn. It is higher for Cr, that is, from 2.4% for SEM EDS and 22% for TEM EDS respectively. The higher deviations for SEM and TEM EDS may be due to different sample conditions whereby, the amount of sample in the SEM is much more than the amount of sample in TEM. The thickness of samples in TEM is also much smaller than in SEM. Therefore, averaging is more accurate in the SEM EDS. The SEM EDS equipment model used here is also newer and has slightly higher resolution giving more accurate results. Although oxygen counts are not considered very accurate for reasons discussed above, they are evaluated here to gauge the accuracy of results (when they are taken into account). It is remarkable that the oxygen content from the EDS measurements are quite close to the calculated value as can be seen in Table 1. Of course values from the measurements show higher content than calculated values but this is expected because although the vacuum environment eliminates most of the oxygen in the sample environment, there are still a number of O_2 molecules present that contributes to the higher oxygen counts measured. There is also adsorbed oxygen on the surface of the particles which will increase the count. The determination of Li in the sample cannot be done as the electron energies involved is out of range of the detector. It is always a problem to study lithium in cathode materials for Li-ion batteries. Stoichiometric values of transition metals in cathode materials, however, can be determined quite accurately by using EDS.

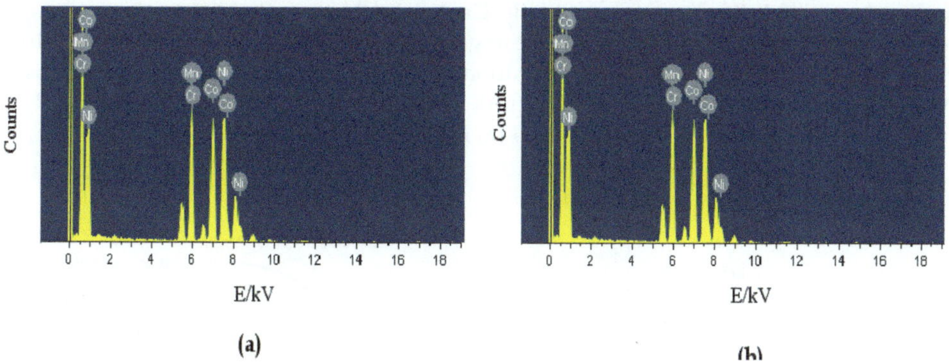

Fig. 3. EDS results obtained from (a) SEM EDS and (b) TEM EDS

Element	Calculated atomic per cent without O (%)	EDS SEM Atomic per cent without O (%)	EDS TEM Atomic per cent without O (%)	Calculated atomic per cent with O (%)	EDS SEM Atomic percent with O (%)	EDS TEM Atomic percent with O (%)
Cr	10.0	10.40	8.43	3.33	3.41	2.59
Mn	30.0	29.97	30.80	10.0	9.97	9.47
Co	30.0	30.30	31.25	10.0	10.13	9.60
Ni	30.0	29.33	29.25	10.0	9.25	9.07
O	-	-	-	66.67	67.24	69.27

Table 1. Comparisons between calculated synthesized values of $LiCo_{0.3}Ni_{0.3}Mn_{0.3}Cr_{0.1}O_2$ material and EDS results from SEM and TEM (O - oxygen). The atomic percentages are compared for EDS measurements with and without oxygen.

4.2 X-ray and neutron powder diffraction

The technique of X-Ray diffraction (XRD) or more precisely in this case, X-Ray powder diffraction (XRPD), is a very well established method used for material studies (Klug, et al. 1954; Kittel 2005). When X-Rays hit a crystalline target, the phenomenon of diffraction occur producing interference fringes as shown in Fig. 4 (a). This arises from a scattering process by which the X-Rays are scattered by the target material without change in wavelength (also called Bragg scattering). Diffraction occurs only when certain conditions are satisfied according to Bragg law or Laue equation (Klug, et al. 1954). The Bragg equation is given by,

$$2d \sin\theta = n\lambda \qquad (1.0)$$

where d - interplanar spacing between successive atomic planes in the crystal

- θ - angle between the atomic plane and the reflected beam (also called Bragg angle)
- n - integer
- λ - wavelength of X-Rays

Equation (1.0) is actually the condition for constructive interference (Christman 1988). The incident beam of X-Rays used is normally Cu K_α. In the so-called θ-θ optical set-up, the incident X-Ray beam moves over the same angle as the detector, as shown in Fig 4 (b). In Fig 4 (a), S_1 and S_2 are slits and L is the position of the detector. The 2θ angle, which is the angular value given with corresponding peak intensities in the XRD pattern, is actually the angle of deviation and can be seen in Fig. 4 (a) marked δ. The value of δ is 2θ.

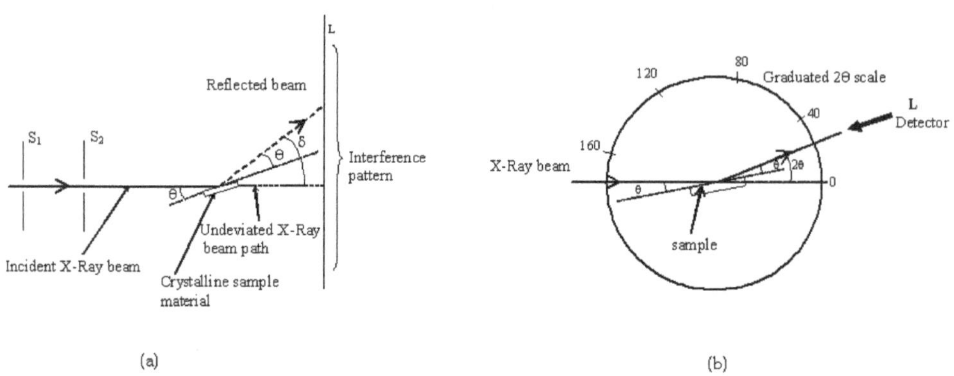

Fig. 4. X-Ray beam paths in an X-Ray diffractometer for (a) incident X-Rays on a crystalline sample undergoing diffraction and producing an interference pattern and (b) schematic of a θ-θ optical set-up. The X-Ray beam moves in conjunction with the detector.

In order to use XRD to elucidate stoichiometries of materials, quantitative analysis has to be done. There are several limitations and points to consider. First and foremost is that the diffractometer must be well aligned and calibrated. The optical settings should be suitable for good resolution and high counts. As a guide, the highest peak for the sample should be around 10,000 counts for good statistics and the 2θ angle should be as high as can be measured by the goniometer. The step size must be appropriately chosen to reflect suitable number of data points in a peak. An accepted rule is that there should be 10-20 individual data points above the full width at half-maximum (fwhm) (Jenkins et al. 1996). The data can then be considered good and suitable for quantitative analysis. If the crystals in the sample show plate-like or rod-like shapes, then there are preferred orientation issues that have to be addressed. To minimize this effect, the sample preparation should be done by the back-loading method and the spinning mode used during measurement. In extreme cases, the sample has to be loaded in glass capillaries and the measurements should be done using a Debye-Scherrer optical geometry. The spinning mode is also used for this optical set-up. In most cases, data is acquired using the Bragg-Brentano geometry.

Neutron diffraction is a technique only available at large neutron research facilities such as in the HFIR (Oak Ridge National Lab, United States of America), J-Parc (Japan), ILL (France) and OPAL (Australia). Beam time allocation is not easily available and can only be obtained *via* a strictly reviewed proposal system. There are two types of neutron diffractions available, that is, the constant wavelength and the time of flight (tof) neutrons (David et al. 2002). Neutrons interact differently with matter compared to that of X-Rays. X-Rays interact with the electron cloud surrounding the nucleus of an atom but neutrons interact with the nucleus directly. Therefore, X-Ray scattering cross section is dependent on the atomic number (Z) of atoms whereas neutron scattering cross section is a more complex process described by the so called Fermi pseudopotential. However, one advantage of neutrons is that it can be scatterered by light atoms relatively well and thus, is an advantage for materials containing elements such as lithium and hydrogen in the presence of heavier elements such as Cu, Co, Mn, etc. (Kamarulzaman et al. 2009).

Quantitative XRD and ND data analysis of the $LiCo_{0.3}Ni_{0.3}Mn_{0.3}Cr_{0.1}O_2$ material is done by the Rietveld refinement method. For the XRD in this work, the diffractometer used is the PanAlytical Xpert Pro MPD with the Bragg-Brentano optical configuration and a solid-state detector, called the Accelerator. The structure refinement software used is the Xpert Highscore Plus using a reference structure, ICSD 99891. Neutron data is acquired from the OPAL reactor facility, at the Australian Nuclear Science and Technology Organization (ANSTO) using constant wavelength neutrons. The optical set-up is the Debye-Scherer configuration with position sensitive detectors. The data analysis was done using the GSAS (Larson et al. 2004) software with EXPGUI (Toby 2001). Stoichiometry of the cathode material can be obtained by performing a Rietveld refinement on the X-Ray and Neutron data. Although Rietveld refinement is used primarily to study the structure of materials, stoichiometric information can be elucidated by refining the sight occupany factor (s.o.f.) of the atoms in their crystallographic sites in the structure (Giacovazzo 2006). The stoichiometry of the materials can then be obtained. Another advantage of using ND is that, the Li content can be elucidated and quantitative values obtained. It is otherwise very difficult to check the Li stoichiometry experimentally. The refined XRD and ND data are shown in Fig. 5 and Fig. 6 respectively. Site occupancy factors of the metals and transition metals are shown in Table 2. It can be seen that the R_w and chi squared values imply high accuracy in the quantification obtained from the refinements. The values obtained from the two diffraction methods agree very well. It is possible to evaluate lithium stoichimetry from neutron data due to the strong interaction of neutrons with light elements and this is an added bonus derivable from neutron diffraction experiments. The stoichiometries obtained for the $LiCo_{0.3}Ni_{0.3}Mn_{0.3}Cr_{0.1}O_2$ material is $LiCo_{0.28}Ni_{0.26}Mn_{0.8}Cr_{0.28}O_2$ from the XRD result and $Li_{0.882}Co_{0.28}Ni_{0.28}Mn_{0.3}Cr_{0.078}O_2$ from the ND result. Both results show that there are slight metal deficiencies in the material except for Mn

(ND result). Another important result is that there is also some Li deficiency in the cathode material which will affect their electrochemical performance.

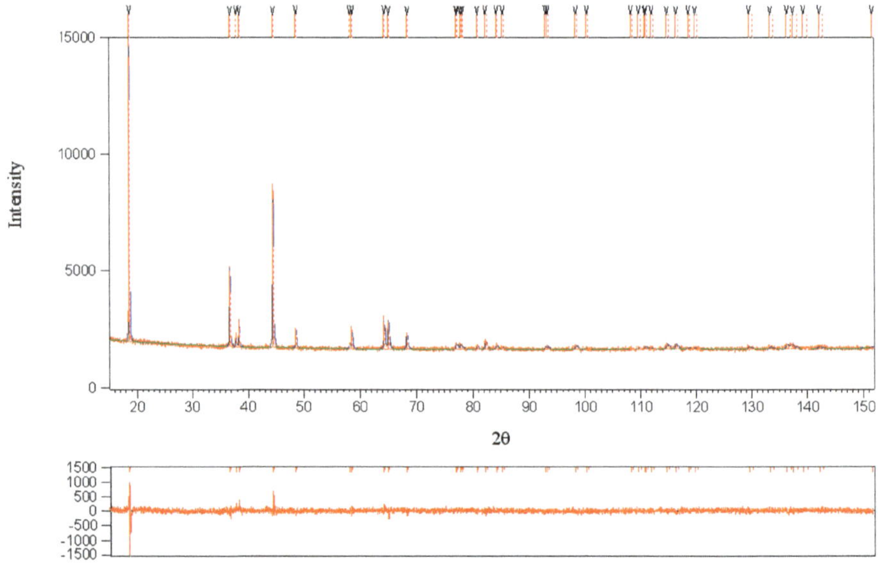

Fig. 5. Refined XRD dataset of $LiCo_{0.3}Ni_{0.3}Mn_{0.3}Cr_{0.1}O_2$

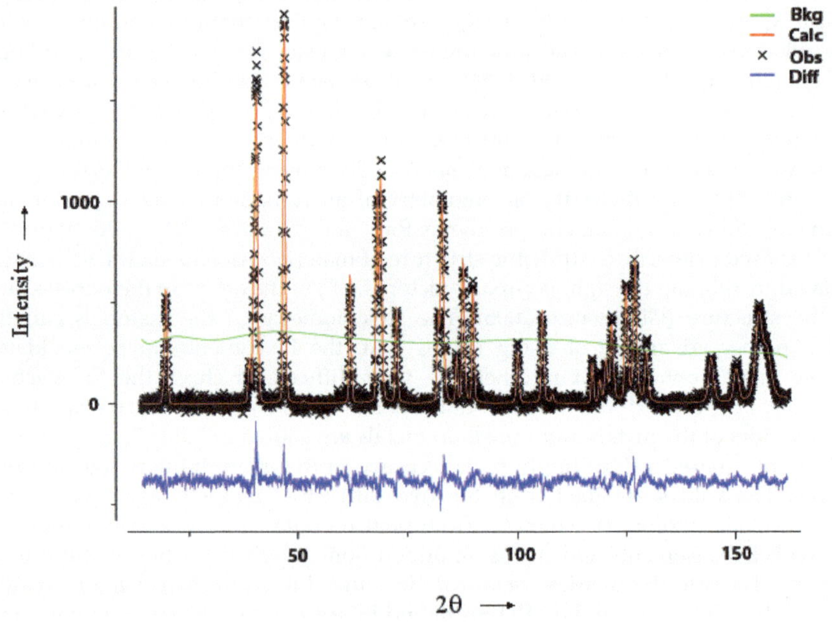

Fig. 6. Refined ND dataset of $LiCo_{0.3}Ni_{0.3}Mn_{0.3}Cr_{0.1}O_2$

Mode	s.o.f of Li	s.o.f of Co	s.o.f of Ni	s.o.f of Mn	s.o.f of Cr	R_w (%)	χ^2 (%)
XRD	1.0000	0.2800	0.2600	0.2800	0.0800	3.1	1.7
ND	0.8820	0.2800	0.2800	0.3000	0.0780	6.2	1.5

Table 2. Site occupancies of Li, Co, Ni, Mn and Cr of the cathode material $LiCo_{0.3}Ni_{0.3}Mn_{0.3}Cr_{0.1}O_2$

4.3 X-ray photoelectron spectroscopy

X-Ray Photoelectron Spectroscopy is a technique used to study the composition of a material, its electronic state and chemical environment. Stoichiometries of materials can be obtained by XPS through the quantitative analysis of the XPS spectrum, that is, from the narrow scans. XPS is a surface technique which can detect the compositional make-up of a sample up to about 10 nm thickness. Care must be taken to perform surface cleaning before the actual XPS measurement is done in order to get rid of surface contaminants such as carbon and oxygen. If careful precautions are taken, XPS measurements can be highly accurate and may be taken to represent the bulk.

The X-Ray Photoelectron Spectrometer is a machine that utilizes photoionization and the detection of the photoelectrons *via* an energy dispersive detector. The principle of the equipment is the ejection of electrons in a material by energetic electromagnetic waves, which is normally X-Rays. The energy of the X-Rays is given by the famous Planck equation (Wehr et al. 1978),

$$E = h\upsilon \tag{2}$$

where h is Planck's constant and υ is frequency. The relationship between the X-Ray energy and the kinetic energy of the photoelectrons ejected out of the orbitals of the elements in the sample material is given by,

$$h\upsilon = E_{KE} + \phi \tag{3}$$

where E_{KE} is the kinetic energy of the photoelectron and ϕ is the work function (also called the binding energy). Equation (3) is the famous Einstein equation by which he successfully explained the phenomenon of the photoelectric effect (Wehr et al. 1978). Thus, XPS measures the kinetic energy of the photoelectrons and calculates the binding energy. Fig. 8 illustrates the principle of the XPS. It shows the absorption of an X-Ray photon by a core electron which escapes from the atom with a specific kinetic energy. The kinetic energy of the electron is then measured by the energy dispersive detector and the characteristic binding energy can then be determined. XPS detects the first 2 to 10 nm of the top layer of a material (depending on the composition of the sample). Its detection limit is about 0.1 percent of total elemental content. A schematic of an XPS is shown in Fig. 9 showing the position of the X-Ray gun, sample chamber and detector.

Several considerations have to be made when acquiring data from an XPS. For a dual beam XPS, the X-Ray guns are usually those of Mg and Al and the best X-Ray source for the sample has to be chosen. For optimized results, peak overlaps due to very close peak positions of the elements, Auger peaks and satellite peaks of the compositional elements in the sample must be taken into account and the right choice of X-Ray source has to be chosen. Stoichiometric information of a sample can be obtained by quantifying an XPS result. For the XPS data shown in Fig. 10, a JOEL JPS 9010 was used and the X-Ray source is Mg. This is because the $LiCo_{0.3}Ni_{0.3}Mn_{0.3}Cr_{0.1}O_2$ material is a very challenging sample to analyze containing elements with their positions very close together in the periodic table. Severe overlapping of peaks will occur if the Al source is used. The Mg k_{α} source produces cleaner and better resolved spectrum giving more accurate results.

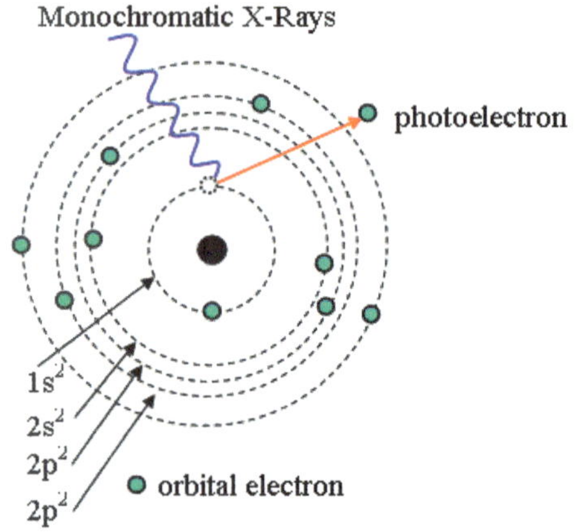

Fig. 7. X-Ray photon being absorbed by an orbital electron and escaping out of the atom with a certain kinetic energy

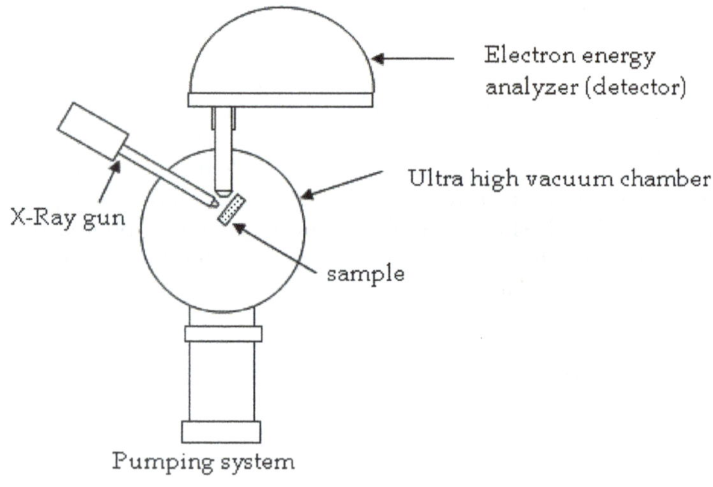

Fig. 8. Schematic of an X-Ray Photoelectron Spectrometer

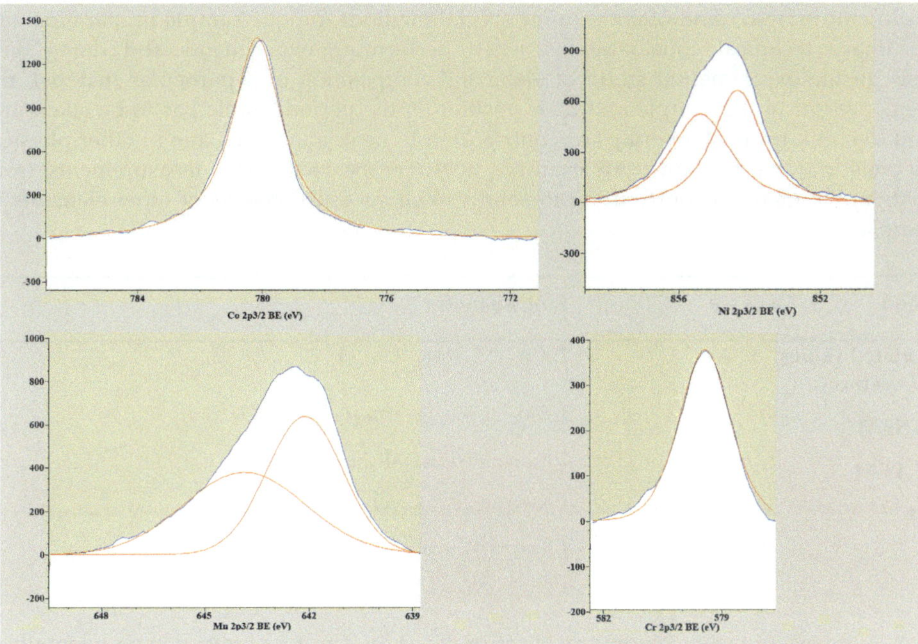

Fig. 9. XPS narrow scan of Co, Ni, Mn and Cr for $LiCo_{0.3}Ni_{0.3}Mn_{0.3}Cr_{0.1}O_2$ material

The data processing done for the results shown in Fig. 10 is background subtraction (due to ineleastic scattering) using the Shirley method, smoothing using the Savitzky-Golay method and deconvolution for the Ni and Mn peaks. Results indicate the presence of two oxidation states for Mn and Ni and one oxidation state for Co and Cr. The stoichiometry elucidated from the XPS results based on the $2p_{3/2}$ orbital of the transition metals for the synthesized $LiCo_{0.3}Ni_{0.3}Mn_{0.3}Cr_{0.1}O_2$ material is $LiCo_{0.35}Ni_{0.21}Mn_{0.29}Cr_{0.15}O_2$. The result obtained for the Cr and Co content is a little lower than those obtained from the XRD and ND data but the Mn content seems to be spot on. This may be due to some surface effects of the XPS data. The slightly higher count for Cr obtained from XPS seem to agree with the TEM EDS results implying that there may be surface effects whereby Cr is hypothesized to migrate towards the surface of the sample. The surface conditions of the material may therefore, not be the same as the interior. Results however demonstrate that XPS is an acceptable method for obtaining quantitative stoichiometric values experimentally for synthesized samples with complex composition of transition metals sitting side-by-side in the periodic table. It can also give further information on the oxidation states of the transition metals which have important implications on the electrochemical behaviour of the cathode material.

5. Conclusions

For synthesized materials, it is very important to experimentally validate their stoichiometries. Several methods are available but all of them have limitations and these limitations must be considered when taking measurements. Specific techniques for some materials must be used depending on sample type and composition. Light elements are more difficult to detect and often requires special methods, unique sample preparation and extraordinary technical considerations when performing experiments and doing data analysis. In the experimental study of elemental composition of a particular material, the elements present in the sample and their position in the periodic table has to be taken into account to check for peak overlaps or contribution to peak intensities due to other physical interactions and mechanisms. An example is Auger peaks in XPS measurements from elements in the material contributing to counts of characteristic peaks of other elements in the sample.

Method	Stoichiometry
Calculated values from synthesis	$LiCo_{0.3}Ni_{0.3}Mn_{0.3}Cr_{0.1}O_2$
EDS SEM	$LiCo_{0.3030}Ni_{0.2933}Mn_{0.2997}Cr_{0.1400}O_2$
EDS TEM	$LiCo_{0.3125}Ni_{0.2925}Mn_{0.3080}Cr_{0.0843}O_2$
XRD	$LiCo_{0.2800}Ni_{0.2600}Mn_{0.2800}Cr_{0.0800}O_2$
ND	$Li_{0.8820}\,Co_{0.2800}Ni_{0.3}Mn_{0.300}Cr_{0.0780}O_2$
XPS	$LiCo_{0.35}Ni_{0.21}Mn_{0.29}Cr_{0.15}O_2$

Table 3. Summary of the stoichiometry of $LiCo_{0.3}Ni_{0.3}Mn_{0.3}Cr_{0.1}O_2$ obtained experimentally from different methods and compared with calculated values

A summary of stoichiometric measurements using different methods of EDS, XRD, ND and XPS is given in Table 3. Each method has its own merits, advantages, disadvantages and limitations and the scientist sometimes has to use more than one method in order to elucidate the stoichiometry of their synthesized products. It can be said that overall, for the techniques of EDS, XRD, ND and XPS, the experimental values are reasonably close to the calculated values and to each other. More importantly, the methods used are successful in analyzing such a complex material as $LiCo_{0.3}Ni_{0.3}Mn_{0.3}Cr_{0.1}O_2$ containing a number of elements sitting side-by-side in the periodic table. Experimental determinations of the metal and transition metal composition generally indicate deficiencies from all methods except XPS which was attributed to the very narrow thickness of evaluation volume that may not be indicative of the bulk. Of the five methods used, only neutron diffraction is capable of giving Li content in the sample. Thus, experimental methods, special techniques and sample considerations are of paramount importance for the study of stoichiometries of synthesized cathode materials. Calculated stoichiometries must not be presumed and must be closely derived from experimental measurements in order to obtain reliable quantitative values of elemental compositions.

6. Acknowledgments

The authors are very grateful to Dr Max Avdeev and Mr Ned Blagojevic of ANSTO for the help in the ND experiments. The project was internally funded by the Institute of Science, Universiti Teknologi MARA and grants from the Misistry of Science, Technology and Innovation (MOSTI: 03-01-01-SF0076) and Ministry of Higher Education (MOHE), Malaysia.

7. References

Amatucci, G.; Tarascon, J.M.; Larcher, D. and Klien, L.C. (1996). *Synthesis of electrochemically active $LiCoO_2$ and $LiNiO_2$ at 100 °C*, Solid State Ionics, Vol.84, pp.169-180

Antolini, E. (2004). *$LiCoO_2$: formation, structure, lithium and oxygen nonstoichiometry, electrochemical behaviour and transport properties*, Solid State Ionics, Vol.170, pp.159-171

David, W.I.F.; David, W.I.F.; Shankland, K.; McCusker, L.B. and Baerlocher Ch. (Eds), (2002). *Structure Determination from Powder Diffraction Data*, Oxford, ISBN 0-19-920553-1

Ding, N.; Ge, X.W. and Chen, C.H. (2005). A new gel route to synthesize $LiCoO_2$ for lithium-ion batteries, *Mat. Res. Bull*, Vol.40, pp.1451-1459

Echlin, P. ; Fiori, C.E.;Goldstein, J.I.; Joy, D.C.; Lyman, C.E. ; Lifshin, E. ; Newbury, D.E. and Romig, A.D. (1992). *Scanning electron microscopy and X-ray microanalysis*, Plenum Press ISBN 0-306-44175-6

Fey, G.T.K.; Chen, J.G. and Subramanian, V. (2003). Electroanalytical and thermal stability studies of multi-doped lithium nickel cobalt oxides, *Journal of Power Sources*, Vol. 119–121, pp. 658–663

Christman J.R. (1988). Fundamentals of Solid State Physics, John Wiley & Sons, ISBN 0-471-81095-9

Giacovazzo, C.; Monaco, H.L.; Artioli, G.; Viterbo, D.; Ferraris, G.; Gilli, G.; Zanotti, G. and Catti, M. (2006). *Fundamentals of Crystallography*, Oxford Sc. Pub., ISBN 0-19-850958-8

Goldstein, J.I.; Newbury, D.E. and Williams, D.B. (Ed.) (1995). *X-Ray spectrometry in electron beam instruments*, Plenum Press, ISBN 0-38-776502-6

Huang, K.; Peng, B.; Chen, Z. and Huang, P. (2000) Preparation, structure and electrochemical properties of spinel $Li_{1-x}Mn_{2-y}O_4$ cathode material for lithium ion batteries, *Solar Energy Mater. & Solar Cells*, Vol. 62, pp.177-185

Jenkins R. And Snyder R.L. (1996). *Introduction to X-Ray Powder Diffractometry*, John Wiley & Sons, ISBN 0-471-51339-3

Joy, D.C. (1995). *Monte Carlo modeling for electron microscopy and microanalysis*, Oxford University Press, ISBN 0-19-508874-3

Kamarulzaman, N.; Yusoff, R.; Kamarudin, N.; Shaari, N.H.; Abdul Aziz, N.A.; Bustam, M.A.; Blagojevic, N.; Elcombe, M.; Blackford, M.; Avdeev, M. and Arof, A.K. (2009) Investigation of the cell parameters, microstructures and electrochemical behaviour of $LiMn_2O_4$ normal and nano powders, *J. Power Sources*, Vol.188, pp.274-280

Kittel, C.(2005). *Introduction to Solid State Physics*, John Wiley & Sons Inc, New York, ISBN 0-471-68057-5

Klug, H.P. and Alexander, L.E. 2nd Ed. (1974). *X-Ray Diffraction Procedures*, John Wiley & Sons Inc, New York ISBN 3-900051-07

Larson, A.C. and Von Dreele, R.B. GSAS, Los Alamos National laboratory Report Laur 86-748 (2004)

Li, G.; Iijima, Y.; Kudo, Y. and Azuma, H. (2002). Structural Changes of Manganese Spinel at Elevated Temperatures, *Solid State Ionics*, Vol.146, pp.55-63

Li, G.; Yamada, A.; Fukushima, Y.; Yamaura, K.; Saito, T.; Endo, T.; Azuma, H.; Sekai, K. and Nishi, Y. (2000). Phase segregation of $Li_xMn_2O_4$ ($0.6<x<1$) in non-equilibrium reduction processes, *Solid State Ionics*, Vol.130, pp.221-228

Lide D.R.; Chief Ed. (1995) Handbook of Chemistry and Physics, CRC Press, ISBN 0-8493-0476-8

Lu, C.H. and Saha, S.K. (2001). Morphology and electrochemical properties of $LiMn_2O_4$ powders derived from the sol–gel route, *Mater. Sc. and Eng.*, B79, pp.247-250

Ohzuku, T.; Kitagawa M.; and Hirai, T. (1990) Electrochemistry of manganese dioxide in lithium nonaqueous cell. III. X-ray diffractional study on the reduction of spinel-related manganese dioxide, *J. Electrochem. Soc.*, Vol. 137, No. 3, pp.769-775

Ohzuku, T.; Takeda, S. and Iwanaga, M. (1999). Solid-state redox potentials for $Li[Me_{1/2}Mn_{3/2}]O_4$ (Me: 3d-transition metal) having spinel-framework structures: a series of 5 volt materials for advanced lithium-ion batteries, *J. Power Sources*, Vol. 81-82, pp.90-94

Oxford EDS manual

Park, S.C.; Kim, Y.M.; Kang, Y.M.; Kim, K.T.; Lee, P.S. and Lee, J. Y. (2001). Improvement of the rate capability of $LiMn_2O_4$ by surface coating with $LiCoO_2$, *J. Power Sources*, Vol. **103**, pp.86-92

Shin, Y. and Manthiram, A. (2004). Factors Influencing the Capacity Fade of Spinel Lithium Manganese Oxides, *J. Electrochem. Soc.*, Vol.151, (2), pp.A204-A208

Tanaka, T.; Ohta, K. and Arai, N. (2001). Year 2000 R&D Status Of Large Scale Lithium Ion Secondary Batteries In the National Project Of Japan, *J. of Power Sources*, Vol.97-98, , pp. 2-6

Toby, B.H. (2001). EXPGUI, a graphical user interface for GSAS, *J. Appl Cryst* Vol.34, pp.210-213

U.S. Patent 2,530,110 (Filed, 1944, granted 1950)

U.S. Patent 2,631,356 (Filed June 15, 1950. granted1953)

Veluchamy, A.; Ikuta, H. and Wakihara. (2001). Boron Substituted Manganese Spinel Oxide Cathode For Lithium Ion Battery, *Solid State Ionics*, Vol. 142, pp.161-171

Wehr, M.R.; Richards, J.A.; Adair, T.W. (1978). *Physics of the Atom*, Addison-Wessley, Reading, Mass.

Wu, C.; Wang, Z.; Wu, F.; Chen, L.; Huang, X. (2001). Spectroscopic studies on cation-doped spinel LiMn2O4 for lithium ion batteries, *Solid State Ionics*, Vol.144, pp.277-285

Yamaguchi, R.; Ikuta H.; Wakihara M. (1999) Heat of Formation for $LiM_yMn_{2\text{-}y}O_4$ (M = Co, Cr, Li, Mg,Ni) Spinel Solid Solution, *J. Therm. Anal. Calorim.*, Vol. 57 pp.797-806

Zou, M.; Yoshio, M.; Gopukamar, S.; Yamaki, J. (2005). Performance characteristics of $Li//Li_{1\pm x}CoO_2$ cells, *Mater. Res. Bull*, Vol.40, pp.708-714

11

A Study on Hydrogen Reaction Kinetics of Pt/HfO_2/SiC Schottky-Diode Hydrogen Sensors

W.M. Tang, C.H. Leung and P.T. Lai
Department of Electrical and Electronic Engineering,
The University of Hong Kong, Hong Kong,
China

1. Introduction

Hydrogen has a wide range of applications and is an essential raw material in many industries such as chemical, food, metallurgical, electronics and others. It can be produced by a variety of methods, including acid and iron methods, steam-reforming production method, coal and electrolysis methods, biological processes, as well as chemical decomposition of hydrogen-containing compounds. Most of the hydrogen produced in the United States is used to synthesize ammonia by the Haber process which can then be used as a soil fertilizer. Hydrogen is also used to produce synthetic methyl alcohol and synthetic petroleum by the Patart method and the Fischer-Tropsch process respectively (Almqvist, 2003). Hydrogen can also be used in the food processing industry to convert vegetable oils into hard and white fats that can be used to make margarine. This process is called hydrogenation, which is the major chemical application of hydrogen. Some oil companies use hydrogenation process to do desulfurization and hydro-cracking. In addition, hydrogen has many industrial applications such as metal smelting, welding, cutting, cooling in power station, electronic fabrication process, fiber optics manufacturing, medical installation and corrosion prevention in nuclear reactors.

The world population rapidly increases nowadays and there is an increasing demand for energy. Fossil fuels are being quickly used up and are not renewable. It has been estimated that the known reserves of petroleum and natural gas will be used up by 2090. Coal deposits are more plentiful but they may last for less than 300 years. It is apparent that mankind is heading toward an energy crisis that can be averted only through conservation coupled with a renewable energy technological revolution. The world must turn to alternative energy sources to meet future energy needs. There are many alternative energy carriers such as solar, wind, tidal, nuclear, etc. Of this wide range of energy carriers, hydrogen is surely one of the most attractive energy carriers. Hydrogen is a clean renewable energy source because it reacts with oxygen producing water only. It causes no air and water pollutions and has a very high fuel value (142 kJ/g). It will be an alternative to fossil fuels in the near future. In most cases, hydrogen is used as a fuel involving the use of fuel cells. Hydrogen can be liquefied at 20.3 K by a multi-stage process, and then liquid hydrogen is easier to store and transport. It can be used for scientific cooling applications and as a rocket fuel for space flights.

Although hydrogen has a lot of useful applications, it has some disadvantages. It can make various metals such as high-strength steels, aluminum, and titanium alloys become brittle and crack. The mechanism starts with hydrogen atoms diffusing through the metal during various manufacturing operations or operational use. These hydrogen atoms will recombine to form hydrogen molecules in the tiny voids of the metal creating a high pressure inside the metal to reduce its ductility and tensile strength. When the pressure is too high, the metal will crack eventually. This process is called hydrogen embrittlement or hydrogen grooving which can pose a serious engineering problem. Hydrogen is also a very dangerous gas. It is highly flammable with a high rate of flame propagation. Its ignition temperature is 500 ^{o}C and the explosion range is 4 % to 75 %. The explosion range indicates the amount of hydrogen mixed with air that will cause serious explosions when ignited with a sufficient amount of heat. If hydrogen mixes with air and the concentration of hydrogen is less than 4 %, the mixture is not explosive because there is not enough fuel. The mixture is also not explosive if the mixture contains more than 75 % hydrogen because there will be too much fuel but not enough oxygen for combustion. Only within the lower explosion limit (LEL) and the upper explosion limit (UEL), the H_2-air mixture burns when ignited, causing significant flame propagation. There is a point within LEL and UEL at which the flame propagation is the greatest (Korver, 2001). As hydrogen poses potential danger, storage and use of hydrogen need special care and attention. In order to avoid the leakage of hydrogen that can cause serious explosion, sensors that can detect hydrogen quickly and accurately are very important and essential in practical applications. Apart from leak detection, monitoring of the hydrogen or hydrocarbon concentration in fuel cell, hydrogen driven vehicle, car exhaust, flue and biomedical and chemical industries is also an important application of hydrogen sensors.

The organization of this chapter is as follows. First, it will briefly review the type of hydrogen sensors and the evolution of Schottky-diode hydrogen sensor. Then, the working principle of Metal-Insulator-Semiconductor (MIS) Schottky-diode hydrogen sensor will be presented. The physical and chemical mechanisms responsible for hydrogen detection will be discussed. Then, it will describe the fabrication procedure of the hydrogen sensors and the measurement methodology. The steady-state and transient behaviour of the Pt/HfO_2/SiC Schottky-diode hydrogen sensors are then analysed. The sensor response of the device is found to increase with annealing time because longer annealing time can enhance the densification of HfO_2 film and improve the oxide stoichiometry. The hydrogen reaction kinetic and the enthalpy change are then investigated. Finally, the results obtained in this study are summarized.

2. Types of hydrogen sensor

There are various types of hydrogen sensors which use different sensing mechanisms to detect hydrogen. Many of them use catalytic metals such as palladium (Pd) or platinum (Pt) as a hydrogen trap because these metals have high hydrogen solubility and can selectively absorb hydrogen gas. The absorption of hydrogen in the catalytic metals depends on temperature and hydrogen concentration. The adsorbed hydrogen gas molecules can dissociate at the Pd or Pt surface to form hydrogen atoms. According to Lundstrom (Lundstrom, 1981), a linear relation can be assumed between the measured response signal (which can be voltage, pyroelectric signal, frequency shift, optical signal, etc.) and the coverage of hydrogen atoms. Six types of hydrogen sensors are introduced in this section. They are pyroelectric sensors, piezoelectric sensors, fiber-optic sensors, electrochemical sensors, and semiconductor sensors.

2.1 Pyroelectric sensor

The first pyroelectric chemical sensor (Pd-$LiTaO_3$) was reported in 1981 (Zemel et al., 1981). It makes use of the unique property of pyroelectric material together with a catalyst metal to detect hydrogen. A pyroelectric material can generate an electric potential when it is heated or cooled. A schematic diagram of a pyroelectric hydrogen sensor is shown in Fig. 1. It consists of three main parts: (1) a single-crystal $LiTaO_3$ wafer which is a pyroelectric material, (2) two NiCr electrodes, one of them is coated with a catalyst material (Pd in this case) and (3) a heater. When hydrogen is introduced into the test cell, it is absorbed by Pd, causing an extra heat gain in the Pd-NiCr electrode and hence a voltage difference between the Pd-NiCr and the reference NiCr electrodes. This type of sensor is not popular because the fabrication procedures are complex and the sensor can be easily influenced by temperature fluctuation due to environmental factors. Besides, the response of the sensor is slow due to the diffusive transport of thermal energy from the Pd electrode to the reference electrode.

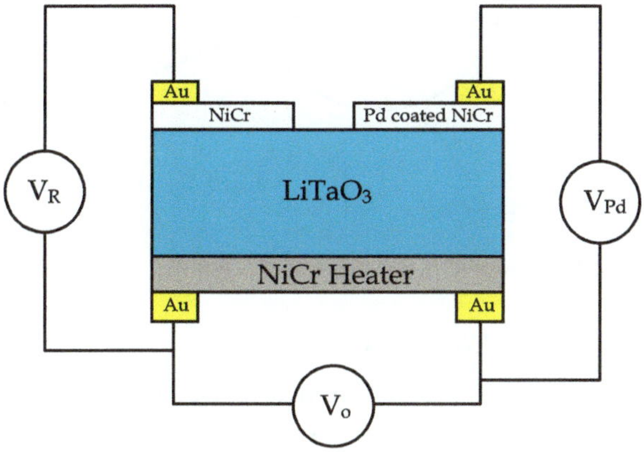

Fig. 1. Schematic diagram of a pyroelectric hydrogen sensor (Christofides et al., 1990).

2.2 Piezoelectric sensor

Piezoelectric sensor can be divided into two groups: piezoelectric quartz-crystal microbalance (PQCMB) sensor and surface acoustic wave (SAW) sensor.

2.2.1 Piezoelectric quartz-crystal microbalance (PQCMB) sensor

A coated piezoelectric crystal (e.g. quartz) is very useful for gas detection. The sensing principle was described in 1959 (Christofides, 1990). The setup of the PQCMB sensor is shown in Fig. 2. It consists of an oscillator powered by a regulated power supply and a frequency meter for measuring the frequency output of the oscillator. The response of the coated PQCMB sensor could be noticed from the change in the resonance frequency of the quartz crystal. The oscillation frequency of the quartz crystal depends on its total mass and the mass of the chemically sensitive layer. When hydrogen molecules are absorbed in the Pd coating layer, the resonance frequency decreases in proportional to the quantity of dissolved molecules. Hence, by observing the change in resonance frequency, the concentration of hydrogen gas can be measured.

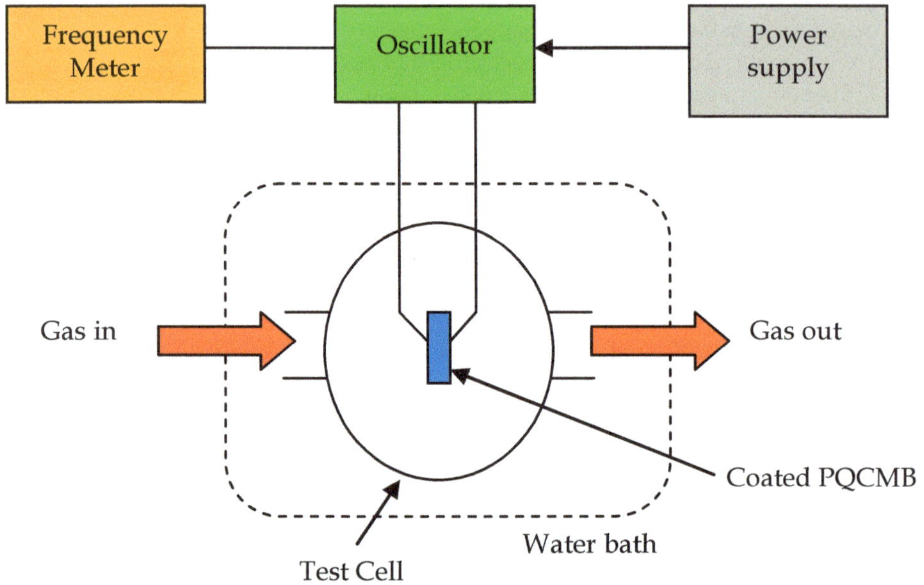

Fig. 2. Setup of the PQCMB sensor (Christofides et al., 1990).

2.2.2 Surface acoustic wave (SAW) sensor

The first SAW hydrogen sensor was demonstrated in 1982 (Amico, 1982/83; Amico, 1982), and its sensing mechanism is simple. When hydrogen molecules are absorbed in a Pd or Pt layer, the SAW characteristics such as amplitude, phase, velocity, etc will be changed. A schematic diagram of a SAW hydrogen senor is shown in Fig.3. The sensor is fabricated on a piezoelectric substrate YZ-$LiNbO_3$. Surface acoustic waves are generated by an input transducer T and then collected by two output transducers T_R and T_S. There are two propagation paths in the device. One path (l_S) is coated with a catalyst metal layer (Pd in this case), while the other path (l_R) is uncoated as a reference. A double-balance mixer connects the two voltage outputs (V_S and V_R) through inductances L_S and L_R. The output from the mixer then goes through a low-pass filter and the differential output voltage ΔS (V_R - V_S) is filtered out and recorded. When absorption of hydrogen takes place at the Pd surface, a voltage change can be detected. Comparing the two groups of piezoelectric sensor, the SAW sensor has a higher sensitivity than the PQCMB sensor because the former can have a resonance frequency a few hundred times higher.

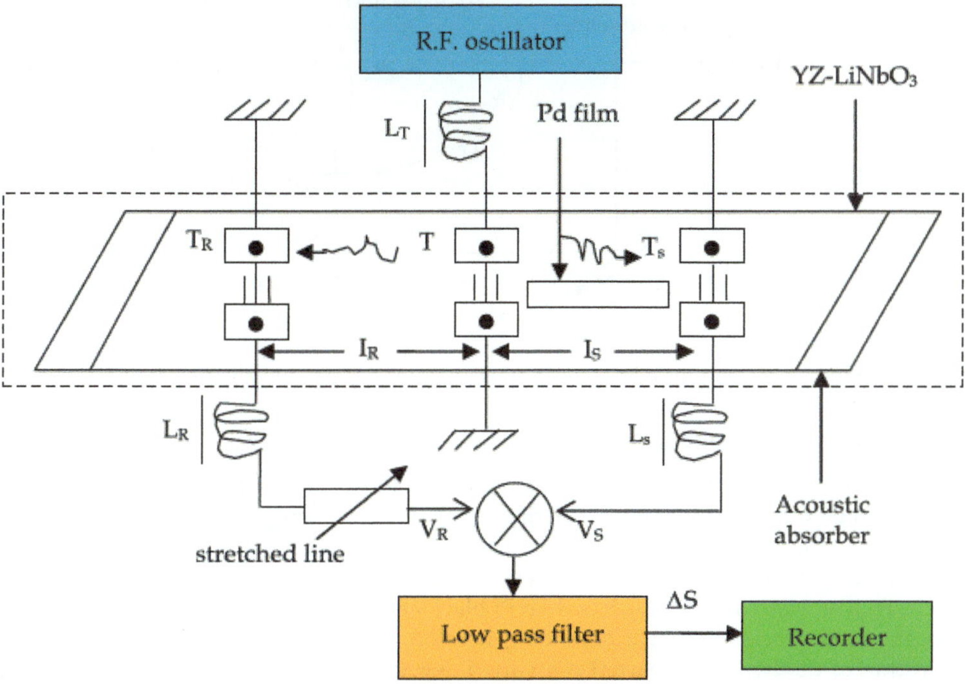

Fig. 3. Schematic diagram of a SAW hydrogen senor (Amico et al., 1982/83).

2.3 Fiber-optic sensor (FOS)

The first FOS to detect hydrogen was developed by Butler in 1984 (Butler, 1984). A schematic diagram of a Pd-coated fiber-optic hydrogen sensor is shown in Fig. 4. Interferometry is applied in this type of sensor. The sensing mechanism of the device is simple and straightforward. When hydrogen molecules are absorbed in the chemically sensitive Pd coating, they will alter the optical properties such as absorbance, reflectance, luminescence, or scattering of the coating. The laser is split into two paths by a beam splitter: one through the Pd-coated fiber and the other through the uncoated fiber. Both ends of the fibers are connected to a quartz plate. When the phase difference between the two interfering beams is $2n\pi$ (n is a positive integer), a maximum fringe pattern will appear. A transverse shift in the position of the fringe pattern can be observed when the fibers are exposed to hydrogen gas. The amount of fringe shift is related to the quantity of absorbed species (gas molecules or atoms) in the optical path. The FOS is very useful in environments with considerable electromagnetic activity because the optical nature of the signal does not introduce any electrical interference. However, one drawback for FOS is that the ambient light could interfere with background noise, causing FOS to be preferably used in dark environment for optimal operation and noise minimization.

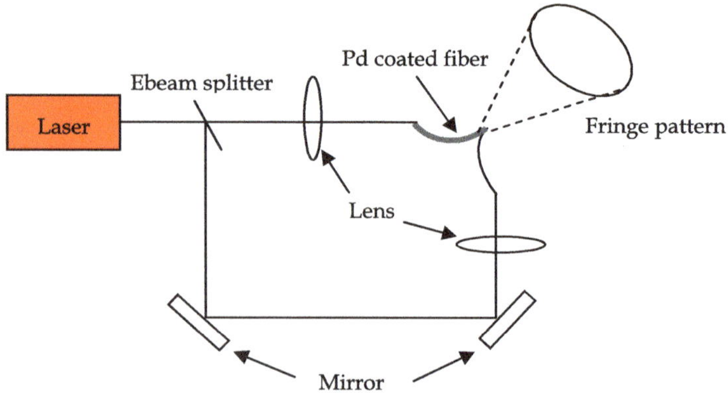

Fig. 4. Schematic diagram of a Pd-coated fiber-optic hydrogen sensor (Butler, 1984).

2.4 Electrochemical sensor

The first electrochemical hydrogen sensor was developed in 1978 (Childs et al., 1978). This sensor can be used for the study of hydrogen evolution during the corrosion of copper in pure water. Fig. 5 shows the schematic diagram of the electrochemical hydrogen sensor. The cell consists of a thin reference electrode, a working electrode made of Pt black, and an electrolyte. Hydrogen gas reacts with Pt black at the anode to give out hydrogen atoms and electrons. The potential difference generated in the cell increases linearly with the logarithm of the hydrogen partial pressure. Long response time is a disadvantage of this sensor.

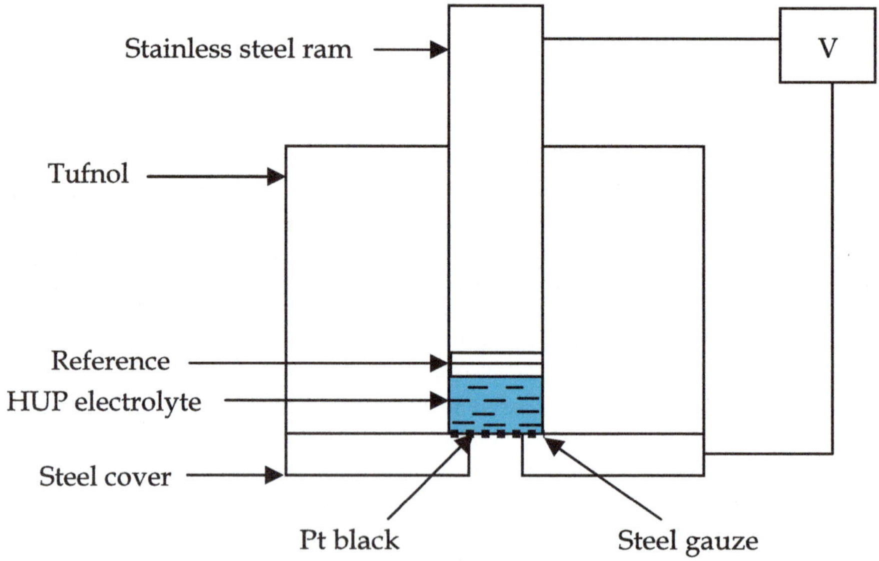

Fig. 5. Schematic diagram of the electrochemical hydrogen sensor (Kumar et al., 1988).

2.5 Semiconductor sensor

The detection of hydrogen gas with semiconductor sensors is primarily done with metal-oxide semiconductors. The initial work on this type of sensor can be traced back to the work of Brattain and Bardeen in 1953 (Brattain et al., 1953). The sensing properties of this sensor are based on reactions between oxide semiconductors and gases in the atmosphere. These reactions can change the electrical characteristics of the device. Oxygen in the atmosphere can be absorbed on the semiconductor surface and dissociates to form different species such as O_2^-, O^{2-} and O^-, where the electrons are extracted from the semiconductor. This electron extraction increases the resistance of the device (assuming an n-type semiconductor). When the sensor is exposed to a reducing gas such as hydrogen, the hydrogen reacts with the absorbed oxygen species to form water and the electron is re-injected into the semiconductor to reduce the resistance. Several metal-oxide semiconductors have been used as gas-sensing materials such as tin oxide (SnO_2) (Korotcenko et al., 2001), tungsten trioxide (WO_3) (Kim et al., 2006), titanium oxide (TiO_2) (Smith et al., 1993) and zinc oxide (ZnO) (Tomchenko et al., 2003). In order to accelerate the reaction rate and increase the sensitivity, a catalyst Pd or Pt is usually deposited on the surface of the oxide semiconductor. Apart from using catalyst, the sensor performance can be also improved by utilizing nano-structured materials, which have high surface-to-volume ratio for gas absorption. However, expensive equipment, materials and extensive training required to produce nanoscale structures increase the production cost of the sensor and hence limit their practical viability in industrial and commercial applications.

There is another category of semiconductor hydrogen sensor in which the species of interest are adsorbed at the electrode/insulator interface and induce interfacial polarization. The first hydrogen sensor under this category was developed in 1975 (Lundstrom et al, 1975) and was an MOS structure using silicon as substrate, silicon dioxide as gate insulator and palladium as gate electrode. Since then a lot of research on such hydrogen sensors has been done. Different materials and fabrication methods have been used to make the hydrogen sensors (Lundstrom et al., 1989; Spetz et al., 1999; Kim et al., 2000). There are many structures that can be used to make this type of hydrogen sensors, such as metal-oxide-semiconductor field-effect transistor (MOSFET) (Lundstrom et al., 1975), ion-sensitive FET (ISFET) (Bergveld et al., 1972), MOS capacitor (Steele et al., 1976a), Schottky diode (Steele et al., 1976b), etc. When the device is exposed to hydrogen, dissociated hydrogen atoms adsorbed at the electrode-insulator interface are polarized due to the displacement of their negatively-charged electron cloud relative to their positively-charged nucleus by an external applied field. This dipole layer causes a shift of the electrical (*I-V* or *C-V*) characteristics of the device.

3. Evolution of Schottky-diode hydrogen sensor

Among the semiconductor hydrogen sensors, Schottky diodes are preferably selected as gas sensors due to its much simpler fabrication procedures and electronic circuitry required for operation. The early Schottky-diode hydrogen sensors often used silicon as substrate because high-quality Si is commercially available and its cost is very low. Silicon is the second most abundant element in the world and is the dominant semiconductor used in the solid-state electronics industry. The properties of silicon have been widely studied and the silicon technology is the most mature and advanced among all semiconductor technologies.

Si-based hydrogen sensors can perform very well at room temperature (Fang et al., 1997) but they can only operate in environment below 250 °C due to the relatively small bandgap of Si (1.12 eV). In order to make these sensors function in harsh environments such as hot-engine and car exhaust, a cooling system is needed. However, this cooling system increases not only the operating and maintenance costs, but also the weight of the whole system, and leaves less space to integrate other electronic components. As a result, the reliability and efficiency of the system become lower. Therefore, development of hydrogen sensors that can directly operate in a high-temperature environment is very useful and important. To achieve this, a new sensor technology based on high-temperature semiconductor materials was then created.

High-temperature semiconductor materials are compound semiconductors such as SiC, CdS, GaP, AlN, GaN, etc with larger bandgap and thus lower intrinsic carrier concentration, which enables them to operate at much higher temperatures than silicon. Devices made of wide-bandgap materials can operate potentially up to 1000 °C. Thus, they are very suitable for making hydrogen sensors for aerospace and automobile applications, process-gas monitoring and leak detection. Uncooled operation of these sensors enables significant reductions in aircraft weight, operating and maintenance costs, overall system size, pollution, and also increases in operational safety, efficiency and reliability. Among the wide-bandgap materials, silicon carbide is by far the most developed for use in high-temperature, high-power, and high-radiation conditions. It is because sublimation growth of SiC wafer has been developed and high-quality 4H- and 6H-SiC wafers have been commercially available by Cree Research since 1989. These greatly facilitate the development of SiC semiconductor electronics. Besides, SiC can make good ohmic and Schottky contacts with a wide range of metals. It can also allow high-quality insulator (SiO_2) grown on top of it when it is put in an oxygen environment at high temperature. In addition, patterned etching of SiC has been developed for device fabrication. The electrical, optical, and physical properties of SiC have been extensively studied. With a wider bandgap, SiC not only can make gas sensor for emissions monitoring applications (Hunter et al., 1998; Spetz et al., 1997) but also has the capability to make blue pn-junction light-emitting diodes (LEDs) (Kong et al., 1997). High breakdown field, high saturation velocity together with high thermal conductivity allow SiC-based devices to function properly in high-power-density and high-frequency environments. Moreover, SiC has excellent mechanical properties because it is extremely hard, refractory and rustless. The first SiC Schottky-diode hydrogen sensor was developed by Hunter at NASA Lewis Research Center in 1992 (Hunter et al., 1992). It was a simple Schottky-diode structure, which had a catalytic metal Pd directly deposited on SiC. This sensor structure could sensitively detect hydrogen but was not thermally stable (Hunter et al., 1995; Chen et al., 1996). After this structure was heated for a long time, there were interfacial reactions between Pd and SiC, forming Pd silicides (Hunter et al., 1995) and causing a shift in sensor properties (Hunter et al., 1997). To provide long-term high-temperature stability, different materials, fabrication techniques and sensor structures were developed. NASA Lewis Research Center and Case Western Reserve University used alloys such as palladium chrome as the catalytic metal (Hunter et al., 1998) while some researchers employed more thermally stable metallization such as tungsten. Diamond, amorphous and polycrystalline barium strontium titanate (BST) have been used as a substrate to reduce the interfacial reaction (Kang et al., 1995; Chen et al., 2000; Dietz et al., 1997). A Metal-Insulator-Semiconductor (MIS) structure involving the use of a gate insulator was developed to suppress the interfacial diffusion between the electrode and the substrate, and hence can make the device more stable for gas sensing. Several gate insulators such as SnO_2 (Hunter et al., 2000) and SiO_2 (Chen et al., 1997; Zangooie et al., 2000; Tobias et al., 1997) have been used. These gate insulators provide some improvements on the sensor performance but they are still not very stable after long-duration high-temperature operation (Hunter et al., 2000). Since the sensitivity and stability of the sensor largely depend on the gate insulator, development of a high-quality gate insulator for MISiC Schottky sensors has become an essential issue. SiO_2 growth in NO gas was then developed to improve the oxide reliability and reduce the interface states and oxide charges (Xu et al., 2003). A technique of growing the oxide layer in O_2 and trichloroethylene (TCE) was also employed to fabricate a hydrogen sensor (Tang et al., 2005). This sensor demonstrated excellent H_2-sensitive properties due to the passivation effect of Cl_2 or HCl at the SiO_2/SiC interface. Researchers are continually seeking new materials and methods to produce more sensitive, stable and reliable sensors, which can function at higher temperature for a longer period of time.

4. Working principle of MIS Schottky-diode hydrogen sensor

A schematic representation of a MISiC Schottky-diode hydrogen sensor is shown in Fig. 6. It consists of three main parts: (1) the substrate, (2) the gate insulator and (3) the front electrode.

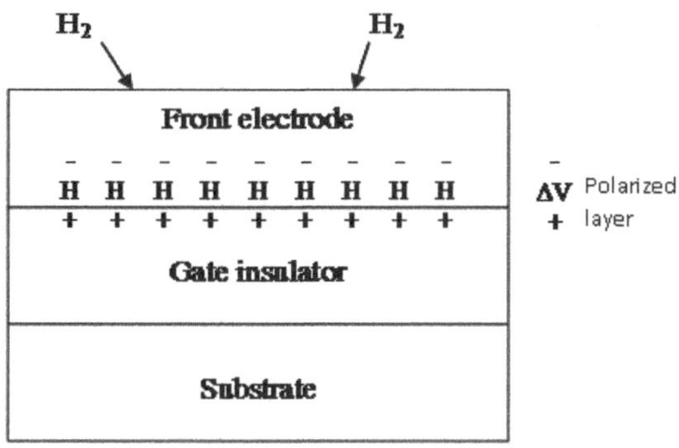

Fig. 6. Schematic representation of a MISiC Schottky-diode hydrogen sensor.

When hydrogen-containing molecules come to the catalytic gate metal, they dissociate on the surface of the metal and form hydrogen atoms. These hydrogen atoms then diffuse through the metal in the order of nano- to microseconds to the insulator surface, and form a polarized layer (see Fig. 6) at the electrode-insulator interface (Ekedahl et al., 1998). This polarized layer reduces the potential barrier at the electrode-insulator interface, and hence leads to a shift in the *I-V* curve of the Schottky diode. The amount of voltage shift at a fixed current is assumed to be proportional to the atomic hydrogen concentration at the interface or the hydrogen coverage θ_i. The maximum voltage change ΔV_{max} occurs when each hydrogen absorption site at the interface is occupied by one hydrogen atom, i.e. the hydrogen coverage is one. The hydrogen coverage at a certain hydrogen concentration depends on two processes. One process is hydrogen dissociation, by which hydrogen atoms diffuse through the metal and are blocked at the interface. The other process is the way hydrogen flows out back to the ambient. Hydrogen atoms at the metal surface can recombine together to form hydrogen molecules or react with oxygen species to form water molecules. Hence, at steady state, the actual coverage of hydrogen at the interface is a result of a balance between these two processes. Fig. 7 shows the schematic energy-band diagram for a MISiC Schottky-diode hydrogen sensor. Electrons in the conduction band of the substrate (SiC) have to overcome an energy barrier ϕ_b to reach the gate (Pt). The polarized hydrogen layer can increase the electron concentration of the metal and hence its Fermi level near the insulator layer. As a result, the lower energy barrier at the metal-insulator interface increases the forward current of the device. The above sensing mechanism is a reversible process because the hydrogen atoms adsorbed at the electrode/insulator interface can diffuse out of the interface to the surface of the electrode and finally recombine together to form hydrogen gas. Thus, when the introduction of hydrogen to the sensor is stopped, the sensor will return to its initial condition after some time. The reactivity and solubility of hydrogen are different for different catalytic metals (Lechuga et al., 1992). MISiC sensors with porous catalytic metal have higher sensitivity to detect gases because they have larger active surface areas available for hydrogen adsorption (Lechuga et al., 1992; Spetz et al., 2000).

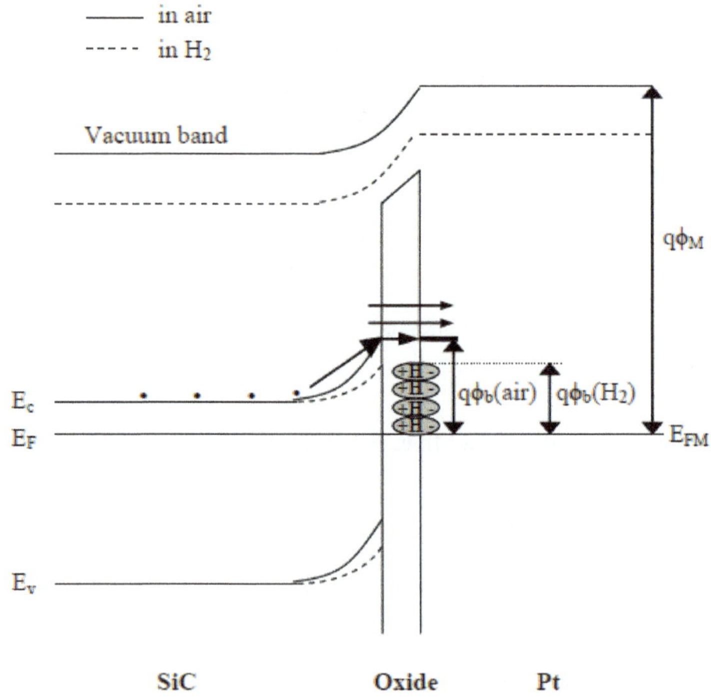

Fig. 7. Schematic energy-band diagram for a MISiC Schottky-diode hydrogen sensor.

5. Experimental details

5.1 Fabrication procedures of the hydrogen sensors

N-type (0001) Si-face 4H-SiC wafer, manufactured by CREE Research, was used in this study. The SiC wafer had a 5-μm epitaxial layer grown on heavily doped substrate. The doping level of the epitaxial layer was 5-6 x 10^{15} cm^{-3}. The wafer was cleaned using the conventional Radio Corporation of America method followed by a 60 sec dipping in 5% hydrofluoric acid to remove the native oxide. The wafer was then loaded into a Denton vacuum LLC Discovery 635 sputterer, which was then pumped down to 2.67 x 10^{-4} Pa. The most promising high-k dielectric for Si technology, hafnium oxide HfO_2, was chosen as the gate insulator of the MISiC sensor as such sensor showed much higher sensitivity than its SiO_2 counterpart (Tang et al., 2008). HfO_2 was deposited at room temperature by DC sputtering of hafnium metal (99.99 % purity) with a sputtering power of 17.52 W in a mixed Ar/O_2 ambient (Ar to O_2 ratio 4:1) for 20.9 min. An electrode consisting of 100-nm Pt with a diameter of 0.5 mm was then deposited on the wafer by DC-magnetron sputtering through a stainless steel shadow mask. The sample then underwent an annealing by loading it into a furnace at 650 °C in N_2 (1000 ml/min) for 5 min (denoted as H05 sample). In order to investigate the effects of annealing time on the sensor performance, two more hydrogen sensors annealed under the same annealing gas but with different annealing durations were fabricated. The sample annealed at 650 °C for 10 min was denoted as H10 sample while the one annealed for a longer time (60 min) was denoted as H60 sample. The oxide at the back of the wafers was removed using 20 % HF solution. The back of the SiC wafer was then pasted using silver epoxy on a gold-coated header with one pin as the back contact. The header was then put into an oven at 200 °C for half an hour to harden the silver epoxy. Lastly, a gold wire was connected between the Pt electrode and one of the pins of the header using a hybrid wedge bonder. The hydrogen sensor fabricated in this study is shown in Fig. 8.

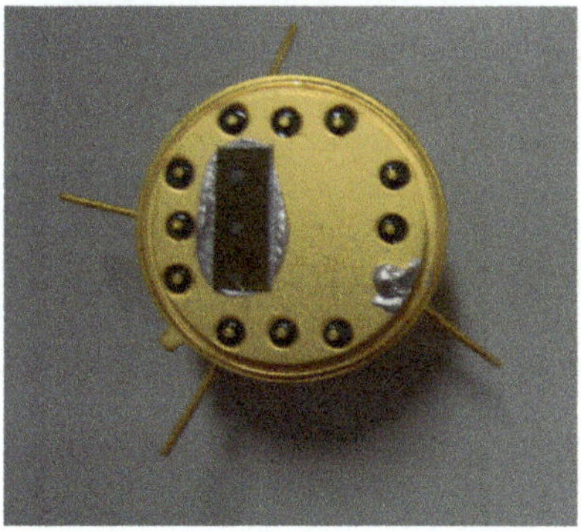

Fig. 8. Photograph of three hydrogen sensors bonded on a header.

5.2 Measurement methodology

The hydrogen-sensing properties of the samples were compared with each other by taking measurements under various temperatures and hydrogen concentrations using a computer-controlled measurement system (See Fig. 9). This measurement system mainly consisted of two parts: the gas-mixing supply system and the parameter-testing system. The test sample was placed in a stainless steel reaction chamber enclosed by the thermostat and gases were injected into the reaction chamber through the digital gas flow controllers (DFCs). The thermostat, the HP 4145B semiconductor parameter analyzer and the DFCs were all connected to a computer and were controlled by computer programs. The computer programs could provide different measurement conditions and the measurement results were automatically saved in the computer.

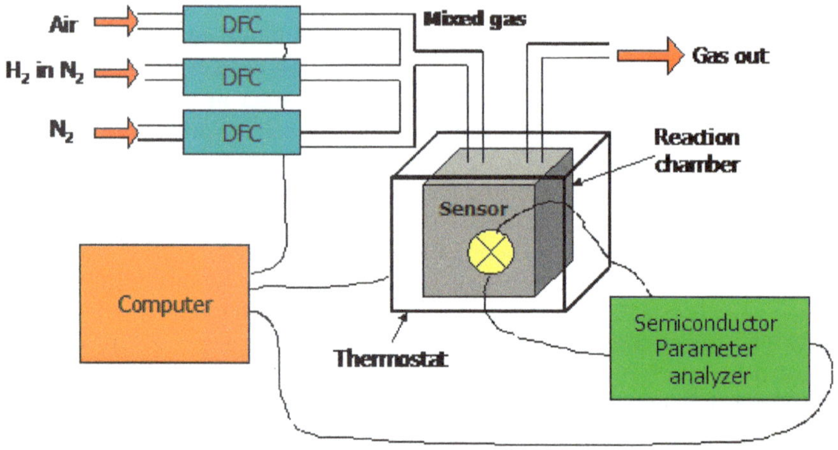

Fig. 9. Computer-controlled measurement system.

6. Results and discussion

Fig. 10 shows the *I-V* curves of the H60 sample measured in air and in different hydrogen concentrations (238 ppm, 429 ppm and 619 ppm) at 50 °C and 150 °C. The *I-V* curve shifts to the left as temperature increases. The hydrogenation effect becomes more obvious at 150 °C. The *I-V* curves shift to the left as hydrogen concentration increases. The shift toward a lower voltage is due to the formation of a polarized layer at the electrode-insulator interface. The *I-V* curves of the H05, H10 and H60 samples measured in air and in 48 ppm H_2 in N_2 at 450 °C are compared in the inset of Fig. 10. Among the three studied devices, the H60 sample has the largest current variation (ΔI) measured at a fixed bias voltage.

Fig. 11 depicts the sensor response of the H60 sample upon exposure to different H_2 concentrations in N_2 at several operating temperatures. The sensor response is defined as $(I_{H2} - I_{air})/I_{air}$, where I_{H2} and I_{air} are currents measured under hydrogen environment and air respectively. The response signal of the H60 sample increases with hydrogen concentration and temperature. When temperature increases, the hydrogen under higher pressure will bombard the surface of the electrode more frequently. Hence, more hydrogen molecules can adsorb at the surface of the electrode and decompose into hydrogen atoms. More hydrogen atoms diffuse through the metal layer to the metal/insulator interface to form a polarized layer. This stronger polarized layer gives a larger barrier-height reduction, thus increasing the sensor response. Increasing H_2 concentration can also cause more hydrogen atoms adsorb at the metal-insulator interface to produce a stronger dipole layer, resulting in more barrier-height lowering, and thus higher sensor response. The inset of Fig. 11 compares the sensor response of the H05, H10 and H60 samples upon exposure to different H_2 concentrations in N_2. It is found that extending the annealing duration can increase the sensor response of the hydrogen sensors because it can enhance the quality of the HfO_2/SiC interface and improve the dielectric properties of the HfO_2 film to produce larger barrier-height modulation and to give smaller I_{air}.

Fig. 12 shows the adsorption transient behaviour of the H60 sample upon exposure to 800-ppm H_2 in N_2 at 50 °C, 150 °C and 450 °C for a bias voltage of 2.5 V. The current shift increases with hydrogen concentration and temperature because more hydrogen-containing molecules adsorbed at the electrode-insulator interface can form a stronger polarized layer to give a larger current shift. The response time which is defined as e^{-1} times the final steady-state value can be obtained from the transient-response curve. The response time of the H60 sample at 50 °C, 150 °C and 450 °C is 25, 13 and 9 sec respectively. The response time of the hydrogen sensor decreases with the increasing operating temperature because high temperature enables faster hydrogen decomposition and diffusion, and hence faster response. The current–time (*I-t*) characteristics of the H05, H10 and H60 samples at 450 °C are shown in the inset of Fig. 12. These curves can be used to obtain the current shift (I_{H2} - I_{air}) and thus sensitivity of each sample. The measured data shows that the H60 sample has the highest sensitivity while the H10 sample has the lowest.

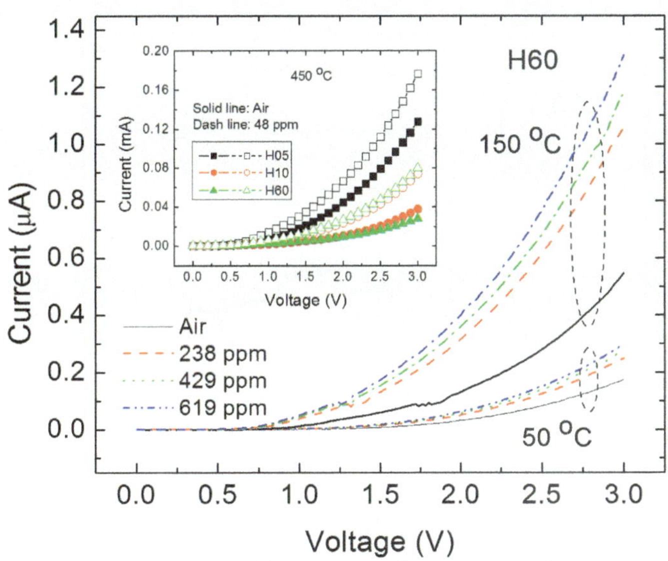

Fig. 10. *I-V* curves of the H60 sample measured in air and in different hydrogen concentrations at 50 °C and 150 °C. The inset shows the *I-V* curves of the H05, H10 and H60 sample at 450 °C.

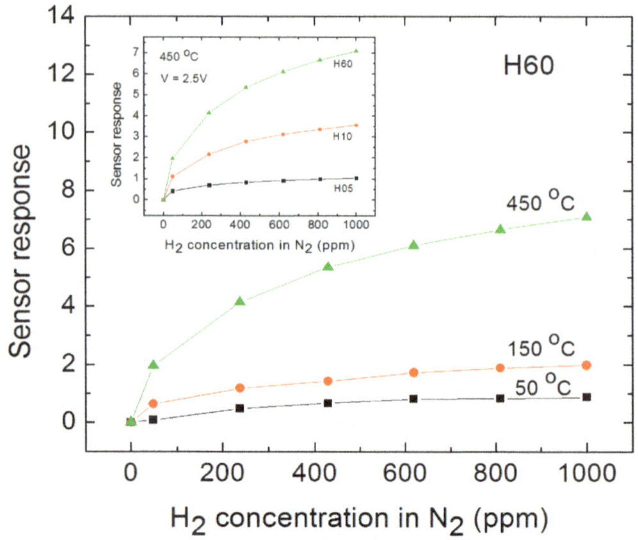

Fig. 11. Sensor response of the H60 sample upon exposure to different H_2 concentrations in N_2 at several operating temperatures (bias voltage V = 2.5 V). The inset compares the sensor response of the H05, H10 and H60 samples at 450 °C.

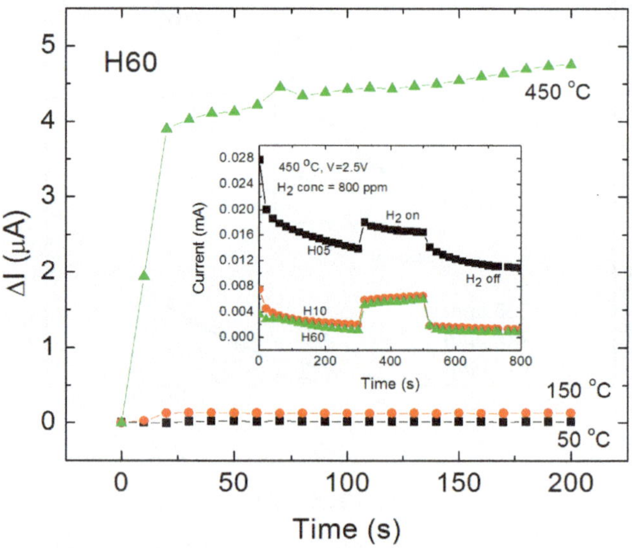

Fig. 12. Adsorption transient behaviour of the H60 sample upon exposure to 800-ppm H_2 in N_2 at 50 °C, 150 °C and 450 °C. The *I-t* characteristics of the H05, H10 and H60 samples at 450 °C are shown in the inset.

According to the hydrogen reaction kinetics (a mathematical model that describes the characteristics and mechanism of chemical reactions happened to hydrogen moving from the electrode surface to the electrode/insulator interface), the hydrogen coverage θ at the interface under steady-state condition can be written as (Johanssan et al., 1998)

$$\theta/(1 - \theta) = k_o\,(P_{H2})^{1/2} \quad (1)$$

where k_o is a temperature-dependent constant which depends on the difference of adsorption between the surface and interface and P_{H2} is the hydrogen partial pressure. The change in voltage across the hydrogen dipole layer ΔV is proportional to the hydrogen coverage (i.e. $\Delta V \propto \theta$) and the proportionality constant is ΔV_{max} (maximum voltage change at a fixed temperature). By substituting $\theta = \Delta V/\Delta V_{max}$ into equation (1),

$$1/\Delta V - 1/\Delta V_{max} = [k_o \Delta V_{max}(P_{H2})^{1/2}]^{-1} \quad (2)$$

By using the equation $\ln I = \ln I_o + qV/(nkT)$, where k is the Boltzmann constant, T the temperature in K, n the ideality factor, I_o the saturation current, and q the elementary charge, equation (2) can be written as

$$1/\ln(I_{og}/I_o) = [\ln(I_{ogmax}/I_o)]^{-1}\{1+[k_o(P_{H2})^{1/2}]^{-1}\} \quad (3)$$

where I_{og} and I_{ogmax} are respectively the saturation current and maximum saturation current of the sensor in hydrogen environment. Under air ambient, the partial pressure of oxygen is 21.4 kPa. As can be seen in Fig. 13, the plot of $1/\ln(I_{og}/I_o)$ versus $(1/P_{H2})^{1/2}$ is a straight line, thus confirming the hydrogen reaction kinetics in the H05, H10 and H60 samples. From the slope and y-intercept of the plot, k_o is calculated and shown in the inset table of Fig. 13. It is found that the H60 sample has the largest k_o. Fig. 14 shows the plot of $1/\ln(I_{og}/I_o)$ vs $(1/P_{H2})^{1/2}$ for the H60 sample at different temperatures (50 °C, 150 °C, and 450 °C). The k_o value increases with increasing operating temperature. According to the van't Hoff equation (Silbey et al., 2001)

$$\ln k_o = -\Delta H^o/RT + \Delta S_o/R \quad (4)$$

where ΔH^o is the enthalpy change; ΔS_o is the entropy change; and R is the gas constant (8.314472 $JK^{-1}mol^{-1}$). Fig. 15 illustrates a plot of $\ln k_o$ versus $1/T$ for the H60 sample. From the slope of the plot, the enthalpy change is determined as 1.18 kJ/mol. Since the hydrogen adsorption process is endothermic, it is favorable for high-temperature detection.

7. Conclusion

MISiC Schottky-diode hydrogen sensors with HfO_2 gate insulator annealed in N_2 at 450 °C for different durations have been fabricated and studied. Steady-state and transient-response measurements at different temperatures and hydrogen concentrations are carried out using a computer-controlled measurement system. Measured data are then used to investigate the sensitivity and response speed of the sensors. By analyzing the data, the hydrogen-reaction kinetic of the devices is deduced. Experimental results show that the sensitivity increases with the N_2-annealing time. Longer annealing duration can enhance the

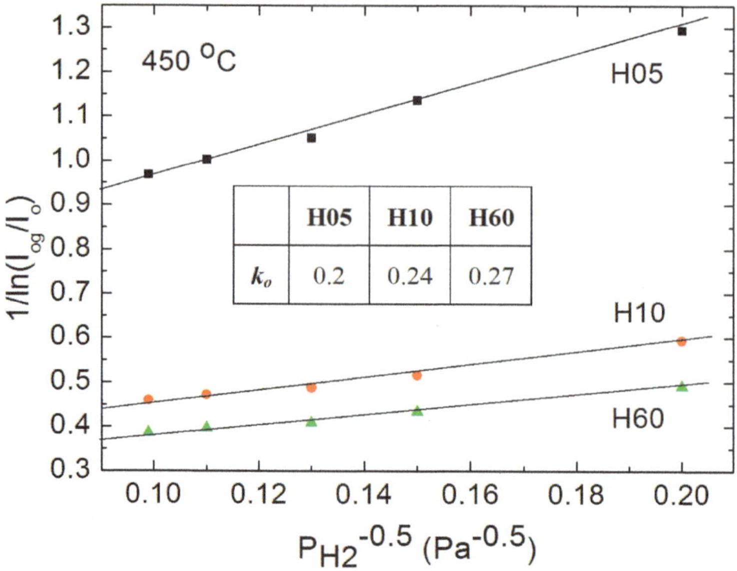

Fig. 13. Steady-state reaction kinetic analysis for hydrogen absorption of H05, H10 and H60 samples at 450 °C.

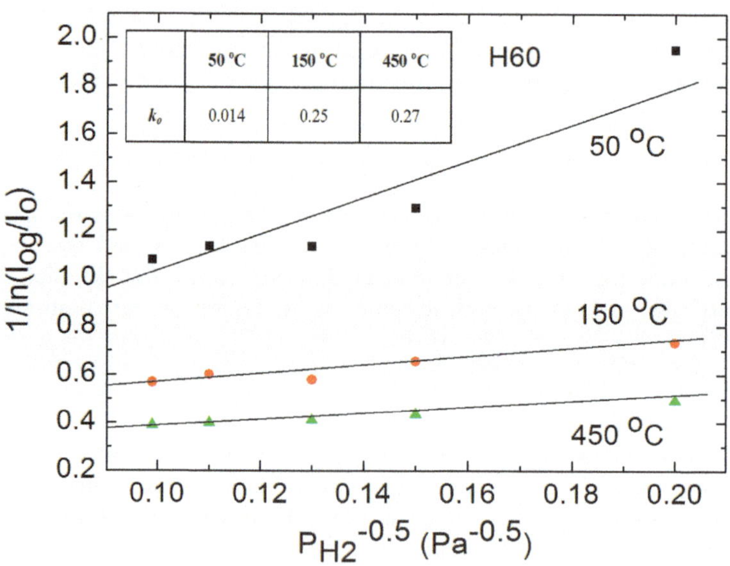

Fig. 14. Plot of $1/\ln(I_{og}/I_o)$ vs $(1/P_{H2})^{1/2}$ for the H60 sample at different temperatures (50 °C, 150 °C, and 450 °C).

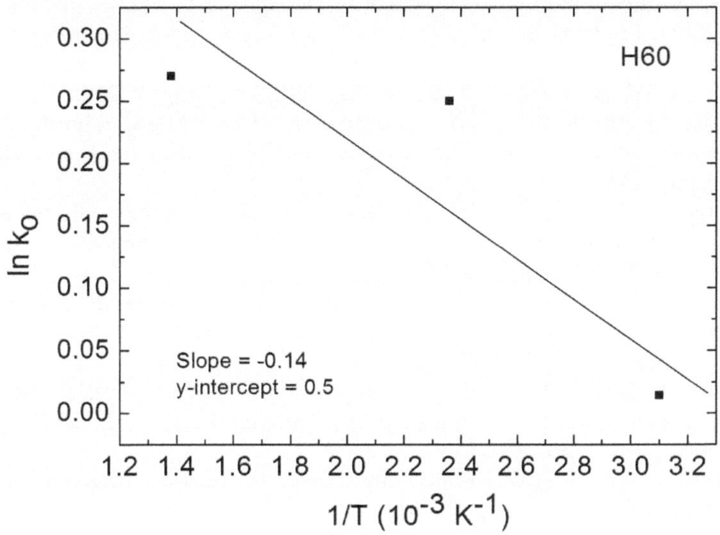

Fig. 15. Plot of ln k_o versus $1/T$ for the H60 sample.

densification of the HfO_2 film; improve the oxide stoichiometry; and facilitate the growth of the interfacial layer, thus causing a remarkable reduction of I_{air} and thus higher sensitivity. It is also found that the response time of the hydrogen sensor decreases by 64 % as the operating temperature increases from 50 °C to 450 °C. In addition, according to the hydrogen kinetic analysis, the temperature-dependent constant of the Pt/HfO_2/SiC Schottky diode hydrogen sensor decreases as the operating temperature increases. This indicates that the atomic hydrogen concentration at the interface or the hydrogen coverage θ_i at the Pt/HfO_2 interface increases at higher temperature. Moreover, the positive enthalpy change obtained from the kinetic analysis implies that the hydrogen adsorption process is endothermic.

8. Acknowledgment

We would like to acknowledge the University Development Fund (Nanotechnology Research Institute, 00600009) of the University of Hong Kong and RGC of HKSAR, China (Project No. HKU 713310E).

9. References

Almqvist, E. (2003). *History of Industrial Gases*, Kluwer academic/plenum publishers

Bergveld, P. (1972). Development, operation, and application of the ion-sensitive field-effect transistor as a tool for electrophysiology. *IEEE Trans. Biomed. Eng.*, Vol. BME-19, pp. 342-351

Brattain, W. H. & Bardeen, J. (1953). Surface properties of germanium, *Syst. Tech. J.*, Vol. 32, No. 1, pp. 1-41

Butler, M. A. (1984). Optical fiber hydrogen sensor. *Appl. Phys. Lett.*, Vol. 45, pp.1007-1009

Korver, W.O.E. (2001). *Electrical Safety in Flammable Gas/Vapor Laden Atmospheres*, William Andrew

Chen, L.-Y., Hunter, G.W., Neudeck, P.G., Bansal, G., Petit, J.B., Knight, D., Liu, C.-C. & Wu, Q.H. (1996). Electronic and interfacial properties of Pd/6H-SiC Schottky diode gas sensors, *Proceedings of the Third International High Temperature Electronics Conference*, Albuquerque, NM

Chen, L.-Y., Hunter, G.W., Neudeck, P.G. (1997). Comparison of interfacial and electronic properties of annealed Pd/SiC and Pd/SiO_2/SiC Schottky diodes. *J. Vac.Sci.Technol.A*, Vol.15, pp. 1228-1234

Chen, X. F., Zhu, W. G. & Tan, O. K. (2000). Microstructure, dielectric properties and hydrogen gas sensitivity of sputtered amorphous $Ba_{0.67}Sr_{0.33}TiO_3$ thin film. *Materials Science and Engineering B*, Vol. B77 pp. 177-184

Childs, P. E., Howe, A. T. & Shilton, M. G. (1978). Battery and other applications of a new proton conductor: hydrogen uranyl phosphate tetrahydrate, $HUO_2PO_4.4H_2O$. *J. Power Sources*, Vol. 3, pp. 105-114

Christofides, C. & Mandelis, A. (1990). Solid-state sensors for trace hydrogen gas detection. *J. Appl. Phys.*, Vol. 68, No. 6, pp. R1-30

D'Amico, A., Palma, A. & Verona, E. (1982/83). Surface Acoustic-Wave Hydrogen Sensors. *Sensors and Actuators*, Vol. 3, pp. 31-32

D'Amico, A., Palma, A. & Verona, E. (1982). Palladium-surface acoustic wave interaction for hydrogen detection. *Appl. Phys. Lett.*, Vol. 41, pp. 300-301

Dietz, G. W., Schumacher, M., Waser, R., Streiffer, S. K., Basceri, C. & Kingon, A. I. (1997). Leakage currents in $Ba_{0.67}Sr_{0.33}TiO_3$ thin films for ultrahigh-density dynamic random access memories, *J. Appl. Phys.* vol. 82 pp. 2356

Edmond, J., Suvorov, A., Waltz, D. & Carter, C. (1997). 6H-Silicon Carbide Light Emitting Diodes and UV Photodiodes. *Phsyica Status Solidi (a)*, Vol. 162, p. 481

Ekedahl, L.-G., Eriksson, M., Lundstrom, I. (1998). Hydrogen sensing mechanisms of metal-insulator interfaces. *Acc. Chem. Res.* Vol. 31, No. 5, pp. 249-256

Fang, Y. K., Hwang, S. B., Lin, C.Y. & Lee, C.C. (1997). Trench Pd/Si metal-oxide-semiconductor Schottky barrier diode for a high sensitivity hydrogen gas sensor. *J. Appl. Phys.*, Vol. 82, pp. 3143-3146

Hunter, G. W., Neudeck, P. G., Jefferson, G. D. (1992). The development of hydrogen senors technology at NASA Lewis Research Centre, *4th Ann. Space System Health Management Technology Conference*

Hunter, G.W., Neudeck, P.G., Chen, L.-Y., Knight, D., Liu, C.-C. & Wu, Q.H. (1995). Silicon Carbide-Based Hydrogen and Hydrocarbon Gas Detection, *1st Joint Propulsion Conference and Exhibit*, San Diego, CA

Hunter, G.W., Neudeck, P.G., Chen, L.-Y. (1997). SiC-based Schottky diode gas sensor. *7th Int. conf. Silicon Carbide, III-Nitrides and Related Materials*, Stockholm, Sweden, Trans.Tech., Vol.2, pp. 1093-1096

Hunter, G.W., Neudeck, P.G., Chen, L.Y., Knight, D., Liu, C. C. & Wu, Q.H. (1998). Microfabricated chemical sensors for safety and emission control applications. *Proc. 17th Digital Avionics Systems Conf.*, Vol. 1, pp. D11/1-D11/8

Hunter, G. W., Neudeck, P. G., Chen, L. Y., Knight, D., Liu, C. C. & Wu, Q. H. (1998). SiC-Based Schottky Diode Gas Sensors, *Silicon Carbide, III-Nitrides, and Related Materials. Materials Science Forum*, 264-268

Hunter, G.W., Neudeck, P.G., Gray, M. (2000). SiC-based gas sensor development. *Materials Science Forum*, Vol. 338-342, pp. 1439-1442

Johanssan, M., Lundstrom, I., Ekedahl, L.-G. (1998). Bridging the pressure gap for palladium metal-insulator-semiconductor hydrogen sensors in oxygen containing environments. *J. Appl. Phys.*, Vol. 84, No. 1, pp. 44-51

Kang, W. P., Gurbuz, Y., Davidson, J. L. & Kerns, D. V. (1995). A polycrystalline diamond thin-film-based hydrogen sensor. *Sensors and Actuators B*, Vol 24-25, pp. 421-425

Kim, C.K., Lee, J.H., Chow, N.I. & Kim, D.J. (2000). A study on a platinum-silicon carbide Schottky diodes as a hydrogen gas sensor. *Sensors and Actuators B*, Vol. 66, pp. 116-118

Kim, H. S., Moon, W. T., Jun, Y. K. & Hong, S. H. (2006). High H_2 sensing performance in hydrogen trititanate-derived TiO_2. *Sensors and Actuators B*, 120, pp. 63-68

Korotcenko, G., Brinzari, V., Schwank, J., DiBattista, M. & Vasiliev, A. (2001) Peculiarities of SnO_2 thin film deposition by spray pyrolysis for gas sensor application. *Sensors and Actuators B*, Vol. 77, pp. 244-252

Kumar. R.V. & Fray, D.J. (1988). Development of solid-state hydrogen sensors. *Sensors and Actuators*, Vol. 15, No. 2, pp. 185-191

Lechuga, L. M., Calle, A., Golmayo, D., Briones, F. (1992). Different catalytic metals (Pt, Pt, and Ir) for GaAs Schottky barrier sensors. *Sensor and Actuators B, Chem.*, Vol. 7, pp. 614-618

Lundstrom, I., Shivaraman, S., Svensson, C. & Lundkvist, L. (1975). A hydrogen-sensitive MOS field effect transistor. *Appl. Phys. Lett.*, Vol.26, pp. 55-57

Lundstrom, K.I. (1981). Hydrogen sensitive MOS-Structures: Principles and Applications. *Sensors and Actuators*, Vol. 1, pp. 403-426

Lundstrom, I., Armgarth, M. & Petersson, L.-G. (1989). Physics with catalytic metal gate chemical sensors. *CRC Crit. Rev. Sol. State Mater. Sci.*, Vol. 15, No. 2, pp. 201-278

Silbey, J., Alberty, R.A. (2001). *Physical Chemistry*. New York, Wiley

Smith, D., Falconer, R., Wittman, E. L. & Vetelino, J. F. (1993). Stability, sensitivity and selectivity of tungsten trioxide films for sensing applications. *Sensors and Actuators B*, Vol. 13-14, pp. 264-268

Spetz, L., Baranzahi, A., Tobias, P. & Lundstrom, I. (1997). High Temperature Sensors Based on Metal-Insulator-Silicon Carbide Devices. *Physica Status Solidi (a)*, Vol. 162, p. 493

Spetz, A.L., Tobias, P., Baranzahi, A., Martensson, P. & Lundstrom, I. (1999). Current status of silicon carbide based high-temperature gas sensors. *IEEE Trans. Electron Devices*, Vol.46, No. 3, pp. 561-566

Spetz, A. L., Tobias, P., Uneus, L., Svenningstorp, H., Ekedahl, L.-G. (2000). High temperature catalytic metal field effect transistors for industrial application. *Sensors and Actuators B*, Vol. 70, pp. 67-76

Steele, M. C., Hile, J. W. & MacIver, B. A. (1976). Hydrogen-sensitive palladium gate MOS capacitors. *J. Appl. Phys.*, Vol. 47, pp. 2537-2538

Steele, M. C. & MacIver, B. A. (1976). Palladium/cadmium-sulfide Schottky diodes for hydrogen detection. *Appl. Phys. Lett.*, Vol. 28, pp. 687-688

Tang, W. M., Lai, P. T., Xu, J. P., Chan, C. L. (2005). Enhanced hydrogen-sensing characteristics of MISiC Schottky-diode hydrogen sensor by trichloroethylene oxidation. *Sensors and Actuators A*, Vol. 119, pp. 63-67

Tang, W.M., Leung, C.H., Lai, P.T. (2008). Improved sensing characteristics of MISiC Schottky-diode hydrogen sensor by using HfO_2 as gate insulator. Microelectronics Reliability, Vol. 48, pp. 1780-1785

Tobias, P., Baranzahi, A., Spetz, A.L. (1997). Fast chemical sensing with metal-insulator silicon carbide structures. *IEEE Electron Device Lett.*, Vol.18, No. 6, pp. 287-289

Tomchenko, A. A., Harmer, G. P., Marquis, B. T. & Allen, J. W. (2003). Semiconducting metal oxide sensor array for the selective detection of combustion gases. *Sensors and Actuators B*, Vol. 93, pp. 126-134

Xu, J.P., Lai, P.T., Zhong, D.G., Chan, C.L. (2003). Improved hydrogen-sensitive properties of MISiC Schottky Sensor with thin NO-Grown oxynitride as gate insulator. *IEEE Trans. Electron Devices*, Vol. 24, No. 1, pp. 13-15

Zangooie, S., Arwin, H., Lundstrom, I. (2000). Ozone treatment of SiC for improved performance of gas sensitive Schottky diodes. *Materials Science Forum*, Vol. 338-342, pp. 1085-1088

Zemel, J.N., Keramati, B., Spivak, C.W. & D'Amico. A. (1981). Non-FET Chemical Sensors. *Sensors and Actuators*, Vol. 1, pp. 427-473

Part 6

Stoichiometry Driven Solid Phase Synthesis

12

The Solid-Phase Synthesis of the Inorganic Non-Stoichiometric Compounds-Fibrous Fluorosilicates

Naira B. Yeritsyan and Lida A. Khachatryan
Name Institute of General and Inorganic Chemistry, National Academy of Science Republic of Armenia

1. Introduction

The synthesis of inorganic non-stoichiometric compounds such as thermostable fibrous fluorosilicates of amphibole group shaped in form of fibrous-acicular crystals represents one of the perspective directions in the modern materials science. The importance of fibrous amphiboles, including synthetic fibrous fluorosilicates (FFS) derives from multifunctionality and specificity of their properties. They are elastic, flexible, and resistant to aggressive media, have high mechanic strength, low electric and heat conductance, heat and fire resistance, express an ability to absorb certain gases and liquids, etc. By present, fibrous amphiboles and their synthetic counterparts in pure state or composed with other substances (pitch, rubber, cement, ceramics, carbon fluoride) are widely used in various fields of modern industry and technique. The main consumers of fibrous silicates and fibrous-based composites are chemical and electro-technical industries, the manufacturing of fillers for electrolytic chambers and for acid dampers, the production of electrical wires, plastic and rubber for special purposes, paints, clays, varnishes, lubricants in the motor-car, aircraft, ship-building, and rocket plants. In atomic power engineering these compounds are used as structuring and insulting materials (Khachatryan, 1969; Hawkins, 1971; Grigor'eva et al., 1975; Li, 1992, Tareev, 2002-2011).

The existing evidence suggests that these thermostable inorganic polymers are conventionally obtained under hydrothermal conditions in the temperature range of 300-900°C and water vapour pressure of 20-300 MPa as well as under pyrogenic conditions by melt crystallization at 1300-1450°C, and by solid-phase synthesis (Khachatryan, 1969; Grigor'eva et al., 1975; Goncharov and Kovalenko, 1973). The latter method, when compared with others, offers the chance to synthesize fibrous asbestos-like fluoramphiboles at relatively lower temperature (in the range of 900-1100°C) and under atmospheric pressure. In this process the chemical reactants and natural pure minerals, both of stoichiometric composition, are used as the raw materials. Nevertheless, the intensification and reduction in energy costs of processes are required to further optimize the conditions of synthesis. This can be achieved by widening the raw material base and determining new complex feedstocks of non-stoichiometric composition, which should be inexpensive and readily available. Following this general trend, we have proposed to use the mining rocks.

This approach, together with its practical significance, would allow the covering of a number of theoretical questions in the fields of inorganic chemistry, physical chemistry, and silicate crystallochemistry, solid-state physics, ecology, materials science and other adjacent disciplines.

For the first time, the artificial analogues of fibrous amphiboles from mining rocks under hydrothermal (with end products hydroxyl- and hydroxyl-fluoramphiboles) and pyrogenic (from the melt and by solid-phase synthesis, the end product: fluoramphiboles-fluorsilicates) conditions were obtained (Khachatryan, 1969; Yeritsyan and Khachatryan, 2003, 2005).

In this chapter we provide a comparative analysis of the existing evidence with regard to the method of solid-phase synthesis of fibrous fluorosilicates from different blends based either on chemical reactants, pure minerals or mining rocks of magnesium-silicate composition, namely serpentinite, dunite, peridotite and pyroxenite. We summarize the results of our experiments aimed to unveil the impact of different compositions of initial blends and thermal-temporal conditions of synthesis on the yield of FFS as well as on the morphology and size of their crystals. To better understand the mechanisms of phase and chemical modifications in the course of FFS-synthesis the feedstock mining rocks are thoroughly characterized using the methods of microscopic, thermal, X-ray at room and high temperatures, and chemical assays. Finally, the chemical composition, structural and optical features of yield FFSs are followed together with their chemical, thermal, and adsorptive properties.

The results of the systemic investigations presented in the chapter highlight some aspects of phase-formation inside the complex multicomponent fluorosilicate systems and discuss the principles of fibrous fluorosilicate morphogenesis in the course of solid-phase synthesis from mining rocks.

2. Chemical composition and structure of amphibole

Amphiboles refer to the class of silicates with chain-ribbon structure. They crystallize in monoclinic and rhombic crystal systems. The long-prismatic crystals of these compounds possess an absolute cleavage.

Amphiboles are characterized by complex chemical composition with a general cristallochemical formula $AX_2Y_3[Z_8O_{22}](OH, F, Cl)_2$, where the cations Na^+, K^+, Ca^{2+} or some vacancies are placed in the position A; Li^+, Na^+, Ca^{2+}, Mg^{2+} in the position X; Mg^{2+}, Fe^{2+}, Fe^{3+}, Mn^{2+}, Al^{3+}, Cr^{3+} in the position Y; and Si^{4+}, B^{3+}, Al^{3+}, Fe^{3+}, Ti^{4+}, Ga^{4+} in the position Z. The structure of amphiboles includes double chains of silico-oxygenic tetrahedrons $[Si_4O_{11}]^{6-}{}_{\infty}$, between which the cations Y are located in the octahedral positions M1, M2, M3 (Fig.1). The position M4 on the edges of octahedral cation-containing chains is occupied by cations X. The surplus (more than two) cations X are located in the channels between the chains (the position A). The anionic polymer chains, which are extended along the crystallographic axis *c* in the amphiboles, represent the structural prerequisite for the formation of fibrous-acicular crystals. Nevertheless, it is important to underline that fibrous crystals are not the only, and not even the most widespread products of crystallization under either natural or experimental conditions.

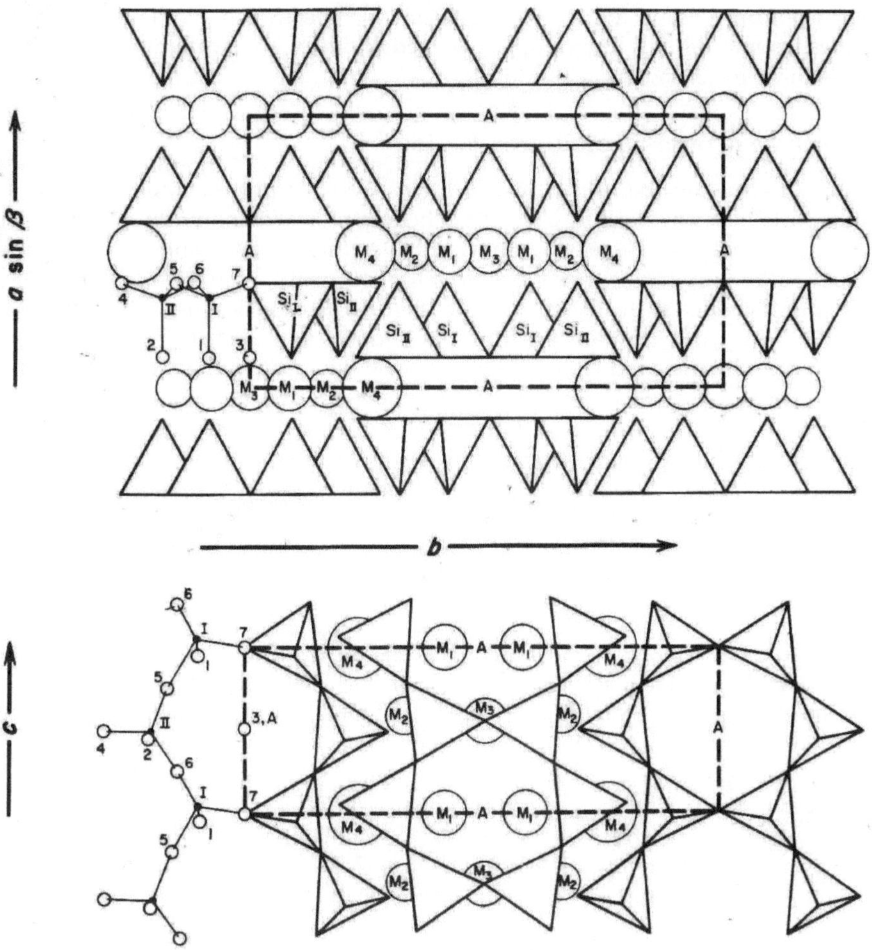

Fig. 1. The structure of monoclinic amphibole in the plane (001) and (100) after Ernst (1968).

The structure of these compounds can undergo a wide range of isomorphic substitutions. The latter determines the formation of different impurities in the composition of the natural amphiboles and provides the variability of their properties, the features, which restrict the use of FFS for a number of important objectives. Therefore, the yield of synthetic fibrous amphiboles with goal-defined stable characteristics greatly attracts the attention of many researchers in the modern materials science.

The presence of vacancies and the potential capability for iso- and heterovalent substitutions in the different structural positions of these non-stoichiometric compounds enables the purposeful development of their properties, and thereby, the synthesis of new inorganic polymers with better physicochemical and technical properties. The yield of thermostable fibrous fluorosilicates, for instance, is achieved by the replacement of hydroxyl group $(OH)^-$ by F^-.

The high isomorphic capacity of FFS provides a perspective to synthesize novel fibrous amphiboles from affordable and inexpensive feedstocks of non-stoichiometric composition, e.g. from mining rocks and waste products.

3. Solid-phase synthesis of fluorosilicates

The yield of fibrous (asbestos-like) fluorosilicates by solid-phase synthesis was described first in the early 1930s by german scientists Scheumann (1932) and Lüdke (1933, 1943). This approach was further developed by the works after Eitel (1952), Espig (1962), Gibbs et al. (1962), and Saito et al. (1965). Since 1963 this topic became a subject of great interest and detailed investigation performed by Fedoseev A. and colleagues (1963-1975). These researchers used chemical reactants (oxides, fluorides, carbonates) and natural pure minerals (talc, chlorite, rodusite, phlogopite, olivine, serpentine, enstatite) as initial substrates. The use of these compounds allowed the yield of fluoramphibole at 900-1100°C under atmospheric pressure. The synthesis was carried out in tightly closed ceramic, graphitic, and platinic bowls or airtight capsules from carbon steel in furnaces with aerial, neutral or reconstructive media. After prolonged isothermic exposure required for the solid-phase reactions to proceed, the temperature was lowered. Although the chemical modifications occurred mainly due to interactions between the particles of initial batches, the concomitant gaseous and liquid phases appeared to alter drastically the rate of the process as well as the composition and shape of the yield products.

Irrespective to the nature of initial components the yield products of solid phase synthesis were represented by dense poriferous cake, which consisted of radial aggregates and hetero-oriented needles or fibres of fluoramphibole having 0.5-1.0 mm length and 0.1-10 μm thickness (Grigor' eva et al., 1975). Depending on the raw stuff used, they contained 25-95% of fluoramphiboles. Fluorides, crystobalite, tridymite, forsterite, humite minerals, pyroxens, mica, glass, etc, were mentioned as conventional impurities. Fluoramphibole often formed 'brush'-like monomineral structures on the surface of yield products (Espig, 1962; Chigareva et al., 1966; Fedoseev et al., 1966; Grigor'eva et al., 1975). The fibres in the surface 'brush' had a length around 5-25 mm and thickness 1-50 μm and constituted 1-5% of product weight.

Altogether, the accumulated evidence suggested that the method of solid-phase synthesis enables the yield of fibrous fluoramphiboles of different chemical compositions and with a variety of physicochemical properties (Fedoseev et al., 1966; Chigareva, 1967; Grigor'eva et al., 1975). For instance, two such compounds, NaMg-fluorrichterite and NaMg-fluorarfvedsonite do not concede to their natural analogues richterite and arfvedsonite by their mechanical strength, chemical resistance, and elasticity, but express higher thermostability. This occurs due to the energy of breaking the chemical bond for Mg-OH (56±5 kcal/mol) in natural minerals being significantly lower than that the one for the Mg-F bond (120±20 kcal/mol) in their synthetic fluor-containing counterparts. Given the restriction in the natural sources of richterite and arfvedsonite, the synthesis of their artificial counterparts is of huge interest, since they can be used as reinforcing fillers in the development of new commercial materials.

In the next section we describe the process of solid-phase synthesis of NaMg-fluorrichterite and NaMg-fluorarfvedsonite from raw stuff of stoichiometric (chemical reactants and minerals) and non-stoichiometric (mining rocks) compositions. The physical-chemical parameters of the synthesis and properties of the yield product will be compared with respect to the different initial batches.

3.1 Solid-phase synthesis of fluorosilicates from raw material of stoichiometric composition

The initial batches used to synthesize fibrous fluorosilicates of non-stoichiometric composition were first described to consist of stoichiometric chemical reactants and pure minerals (Fedoseev et al., 1966, 1970; Chigareva, 1967; Goncharov and Kovalenko, 1973; Grigor'eva et al., 1975; Tresvyatskiy, 1982). These batches corresponded to the compositions of NaMg-fluorrichterite and NaMg-fluorarfvedsonite with variable excess of fluorine, and contained air-dry (dried and burnt in advance) reactants such as SiO_2, MgO, MgF_2, NaCl, Na_2CO_3, Fe_2O_3, etc, and pure minerals olivin, chlorite, serpentine, phlogopite, enstatite, talc, roducite, etc.

The formation of magnesio-fluorrichterite ($Na_2Mg_6[Si_8O_{22}]F_2$) most productively proceeded in the SiO_2, MgO, Na_2CO_3-containing initial batches supplemented with the mixture of NaF and MgF_2. Fluoramphiboles, including NaMg-fluorrichterite, were obtained by solid-phase synthesis in the system of NaF-MgF_2-MgO-MeO-SiO_2 (Me = Ni^{2+}, Co^{2+}, Cr^{3+}) within the temperature range 600-1150°C. Fibrous fluoramphiboles of NaMg-fluorarfvedsonite ($Na_{2.5}Mg_{4.5}Fe^{3+}[Si_8O_{22}]F_2$, $Na_3Mg_4Fe^{3+}[Si_8O_{22}]F_2$) compositions were yield at 800-1050°C.

The synthesis was reported to take place in open and closed systems represented by not tightly closed platinic bowls and by sealed off quartz capsules respectively. Within the open system the amorphous silica turned to tridimite and the formation of forsterite (Mg_2SiO_4) occurred at 700°C. Fluornorbergite ($Mg_2SiO_4{\cdot}MgF_2$), which was initially formed at the edges of forsterite grains crystallized at 800°C. Fluorides were shaped as caked granules (n_m=1.406) within the temperature range of 650-810°C. Neiborite ($NaMgF_3$) and the local foci of eutectic crystallization of fluoride (NaF+$NaMgF_3$) emerged along with these granules upon heating above 800°C. Magnesio-fluorrichterite was yield at 900°C. The characteristics, sequence, and thermal ranges of mentioned phases were not much different when using the closed system. However, the concentration fields of crystallization for all fluorine-containing compounds were shifted toward lower content of MgF_2.

In powder-like crystalline reagents mixtures, the transformation proceeded due to direct interactions between the particles of initial substances with the contribution of liquid and gaseous phases. The latter substantially enhanced the rate of the processes. Several studies were undertaken to establish the conditions enabling high rates of solid-phase reactions in fluoramphibole formation. For instance, attributing importance to the impact of gaseous phase, Scheumann (1932) and Lüdke (1933) used the pneumatolytic approach to synthesize fluoramphiboles in the stream of water vapour. Further, gas-forming supplements were introduced into the initial batches by other researchers (Espig, 1962; Saito et al., 1965; Fedoseev et al., 1970; Goncharov et al., 1973). In some works the synthesis was initiated in the reducing media (Goncharov et al., 1973) or in the presence of mineralizing fluxes (Espig, 1962, Fedoseev et al., 1966, 1970).

The experimental evidence gained in this field allowed an assumption that the formation of NaMg-fluorrichterite from oxides and fluorides in the course of solid-phase synthesis using the open system approach resulted from the reactions (1)-(4). The formation of forsterite in the first stage of these reactions was considered to be thermodynamically preferential, as it occurred in the magnesium-silicate systems different by their component ratio (Budnikov and Ginstling, 1971).

$$NaF + MgF_2 \rightarrow \underset{\text{neiborite}}{NaMgF_3} \quad (1)$$

$$2MgO + SiO_2 \rightarrow \underset{\text{forsterite}}{Mg_2SiO_4} \quad (2)$$

$$Mg_2SiO_4 + MgF_2 \rightarrow \underset{\text{fluornorbergite}}{Mg_2SiO_4 \cdot MgF_2} \quad (3)$$

$$9\,SiO_2 + MgF_2 + Mg_2SiO_4 \cdot MgF_2 + NaMgF_3 \rightarrow \underset{\text{NaMg – fluorrichterite}}{Na_2Mg_6[Si_8O_{22}]F_2} + 2SiF_4 \uparrow \quad (4)$$

These interactions might also develop when natural minerals from the group of orthosilicates (olivine) or magnesium hydrosilicates (chlorite and serpentine) were introduced to the initial batches. These minerals were reported to decay upon heating with the yield of forsterite (Grigor' eva et al., 1975). The formation of fluoramphibole from dehydrated phlogopite proceeded through the generation of amorphous cluster of rough composition $3MgO \cdot 2SiO_2$, in accord to the reaction (5). An inclusion of natural enstatite or the hydrosilicates (talc and roducite), the dehydration process of which usually ends up with the yield of pyroxens, was thought to influence the course of the reaction (6):

$$4(3MgO \cdot 2SiO_2) + 19SiO_2 + 6MgF_2 + 6NaF \rightarrow \underset{\text{NaMg – fluorrichterite}}{3Na_2Mg_6[Si_8O_{22}]F_2} + 3SiF_4 \uparrow \quad (5)$$

$$4MgSiO_3 + 5SiO_2 + 2MgF_2 + 2NaF \rightarrow \underset{\text{NaMg – fluorrichterite}}{Na_2Mg_6[Si_8O_{22}]F_2} + SiF_4 \uparrow \quad (6)$$

Main parameters of synthesis, i. e. the nature of initial reagents, the content of fluorine in the initial batch, thermal conditions, cooling rate and duration of the experiments, highly influence the completeness of interactions, the rate of the process and the morphology of fluoramphibole crystals. As it was demonstrated in studies of the kinetics of NaMg-fluorrichterite formation in the isothermic regimen at 1000°C, the initial stage of the process was limited by chemical reaction, while its further development was restricted by diffusion. The yield of fluoramphibole synthesized from chemical reactants was of 75-95%, whereas this percentage was markedly lower in the end product obtained from pure minerals (around 30-50%).

The maximal yield of magnesio-fluorrichterite (85-95%) was registered at 950-1050°C. The fibres of this silicate were shaped at 900-950°C. Above 1000°C a prismatic appearance of crystals was reported (Grigor' eva et al., 1975). Monomineral clusters, which contain fluoramphibole fibers and needles having 5-30 mm in length, were crystallized on the surface of yield product (Fig. 2). This took place in the initial batch under the condition of the fluorine excess, which constituted 5-10% from the stoichiometric composition of synthesized fluoramphibole. An additional requirement for the cluster formation was the thermal gradient of around 10°C/cm ascertained along the whole height of bowl.

Then fluorides characteristics specified the composition of local foci of eutectic melt within the reaction mixture and determine the temperature of their generation. Such foci greatly intensified the process of NaMg-fluorrichterite formation. The supplementation of the initial batches with fusible salts, e.g. NaCl, Na_2CO_3, Na_2SO_3, etc, promoted the appearance of local clusters of melt at 680-740°C. The use of mineralizing compounds in the solid-phase synthesis under certain conditions enabled higher yield of fluoramphibole and purposeful adjustment of crystal habits from fibrous to short-prismatic.

Thus, the process of solid-phase synthesis of fluoramphibole from either chemical reactants or pure natural silicates and hydrosilicates, i.e. compounds of stoichiometric compositions, was represented by multistep physicochemical interactions conventional for the formation of non-stoichiometric inorganic compounds (Rabenau, 1970). The mechanisms of this process were estimated by the properties of new phases emerged during the decomposition of hydrosilicates.

Fibrous fluoramphiboles of NaMg-fluorarfvedsonite composition were obtained by the solid-phase synthesis from the initial batches containing chemical reactants SiO_2, MgO, Fe_2O_3, Na_2CO_3, and NaCl treated at 800-1000°C as well as from the blend consisting of these reactants additionally supplemented with pure mineral roducite at 800-1050°C (Chigareva et al., 1966, 1971). The content of fluoramphiboles of $Na_{2.5}Mg_{4.5}Fe^{3+}[Si_8O_{22}]F_2$ and $Na_3Mg_4Fe^{3+}[Si_8O_{22}]F_2$ compositions in the final products constituted around 95% and 60% respectively. The crystals of fluoramphiboles were fibrous or fine-acicular, having 0.05-4 mm in length and 1-5 μm in thickness.

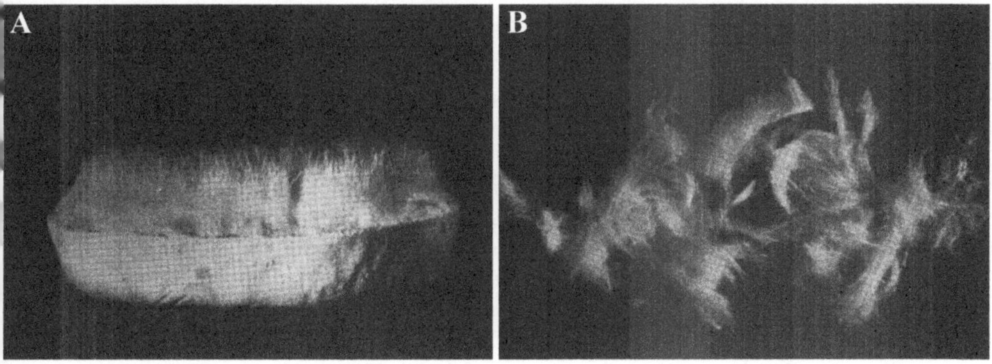

Fig. 2. Synthetic fibrous fluoramphibole: general view of the synthesis product (*A*, 0.8x) and fibrous crystals (*B*, 0.7x) (Fedoseev et al., 1970).

The crystallochemical formula of obtained NaMg-fluorrichterite corresponded to $Na_{2.12}Mg_{5.85}Fe^{3+}{}_{0.01}[Si_8O_{21.88}]F_{2.09}$ (to compare: the theoretical composition of this silicate was $Na_2Mg_6Si_8O_{22}F_2$). For NaMg-fluorarfvedsonite it is expressed as $Na_{2.58}Ca_{0.10}Mg_{4.49}Fe^{3+}{}_{0.97}[(Si_{7.84}Al_{0.03})O_{22}](O_{0.10}F_{1.70}Cl_{0.20})$ vs. theoretical composition $Na_{2.5}Mg_{4.5}Fe^{3+}Si_8O_{22}F_2$, and $Na_3Ca_{0.16}Mg_{3.89}Fe^{3+}{}_{1.25}[(Si_{7.73}Al_{0.05})O_{21.04}](F_{1.70}Cl_{0.36})$ vs. theoretical formula $Na_3Mg_4Fe^{3+}Si_8O_{22}F_2$. The findings of crystallooptical and X-ray assays, which describe the refractive index and parameters of crystal lattice, as well as physical-chemical properties of synthesized fluoramphiboles are summarized in the tables 1 and 7, according to Grigor'eva et al. (1975).

Fluoramphibole composition	Parameters of crystal lattice				Optical constants			
	a, nm	b, nm	c, nm	β	Ng	Np	Ng -Np	cNg
NaMg-fluorrichterite	0.967	1.792	0,527	103°00'	1.596	1.577	0.019	12°
NaMg-fluorarfvedsonite	0.974	1.787	0,528	103°36'	1.630	1.618	0.012	20°

Table 1. Parameters of crystal-lattice and optical characteristics of NaMg-fluorrichterite and NaMg-fluorarfvedsonite.

To further develop technological procedures of NaMg-fluorrichterite synthesis in term of the estimating the optimal composition of initial blends, thermal and temporal parameters of the solid-state reactions enabling the maximal yield of the fibrous amphibole with minimal content of impurities, and testing the reproducibility of cycles, Goncharov and Kovalenko (1973) used burnt and dried in advance chemical reactants (oxides, fluorides, chlorides, etc). The synthesis was carried out in the airtight bowls from low-carbon steel in the furnace with hydrogenous protective atmosphere, i.e. in the reducing media. The supplementation of flux sodium chloride (NaCl) in quantity of 25% from the weight of the initial blend changed the habitus of fluoramphibole crystals from acicular and prismatic to fibrous. The basic parameters of the synthesis were the following: the temperature of the synthesis - 1050-1100°C, retention time at this temperature - 48 h, and a composition of the blend - $2.5NaF \cdot 0.5MgF_2 \cdot 5.5MgO \cdot 8SiO_2$ + 25% NaCl. The fluoramphibole fibres were of 0.05-0.15 mm lenghth and of 5-10 μm in thickness. The chemical analysis revealed, that the synthesized compound had the chemical formula $(Na_{1.89}K_{0.02})_{1.91}Mg_{5.9}[Si_8O_{22.14}]F_{1.86}$.

The described approaches of solid-phase synthesis, i.e. from chemical reactants and pure minerals, reveal a number of qualities restricting their use as conventional methods to obtain fibrous fluroamphiboles of NaMg-fluorrichterite and NaMg-fluorarfvedsonite composition. They consume expensive feedstocks (chemical reactants and pure minerals), which require additional treatment of burning and drying. This treatment by heating the substances upon 300-800°C, along with high temperature of the main process (above 950-1050°C) sets demands for a complex and labour-consuming technology. Therefore, we investigated the process of solid-phase synthesis of fibrous fluoramphiboles from affordable and inexpensive complex raw materials - mining rocks, which mainly contain the minerals of magnesium-silicate composition.

3.2 Solid-phase synthesis of fluoramphiboles from raw material of non-stoichiometric composition – Mining rocks

3.2.1 Properties of the mining rocks

We used as feedstocks the ultra-alkaline mining rocks from local mines: serpentinite (S), serpentinite calcinated at 700°C to constant weigt (S_{calcin}), dunite (D), peridotite (P), and pyroxenite (Px). The mining rocks were utilized in form of fine grinded powder with the granule size less than 50 μm. The rocks were studied by chemical (Table 2), microscopic (Table 3), X-ray diffraction at room and high temperatures (Fig. 7*A*), and differential thermal (derivatograph OD-103, Fig. 6, *curve 1*) assays. From the analysis of chemical and mineralogical compositions it became evident that dunite, peridotite and pyroxenite were substantially serpentinized and represented the classical samples of ultra-alkaline rocks.

They belonged to the non-stoichiometric compounds consisted of several minerals and were used here to donate the oxides MgO, SiO_2, Fe_2O_3 into the composition of initial blends.

Oxydes	Rocks, wt %			
	Serpentinite	Peridotite	Dunite	Pyroxenite
SiO_2	40.05	35.30	38.05	40.97
Al_2O_3	2.50	0.38	0.43	1.68
Fe_2O_3	4.12	5.54	4.66	8.15
FeO	2.77	1.87	3.22	-
TiO_2	Traces	not found	not found	traces
CaO	3.19	0.95	1.01	6.30
MgO	35.70	37.59	38.84	32.60
MnO	-	0.09	0.10	-
SO_3	Traces	0.16	0.16	traces
CO_2	-	1.42	0.96	-
Na_2O	0.07	0.03	0.16	0.09
K_2O	0.08	0,03	0.03	0.06
Cr_2O_3	0.60	0.31	0.33	0.25
Loss	11.10	14.44	11.16	9.40
Moisture of the rock	0.19	2.26	1.23	0.5
Σ	100.35	100.33	100.29	99.25

Table 2. Chemical composition of initial rocks.

Rock	Minerals	Content, wt %
Serpentinite	Serpentine (antigorite) Monoclinic pyroxenes Magnetite	~80 ~10–15 ~5–7
Peridotite	Olivine (serpentinous, from which ~10 % invariable olivine) Rombic pyroxene(enstatite) Monoclinic pyroxenes (diopside, diallag) Magnetite (secondary), chromospinelide Carbonates	~70 ~15–20 5–10 3–5 1.5–2.5
Dunite	Serpentines (chrysotile, antigorite) Olivine Pyroxenes Magnetite and chromospinelide Carbonates	~ 40–45 ~ 40–45 ~5 ~ 3–8 ~2
Pyroxenite	Serpentine (described as chrysotile, antigorite, serpofite, from which ~5% chlorite) Enstatite Monoclinic pyroxenes (diopcide, diallag) Magnetite	~ 40 5-10 40-50 5-7

Table 3. Mineralogical composition of initial rocks.

The composition of initial blends was calculated from the theoretical formula of the NaMg-fluorarfvedsonite ($Na_3Mg_4Fe^{3+}[Si_8O_{22}]F_{2n}$), where n represented an excess digits of fluorine atoms. The excessive content of fluorine in the blend varied within the range of 4.54-27.24 wt%, that constituted n=1, 2, 3, 4, 5, 6 atoms per formula unit of the amphibole. Initial blends together with the mining rocks contained supplementing stoichiometric chemical reactants Na_2SiF_6, NaF, NaCl, and Na_2CO_3, which were used without additional treatment (burning, drying, or powdering).

We examined seven reaction blends of the following compositions: S- and S_{calcin} – NaF (I, II), S- and S_{calcin} - Na_2SiF_6 (III, IV), D - Na_2SiF_6 (V), P - Na_2SiF_6 (VI), and Px - Na_2SiF_6 (VII). Sodium carbonate and sodium chloride were introduced as fluxes (10 - 30wt %). The amount of supplemented SiO_2 was equal to 3–10 wt%. The fibrous fluorosilicates were synthesized in fluorine-steady vessels in the electrical resistance furnaces.

The phase composition in the products of rock treatment and the synthesized compounds were examined by the crystal optical, differential thermal (DTA, OD-103), X-ray diffraction with use of powder and single-crystal methods (DRON-2 and URS-70 diffractometers, CuK_α-radiation, Ni-filter), and electronnomicroscopic (Tesla BS-242) assays.

3.2.2 Results and discussion

Initial mixtures were heated upon 500-1000°C during 1-60 h. The synthesized products were mainly represented by an array of tangled fibrous texture with surface monomineral 'brush' of fluorosilicate fibers (see Fig. 10)

The phase composition of synthesized samples, the percentage of FFS yield (i.e. the rate of rock transformation and fluorosilicate content), the mechanisms and rate of the processes, the morphology and size of FFS crystals were greatly influenced by the parameters of synthesis (the properties and the quantity of mining rocks, fluorine-introducing components NaF and Na_2SiF_6, and fluxes NaCl and Na_2CO_3; thermal-temporal conditions of the process, etc.).

Fluorine content was manipulated by supplementation of a fluorine-introducing agent, e.g. Na_2SiF_6 in the blends III, IV, V, VI, VII, and NaF in the blends I and II. The amount of fluorine varied from 4.5 to 27 wt%, i. e. with an excess in regard to the theoretical composition of synthesized amphibole. The experiments carried out at 900-920°C and exposure of 48 h revealed that when using NaF the rate of S and S_{calcin} modification into fluorosilicate was 25-30% and 65-70% respectively. In these series we failed to obtain monosized and long fibres on the surface of the product. Distinct fibres in the 'brush' were around 1-3 mm in length. The process of rock treatment proceeded more actively in the reaction blends supplemented with fluorine through Na_2SiF_6. The optimal amount of fluorine in the initial blends to obtain the product with high content of fine-fibrous crystals and well developed 'brush' on the surface was found to constitute 7.00-13.50 wt% (3-6 fluorine atoms per formula unit of the amphibole, Fig. 3).

In case of low (≤ 2 fluorine atoms per formula unit) and high (≥10 fluorine atoms per formula unit) fluorine content the amphibole was crystallized in form of thick and prismatic crystals.

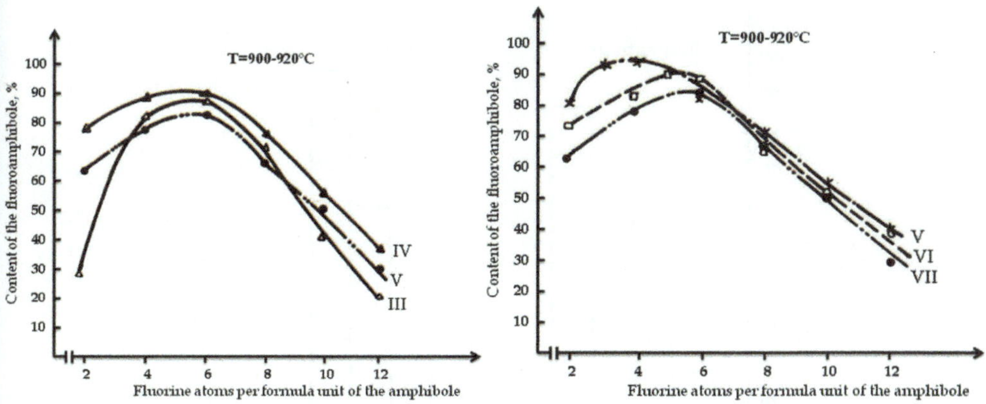

Fig. 3. The dependence of the yield of fluorosilicates on the content of fluorine in the initial blends containing: III -serpentinite, IV - calcinated serpentinite, V - dunite, VI - peridotite, and VII - pyroxenite.

The properties and the amount of fluorides determined the composition and the temperature of formation for the gaseous phase and for the local eutectic melt in the reaction blends. In its turn, the latter factors intensified the process of magnesium-arfvedsonite formation. The supplementation of the initial blends with fusible salts such as NaCl and Na_2CO_3 promoted the generation of local foci of the melt, i.e. the formation of the reaction-capable liquid phase, at relatively low temperature (≤450°C). This extended the thermal capacity of the fluoramphibole formation toward rather low temperatures. The introduction of Na_2CO_3 resulted in strong caking of the treatment product, which contained 30-40% of fluorosilicate in form of short prismatic crystals. The introduction of NaCl into the blends led to the yield of asbestos-like FFS, which constituted in average 75-92% in case of 10-30 wt% of NaCl supplement. An increase in the content of fluxes from 30 to 70 wt% had no effect on rock transformation rate, but substantially influenced the morphology of fluorosilicate crystals by decreasing the amount of fibrous crystals in favour to the particles with acicular-prismatic habits.

Thus, the use of mineralizing compounds in solid-phase synthesis enabled the yield of the product with high concentration of fluorosilicate as well as the purposeful adjustment of crystal habits from fibrous toward short prismatic one.

Further, to clarify how a dispersion of raw components influence the process of synthesis we studied the impact of differently sized granules of monomineral rock serpentinite and of multimineral dunite. The size of granules varied from 50 to 600 μm. Variation in dispersivity of both rocks altered the yield of the fluorosilicate and the morphology of the crystals. Nevertheless, the effect was stronger expressed in case of serpentinite (Fig. 4).

The use of serpentinite granules of 400-600 μm resulted in the prevalence of dissociated fibrous aggregate of fluoramphibole in the end products. Acicular and prismatic crystals were formed when testing granule size 400-90 μm. A decrease in size less than 50 μm led to the formation of detached fine fibres and needles.

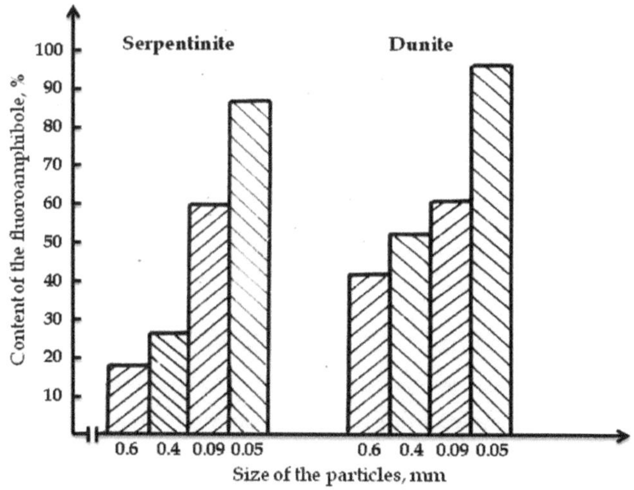

Fig. 4. The influence of the granule size of raw rocks serpentinite and dunite on the content of fluorosilicate in the end product.

The study of thermal conditions in the course of the synthesis showed that the most complete modification of rocks happen in the temperature range 850-920°C (Fig. 5).

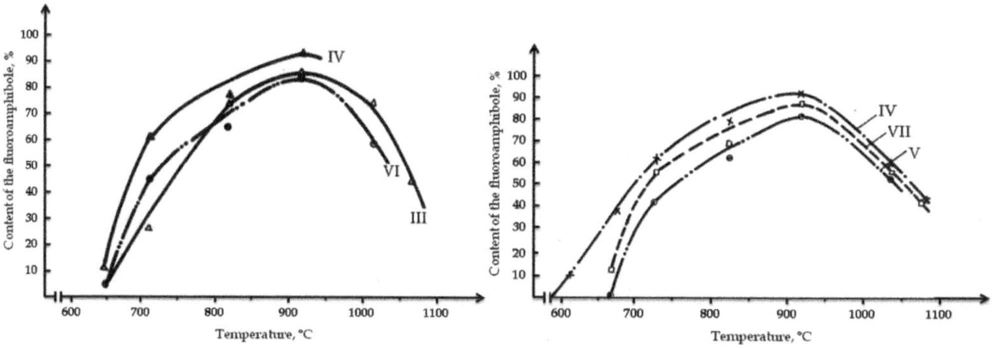

Fig. 5. The effect of heating on the yield of the fluoramphibole (exposure time 48 h) in the reaction blends with: III –serpentinite, IV – calcinated serpentinite, V - dunite, VI - peridotite, and VII – pyroxenite.

The starting temperature for fluoramphibole formation was lower in the blends, which were based on the rocks containing substantial amounts of orthosilicate (blends V and IV) and enstatite (blends VI and VII). The process was defined to have start temperature higher by 50-70°C within the blend (III) with coarse serpentine. The crystallization temperature was elevated in the blend VII owing to the presence of monoclinic pyroxens, e.g. diopside and diallage.

The formation of FFS at 700-720°C was observed on the surface of end product obtained from dunite (blend V), and at 800-820°C in the reaction blends III, IV, VI, VII. In all blends heating upon 850-920°C led to elongation and thickening of fibres/needles in both surface 'brush' and basic mass of the product. At 1000-1100°C the fluoramphibole was crystallized in form of short-acicular and fine-prismatic crystals.

More intensively the solid-phase reactions proceeded at 900-920°C, when the exposure lasted 24-36 h (Fig. 6). The processes were in fact accomplished by the exposure time of 36-48 h.

The fibres were elongated from 0.01 to 0.4 mm by the time of exposure from 1 to 6 h. The further prolongation of exposure by 12-48 h caused an increase in length of fibres within the basic mass and on the surface of the synthesized product.

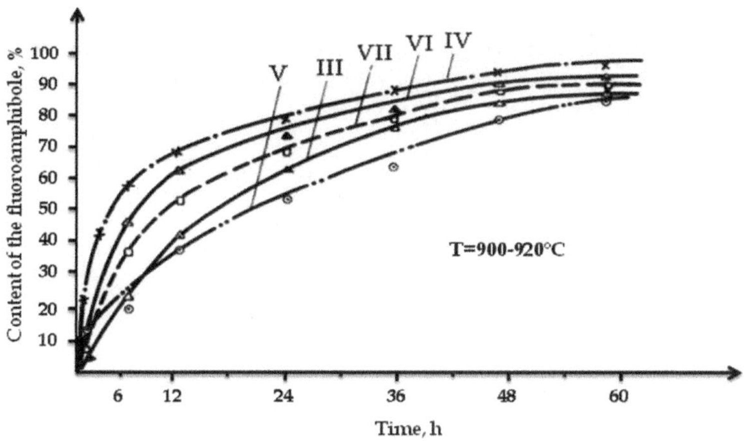

Fig. 6. The effect of the exposure time on the yield of the fluoramphibole in the reaction blends with: III –serpentinite, IV – calcinated serpentinite, V - dunite, VI - peridotite, and VII – pyroxenite.

In summary, synthesized samples contained around 85-95% fluorosilicates in form of fibrous and fine-prismatic crystals. Their length in the basic mass of the product constituted 0.1-1.0 mm, and thickness – 0.01-5.00 μm. Under certain conditions 5-30 mm long and 0.05-8.0 μm thick fluoramphibole fibres form a monomineral 'brush' on the surface of synthesized product (Fig. 7).

The results of crystal-optical and X-ray phase assays demonstrated that the fluorosilicate in shape of acicular-fibrous crystals was a dominant phase in the samples obtained from the reaction blends containing high dispersed mining rocks (with granule size ≤50 μm) supplemented with around 10-30 wt% fluxes (NaCl) at 850-920°C, and within the exposure time 36-48 h. The accompanying minerals in the products of synthesis were mainly represented by galite and sodium fluoride, as well as by CaF_2, MgF_2, $NaMgF_3$, $NaFeF_4$, glass, etc, in markedly lower concentrations.

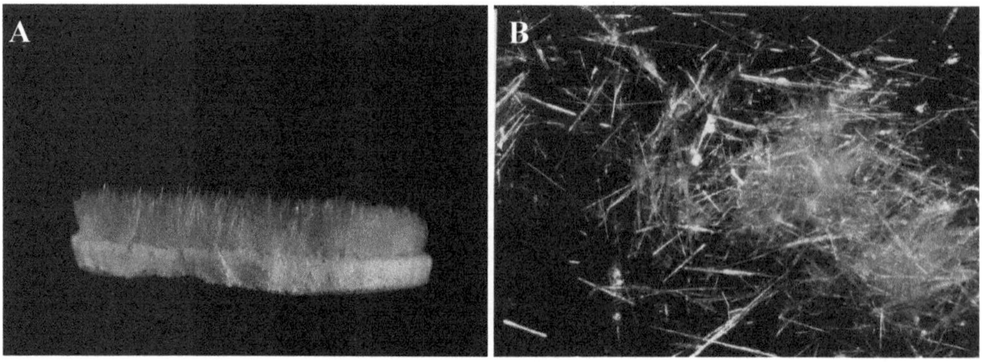

Fig. 7. Synthetic fibrous fluoramphibole from mining rock serpentinite: general view of the synthesis product (*A*, 0.5x) and fibrous crystals (*B*, 0.5x).

To better understand the features of solid phase synthesis from mining rocks we investigated some mechanisms of chemical and phase transformations occurring in the reaction blends as well as the kinetics of the process.

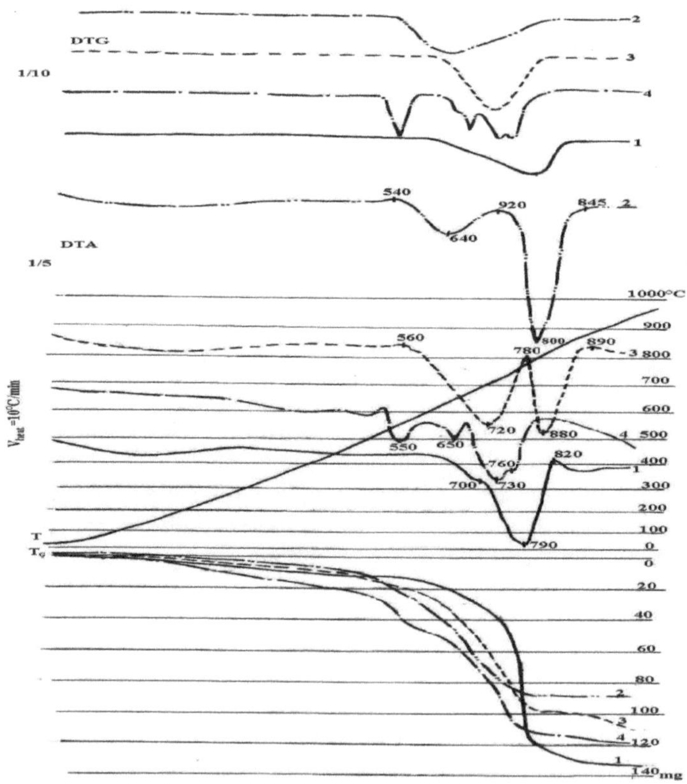

Fig. 8. DTG and DTA curves for the blends containing serpentinite (S, 1), S-NaF (2), S-NaCl (3), and S-Na_2SiF_6 (4).

The formation of asbestos-like ribbon–chain fluorosilicates (amphiboles) in the studied blends appeared to depend substantially on the chemical nature and crystal chemical characteristics of the initial reactants and, consequently, on the physicochemical features of their behaviour upon heating. In turn, this determines the mechanisms responsible for chemical interactions between precursors (in particular, for the character and sequence of chemical transformations) and for the characteristics of final products.

As discussed above, to reveal a correlation between the physic-chemical parameters of initial reactants and chemical and phase transformations during the synthesis of FFS from mining rocks, we studied the phase compositions of the products obtained by the treatment of the reaction blends I-VII under different thermal-temporal conditions. The intermediate and final products of heating were identified by X-ray powder diffraction and crystal-optical assays. Moreover, for a more exact interpretation of the data, the sequence of serpentinite transformations and the dinamics of chemical interactions between S and sodium fluorosilicate (Na_2SiF_6), halite (NaCl), and sodium fluoride (NaF) upon heating were investigated using thermal (Fig. 8, curves 1- 4), X-ray diffraction at room and high temperatures (Fig. 9), crystal-optical, and chemical assays.

The data on blend composition together with the results of these analyses are summarized in the table 4. The transformations of the sodium fluorosilicate and its hardened samples, obtained by heating upon 600°C with exposure time 10 and 30 min were described using the crystal-optical, X-ray powder diffraction and thermal analyses (Khachatryan, 1969, Khachatryan and Karachanyan, 2000).

In the reaction blends within the temperatures range of 90-750⁰C dehydration and dehydroxylation of serpentines chrysotile and antigorite (Fig. 8, curve 1), decomposition of olivine, and the transformation of magnetite (see Table 3) were found to proceed according to the reactions (7)-(10).

$$\underset{\text{chrysotile}}{3MgO{\cdot}2SiO_2{\cdot}2H_2O} \xrightarrow[400-620^\circ C]{-2H_2O} \underset{\text{metaserpentine}}{3MgO{\cdot}2SiO_2} \xrightarrow{\geq 700^\circ C} \underset{\text{forsterite}}{Mg_2SiO_4} + MgO + SiO_2 \xrightarrow{>840^\circ C} Mg_2SiO_4 + MgSiO_3 \tag{7}$$

$$\underset{\text{antigorite}}{3MgO{\cdot}2SiO_2{\cdot}2H_2O} \xrightarrow[\sim 700^\circ C]{-2H_2O} \underset{\text{forsterite}}{Mg_2SiO_4} + MgO + SiO_2 \longrightarrow Mg_2SiO_4 + \underset{\text{enstatite}}{MgSiO_3} \tag{8}$$

$$\underset{\text{olivine}}{(Mg,Fe)_2SiO_4} \xrightarrow{<800^\circ C} 2MgO{\cdot}SiO_2 + Fe_2O_3 + \underset{\text{amorphous}}{SiO_2} \tag{9}$$

$$\underset{\text{magnetite}}{FeO{\cdot}Fe_2O_3} \xrightarrow{200-300^\circ C} \underset{\text{maghemite}}{Fe_2O_3} \xrightarrow{530-800^\circ C} \underset{\text{hematite}}{Fe_2O_3} \tag{10}$$

We failed to observe the formation of complete X-ray amorphous phase during the decomposition of serpentinite under the conditions studied (Table 4, Fig. 9*A*), that was in line with the data obtained by Koltermann (1965).

Initial Blend	Conditions of treatment			Products of treatment
	temperature, °C	exposure, h	supplements, wt%	
S	550	2	-	Partially modified serpentine
S	650	2	-	Partially modified serpentine, forsterite
S	800	2	-	Forsterite, hematite
S - NaCl	550	2	20	Serpentine, forsterite, ore mineral, glass
S - NaCl	800	2	20	Forsterite, enstatite, ore mineral, glass, amphibole
S-Na_2SiF_6	700	2	20	Serpentine, NaMg-threechain silicate, amphibole, forsterite, enstatite, fluorides, ore mineral
S-Na_2SiF_6	800	2	20	Amphibole, NaMg-threechain silicate, forsterite, enstatite, fluorides, ore mineral
S-Na_2SiF_6	900	2	20	Amphibole, fluorides, ore mineral

Table 4. The conditions and products of the thermal treatment of serpentinite (S) and the blends S-NaCl, S- Na_2SiF_6 within the closed system.

The supplementation of the initial blends with fluorides Na_2SiF_6 and NaF or flux NaCl substantially affects the DTA curve of serpentinite. In this case, the introduction of Na_2SiF_6, NaF or NaCl to the serpentinite-containing blend reduced both the dehydration temperature and the temperature of maximal rate of dehydroxylation by 60, 90 and 150°C. This lowered the energy required for the activation of the decomposition process and widened the thermal field of the dehydration for this rock. The fine-fibrous phase of three-chain fluorosilicate $Na_2Mg_4[Si_6O_{16}]F_2$ was formed in the closed bowl when heating the blend S- Na_2SiF_6 upon 700°C. The amount of this silicate was significantly reduced at 800°C.

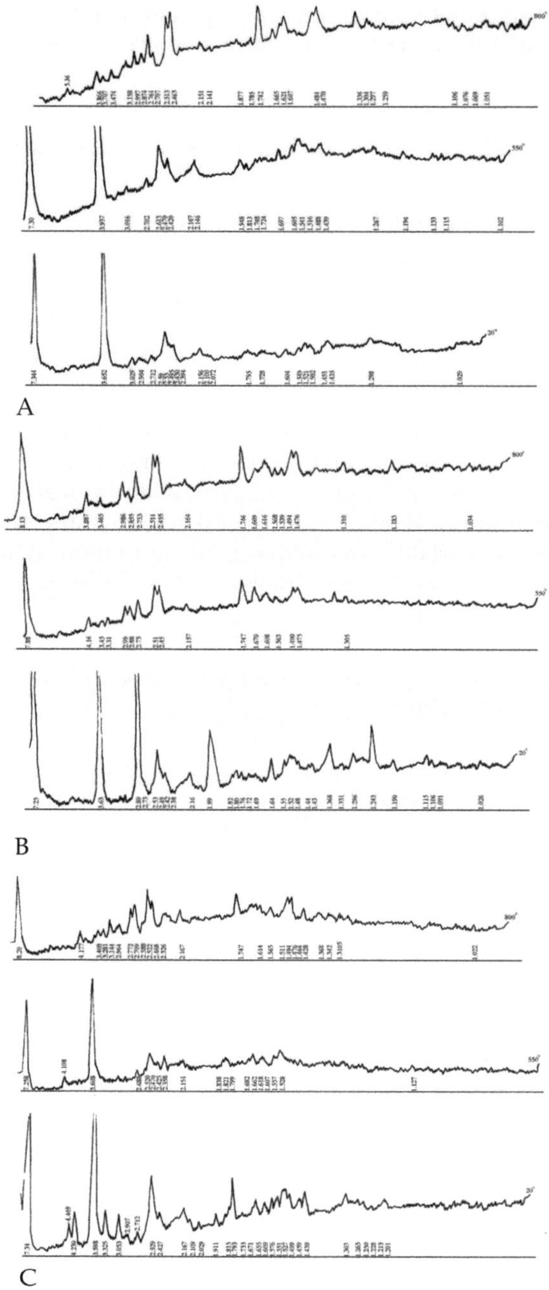

Fig. 9. Thermal X-ray diffraction patterns of serpentinite (S, ***A***) and the blends containing S-NaCl (***B***) and S- Na_2SiF_6 (***C***). The abscissa shows the value of d/n. Temperature (°C), time (in min): 20; 550, 30; 800, 30.

The decomposition of sodium fluorosilicate, which parallels the processes (7)-(10) occurring in the initial blends III-VII at the temperature above 400°C, can be represented by the reactions (11)-(13):

$$.\,Na_2SiF_6 \xrightarrow{540-560°C} 2xNaF + (1\text{-}x)Na_2SiF_6 + xSiF_4\uparrow . \tag{11}$$

$$SiF_4 + 2H_2O \xrightarrow{\geq 420°C} SiO_2 + 4HF\uparrow \tag{12}$$

$$3SiO_2 + 4NaF \xrightarrow{570-620°C} 2Na_2SiO_3 + SiF_4\uparrow \tag{13}$$

The further increase in the temperature (≥ 600°C) was reported to result in the complete decomposition of Na_2SiF_6 with the formation of NaF (Khachatryan, 1969, Istomin, 1980).

Water vapour formed in the reactions (7) and (8) obviously exerted a catalytic effect on the chemical transformations occurring in the blends (Simons, 1950; Prodan et al., 1976; Tret'yakov, 1978). The hydrogen fluoride HF released in the reaction (12) facilitated the decomposition of initial minerals and products of thermal treatment of mining rocks, e.g. metaserpentine, forsterite, enstatite, amorphous SiO_2, and intermediate compounds (Ryss, 1956; Butt et al., 1965; Kogarko and Krigman, 1981). The possible interactions are summarized in the reactions (14)-(17).

$$\underset{\text{metaserpentine}}{3MgO\cdot 2SiO_2} + 6HF \longrightarrow 3MgF_2 + 2SiO_2 + 3H_2O\uparrow \tag{14}$$

$$\underset{\text{amorphous}}{SiO_2} + 4HF \longrightarrow SiF_4 + 2H_2O\uparrow \tag{15}$$

$$2MgF_2 + SiO_2 \longrightarrow 2MgO + SiF_4\uparrow \tag{16}$$

$$MgF_2 + SiO_2 + H_2O_{vapour} \longrightarrow MgSiO_3 + 2HF\uparrow \tag{17}$$

The presence of NaCl in the mixture decreased the elimination of SiF_4 (Istomin, 1980) through the formation of the unstable intermediate compound $Na_2SiF_4Cl_2$, as shown in the reaction (18):

$$2NaCl + SiF_4 \longrightarrow Na_2SiF_4Cl_2 \tag{18}$$

In the products of treatment of the initial blends at 650-800°C, one of the accompanying minerals was fluornorbergite. Additionally to the reaction (3), this mineral could be formed in the reaction (19):

$$\underset{\text{enstatite}}{MgSiO_3} + MgF_2 + MgO \longrightarrow \underset{\text{fluornorbergite}}{Mg_2SiO_4\cdot MgF_2} \tag{19}$$

The compounds formed in these reactions (metaserpentine, forsterite, enstatite, threechain fluorsilicate, fluornorbergite, SiO_2, fluorides and etc.) possessed high reactivity. Forsterite and fluornorbergite contained certain elements of the fluoramphibole structure; enstatite had a structure similar to the fluoroamphibole, whereas NaMg-threechain fluorosilicate resembled the fluoramphibole by its structure and chemical composition. These factors likely favoured the formation of fluoramphibole at relatively low temperatures (~600–750°C). The fluoramphibole formed at these temperatures was defined by low content of iron. Further rise in the temperature promoted the formation of intermediate compounds: Na_2FeF_4, Na_3FeF_6, $NaMgF_3$, CaF_2, etc (Simons, 1950; Ryss, 1956). It is to note, that the formation of fluorosilicates occurred uniformly throughout the whole volume of the blends within relatively wide range of temperatures (600-950°C).

The prolongation of treatment from 1 to 48 h at 900-920°C and the elevation of the temperature from 600 to 920°C led to the elongation and thickening of fluoramphibole fibres, i.e. formation of needles, and to the increase in the refractive index of fibres and needles from n_m=1.582 to n_m=1.618. These changes might result from the complexification of chemical compositions of fluoramphiboles, e.g. their enrichment with Fe^{2+}, Fe^{3+}, Ca^{2+},Cr^{3+},Al^{3+} and other cations, which promotes the yield of more isometric crystals of the product. Furthermore, the recrystallization of the fluoramphibole particles originated from olivine, forsterite, enstatite, and other minerals in the composition of D, S_{calcin}, Px, and P at low temperatures at the initial stages of solid-phase synthesis, might influence the final shape of crystals. Altogether, this indicated, that the structures of fluoramphibole were formed in the course of sequential transformation of initial and intermediate compounds and that the elevation of process temperature together with prolongation of synthesis favoured an improvement of the crystal lattice in fluorosilicates (Godovikov, 1973).

Crystal-optic investigation of the products hardened at 550°C estimated the presence of melt in the studied systems. The generation of liquid phase in these conditions apparently was facilitated by the melting of $FeCl_3$ released in the schematic reaction (20):

$$Fe_3O_4 + NaCl \longrightarrow FeCl_3 + Na_2O \qquad (20)$$

An increase in the amount of liquid phase was associated with the further elevation in the temperature and occurred due to the melting of solid constituents of the initial mixtures and formed eutectic blends (Sharma and Johnson, 1962; Levin et al., 1964; Kogarko et al., 1968; Kuvakin and Rod'kina, 1971).

Thus, within the temperature range of 400-1000°C the solid-phase interactions proceeded with the participation of both gaseous phase and melt, i.e. the crystals of fibrous fluorosilicates grew by 'feeding' from the melt and the gaseous phase. Depending on the conditions of synthesis the gaseous phase may contain SiF_4, HF, and H_2O (see the reactions (11)-(17)). The mole percentage of SiF_4 in the initial blends and in the gaseous phase was raised by the growth of the ratio F/O (Kogarko et al., 1968). The presence and, moreover, the excess of Cl^- intensified the participation of gaseous phase in the formation of fluor-containing compounds ($NaMgF_3$, $MgSiF_6$, SiF_4, HF) and Na_2SiCl_6 as the product of the sequential reactions:

$$NaCl + HF \longrightarrow NaF + HCl \uparrow \quad (21)$$

$$\underset{\text{amorphous}}{SiO_2} + 4HCl \longrightarrow SiCl_4 + 2H_2O \uparrow \quad (22)$$

$$2NaCl + SiCl_4 \longrightarrow Na_2SiCl_6 \quad (23)$$

In gaseous phase Na_2SiCl_6 was stable under the described conditions (see also Wartenberg, 1953). Mass-spectrometric analysis of products evaporated from the alloys in the system NaF-MgF_2 determined the presence of NaF, Na_2F_2, and Na_3F_3 at 800°C (Nikitin et al., 1973). The appearance of sodium in the gaseous phase might also result from the evaporation of NaCl at 750°C in open air and at 700°C in the stream of water vapour (Spitsyn et al., 1952).

The complex magnesium fluoride $MgSiF_6$, which released in the reaction (24) (Kogarko and Krigman, 1981), expressed significant vapour pressure at 750-1100°C.

$$MgF_2 + SiF_4 \longrightarrow MgSiF_6 \quad (24)$$

High vapour pressure seems to be the characteristic property of the complex fluorides ($MgSiF_6$, Na_2FeF_4, Na_2FeF_6, $NaMgF_3$, etc) under the mentioned thermal conditions (Eitel et al., 1956). Using the mass-spectrometric assay a number of researchers have distinguished the silicium oxyfluoride $SiOF_2$ in the gaseous phase of fluorosilicate systems (Nikitin et al., 1973; Novoselova et al., 1964; Novoselova and Azhikina, 1966).

Thus, all cations required for fibrous fluoramphibole synthesis were represented in the gaseous phase. The concentration of these cations depended on the composition of the reaction blend and on thermal-temporal parameters of synthesis. This was confirmed by the formation of fibrous fluoramphiboles on the surface of the main treatment product at ~700-720°C (blend V) and at ~800-820°C (blends III, IV, VI, VII). Furthermore, an increase in the number, length, and thickness of fibers and needles forming a dense 'brush' accompanied the temperature raise up to 850-920°C and the prolongation of the synthesis up to 24-36 h.

The formation of fibrous fluoramphiboles at above 800°C proceeded apparently with the participation of the gaseous phase characterized by higher content of iron. The putative chemical interactions can be reflected by the reactions (25)-(28):

$$\underset{\text{norbergite}}{Mg_2SiO_4 \cdot MgF_2} + Na_3FeF_6 + MgSiO_3 + 7SiO_2 \longrightarrow \underset{\text{NaMg-fluorarfvedsonite}}{Na_3Mg_4Fe^{3+}[Si_8O_{22}]F_2} \quad (25)$$

$$2(\underset{\text{forsterite}}{Mg_2SiO_4}) + NaFeF_4 + 2NaF + 4SiO_2 \longrightarrow Na_3Mg_4Fe^{3+}[Si_8O_{22}]F_2 \quad (26)$$

$$\underset{\text{enstatite}}{Mg_2SiO_3} + 2NaF + NaFeF_4 + 2MgO + 7SiO_2 \longrightarrow Na_3Mg_4Fe^{3+}[Si_8O_{22}]F_2 \quad (27)$$

$$\underset{\text{threechain fluorsilicate}}{Na_2Mg_4[Si_6O_{16}]F_2} + NaFeF_4 + 2SiO_2 \longrightarrow Na_3Mg_4Fe^{3+}[Si_8O_{22}]F_2 \quad (28)$$

With the temperature raise above 1150°C fluoroamphibole partially decayed with the formation of magnesium- ortosilicate, spinel, tridimite, and ore mineral.

The differences during fluorosilicate synthesis observed for serpentinite-contatining blends supplemented with either NaF or Na_2SiF_6, might originate from the circumstance that in the NaF-added blends thermal-temporal parameters of dehydration of initial minerals do not match with those of NaF hydrolysis and with conditions, under which the content of NaF in the gaseous phase reached conspicuous values.

Our results demonstrated the complex character of interaction between the precursors in the studied blends during the synthesis of asbestos-like fluorosilicates. In the course of solid-phase reactions, the gaseous phase participated actively within the temperatures range of ~400-900°C, whereas the liquid phase (the melt of mainly halogen-containing composition) played the leading role at temperatures above ≥600°C. The reactions of fibrous fluoroamphibole formation proceeded with a considerable rate at relatively low temperatures (850-920°C) owing to the presence of the gaseous and liquid phases. The high reactivity of the blends could be explained by mineralogical composition and texture-structure features of the mining rocks: S, D, P, and pyroxenite.

In summary, we suggest that the process of fibrous fluorosilicate formation from mining rocks involved a sequence of complex phase transformations and chemical reactions in combination with structural transformations and recrystallization.

The treatment of magnesium silicate-containing rocks under pyrogenic conditions led to the crystallization of fluorosilicates in the form of fibrous-, thin-, and thick- prismatic crystals of the various length (Fig. 10).

Our investigations indicated that the morphology of fluoramphibole crystals was highly influenced by the following factors: the mineralogical composition of initial rocks, content and character of fluorine-containing components and flux in the initial mixtures, the dispersion rate of rocks, and the temperature and duration of the synthesis.

Morphogenetic patterns of the crystallization process, which take place at the end stage of solid-phase interactions, were defined by succession of structural components of crystalline compounds involved in the reaction. In solid-phase reactions the fluoroamphibole was shaped in form of fibers in case of emergence of intermediate products such as forsterite and fluornorbergite, which contain the oxygen octahedrons of magnesium. The latter by their size and symmetry resemble polyhedrons M (1-3) in the structure of amphiboles. Another intermediate product - threechain FFS, bears as a basic structural component the ribbons of $[Si_6O_{16}]^{8-}{}_{\infty}$ composition, which are represented by three polar tetrahedron (pyroxen) chains connected by common oxygen atoms. Collectively, the fibrous shape of fluoramphibole was thereby determined by a complex of intermediate products expressing proximity of structures and compositions.

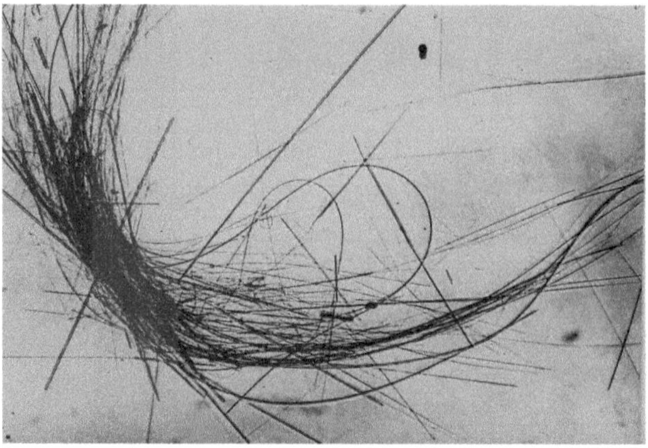

Fig. 10. Micrograph of synthetic fibrous fluorosilicates (x 110).

Chemical composition, wt%	Fluorosilicates obtained from the blends containing:			
	Serpentinite	Peridotite	Dunite	Pyroxenite
SiO_2	58.36	58.20	58.7	55.11
Al_2O_3	0.30	1.25	0.68	2.10
Fe_2O_3	3.08	2.83	3.83	6.05
FeO	0.05	0.12	0.05	traces
Cr_2O_3	0.20	0.08	0.16	0.4
MgO	26.31	28.00	25.1	20.15
CaO	1.86	0.88	0.87	1.59
Na_2O	7.12	5.83	7.10	11.39
H_2O^+	-	-	-	-
H_2O^-	0.28	-	0.61	-
F	4.09	4.70	4.84	4.52
Sum	101.65	101,33	101.94	101.53
$F=O_{equivalent}$	1.72	1.92	2.00	1.9
Σ	99.93	99.11	99.94	99.63

Table 5. The chemical composition of obtained fibrous fluorosilicates.

Synthesized FFS by their chemical composition (Table 5), structure (Table 6, Fig 11), and physical-chemical properties (Table 7) related to NaMg-arfvedsonite asbestos (Hodgson, 1965; Chigareva, 1967; Grigor'eva et al., 1975). Crystallochemical formulas of fibrous fluorosilicates synthesized in serpentinite-, dunite-, peridotite- and pyroxnite-containing blends corresponded respectively to:

$Na_{1.90}Ca_{0.30}Mg_{5.40}Fe^{3+}{}_{0.32}Fe^{2+}{}_{0.006}Cr^{3+}{}_{0.02}Al_{0.05}Si_8O_{22}(F_{1.8}O_{0.34})$,

$Na_{1.55}Ca_{0.13}Mg_{5.70}Fe^{3+}{}_{0.28}Fe^{2+}{}_{0.002}Cr^{3+}{}_{0.01}Al_{0.09}Si_8O_{22}F_2$,

$Na_{1.90}Ca_{0.12}Mg_{5.10}Fe^{3+}{}_{0.40}Cr^{3+}{}_{0.01}Al_{0.11}Si_8O_{22}F_2$,

$Na_3Ca_{0.24}Mg_{4.24}Fe^{3+}{}_{0.67}Cr^{3+}{}_{0.04}Al_{0.35}Si_{7.71}O_{22}F_2$.

Fluoramphibole composition	Parameters of crystal lattice				Optical constants			
	a, nm	b, nm	c, nm	β	Ng	Np	Ng -Np	cNg
$Na_{1.90}Ca_{0.30}Mg_{5.40}Fe^{3+}_{0.32}Fe^{2+}_{0.006}Cr^{3+}_{0.02}Al_{0.05}Si_8O_{22}(F_{1.8}O_{0.34})$	0.972	1.788	0.530	103°44'	1.615	1.600	0.015	24°
$Na_{1.55}Ca_{0.13}Mg_{5.70}Fe^{3+}_{0.28}Fe^{2+}_{0.002}Cr^{3+}_{0.01}Al_{0.09}Si_8O_{22}F_2$	0.968	1.770	0.522	103°30'	1.594	1.583	0.011	24°
$Na_{1.90}Ca_{0.12}Mg_{5.10}Fe^{3+}_{0.40}Cr^{3+}_{0.01}Al_{0.11}Si_8O_{22}F_2$	0.966	1.784	0.523	103°11'	1.618	1.600	0.018	24°
$Na_3Ca_{0.24}Mg_{4.24}Fe^{3+}_{0.67}Cr^{3+}_{0.04}Al_{0.35}Si_{7.71}O_{22}F_2$	0.964	1.783	0.524	103°7'	1.612	1.594	0.018	24°

Table 6. Parameters of crystal lattice and optical characteristics of synthesized fibrous fluorosilicates.

Fibrous fluorosilicates	Thermal properties, T, °C		Chemical resistance: weight loss (in wt %) after boiling (4 h) in:		Adsorption of water vapour: weight gain (in wt %) at air water vapour pressure 16.4 mm Hg	
	Decomposition	Melting	HCl	KOH	in 24 h	in 30 days
From serpentinite	980-1070	1180-1230	5.40	0.80	3.20	3.85
From dunite	960-1000	1160-1180	-	-	2.40	2.70
From peridotite	970-1050	-	4.75	0.80	3.00	3.60
Crocidolite *Hodgson, 1965*	550-700	800	8.0	-	-	-
NaMg-fluorrichterite *Grigor'eva et al., 1975*	950-1000	1170-1190	8.6	0.1	-	-
NaMg-fluorarfvedsonite *Grigor'eva et al., 1975*	940-1000	1000-1060	8.40	1.50	-	-

Table 7. Physical-chemical properties of fibrous fluorosilicates obtained from mining rocks and chemical reactants.

Fluoramphiboles were crystallized mainly in the form of fibres and needles, rarely as dissociated aggregates, which were formed by non-splited bundled fibres (Fig. 10, 11). By their size the fibres and needles in the basic mass were inferior to the crystals in the 'brush'. Thinner crystals were obtained from the blend with serpentinite, while the thicker ones – from pyroxenite.

Electron-microscopic studies revealed that fluorosilicate microcrystals had the form of a ribbon extended by axis '*c*'.

The fibrous fluorosilicates obtained in our study from mining rocks did not concede by their properties to the fibrous fluoramphiboles synthesized from chemical reactants. Furthermore, the thermal properties and chemical resistance of fluorosilicates from mining rocks were better than those of natural asbestos crocidolite. The decomposition temperature of fluorosilicates from mining rocks was by ~120-180°C higher compared with the natural and synthetic hydroxyl amphyboles (Hodgson, 1965; Khachatryan, 1969; Grigor'eva et al., 1975).

Fig. 11. Electron-microscopic images of fibrous fluorosilicate synthesized from serpentinite in form of fibres (*A*, 18000x) and dissociated aggregate (*B*, 25000x).

4. Comparative analysis of solid-phase synthesis from mining rocks and chemical reactants

The described approach to obtain fibrous fluorosilicates of NaMg-fluorarfvedsonite composition by solid-phase synthesis from mining rocks revealed a line of advantages. First, it utilizes less expensive initial blends, which contain inexpensive and affordable mining rocks instead of expensive chemical reactants and pure mineral roducite. Second, the mining rocks and other supplements in our method can be used *per se*, i.e. do not require preliminary treatment such as burning or drying, in contrast to the solid-phase synthesis from reactants and minerals. Third, the use of mining rocks enables the temperature of synthesis to be ~100-130°C lower when compared with other conventional methods. Forth, the latter simplifies the technology and reduces the costs of synthesis. Fifth, the fibrous fluorosilicates synthesized from mining rocks express better physical-chemical properties, i.e. they are more thermostable and have higher chemical resistance.

5. Practical outcome and perspectives of solid-phase synthesis of fibrous fluorosilicates from mining rocks

We developed the method of synthesis of new fibrous crystalline materials, i.e. fluorosilicates of amphibole group, from natural feedstocks. By their texture these materials resemble natural amphibole asbestos, by chemical resistance in acidic and alkaline media, and by sorptive properties they do not concede to natural asbestos and synthetic fibrous fluoramphiboles obtained from reactants and pure minerals. By thermostability they substantially rank over natural amphibole asbestos (see decomposition temperatures in table 7).

This experimental approach was completely reproducible in large-scale laboratory set-ups, where the scales of the experiments were 20-30 times higher. Further, we plan to test this approach in preindustrial environment, as well as to employ the obtained fibrous fluorosilicates as fillers for creation of new composite materials based on organic, silicorganic, alumophosphate, etc, matricies. Such composite materials are utilized in the different fields of the modern technique to obtain thermostable clays, rubber, new pressed materials with valuable physic-mechanical properties, etc. Synthesized fibrous fluorosilicates can be also used as adsobtive substances for the extraction of heavy metal cations from acidic and alkaline water solutions and for the regeneration of specific elements during the sewage treatment in industry.

6. Conclusion

The possibility to synthesize inorganic non-stoichiometric fibrous fluorosilicates from inexpensive and readily affordable complex non - stoichiometric raw material such as magnesium silicate-containing mining rocks has been shown for the first time using the approach of the solid-phase synthesis. We determined the optimal conditions of the synthesis from serpentinite, calcinated serpentinite, dunite, pyroxenite, and peridotite. The fibrous fluorosilicates of NaMg-fluorarfvedsonite composition ($Na_3Mg_4Fe^{3+}[Si_8O_{22}]F_2$ were synthesized by employing precursors with different chemical and mineralogical compositions. Obtained silicates displayed different rate of Na+ → Ca^{2+}, Mg^{2+}; Mg^{2+} → Fe^{2+}; Fe^{3+}→ Cr^{3+}, Al^{3+}; Si^{4+} → Al^{3+} substitutions.

We estimated that mineralogical composition of initial rocks affected the lower limit of the temperature field in the course of fluoramphibole formation and the intensity of the process, as well as determined the maximal rate of rock transformation to the fluorosilicates and the required content of fluorine in the initial blend.

The reaction blends based on the mining rocks possessed a high reactivity, owing to which the fibrous fluorosilicates were synthesized at lower temperature (around 100–130°C) and the duration of the synthesis was shortened several fold. The capability of the investigated rocks to be transformed into the fluoramphibole decreased within the following line: dunite - flamed serpentinite - peridotite - serpentinite - pyroxenite. The same capability of minerals included in the composition of mining rocks became extinct within the line: olivine - enstatite –serpentine- monoclinic pyroxenes (diopside, diallag). The maximal yield (~ 92%) of fibrous fluorosilicates was observed in the reaction blend based on the dunite rock and supplemented with halogen-containing constituents Na_2SiF_6 and NaCl. It is to note, that growing scale of the experiments resulted in increased yield of long-fiber fluorosilicate.

Fibrous fluorosilicate formation within the temperature range of 600-1000°C was demonstrated to result from complex non-stoichiometric reactions developing through sequential and parallel stages with the contribution of liquid and gaseous phases. In this, the process in the first 3h was limited by proper chemical interaction, while its further development was restricted by diffusion.

Some features of phase-formation were described in the studied polycomponent fluorosilicate systems during solid-phase synthesis. The character of intermediate crystalline compounds substantially affected the morphogenetic patterns of fluoramphibole formation by donating specific structural fragments aimed to be bound with growing borders of fluoramphibole crystals. The higher rate of condensation of silicoxygen tetrahedrons enabled the formation of silicates with more complex structure of anion radical.

Examining the morphogenetic patterns, we noticed that the fluoroamphibole in form of fibrous-acicular crystals was formed near the low temperature border of their stability, and when the mining rock of monomineral composition serpentinite was used at its high dispersivity. The mineralizing substances allowed modifying the properties of intermediate compounds and the physical-chemical parameters of the initiation of solid-phase interactions, thereby, offering a possibility for purposeful changes in the habits of crystals from prismatic to fibrous. The described course of the solid-phase synthesis from mining rocks as well as the properties and characteristics of obtained fibrous fluorosilicates were typical for the compounds of non-stoichiometric composition. Thus, the non-stoichiometric composition of amphibole enabled its yield from inexpensive and more affordable non-stoichiometric raw material.

In comparison with natural and synthetic hydroxyl amphiboles the fluoramphiboles obtained by solid-phase synthesis from mining rocks decayed at higher (120-180°C) temperature. By their acid- and alkaline-resistance as well as by their adsorptive properties the synthesized fluorosilicates did not concede to the top-graded natural amphiboles.

7. Acknowledgment

This research was funded by grants from Ministry of Finances (92-168) and Ministry of Education and Science (01-392, 98-37, 07-38) of Republic of Armenia to L.Kh.. Dr. Dmitry P. Romanov, Dr. Boris V. Volkonsky, and Dr. Galina I. Kosulina, deserved our special thanks for their help with taking the X-ray diffraction, DTA, and elecronmicroscopic images. Dr. Lidia F. Grigor'eva is sincerely acknowledged for her valuable comments. We are very grateful to our spouses Dr. Benik G. Yeritsyan and Gevorg E. Gevorgyan, who provided the support with the graphical illustrations and were always available for helpful scientific discussion; they sustain us in all we do and it is to them that this work is dedicated with love and respect.

8. References

Budnikov, P P. & Ginstling, A.M. (1971). Reactions in the mixtures of solid substances, Publ. by construction, Moscow.

Butt, Yu. I.; Okorokov, S.D.; Sychev, M.M. & Timashev, V. V. (1965). *Technology of cohesive substances*, Advanced School, Moscow.

Chigoreva, O.G. (1967). Synthesis and Investigation of Basic Properities Asbestos-like Fibrous Fluoramphiboles, *Extended Annotation of Cand. Sci. Dissertation*, Institute of Silicate Chemistry after Grebenshchikov, AS USSR. Leningrad.

Chigoreva, O.G. & Fedoseev, A.D. (1966). Synthesis and investigations of some properities of fibrous alkali amphiboles. In: *Investigations of Natural and Technical Mineraloformation*, pp. 169-174, Science, Moscow.

Chigareva, O.G. & Grigor'eva, L.F. (1970). About receipt artifical fluorasbestos from natural talc. *Lectures Academy of Science, USSR, ser. Geology*, Vol. 195, No. 5, pp. 1194-1196.

Eitel, W. (1952). Synthesis of fluorsilicates of the mica and amphibole group. In: *Proc. Intern. Symp. Reactivity of Soids*, Gotheburg. Sweden, pp. 335-347.

Eitel, W.; Hatch, R.A. & Denny, M.V. (1956). Synthetic mica investigations. II. Role of Fluorides in Mica Batch Reaction. *Journal of American Ceramic Society*, Vol. 36, No. 1, pp. 341-348.

Ernst, W.G. (1968). *Amphiboles. Cristal chemistry phase relations and occurence*, New York.

Espig, H. (1962). Beitrag zur Synthese asbestartiger und einiger anderen Silikate. *Silikattechnik*, Bd. 13, No 4, ss. 131-136.

Fedoseev, A.D.; Grigor' eva, L.F. & Chigoreva, O.G. (1966). Die Synthesis faseriger Silikate unter thermischen Bedigungen. *Kristall und Techn.* Bd., H. 2, ss. 231-236.

Fedoseev, A.D.; Grigor' eva, L.F. & Krupenikova, Z.V. (1963). Synthesis and investigation of the properities of amphibole asbestos-like fibrous minerals. In: *Silicates and oxydes in the high temperatures*, pp. 180-190, Science, Moscow.

Fedoseev, A.D.; Grigor'eva, L.F. & Makarova, T.A. (1966). *Fibrous silicates, natural and synthetic asbestos*, Science, Moscow-Leningrad.

Fedoseev, A.D.; Grigor'eva, L.F.; Chigareva, O.G. & Romanov, D.P. (1970). Synthetic fibrous fluoramphiboles and their properties. *Amer. Mineralogist*, Vol. 55, No. 5-6, pp. 854-863.

Fedoseev, A.D.; Grigor' eva, L.F.; Chigareva, O.G.; Romanov, D.P. & Krupenikova, Z. V. (1972). About cationic isomorphism in the synthetic fibrous fluorosilicates. In: *Experimental investigations of mineral-formation in the dry oxyde-silicate systems*, pp. 141-146, Science, Moscow.

Gibbs, G.V.; Miller, J.I. & Shell, H.R. (1962). Synthetic fluor-mangesio-richterite. *Amer. Mineralogist*. Vol. 47, No. 1-2, pp. 75-82.

Godovikov, A. A. (1973). *Introduction to mineralogy*, Science, Novosibirsk.

Goncharov, Yu. I. & Kovalenko, V. S. (1973). Toward a mechanism of formation of amfibolic asbestos. *Geochimia*, No. 5, pp.782-787.

Grigor' eva, L.F.; Makarova, T.A.; Korytkova, E.N. & Chigareva, O.G. (1975). *Synthetic Amfibolic Asbestos*. Science. Leningrad.

Hawkins, J.W. (1971). Asbestos: today, tomorrow, the day after tomorrow. *Sampe Journal*, Dec/Jan, pp. 17-20.

Hodgson A. A. (1965). Fibrous silicates. *W.C.I. lecture series*, London, No. 4, 46 p.

Istomin, S.S. (1980). The influence of inorganic sodium compounds on thermal dissociation of sodium fluorsilicate. *Journal of Applied Chemistry, AS USSR*, Vol. 53. No. 6, pp. 1201-1205.

Khachatryan, L.A. (1969). Asbestos-like synthetic fibrous silicates from Armenian mining rocks. *Extended annotation of cand. sci. dissertation*, Institute of Silicate Chemistry after Grebenshchikov, AS USSR, Leningrad.

Khachatryan, L.A. (2004). Disperse layered silicates in magnesium silicate rocks. *Glass physics and chemistry*, Vol. 30, No. 1, pp. 67-72.

Khachatryan L.A. (2007). Fibrous Fluorosilicates from Mining Rocks. *Chemical Journal of Armenia*, NAS RA, Vol. 60, No. 3, pp. 433-445.

Khachatryan, L.A. & Yeritsyan, N.B. (2003). Nanosized fibrous silicates from mining rocks of magnesium-silicate composition. *Chemical Journal of Armenia, NAS RA*, Vol. 56, No. 4, pp. 12-22.

Khachatryan, L.A. & Yeritsyan, N.B. (2011). Synthesis of layered thin-scaly fluorosilicates from mining rocks. *Proceedings of 11th European Symposium of Eurofillers*, Drezden, Germany, August, 2011.

Kogarko, I.N. & Krigman, L.D. (1981). *Fluorine in silicate melts and magmas*. Science, Moscow

Kogarko, I.N.; Krigman, L.D. & Sharudilo, N.S. (1968). Experimental studies of the effect of melt alkalinity on the detachment of fluorine into the gaseous phase. *Geochemistry*, No. 9, pp. 948-954.

Koltermann, M. (1965). *Ber. Dtsch. Keram. Ges*, Bd. 42, Hf. 10, ss. 373.

Kuvakin, M.A. (1971). Ternary mutual system of sodium and magnesium fluorides and sodium and magnesium chlorides. *Journal of Applied Chemistry, AS USSR*, Vol. 16, No. 12, pp. 3290-3292.

Li, K. (1992). *Feijinshu Kuang-Nen-Metal*. Mines, No. 4, pp. 47-49.

Lüdke, W. (1933). Metrodisches zur Synthese von Silikaten mit leichtfluchtigen Substanzen unter stationären Bedingungen. A. Hornblendesynthesen. *Fortschr. Mineral*, Bd. 18, ss. 29-31.

Lüdke, W. (1943). Verfabren zur Herstellung von Asbestos mit langeren Fasern aus kürzfaserigen synthetischen Asbestos. *Pat. Gemany*.1.95221, Ivb (12i) 38.

Nikitin, B.N.; Litvinova, T. I.; Raichenko, T. F. & Voronov, V.A. (1973). Ternary mutual system from magnesium and silicon fluorides and oxides. *Journal of Applied Chemistry, AS USSR*, Vol. 46, No. 3, pp. 646-648.

Novoselova, A.V. & Azhikina, Yu.V. (1966). Investigation of the equilibrum in the system containing beryllium oxide, silicon dioxide, and sodium fluoroberyllate. *Reports of AS USSR, Inorganic Materials*, Vol. 2, No. 9, pp.1604-1607.

Novoselova, A.V.; Orlova, Yu.V.; Sobolev, B.P. & Sidorov, L.I. (1964). On the mechaanism of formation of beryllium silicate. *Lectures AS USSR, ser. Chemical*, Vol. 159, No. 6, pp. 1338-1341.

Prodan, E. A.; Pavlynchenko, M.M. & Prodan, S.A. (1976). *Regularities of Topochemical Reactions*, Science and technique, Minsk.

Rabenau A. (Ed.). (1970). *Problems of non-stoichiometry*, North-Holland Publishing Company, Amsterdam-London.

Ryss, I.G. (1956). *The chemistry of fluorine and its inorganic compounds*, Goskhimizdat, Moscow.

Saito, H. & Takusagawa, N. (1965). Synthesis of fluor-manganese-richterite asbestos by sintering method. *Journal of Chemical Society of Japan*, Ind. Chem. Soc. Vol. 68, No. 12, pp. 2347-2351.

Scheumann, K.H. (1932). Über Hornbleendesdesynthesen bei niedrigen Drucken. *Fortschr. Mineral*, Bd. 17, s. 447.

Sharma R.A. & Johnson, I. (1969). *Journal of American Ceramic Society*, Vol. 52, p. 612.

Simons, J. (Ed.). (1950). *Fluorine chemistry*, Academic, New York.

Spitsyn, V.A.; Shostak, V. I. & Meerov, M.A (1952). A study of the volatility of alkali metal chlorides at high-temperatures. *Journal of General Chemistry*, Vol. 22, No. 4, pp. 758-766.

Tareev B.M. (2002-2011). Electrotechnological materials. Available from: http://aznku.ru/met-meb/elektromaterial/88/index.html.

Tresvyatskiy, S.G. (Ed.). (1982). *New materials from oxides and synthetic fluorosilicates*, Naukova dumka, Kiev.

Tret'yakov, Yu. D. (1978). *Solid-phase reactions*, Chemistry, Moscow.

Wartenberg, H.V. (1953). *Z. Anorg. Allgen. Chem*, Bd. 273, No 3, ss. 257-268.

Yeritsyan, N.B. & Khachatryan, L.A. (2005). Synthesis of nanosized fibrous silicates from mining rocks in hydrothermal conditions, *Proceedings of 8th European Symposium of Polymer Blends and Eurofillers*, Bruges, Belgium, May 9-12, 2005.

13

Observation of Chemical Reactions in Solid Phase Using X-Ray Photoelectron Spectroscopy

Sergey P. Suprun, Valeriy G. Kesler and Evgeniy V. Fedosenko
Institute of Semiconductor Physics,
Siberian Branch of Russian Academy of Sciences
Russia

1. Introduction

Chemical reactions in solid phase are now widely used in modern technology. All electronic devices are products made on their basis. For example, such manufacturing method as vacuum epitaxy uses a sequence of chemical reactions accompanying removal of intrinsic oxides from the substrate surface and formation of heterointerface during deposition. The temperature (energy) of reagents coming from a source is as a rule much higher than the temperature of the substrate, and the atomically clean surface is highly active, all in all this creates the required reaction medium. At the next stage of layer engineering it is possible to apply electron-beam lithography, which actually is a directional spatially localized reaction in solid phase, activated by radiation, etc.

Modern engineering creates complex nanoscale objects and quantum systems with a number of specific properties, and this requires detailed knowledge about processes taking place during their formation on atomic level. One of the methods providing unique information on physical and chemical state of the surface of a solid body is a widely applied method of X-ray photoelectron spectroscopy (XPS). XPS allows us to register energy positions of deep (core) atomic levels, which is determined by the properties of external valency bonds. Change of valency bond configuration (its hybridization, length, in other words, change of the spatial distribution of charge density), which is connected with the atom environment, causes an energy shift in the underlying atomic levels (Briggs & Seah, 1990; Sun 2007). Sensitivity of this method is high enough, so that it is possible to identify the chemical compound containing the observed element judging by the value of such shift. Since spectral measurements register electrons emitted from deep levels in the course of specimen irradiation with monochromatic X-ray radiation, then it is the electron free path in the material that determines the analysis depth. Therefore the obtained information shows the dependence between total intensity of electrons leaving the layer ≈ 1 - 2 nm thick and energy. Detailed description of the method and its characteristics can be found in a number of studies (Seah et al., 2001), so it is not going to be expounded here again. Let us consider specific examples of XPS application in development of various technologies which use the molecular beam epitaxy (MBE).

As it was mentioned above, small analysis depth causes certain difficulties in preparation of specimens to be examined, since we need to exclude the uncontrollable chemical influence on their surface. To study the interface between two materials we need to have air lock chamber between the technological chamber for epitaxy and analysis chamber of installation in order to avoid contact with the atmosphere between operations. Another indispensable condition is a precisely controlled deposition of layers with target thickness of about 1 nm. With such requirements met, and with the help of XPS method, it is possible to obtain unique data on heterointerface formation and structural metamorphoses on the surface (changes of crystalline state of the layer).

In order to observe changes in solid phase in situ, for example during formation of the heterointerface, it is possible to use the method described below. First, the atomically clean surface is prepared before epitaxy and then the chemical state of all elements on the surface is registered. If we acquire spectra at different angles of the specimen inclination in respect to the axis of analyzer collecting photoelectrons, we will obtain dependence of chemical state of elements on depth. It is due to the fact that effective analysis depth is defined as $d_{ef} = \lambda \times \cos \theta$, where λ is electrons path in the material, θ - is an angle between the analyser axis and the surface of the specimen. Such method makes it possible to distinguish between the surface and bulk components of the same element. Surface atoms with dangling valency bonds differ by the core levels energy position from the bulk atoms with a complete set of bonds (Sun, 2007). At the next stage a layer of another material is deposited onto the substrate at room temperature. Usually, the thickness of the layer must not exceed 1 nm to enable observation of chemical state of elements in the interface zone from the side of the substrate.

After deposition of the material chemical bonds can be formed even at room temperature, because of the dangled bonds existing on the surface and thermal energy of the source material. Therefore the observation of the heterointerface formation begins immediately after deposition of the layer and continues during the substrate heating. Chemical bonds in solid phase will be formed consecutively with the rise in temperature. Since the layer on the substrate surface is thin, diffusion in this case is not a limiting process. In this manner we analyzed different methods of obtaining clean surfaces of Si, InAs, GaAs and creation of heterojunctions on their basis. This method of studying physical and chemical changes in solid phase resulted in considerable reduction of time spent to fine-tune technological modes matching optimum temperatures necessary to produce structures with required properties.

When interpreting the obtained spectra one should remember that the observed peak of an element is a sum of peaks corresponding to different states of such element within analysis depth. All this results in the change of the peak form, its widening (narrowing) with a change in chemical composition of the surface. The number of components in analysis of the observed peak is selected depending on the situation and, strictly speaking, must be justified. The process of spectral components decomposition includes operations of the energy scale shift (to compensating for the possible charging of the specimen), subtraction of the background, taking into account the shape of the peak determined by exciting radiation line width, energy resolution of the analyzer etc. Mathematical aspects of spectra processing is described in a number of specialized works. Further on we describe and discuss specific results obtained in situ using XPS method on specimens prepared according to the procedure described above.

2. Analysis of intrinsic InAs oxide formation in glow-discharge plasma

Interest to metal-insulator-semiconductor structures (MIS-structures) based on A_3B_5 compounds continuously grows owing to the high mobility of their charge carriers. However, despite the long research history, the problem of thermodynamically stable and dielectrically hard oxide formation with a low electronic states density at the interface in such semiconductors and InAs, in particular, has yet to be solved.

Direct deposition of SiO_2 films on InAs surface results in the formation of a transition area with a complex chemical composition. When studying such structures, one may observe a considerable hysteresis on capacitance-voltage (C-V) characteristics, high density of surface states and pinning of Fermi level, which makes it impossible to manufacture MIS-devices on their basis (Schwartz et al., 1971; Kovchavtsev et al., 1986; Gurtov et al., 1986). Referring to the development of a new method, i.e. ALD molecular layering (Atomic Layer Deposition), research began trying to improve the parameters of MIS-structures on InAs basis, using Al_2O_3 insulator with high dielectric permittivity. The first results obtained (Li et al., 2008) showed that such approach can be very promising.

So far the most widespread and the best-studied methods of formation of oxide films on the surface of A_3B_5 semiconductors are the thermal oxidation in oxygen atmosphere, anode oxidation in liquid electrolytes and in oxygen containing plasma. The former (Leandersson et al., 2003), requires heating of the sample to high temperatures (above 350°C). In the case of InAs, it causes stoichiometry disturbances in the near-surface zone of the semiconductor and in the growing oxide films because of the predominant evaporation of arsenic atoms and the processes of oxide crystallization. The ratio between As and In oxide phases in the oxide film, determining its dielectric properties, depends strongly on oxidation temperature. At higher temperature, the relative share of metal oxides increases, which results in a deterioration of the insulator electrophysical parameters. Electrophysical characteristics of MIS-structures obtained by this method are unacceptable for atomic layer engineering. Taking into account the restriction of necessarily performing all technological operations at temperatures lower than 200-250°C, one may say that the process of thermal oxidation has no prospects.

Anode oxidation in liquid electrolytes or in oxygen-containing plasma is one of low-temperature methods. Oxide films obtained by anode oxidation in liquid electrolytes are characterized by a diffuse interface between the insulator and semiconductor (Grigorovich et al., 1981), and by strong dependence of the ratio between As and In oxide phases and elemental arsenic on synthesis conditions (on such parameters as electrolyte acidity, anodic current density, sample temperature etc.). Moreover, there is an essential gradient of chemical compound with respect to film thickness observed (Grigorovich et al., 1981; Sorokin & Gatko, 1985), and the structure of InAs oxides inevitably includes unwanted impurities, i.e. the electrolyte components. MIS-structures made on their basis have a number of negative properties: hysteresis phenomena, high density of surface states, unstable threshold voltage and photo-induced memory effect. The problems associated with such method of oxide layer formation are studied in great detail in a number of original works (Kornyushkin et al., 1996; Valisheva et al., 1999; Kovchavtsev et al., 2001).

There is no data in literature on InAs oxidation in plasma. From previous works (Chang et al., 1978; Ahrenkel et al., 1982) studying dry oxidation of GaAs in oxygen plasma, it follows that the method can be used to obtain intrinsic oxides which are homogeneous in depth and close to stoichiometry in their structure. It was found, that in the described oxidation modes with an increase of oxide thickness one may observe increase of thickness of oxide film-substrate transition area and increase in the amount of elemental arsenic at the interface (Chang et al., 1978). It was noted, that the accumulation of elemental As in oxide layers is accompanied by build up of the surface states. Partial improvement of electrophysical parameters of MIS-structures obtained by GaAs oxidation in oxygen plasma was observed after addition of fluorine to the oxidizing environment and after additional annealing of samples after oxidation (Ahrenkel et al., 1982). Thus, the chemical composition of oxide films on the surface of A_3B_5 semiconductors strongly depends on their production process, which finally determines the electrophysical properties of such MIS-structures.

Further improvement of the intrinsic oxide quality on the surface of binary A_3B_5 type semiconductors is possible using low-temperature oxidations in plasma. Such process can be performed at room temperature, in a vacuum chamber, using high-purity oxidizing gases. It offers a great advantage compared with thermal and anode oxidation methods, making it possible to exclude the effects related to stoichiometry disturbance and contamination of the oxide.

In the next section we describe the results of our study of chemical composition and formation kinetics of super-thin (5-10) nm intrinsic oxide films on the surface of InAs in oxygen plasma, in normal and glow discharge plasma, and the effects of cathode and anode sample polarization and influence of gas atmosphere, obtained *in situ* using the XPS method.

2.1 Preliminary sample preparation

Formation of super-thin oxide layers imposes special requirements for purity and surface relief of the initial samples. Chemical finishing of the surface of InAs plates was carried out using the technique described by (Kesler et al., 2010). Before loading into a vacuum chamber the samples were degreased by boiling in toluene for 10 minutes, then treated with a saturated solution of hydrochloric acid in isopropanol (HCl-IPA) for 5-30 minutes at room temperature. Immediately after chemical processing and rinsing in isopropanol, the samples were loaded into ultrahigh-vacuum chamber with working pressure less than 1×10^{-9} Torr.

Chemical composition and oxide films thickness were studied using the XPS method. For XPS spectra excitation we used a non-monochromatic aluminium source with line energy Al K_α of 1486.6 eV. All XPS-spectra were acquired with analyser resolution $\Delta E = 0.7$ eV, which corresponds to Au $4f_{7/2}$ line width at half-height equalling 1.31 eV. The device was calibrated using Cu $2p_{3/2}$, Cu $3p_{3/2}$ and Au $4f_{7/2}$ lines, which ensured precision of the energy position registration not less than 0.1 eV. Signal electrons were registered at the angle of 20° off the normal to the sample surface. Overlapping photoelectron peaks were resolved into separate components (Gauss and Lorentz functions) using XPSPEAK95 Version 2.0 software. The parameters of the form of photoelectronic lines were determined through measurement of freshly made chemically pure standard reference samples: InAs and anode oxide films on InAs surface formed in the alkaline electrolyte.

XPS-spectra of the samples surface at different stages of their finishing cleaning are shown in Fig. 1. After degreasing in toluene there is an oxide film of the surface of samples about 2 nm thick consisting of In and As oxides. Etching in HCl-IPA solution for 5 minutes results in almost complete removal of oxides, but some traces of hydrocarbon impurities, oxygen and chlorine, still remain on the surface. Prolonged treatment does not lead to substantial reduction of residual impurities concentration and, at the same time, monotonous increase in the concentration of elemental arsenic on the surface of samples (Fig. 2) is observed.

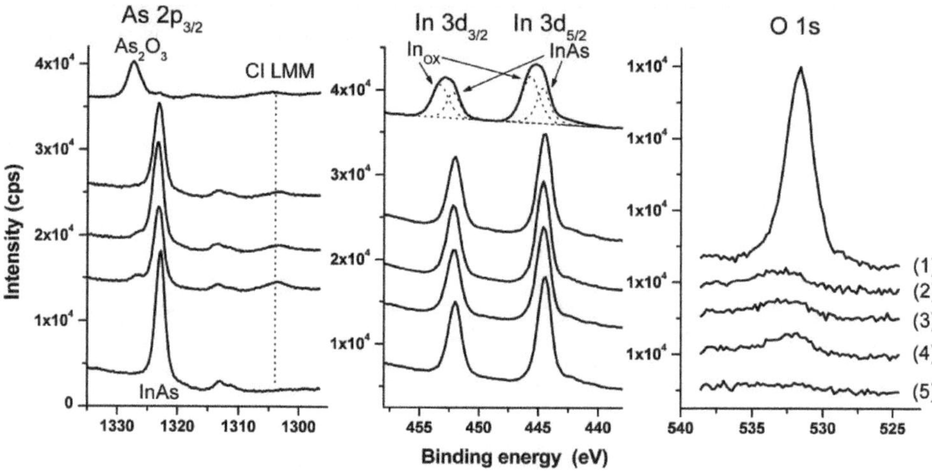

Fig. 1. XPS lines of As, In, and O on the surface of InAs (111)A sample: (1) - after degreasing in toluene, (2), (3), and (4) - after HCl-IPA treatment for 5, 15, and 30 minutes, respectively, (5) - after 60-min of vacuum annealing at 300°C.

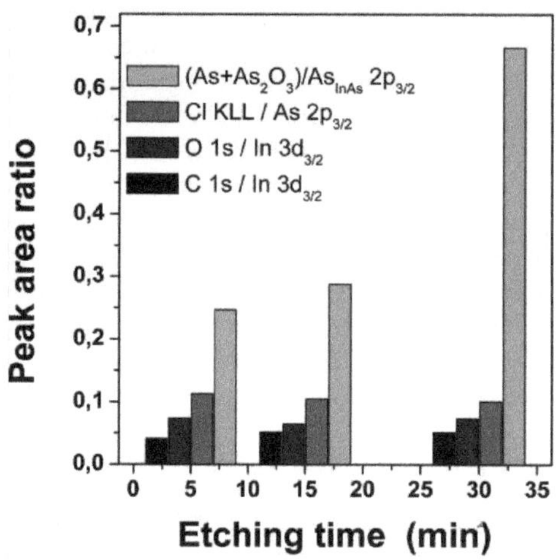

Fig. 2. Peak area ratios observed in the samples treated in isopropanol saturated with hydrochloric acid as a function of etching time.

Such processing results in surface passivation, practically not changing its average roughness value, and also increases lifetime of nonequilibrium charge carriers in the near-surface zone of the samples. However, in this case we may observe formation of local microdefects in the form of "hillocks" consisting mainly of arsenic and its oxides a few nanometers high and several tens of nanometers wide, and with surface density up to 3×10^8 cm^{-2} (Kesler et al., 2010). Chemical reactions which can supply excessive arsenic for formation of such "hillocks" are discussed in (Sun et al., 2008).

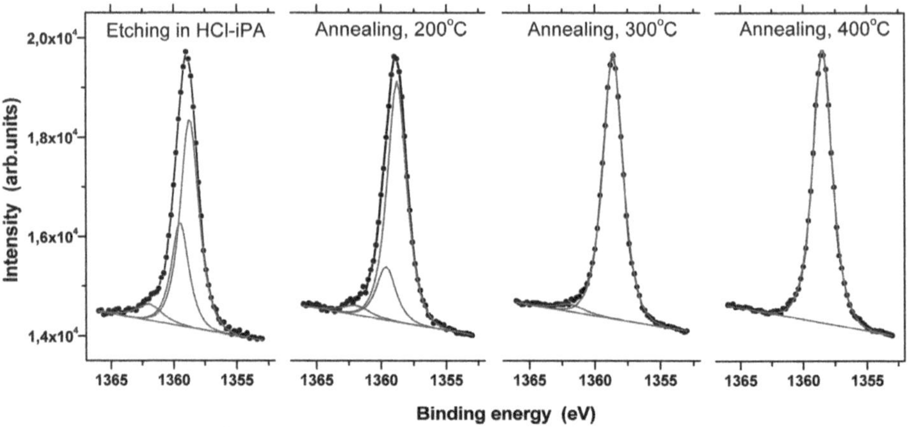

Fig. 3. Detailed spectra of As $2p_{1/2}$ and deconvolution results for the surface of InAs after etching in HCl-IPA solution and vacuum annealing.

Finishing cleaning of the surface of InAs samples (111)A was performed using vacuum annealing. XPS method *in situ* is literally irreplaceable when it comes to monitoring purity of the surface. Since photoelectrons escape depth for As $2p_{1/2}$ line is $\approx$ 0.6 nm, this peak is the most sensitive to the changes in the surface chemical composition. Fig. 3 shows the peak and its decomposition into three components. Main component with binding energy of 1358.7 eV is related to arsenic atoms in InAs volume, and components with energies of 1359.5 eV and 1362.0 eV are related to elemental As and its oxide in the surface layer. The ratio between intensities of the surface and bulk peak components allows us to calculate the thickness of the covering surface layer (Tereshchenko et al., 2006). After the sample is etched in HCl-IPA solution for 30 minutes thickness of the covering arsenic layer is 0.2 nm, which corresponds to 1 monolayer (ML). After 30 minutes of vacuum annealing at 200°C the degree of surface coating with elemental arsenic decreases approximately by 50%, and annealing at 300°C results in complete absorption of excessive As. Complete removal of As oxides and Cl traces is observed at temperatures about 400°C. Hydrocarbon impurities are not completely removed even at a temperature ~ 500°C. Minimal value of carbon peak is 30 % from its initial value, which corresponds to surface coating with carbon ~ 0.1 ML.

Comparison of intensities of As 3d and In 4d photoelectronic peaks characterized by similar photoelectrons escape depth, allows us to estimate changes in stoichiometry of the near-surface zones of the samples. The results of the composition measurements at InAs surface during various stages of vacuum annealing are shown in Fig. 4. Increase in the annealing temperature up to (200 - 500) °C leads to the monotonous decrease of relative content of As and at the same time to a decrease in lifetime of minority charge carriers according to the results of electrophysical measurements.

It is widely thought that the first thing that takes place during the low-temperature annealing of InAs is evaporation of excessive elemental arsenic accumulated at the surface after etching in HCl-IPA. On the other hand, at higher temperatures As is evaporated from the substrate. Vacuum annealing does not result in any significant changes in average surface roughness, but at temperatures exceeding 300°C one may observe complete disappearance of such surface microdefects as "hillocks" (Kesler et al., 2010). Evaporation of elemental arsenic decreases the effects of surface passivation. Similar regularities were found in (Dultsev & Kesler, 2009); the results of studies related to accumulation of As at InAs surface may be found in (Sun et al., 2008; Tereshchenko et al., 2003).

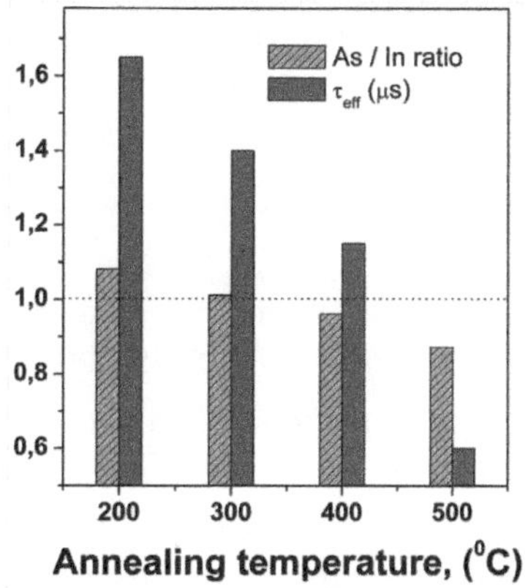

Fig. 4. As / In surface concentration ratio and effective lifetime of minority charge carriers in the subsurface zone of InAs samples vacuum-annealed in the temperature range (200 - 500)°C.

Optimal preparation modes of the surface of InAs (111) are surface etching in HCl-IPA solution for 5-30 minutes and subsequent annealing of samples in vacuum for 30-60 minutes at temperatures about (300 - 350)°C. Higher annealing temperatures cause significant disturbance of surface stoichiometry due to predominant evaporation of As, which leads to the reduction of effective life time of non-equilibrium charge carriers. This method of preliminary preparation ensures high purity, stability and inertness of InAs surface.

2.2 Results of studying InAs oxidation in glow-discharge plasma

Oxidation process was conducted directly in one of high-vacuum modules of the surface analysis installation, which allowed us to control the sample chemical composition immediately after each cycle of plasma processing without any contact with the atmosphere. The examined samples were mounted on a molybdenum holder enabling electric contact to the aluminium auxiliary electrode put under positive/negative electric potential in relation to the frame of the vacuum chamber. During our experiments the sample was put in a stable glow discharge column, which ensured uniformity of surface processing. The experimental procedure is described in detail in (Kesler et al., 2010a; Kesler, 2011).

Total thickness of the films being grown was determined using XPS method by a component of As 3d peak with binding energy of 41.1eV (InAs substrate) assuming exponential reduction of signal intensity with the increase of oxide thickness. In order to estimate the thickness of the film of InAs intrinsic oxide, we used the approach developed in (Seah & Spenser, 2002) based on a measurement of the relation between integral intensities of photoelectronic peaks of oxide films and the substrate:

$$d = \lambda_{ox} \cdot \cos(\theta) \cdot \ln\left(1 + \frac{I_{sub}^{\infty} \cdot I_{ox}}{I_{ox}^{\infty} \cdot I_{sub}}\right)$$

where d is the thickness of the oxide film, I_{sub}^{∞} and I_{ox}^{∞} - integral intensities of As 3d peaks from semi-infinite substrate samples and anode oxide film measured in similar experimental conditions, I_{ox} and I_{sub} - measured integral intensities of As 3d peak components with binding energy ~ 45 eV (oxide film) and ~ 41 eV (substrate), λ_{ox} - (attenuation length) of non-elastic dispersion of 3d As electrons in the film, θ - registered electrons emission angle off the normal to the sample surface. In our calculations we used λ_{ox} = 3.08 nm.

When studying basic regularities and features of InAs oxidation in glow discharge plasma, we analyzed three modes:

1 - the sample was under negative potential in relation to the auxiliary electrode, and voltage in the spark gap was close to glow discharge ignition voltage;

2, 3 - the sample was under positive potential in relation to the auxiliary electrode, and voltage in the spark gap corresponded to the dark and normal glowing discharge, respectively.

Choice of the range from the dark discharge to the beginning of plasma glow is related to the fact that the smallest radiation disturbances of the surface of studied sample are observed in this range. Moreover, plasma of electronegative gases (O_2, CO_2) must show the effect of "sticking" electrons with formation of negative oxygen ions necessary for oxidation of the sample located on the anode.

Using cathode polarization (negative potential on the sample), the sample surface is bombarded by positive oxygen ions with energies determined by a potential difference in the spark gap. Formation of oxide films is accompanied by oxygen ions penetrating deep inside the sample and formation of chemical bonds with the lattice atoms. Fig. 5*a* shows XPS lines of arsenic and indium for InAs (111)A sample after chemical preparation and stage-by-stage processing in oxygen plasma for 10 – 50 minutes.

The surface of the samples is not oxidized in oxygen atmosphere immediately after chemical preparation at pressure of 0.15 torr for a long exposure time (more than 65 minutes) (Kesler et al., 2010a). Binding energy and full width at half-height (FWHM) of As 3d and In 4d peaks is respectively 41.1 (1.7) eV and 17.7 (1.8) eV. After ignition of the glow discharge plasma an additional peak with binding energy 45.4 eV and width ≈ 2.3 eV appears in As 3d spectrum, corresponding to the oxide state of As atoms. Analysis of the peak form shows that it is a superposition of two components with binding energies 45.7 eV and 44.4 eV, identified as states As^{5+} and As^{3+}. At the same time one may observe a shift of In 4d line toward higher binding energies up to 18.2 eV and its widening up to 2.4 eV, indicating appearance of indium oxides. Obtained experimental peaks In 4d can be well described by superposition of just two components with binding energies - 17.7 eV (InAs) and 18.8 eV (indium oxide). Results of our analysis of oxidation kinetics and chemical composition of near-surface zone of the sample during various oxidation stages are shown in Fig. 5*b*. It can be seen, that values of thickness of oxide films on the surfaces calculated from the attenuation of As 3d (substrate) peak and from the ratio between oxidized and non-oxidized arsenic components, are quite consistent. In this case the thickness of oxide films does not really depend on oxidation time and is about 2.6 nm.

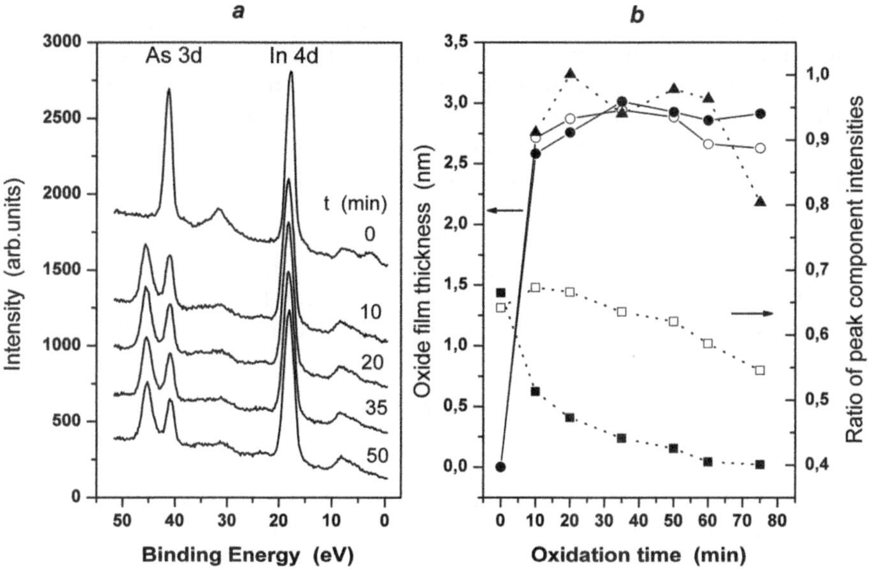

Fig. 5. XPS spectra of As 3d and In 4d in InAs sample during various stages of its processing in oxygen glow discharge plasma at sample voltage of 350 V (a), experimental values of the relation between integrated intensities of peak components and calculated values of thickness of the formed oxide films (b): (□) - total relation of As 3d / In 4d peak components intensities; (■) - relation of peak components intensities of As 3d / In 4d connected with the substrate; (▲) - relation of peak components intensities of As 3d / In 4d connected with the oxide film; (•) – calculation of the total thickness of oxide films from the absolute attenuation of As 3d peak from the substrate; (○) - calculation of the thickness of oxide films from the ratio of As 3d peak components connected with the oxide film and InAs substrate.

Increase in oxidation time results in observed monotonous reduction of the relation between integral intensities of As 3d and In 4d peaks, but the relation between oxide components of those peaks changes only slightly. Most significant reduction in relative arsenic content in relation to indium in the near-surface zone is observed near the boundary between the substrate and the oxide film.

The results of our study of InAs oxidation kinetics when the anode polarization is used (positive potential on the sample) are shown in Fig. 6. In this case the negatively charged oxygen ions take part in oxidation. In the dark discharge mode there is no visible discharge glow, the potential difference in the spark gap is 320V and it is close in its absolute value to the cathode polarization mode described above. In the normal glow discharge mode the potential difference is 500 V, and bright yellow glow is observed.

After discharge ignition, an oxide film begins to form on the surface just as in case of cathode polarization, as evidenced by a change in the acquired XPS spectra for indium and arsenic (Fig. 6 ***a***, ***b***). Peculiarities of the changes in shape and position of As 3d and In 4d lines, observed earlier, are qualitatively repeated. Most evident distinctions are observed in the oxidation process kinetics (Fig. 6 *b*, *d*). In anode polarization mode, with the increase of processing time in plasma, we observe monotonous increase in the total thickness of the film, calculated from the attenuation of As 3d (substrate) peak. Estimations of thickness of intrinsic oxide films on the surface of InAs samples, made by the ratio of oxidized and not oxidized As components, give smaller values. The difference between total thickness of oxide films and thickness of intrinsic InAs oxide grows linearly at the rate of 0.001 $nm \times min^{-1}$ and 0.03 $nm \times min^{-1}$ with time of processing in plasma for discharge voltage +320 V and +500 V, respectively. Growth of intrinsic oxide on InAs films in dark discharge proceeds logarithmically with time (d [nm] $=0.38+0.3 \cdot \ln(t)$), growth of intrinsic oxide film in the glow discharge mode occurs only at initial oxidation stages, then the saturation is attained and an insignificant reduction of oxide thickness is observed. In the glow discharge mode (+320V) the relation between integral intensities of As 3d and In 4d peaks and intensities of components connected with the film and the substrate, does not really depend on duration of plasma processing.

Glow discharge mode (+500V) shows strong dependence between the relation of integral intensities of As 3d and In 4d peaks and their components on oxidation process duration. At the initial stages of InAs oxidation, considerable growth of oxide arsenic component in relation to similar indium component is observed: the relation between As_{ox} / In_{ox} components reaches 2.5. At the same time one may observe a reduction of relative arsenic content in relation to indium in the zone near the substrate - oxide film interface.

Relative share of As^{5+} in oxide goes down during later oxidation stages (Fig. 6*a*), and the ratio As_{ox} / In_{ox} decreases to 1.0. With the increase in InAs processing duration in plasma for over 50 minutes, one may observe the growth of the ratio between arsenic and indium components, related to the substrate, and the increase in As 3d peak width up to ≈3eV, which indicates appearance of arsenic atoms in the oxide film with lower oxidation degree and/or lower content of elemental arsenic. Considerable intensity reduction of this component of As 3d peak with the increase in oxide film thickness in the course of oxidation makes more exact estimations difficult.

Fig. 7 shows XPS spectra of As, In and Al for the samples with an oxide film synthesized using various gases in different plasma modes. Films obtained in oxygen plasma consist of a mixture of indium and arsenic oxides. The contribution of arsenic oxides in them exceeds the contribution of indium oxides, and the contribution of As^{5+} oxide exceeds that of As^{3+} oxide. Addition of NF_3 gas to oxygen plasma leads to a substantial growth of oxidation rate (up to 35 times in case of 100% content of NF_3) and to a strong shift of chemical composition towards formation of indium and arsenic fluorides. All indium in the film is combined with fluorine, and arsenic peak represents superposition of several arsenic oxyfluorides. The composition of the formed film also includes an insignificant amount of nitrogen.

Presence of aluminium peaks in XPS spectra indicates that the material is transferred from the cathode to the anode with the simultaneous oxidation. It explains observed distinctions between the obtained estimates of the total thickness of oxide films and the thickness of

InAs intrinsic oxide (Fig. 6*b*, *d*). Growing film of aluminium oxide acts as a barrier preventing further oxidation during the diffusion of oxygen to the surface of InAs. Deposition rate of aluminium oxide film is strongly suppressed if low discharge voltage is used, and with the increase in gas pressure. In these cases no fast formation of a continuous barrier layer of aluminium oxide is observed, and aluminium is built into the growing oxide film of InAs. This given assumption is confirmed by chemical shift of Al 2p peak (Fig. 7) which depends on the synthesis conditions.

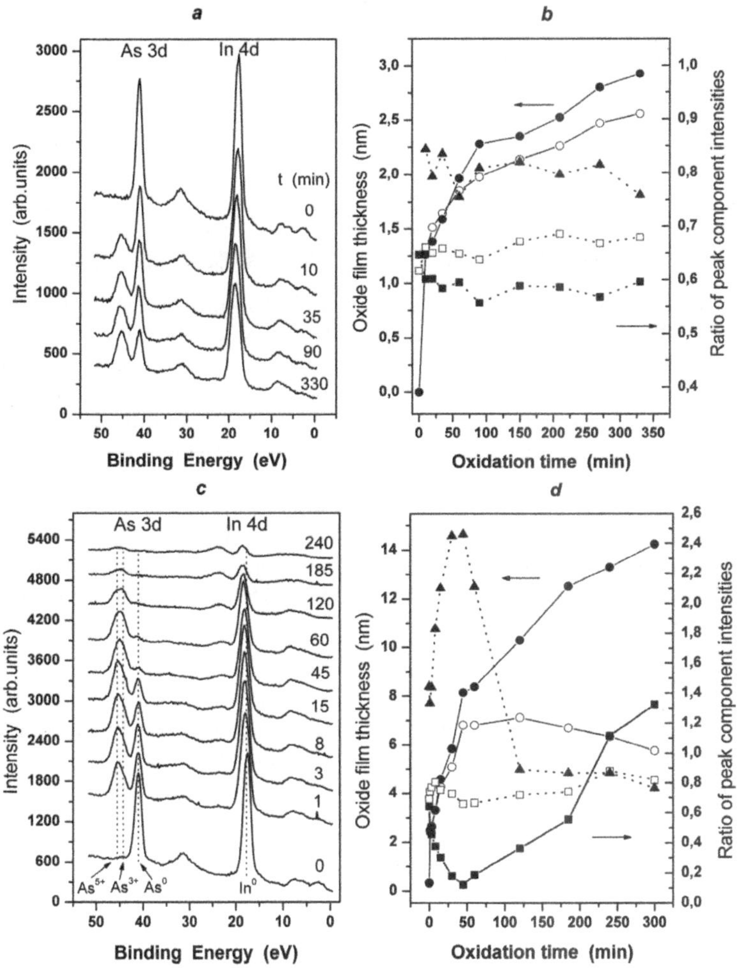

Fig. 6. X-ray photoelectron spectra of As 3d and In 4d from the surface of InAs sample at various oxidation stages in oxygen glow discharge plasma (NF_3 gas content is less than 1 %) at sample voltage +320 V (*a*) and +500 V (*c*). Experimental values of the ratio between integral intensities of peak components and calculated values of oxide film thickness being formed at sample voltage +320 V (*b*) and +500 V (*d*). The designations correspond to those used in Fig. 5.

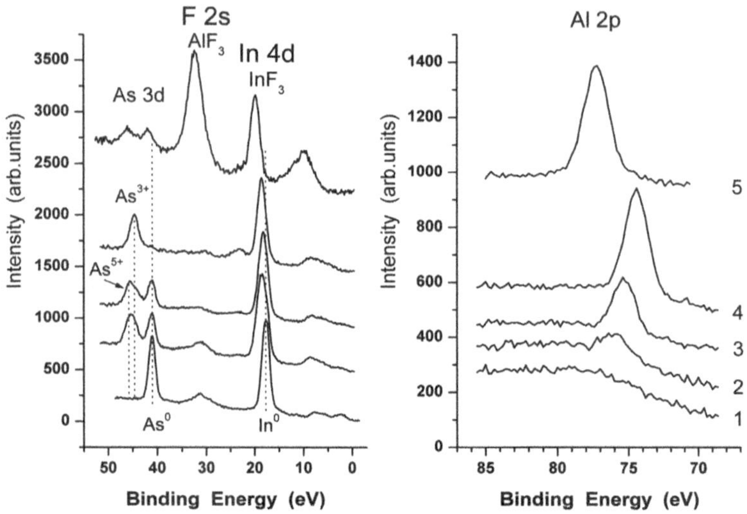

Fig. 7. X-ray photoelectron spectra from the surface of InAs samples: (1) –initial sample after chemical cleaning; (2) – after oxidation in plasma (O_2 + less than 1% NF3) for 330 minutes, at pressure 0.15 Torr, voltage + 320 V, current 0.17 mA; (3) - after oxidation in O_2 plasma (no NF_3 gas in plasma) for 65 minutes, at pressure 0.15 Torr, voltage + 500V, current 1.5 mA; (4) - after oxidation in plasma (CO_2 + less than 1% NF_3) for 2220 minutes, at pressure 0.5 Torr, voltage + 375V, current 1.0 mA; (5) - after oxidation in plasma (O_2 + 7 %NF_3) for 65 minutes at pressure 0.15 Torr, voltage + 500V, current 1.5 mA.

The information on chemical composition and kinetics of oxide films growth on InAs surface, obtained for the first time *in situ* with the help of XPS method, made it possible to find out the following features of oxide films growth in glow discharge plasma using O_2, CO_2 gases and gas mixtures O_2 and NF_3:

1. An oxide film several nanometers thick is formed on the sample surface during cathode polarization as a result of sample bombardment by positive ions of oxygen plasma. Thickness of the oxide film does not really change with increase in duration of surface processing, but the ratio between As/In concentrations both in oxide film and in near-surface zone of the semiconductor decreases. By varying the discharge voltage, it is possible to control the oxide thickness, which indicates the dominating role of ion sputtering in the process.
2. Two-layer oxide films are formed on the surface of InAs during the anode polarization of the sample in a glow discharge mode. The bottom layer, several nanometers thick consisting of arsenic and indium oxides, is formed at the initial stages of oxidation, and later its thickness does not practically change. The top layer – Al_2O_3 serves as a barrier to oxygen diffusion, and its thickness grows linearly with the time of processing in glow discharge plasma. The chemical composition of the growing intrinsic InAs oxide film and the stoichiometry of the near-surface substrate zones depend strongly on the parameters of oxidation process. Voltage reduction in the spark gap and/or increase in gas pressure result in decrease in the growth rate of InAs oxide films, and lead to embedding of cathode material atoms in their oxide form into the film.

3. Use of O_2 + 7 % NF_3 gas mixture or 100 % of NF_3 gas causes sharp increase in oxide film growth rate - 7 and 35 times, respectively. Fluorine causes attenuation of barrier properties of the aluminium oxide protective layer. Indium and arsenic are present in top layers of oxide films even if they are 35-85 nanometers thick.
4. In films generated in oxygen plasma, the contribution of arsenic oxides exceeds that of indium oxides. Addition of NF_3 gas to oxygen plasma results in a strong shift in synthesis towards formation of indium and arsenic fluorides. All indium in the film is combined with fluorine, and the arsenic peak is a superposition of several components –arsenic oxyfluorides. Composition of the generated film also includes nitrogen.
5. Mostly arsenic oxides As_2O_5 are formed at the initial stages of formation of InAs oxide films in O_2 and CO_2 plasma. With a rise in the film thickness, the share of five-valency arsenic oxide decreases, and the share of trivalent arsenic oxide grows.

The conducted research shows the possibility of controllable and precise growth of tunnel-thin oxide films to be used in various microelectronic and nanoelectronic devices. The results described in (Guzev et al., 2008; Kesler et al., 2010a; Kesler et al., 2011) showed that it is probably possible to passivate the surface of InAs with oxide films 3-5 nm thick formed in glow discharge plasma, which opens new prospects for modern device engineering.

3. Formation of platinum silicide on silicon

There are a number of problems in modern production technology of photodetector arrays, sensitive in spectral range not typical for silicon, the primarily used semiconductor material. The systems for signals processing are made of elements mounted on a silicon wafer, and the photodetectors are the areas of other material connected to the substrate in a defined way. Photosensitive elements are sometimes grown in special remaining windows on a wafer, with the electronic circuit already formed on it. And elements need to be formed on an atomically clean substrate surface. It is well known that the intrinsic silicon oxide is removed from the surface only at high temperature, which may deteriorate the properties of the electronic part of the array. One example of similar technology is creation of photodetector arrays on *p*-Si-PtSi Schottky barriers, sensitive in 5-micron range.

The following chemical way of Si processing was suggested, in order to reduce the temperature necessary to obtain atomically clean surface in such windows. The first stage includes operations performed according to the standard RCA technique. The next step is a finishing etching of the rotating wafer, using a solution of fluoric acid in ethanol in the ratio 1:20 in inert gas atmosphere in a hermetic box (Grunthaner et al., 1989). This operation allows "deposition" of hydrogen on dangling silicon bonds on the surface, and the surface is hydrogenised. Then the wafer is loaded into a hermetic container and put into the load lock of the vacuum installation where inert gas is pumped out, and the specimen is transferred to the MBE growth chamber. Presence of hydrogen stops the reaction of silicon oxidation and prevents oxide formation. When such substrate is heated in high vacuum at temperatures about 300°C, Si-H bonds get broken exposing the atomically clean surface ready for further operations.

However, the process of Schottky barrier formation using the above-described way of cleaning, showed strong dependence of electrophysical properties on the time interval between the last operation of chemical treatment and the vacuum processing of the specimen (Grunthaner et al., 1989). Analysis of hydrogenised Si surface showed that Auger and XPS spectra contain peaks corresponding to carbon and oxygen after exposure of the wafer to the atmosphere, but the chemical state of silicon corresponds to a clean material. This means that the presence of Si-H bonds on silicon surface protects it from oxidation, and the registered elements are in a physically adsorbed state. Nevertheless, some aspects of such protection are still unclear, such as the necessity of inert gas after processing, dependence on time of exposure to the atmosphere, etc.

To better understand the mechanism behind this process, we set up the following experiment in our laboratories (Suprun et al., 1995). After hydrogenization of the surface, the specimen was exposed to water vapours, and then loaded into the analysis chamber. With Mg K_{α} line used as the source of radiation, with energy 1253.6eV, XPS spectrum of the initial surface featured peaks corresponding to oxygen and unoxidized silicon. Thus the hydrogen passivation of dangling silicon bonds renders its protective effect on surface. In the subsequent heating of the specimen in analysis chamber Si 2p peak was recorded (Fig. 8). No significant changes in the spectrum were observed up to the temperature about 350°C, and the further heating resulted in oxidation of silicon surface. Fig. 8 shows the beginning and the end at 450°C of the process of Si 2p "flow" from the clean to oxidized state. One may conclude that the desorption energy increases on such "van der Waals" surface stabilized by hydrogen. As a result of specimen heating, the breakdown of Si-H bonds occurs earlier than the evaporation of physically adsorbed water, which leads to oxidation of silicon. It means that in order to obtain clean surface in such a way, it is necessary to exclude the physical adsorption of compounds leading to its oxidation. Moreover, it is recommended to reduce water content in initial reagents as much as possible, for example, by using dehydrated ethanol.

Sensitive elements of photodetectors array on a basis of Schottky barrier *p*-Si-PtSi are formed by deposition of a layer of platinum 2 - 3 nm thick in the prepared windows with subsequent heating of the wafer in high vacuum. Fine-tuning of this technology requires detailed knowledge of the processes taking place during possible formation of platinum silicides - Pt_2Si and PtSi. For this reason we developed a model experiment on silicon substrate, to observe changes in the chemical state of Si-Pt system with heating. Here we need to take into account some features of the XPS method. Photoelectrons during the spectrum acquisition are picked up from an area about 0.5 cm in diameter, and the signal is therefore integrated over that area. This imposes strict requirements over the uniformity of thickness of the deposited layer in the area to reduce additional widening of registered peaks.

To monitor the thickness of nanoscale coatings on monocrystal substrate we used Raman scattering method (RS), registering the spectrum of Raman emission of the substrate (Suprun et al., 1995*a*). Intensity of the emission passing through the deposited layer exponentially falls with its thickness, allowing us to determine its characteristics. The atomic resolution electron microscopy of the structure cross-sections was used for independent calibration.

According to the procedures described above, a Pt layer was deposited on a clean Si surface at room temperature and then the specimen was heated and XPS spectra was acquired. Obtained Pt $4f_{7/2}$ peaks were approximated with Gauss functions and their area, position and width were determined. Such spectrum analysis methodology is rather qualitative, since it does not imply precise decomposition of peaks into components, but nevertheless it is fairly informative.

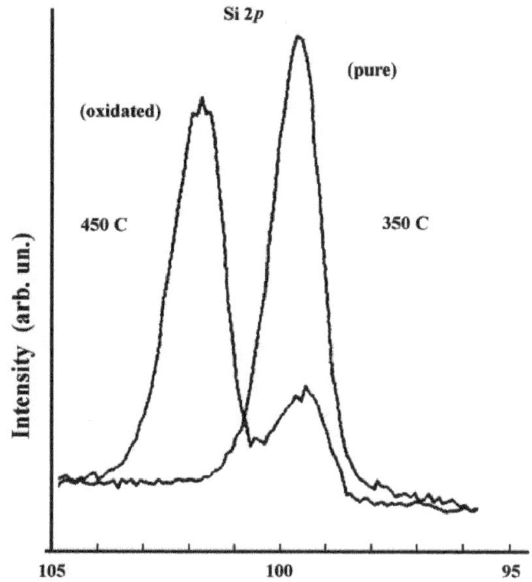

Fig. 8. Changes in XPS spectra of the sample surface during annealing at 450°C.

The inset in fig. 9 shows the evolution of Pt 4f peaks for a specimen with a layer 15.6 nm thick during its heating. It is clearly seen that the spectrum with time (each spectrum acquisition took 6 minutes) is shifting towards lower kinetic energy, which corresponds to an increase in electron binding energy in atom, and peaks intensity is decreasing as well. The figure shows the results of Pt 4f $_{7/2}$ peak processing by the position of its maximum and by its width at half-height. The position of the maximum (black circles, right curve) shifts by more than 1 eV. Peak width here (open circles, left curve) is actually a time derivative of the first curve. The obtained dependences reflect the change of Pt chemical state within the analysis depth, transition of the pure metal into silicide, and the peak widening due to registration of element in two different states.

Fig. 10 shows the behaviour of the area under peaks in the spectra of two specimens with average metal thickness of 4.5 and 15.6 nm during annealing: the first one was annealed at 300°C and the second at 400°C. It can be seen that platinum concentration on the surface falls by half during heating, asymptotically reaching 50±5%, which presumably corresponds to the formation of PtSi compound. These curves have different shape determined by the course of reaction in films of different thickness. Reaction at the interface, as it was demonstrated on layers less than 3 nm thick, begins with Pt deposition on the specimen at room temperature, owing to high temperature of the source - the electron beam evaporator (about 2000°C) and exothermic nature of the reaction.

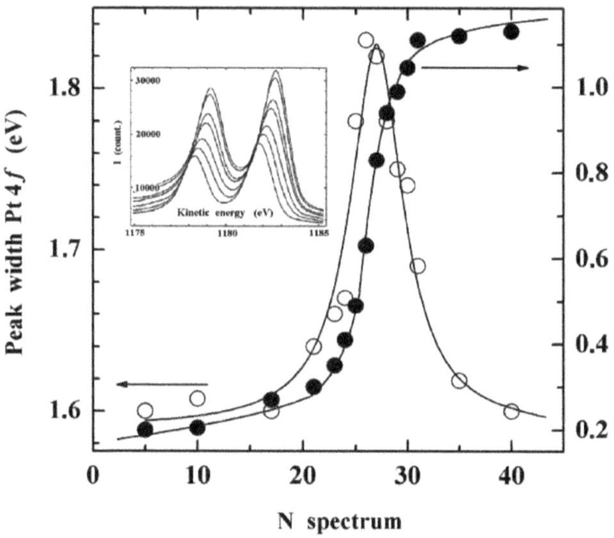

Fig. 9. Changes during the annealing of a 15.6 nm Pt layer at 400°C versus spectrum number: position of Pt 4f peak (right scale, open circles); peak width (left scale, solid circles).

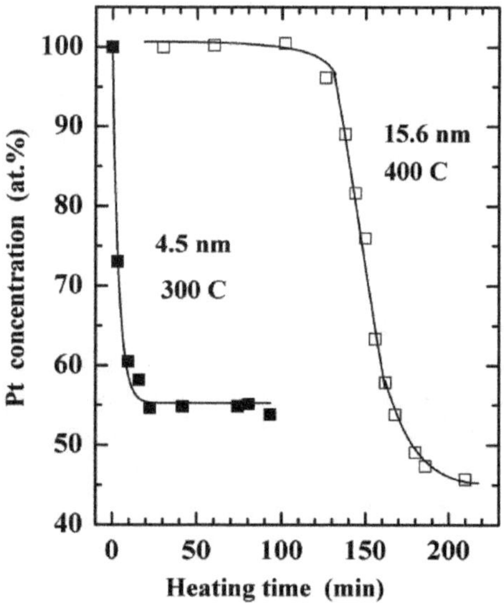

Fig. 10. Platinum surface concentration versus annealing time: 4.5 nm thick Pt layer annealed at 300°C (solid boxes); 15.6 nm layer annealed at 400°C (open boxes).

Subsequent X-ray structure analysis of specimens cross-section showed that resulting polycrystalline film had MnP type orthorhombic structure with the following constants: a = 0.592 nm, b = 0.558 nm, c = 0.360 nm, which corresponds to the lattice parameters of PtSi compound (Suprun et al., 1995).

There is one more interesting moment in this technology, which is not going to be considered here, but should be considered in atomic layer engineering. The point is that the exposition of the obtained compound to the atmosphere changes the chemical composition of its surface due to formation of silicon oxides. Oxygen is a chemically active element, and breaking up of platinum silicide leads to a number of unwanted effects.

Thus, the use of the XPS method gives us a clear idea of chemical reactions on the surface of a solid body, which helps optimise manufacturing techniques of objects with desired properties.

4. Formation of GaAs/Ge heterointerface in the presence of intrinsic GaAs oxides

Application of MBE method in manufacturing of heterostructures means use of chemical reactions in solid phase having very their own specific features. For instance, growth of the epitaxial layer takes place in nonequilibrium conditions. In epitaxy the ratio between the flow of the material from the source, determined by its temperature, and the pressure of the equilibrium vapour of that material over the substrate surface may be more than 10^{10} times higher! One example is the obtaining of Ge epitaxial layers (melting temperature 938°C) on GaAs substrate at substrate temperature (Ts) 200°C. Obviously such estimation of the system deviation from equilibrium conditions is not correct. The matter is that the crystal lattice of Ge is built from sp^3 hybridized atoms, and the life time of such excited state during the evaporation of material according to some estimations reaches 10^{-3} s (Davey et al., 1963). The duration of flight of the adatom or cluster from the evaporator to the substrate in MBE chamber is significantly smaller than this value. Thus, hybridized atoms ready to be embedded into the crystal lattice come to the substrate surface. Their initial energy is not determined by the temperature of substrate, but it will determine the exited state relaxation time. On the whole, stoichiometry of the reaction product in solid phase, which is characterized by crystal perfection of the grown layer, depends on the selected technological mode.

However, there is one more feature related to the formation of the heterointerface. Boundary conditions not only require adherence to stoichiometry, since the excessive component interferes with crystal lattice periodicity, but they also require coordination of atomic dipoles on the boundary (Gleim et al., 2003; Pashley, 1989). All that leads to such effects as "floating" - arrival of not embedded component from the heterointerface to the growing layer surface, spatial degradation of the boundary, which is necessary for coordination of dipole moment, etc. Let us consider reception of heterojunction (HJ) GaAs-Ge as an example.

Nowadays this heterojunction is one of the most exhaustively studied. There is extensive literature reporting results of experimental studies of production processes, properties, and theoretical models describing characteristics of an ideal heterojunction and deviations from them occurring in the presence of different imperfections and defects. Such heterostructure is notable for the same type of symmetry of crystal lattices on both sides of the interface and perfectly matching lattice periods. Distinctions caused by the type of valency bond in GaAs and Ge are studied in great detail both theoretically and experimentally. It was shown that during formation of the heterointerface a transitive mixed composition layer is formed on

GaAs polarized faces, which matches atomic dipole potential. There is a good agreement of theoretical data on the value of the valence band discontinuity in those semiconductors with measured values, with a certain spread in observed data (Franciosi & Walle, 1996; Yu et al., 1992). This may be caused by the fact that when a compound is used as a substrate, the composition on the free surface always differs from the bulk stoichiometry, which complicates the processes of interface formation.

The structure and morphology of the intrinsic GaAs oxide which plays an essential role in the technology of heterojunction production are also studied in great detail. According to the phase diagram, the thermodynamic equilibrium with GaAs under normal conditions is attained by the combination of limiting oxide Ga_2O_3 and arsenic (Torkhov, 2003). In this connection, processes of arsenic oxides reduction and Ga_2O_3 oxide cluster formation (Van Buuren at all., 1991) will be observed in the oxide formed during chemical processing with participation of the monocrystal surface. Substrate heating in vacuum causes desorption of arsenic and its oxides and formation of loose coating with possible lacunas. At the next stage of clearing in the presence of fragments of free GaAs surface one may observe a reaction leading to reduction of valency of Ga in the limiting oxide and desorption of components at temperatures about 500°C (Carli & Bianchi, 1994; Isomura et al., 2007)

$$Ga_2O_3(s) + 4 \cdot GaAs(s) \Rightarrow 3 \cdot Ga_2O \uparrow + 2 \cdot As_2 \uparrow \quad (1)$$

Thus, on GaAs surface during heating one may observe a complex chain of chemical reactions leading to the removal of intrinsic oxides of the material and loss of some stoichiometric arsenic. It was shown that the aforementioned system is very sensitive to the method of oxide preparation and to its initial thickness (Yamada & Ide, 1995). Because of the low pressure of Ga_2O_3 vapours the time (or temperature) necessary for completion of the second stage of surface cleaning, accompanied by reduction of Ga valency, is determined by the initial oxide thickness. It should not be overlooked, that the reaction with participation of GaAs only goes intensively in the presence of free surface which could be not so easily accessible in case of thick oxide. One of the drawbacks of complete removal of GaAs oxides by means of thermal heating in vacuum is certainly coarsening of the surface and its enrichment with gallium. During subsequent deposition of material, it leads to the formation of various defects in the interface zone and "tails" of background impurities penetrating into the growing layer (Kawanaka & Sone, 1990). Therefore the usual procedure is to grow the buffer GaAs layer on the substrate in order to restore the structure and relief of the surface. However, even in this case there still can be a wide spread in electronic properties and electrophysical parameters of the heterojunction, caused by the presence of a superstructure with its own Ga to As ratio on the substrate surface during epitaxy.

Earlier we used a method of scattered laser radiation to study desorption of intrinsic GaAs oxide during its heating in vacuum (Suprun et al., 1997). It was shown that it is possible to obtain island-like Ge films on GaAs substrate in the intrinsic oxide lacunas in case of incomplete surface cleaning. Moreover, using reflection high energy electron diffraction (RHEED), we observed a formation of GaAs-Ge heterointerface under the conditions of incomplete oxide removal from the substrate surface, which is described below.

Standard semi-insulating and n+ GaAs (001) wafers 50 mm in diameter were used as specimens after their chemical processing (Suprun & Fedosenko, 2009). The wafers were fixed on a molybdenum carrier, loaded into the load lock of MBE installation and then vacuumed. After its transfer to the growth chamber the specimen was heated up to 500°C and Ge was deposited at the rate less than 0.5 nm×min^{-1}. RHEED method allowed us to observe formation of the atomically-smooth epitaxial layer with 2×2 superstructure simultaneously with the removal of remaining gallium oxides (Fig. 11). Note that formation of GeO and its desorption do not take place in this case because of the low substrate temperature.

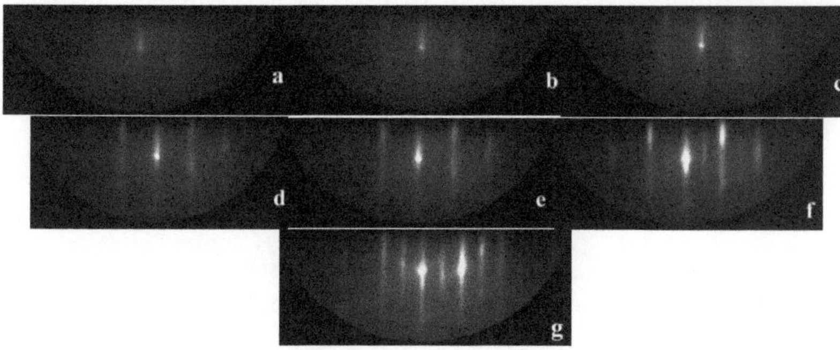

Fig. 11. Successive changes in the diffraction pattern observed during deposition of Ge onto GaAs(001) at 500° C. (*a*) The initial surface in direction [110] before deposition; after the deposition of the Ge layer with a thickness of (*b*) 0.4 nm, (*c*) 1.2 nm, (*d*) 1.5 nm, (*e*) 2.0 nm, (*f*) 2.5 nm, and (*g*) 10.5 nm.

As it follows from Fig. 11, one of the features of the growth process under these conditions is that 2×2 superstructure is formed gradually and finishes when the layer is several nanometers thick. It can be caused by the fact that the growth begins in the oxide lacunas, and it seems as if the picture is "developing" through the diffuse background. By contrast the reconstruction on a completely cleaned GaAs surface is only observed after deposition of 2 germanium monolayers.

Two factors are critical for this mode of epitaxy: the first is the chemical preparation of the specimen during which a thin intrinsic GaAs oxide is formed, and the second is the rate of Ge deposition during formation of heterointerface. Non-compliance with these conditions leads to formation of polycrystalline Ge layer (Suprun & Fedosenko, 2009).

The following experiment was developed to study the mechanisms of the observed process. After chemical treatment, n+GaAs plate (001) was loaded into the analysis chamber. XPS spectra were acquired of GaAs surface before and after heating of the specimen to 500°C. Al K_{α} line with energy of 1486.6 eV was used as the source, spectra were acquired with 0.6 eV resolution and angle of acquisition 45 and 70° off the normal to the plate surface. Analysis depth ratio for these acquisition angles was 2:1, respectively. Figures 12 and 13 show As and Ga spectra, and the distance between peaks, corresponding to bulk components, was 21.9 eV. The following table shows parameters of Ga 3d and As 3d doublets, namely, values of the spin-orbital splitting, components ratio in the doublet, values of full width at half-height of Gauss and Lorentz functions used to fit the experimental spectra, energy shifts of various compounds of Ga and As and their surface components in relation to their bulk components.

Peak	Ga 3d	As 3d
Spin-orbital splitting (eV)	0.45	0.69
Components ratio	1.5	1.4
Gaussian width (eV)	0.6	0.6
Lorentzian width (eV)	0.7	0.7
Surface component S1 (eV)	0.34	-0.43
Surface component S2 (eV)	-0.32	
Chemical shift – As (eV)		0.6
As_2O_3		3.2
Ga_2O	0.4	
Ga_2O_3	1.5	

Table 1.

Values of spin-orbital splitting, components ratio in the doublet and energy shifts for surface components shown in the table were taken from (Lay et al., 1991), and the chemical shifts values ware taken according to (Holloway & McGuire, 1995). Initial state of the surface matches contemporary views on formation of intrinsic GaAs oxide. For example, As 3d peak (Fig. 12 above) can be shown as a combination of peaks of pure arsenic, the component corresponding to As-Ga (***B***) bulk bond, and relatively small peak of As_2O_3 oxide. At the same time initial condition of Ga (Fig. 13 above) is characterized by the presence of Ga_2O, Ga_2O_3 oxides, Ga-As (***B***) bulk components and not so pronounced surface components S_1 and S_2 (this component is not marked in the top part of Fig. 13).

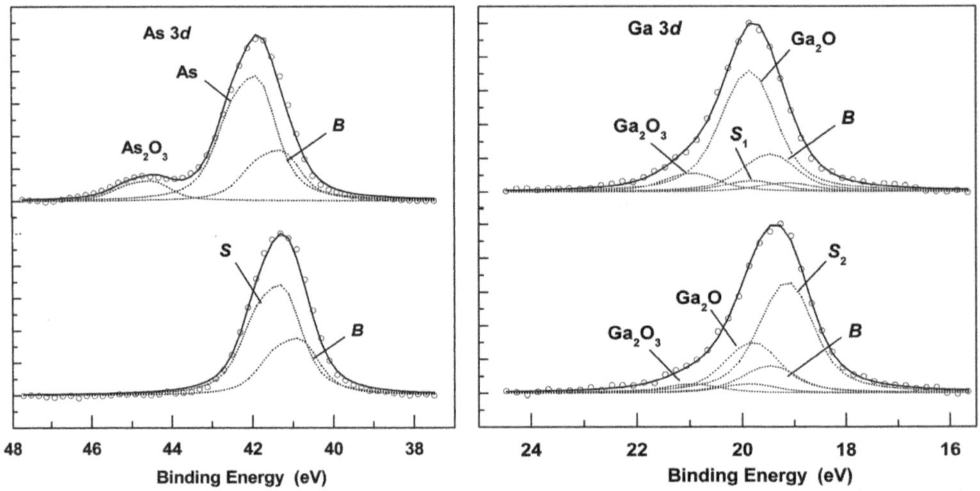

Fig. 12. (right). Spectra of Ga 3d, obtained immediately after chemical processing of the specimen (above), and after its heating to 500°C (below). Symbols on the spectra correspond to the table and Fig. 12.

Fig. 12 (left). Spectra of As 3d obtained immediately after chemical processing of the specimen (above), and after its heating to 500°C (below). Experimental data on the spectra are marked with circles, dashed lines correspond to fitting components, and the continuous line shows the sum of all fitting components. Symbols on the spectra: ***B*** – is a component corresponding to As-Ga bulk bond, ***S*** is a surface component, As is elemental arsenic.

When the specimen is heated to 500°C, one may observe the complete desorption of arsenic and its oxide (Fig. 12 below). The registered peak is the sum of As-Ga volume and surface components. Significant changes are observed in Ga 3d peak (Fig. 13 below). Content of Ga_2O, Ga_2O_3 oxides decreases and surface component S_2 corresponding to low-coordinated Ga becomes dominating in the spectrum. Presence of gallium with dangling valency bonds, hence being in chemically active state, should facilitate the reaction development according to formula (1). If at this time we begin deposition of germanium onto the surface, it, presumably, will lead to binding of released arsenic, which prevents disturbance of surface stoichiometry and its enhancement with gallium.

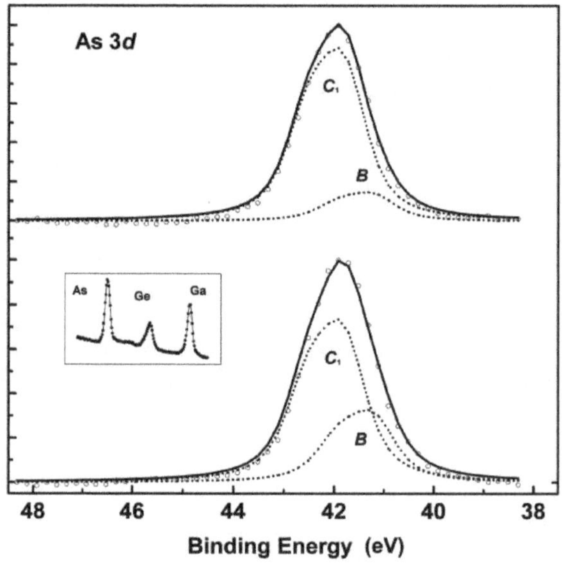

Fig. 13. As 3d spectra obtained after Ge deposition at photoelectrons acquisition angle of 70° (above) and 45° (below) off the normal to the plate surface, the insert shows summarizing spectrum. Components symbols are the same as in figures above.

Analytical installation was equipped with a growth chamber which allowed us to obtain the required structure and to transfer the specimen into the analysis chamber without its contact with the atmosphere. The research *in situ* technique was similar to the one used to study low temperature recrystallization of Ge on ZnSe and is described in (Suprun & Fedosenko, 2007). Ge layer was deposited at room temperature on a substrate with surface structure shown in Fig. 12 and 13 after it was heated up to 500°C. Then the specimen was transferred to the analysis chamber where it was consecutively heated to 500°C, with a step of 100°C, and its surface spectra recorded.

Attention must be paid to significant difference between the situation with the formation of Ge-GaAs heterostructure in MBE installation and the modelling experiment carried out in the analytical installation. Deposition of Ge in MBE installation began at low rate at 500°C after removal of arsenic oxides, and in analytical installation Ge layer about 1 nm thick was deposited on similarly prepared substrate surface at room temperature. Hence, in the second case Ge layer blocked the substrate surface, complicating interaction of the components in the heterosystem and their subsequent desorption. For us it was important to observe basic changes of chemical state of elements in the interface zone and the surface.

Fig. 14 shows As 3d spectra, obtained immediately after deposition of Ge onto the substrate at room temperature. Note, that in contrast to Fig. 12 there is no surface arsenic component, and the integral spectrum only consists of two components. One of them - the bulk - is more pronounced at greater depth of the analysis (angle - 45° off the normal to the surface), another one, (**C_1**) – behaves quite the opposite. Component **C_1** is shifted by 0.6 eV in relation to the bulk component, B, and the observed dependence on the analysis depth indicates that it is located closer to the surface. This component (As on Ge) may be related to a well studied process of Ge surface energy reduction because a part of arsenic comes to the surface from the interface (Monch & Grant, 1980; Bauer, 1983). The reaction is activated due to high energy of Ge adatoms coming from the source during the deposition.

There is one more interesting point: the change of the structure (not the composition) of the surface layer may be traced by As and Ga spectra dependences on temperature: heating of the specimen leads to a sharp increase of volume component at Ts ≈ 300°C if photoelectrons are registered at 45° off the normal to the surface (Fig. 15). This figure shows temperature behaviour of bulk components (***B***) contribution (in relative units) to the total peak area of the respective element at two acquisition angles. Such difference in the behaviour of this component may be due to crystallisation of initially amorphous Ge layer deposited at room temperature. This process enables possible local reduction of Ge thickness at grain boundaries, which leads to increased flux of photoelectrons from the substrate. At a flat acquisition angle – 70° off the normal – this effect is less pronounced due to shadowing of these areas by relief roughness. Subsequent reduction of bulk components contribution to the spectrum reflects smoothing of the surfaces with temperature due to diffusion.

Rise in the specimen temperature occurs simultaneously with the process of reduction of gallium valency in oxide and its desorption from the surface. As can be seen from Fig. 9, the basic component in gallium spectrum is Ga_2O oxide. Registered reduction of this component at 300°C occurs as a result of increased contribution of the bulk component described above. Near 500°C, the process of Ga_2O oxide desorption goes intensively despite the fact that the presence of Ge at the layer surface could complicate the situation.

Fig. 15 (left). Relative area of bulk component (***B***) in spectra of As (circles) and Ga 3d (triangles), measured at 70° (white) and 45° (black) off the normal to the surface as a function of temperature.

Sudden change in the composition of As 3d peak is observed at the same temperature of 500°C: there appears another arsenic compound with germanium. The peak fitting component designated as **C_2**, is shifted by 0.8 eV in relation to the bulk component ***B*** (Fig. 17). The similar state of arsenic is described in (Stucki et al., 1983) for germanium arsenide

(GeAs) monocrystal. It is safe to assume, that the saturation degree of arsenic bonds at GaAs-Ge heterointerface is higher than that on Ge surface, and the value of chemical shift is respectively greater. Absence of the bulk component in the spectrum indicates high thickness uniformity of germanium layer at the final stage of heating.

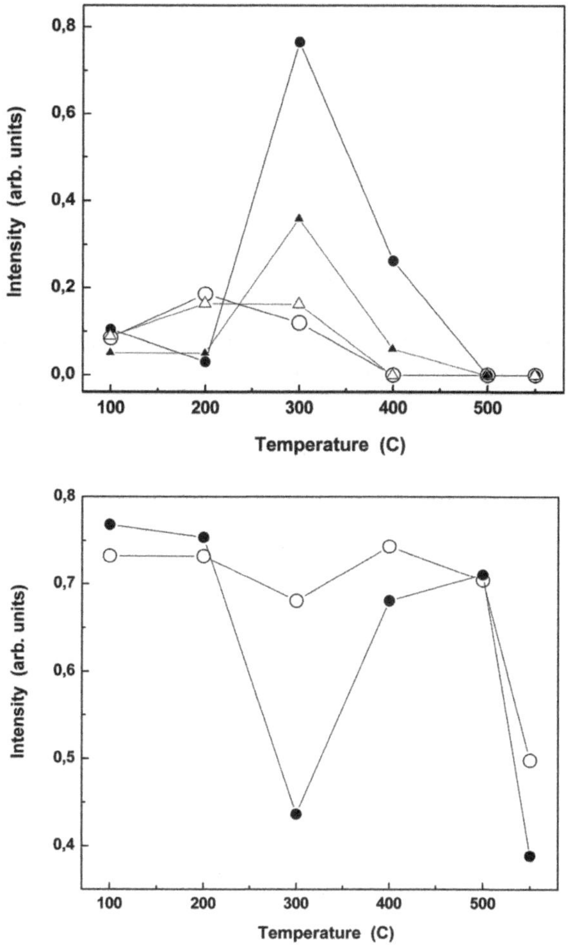

Fig. 14. (right). Relative area of Ga_2O component, measured at 70° (white circles) and 45° (black circles) off the normal to the surface as a function of temperature.

It is informative to compare dependences between charge carriers concentration and thickness of Ge layers grown on the surface of semi-insulating GaAs, stabilized by Ga, and without removal of oxide layer at T_S = 500°C. In the second case, after the formation of the interface zone, the growth rate increased up to 0.3 nm×sec^{-1}. Layers thickness was ≈ 2 micron, then Hall structures were formed on the specimens using photolithography, which were measured during the gradual etching of Ge. Calculations of charge carriers concentration and mobility were made for the etched layer; the results are shown in Fig. 18.

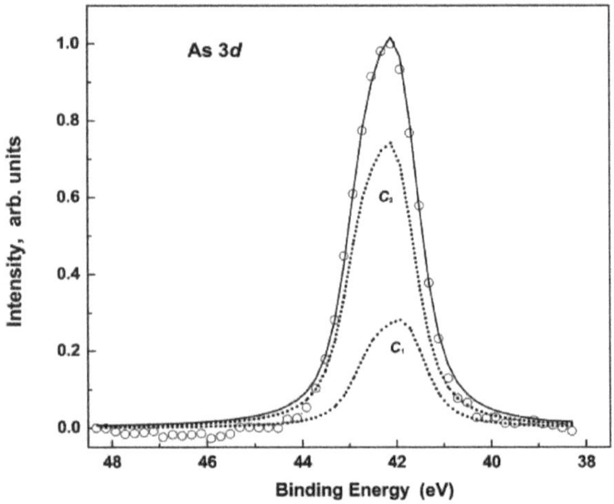

Fig. 15. As 3d spectrum of the surface obtained after deposition of Ge and heating of the specimen up to 500°C, dashed lines show components C_1 and C_2 (As-Ge compounds on the surface and at the heterointerface).

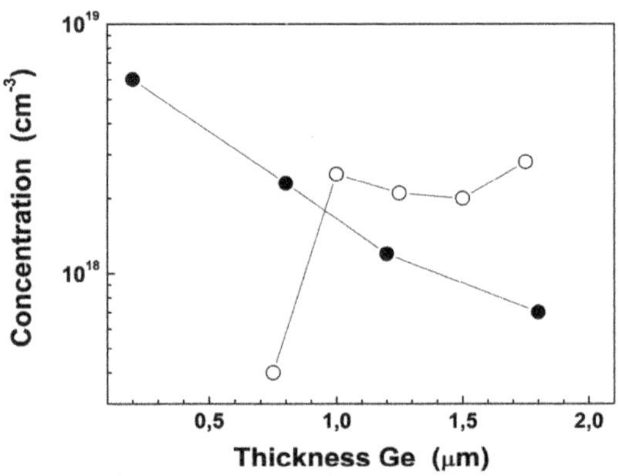

Fig. 16. Dependences between hole concentration and films thickness calculated for etched away layer: black circles - Ge grown on Ga stabilized substrate surfaces, open circles show low temperature epitaxy.

In both cases p-type conductivity layers were obtained, but dependences between concentration and thickness had different behaviour. During the deposition onto substrate surface with the superstructure stabilized by gallium, there is a gradual embedding of the excessive component into the growing layer (Ga solubility limit in Ge is about 10^{20} cm^{-3}). As a result one can observe the impurity concentration profile growing towards the heterointerface. In the second case concentration in the boundary zone is rather small and increases with the film thickness with a rise in the growth rate during epitaxy. This indicates existence of another source of background impurity, which means that the interface cannot obviously be such source. Measured Hall mobility of charge carriers on etched away layer was in the range of 200 - 350 $cm^2 \times V^{-1} \times s^{-1}$ and corresponded to its value in bulk Ge monocrystal with the same impurity concentration. It indicates high structural perfection of the obtained epitaxial films.

Thus, application of XPS method allows us to understand complex phenomena of formation of heterointerface, and to optimize the technological processes. Reliable information often can only be obtained using various research methods supplementing each other. This is why here and further on we will present some results obtained using other techniques, which helps studying the peculiarities of the course of reactions in solid phase.

5. Preparation and analysis of ZnSe/Ge structures on GaAs substrate

Problems similar to those described above also occur during formation of GaAs-ZnSe heterojunction. In our work (Suprun et al., 2009) we described peculiarities of this heterointerface formation, related to ZnSe nutrient being used as a source during the growth. Crystal lattices of both materials in this heterojunction are also of the same type and have very similar constants. The spatial symmetry on both sides of the boundary between GaAs-ZnSe is not discontinued; the coordination number remains the same, it is only the dipole value between the nearest neighbours that changes. This respectively causes formation of a transitive layer from the elements composition (III–VI and II–V), matching the charge distribution at the boundary, as it was described, for example, in (Schmeißer et al., 1999; Pan et al., 2002). In this case the complete lattice period has on average 4 electrons per lattice point. But if there is any deviation from bulk stoichiometry on the substrate surface owing, for example, to the presence of a superstructure, then again there will be a problem with the excessive element embedding.

We used epitaxial layers of ZnSe on GaAs as a substrate in formation of an array of mechanically unstressed quantum dots (QD) of Ge (Neizvestny et al., 2001; Talochkin et al., 2005). XPS method was used for direct observation of recrystallization of germanium amorphous layer. This process is accompanied by the change of coordination number in the examined structure, which characterizes the average number of the dangling valence bonds per atom. In other words, the smaller is the size of the clusters, the greater is the ratio of the surface layer to volume, and the larger is the number of atoms with non-saturated bonds. Recrystallization is usually attributed to the phase changes, but from the XPS method point of view, the observed changes in binding energy of the analyzed element are absolutely similar to those occurring in chemical reactions. We may look at the term stoichiometry in a slightly different way if we mean the saturation degree of valency bonds. For a monocrystal of some compound, for example ZnSe or GaAs, compositional disorders are directly connected with the broken spatial symmetry. For a covalent semiconductor, such as Ge or Si, everything will be the same, though it is not so obvious, since the element composition does not change.

Analysis of physical properties of a material in the form of nanoclusters is now a very topical and often rather complicated problem. With the objects of such a size, the measurement method must be extremely "delicate" in order not to introduce systematic error caused by interaction with the device. In this connection of special interest are works by Chang Sun et al. (Sun, 2007), offering a model of bond-order-length-strength (BOLS) correlation mechanism. BOLS model makes it possible to examine dependences between the size and form of a nanoobject and such values as shift of core levels, width of the band gap, temperature of phase transitions, etc. It is based on the principle that the only factor determining all properties of a solid body is the energy of interatomic interaction; the basic atomic characteristic in the crystal lattice is the coordination number (CN), characterizing the number of nearest neighbours; existence of dangling valency bonds causes CN reduction; the spatial size of the dangling electronic orbital is reduced, which leads to the increase in energy per CN (though the total binding energy of atom falls due to CN reduction). Finally, reduction of the object size leads to higher ratio between volume of the surface shell, being no more than 3 monolayers thick, to the total volume, which respectively increases the role of the surface in the examined physical properties. It is clear, that existence of dangling bonds leads to a change in atom energy in the lattice, including change of the position of internal core levels and all parameters of the lattice caused by the change of binding energy with neighbours. Comparison between calculations and experimental data in those works shows good agreement, and it is offered for a wide range of materials, for which small size clusters were obtained. However, there are still no data for such semiconductor as germanium.

We previously described some methods used to obtain an array of unstrained Ge quantum dots (QD) on heteroepitaxial ZnSe layers. The examined ZnSe/Ge heterosystem is characterized by similar type of the lattice and small mismatch value for their constants, ≈ 0.2%, which excludes occurrence of the mechanical stress in the substrate-nanoobject system. When studying the islets formation in this system, we observed a low-temperature recrystallization of Ge – the transition from the amorphous state to the polycrystalline one, and then into the continuous epitaxial layer (Suprun & Fedosenko, 2007). The temperature was significantly below the melting temperature of the bulk material. Later we conducted experiments *in situ* for direct observation of this process using RHEED and XPS methods, and the results were analyzed in terms of BOLS model.

XPS method was used to acquire the spectra shown in Fig. 19 (the insert shows RHEED patterns, obtained when a similar specimen was heated in MBE chamber). Al K_{α} line was used as the radiation source. From the growth chamber through the air lock the carrier was transferred into the analysis chamber, where spectra were recorded. After deposition of a 1 nm thick germanium layer, the specimen was heated step-by-step, and XPS spectra were recorded for each step. As can be seen from the Fig. 19, the distance between Zn 3d and Se 3d peaks after Ge deposition, and during subsequent heating of the specimen from room temperature to 623 K, changes only slightly, no more than by 0.10 eV, which remains within the error limit of peak position measurements. This finding proves that there is no significant change in chemical bonding between Zn and Se in the presence of Ge. At the same time the initial position of Ge 3d peak after deposition (the curve on Fig. 19, corresponding to the temperature of 293K) is shifted in relation to the position in the bulk material towards higher binding energies. Increase in binding energy of core peaks is typical for nanoscale clusters (Howardet al., 2002; Nanda al., 1999). During further heating of the specimen one may observe significant shift of Ge 3d level by more than 1 eV towards lower binding energies.

To find the dependence of Ge 3d level shift on temperature we chose the position of Se 3d peak as a reference point. Both peaks underwent the background subtraction and were fitted using Gauss function, which allowed us to measure Ge 3d shift with error of ± 50 meV. Fig. 20 shows the dependence of energy difference between Se 3d and Ge 3d peaks on specimen temperature, and the dependence of the value of Ge 3d peak dispersion on temperature. It can be seen, that with a rise in specimen temperature there is a shift of the core level towards lower binding energies, and the curve gets closer to saturation near 450 K. According to the reference data, the difference in binding energy between Se 3d and Ge 3d levels in bulk ZnSe and Ge materials is 25.25 eV. Fig. 20 shows, that peak position even in the final state does not reach the limiting value corresponding to bulk materials, which is due to small thickness of the forming layer.

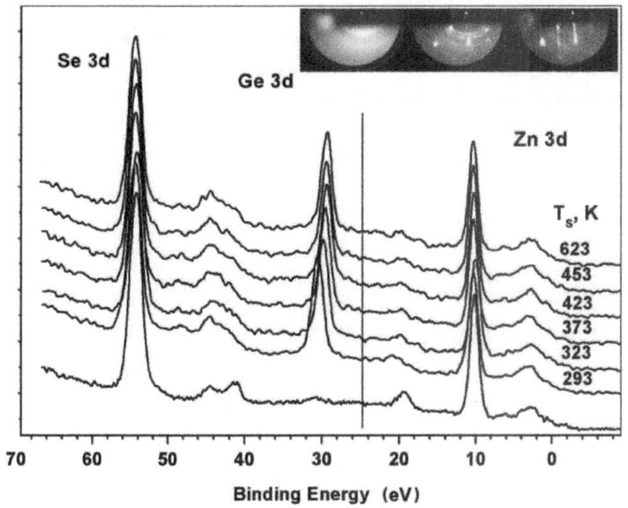

Fig. 17. Changes of XPS spectra of GaAs/ZnSe/Ge surface with temperature, the lowermost - before Ge deposition. On the inset, changes in RHEED patterns are shown, occurring during the heating of the crystal structure of Ge layer ≈ 1 nm thick deposited at room temperature on ZnSe epitaxial film.

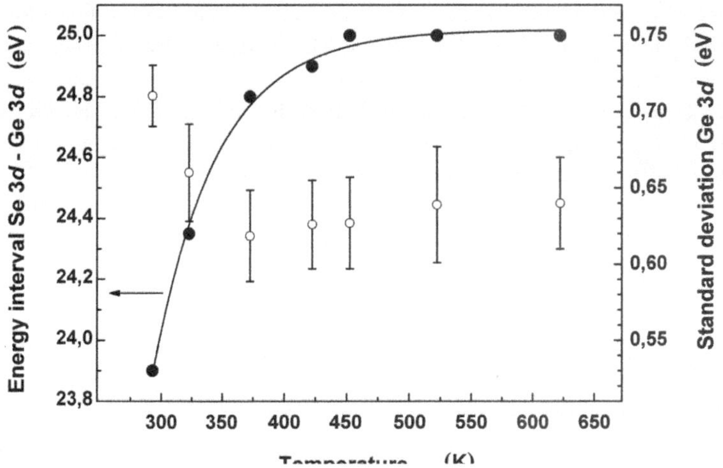

Fig. 18. Energy difference between Se 3d and Ge 3d peaks as a function of the specimen temperature (left), dependence of the value of Ge 3d peak dispersion on temperature (right).

Let's now describe the observed changes of Ge 3d level position in terms of BOLS model, assuming that there are clusters about 1 nm in the amorphous layer and the interaction between the material and substrate is weak.

Value of core level energy shift (Sun et al., 2001; Sun, 2004) can be described by the formula:

$$E_{3d}(\Delta)-E_{3d}(1)=\left[E_{3d}(\infty)-E_{3d}(1)\right](1+\Delta) \tag{2}$$

where $E_{3d}(\Delta)$ is the core level energy for nanoobject; E_{3d} (1) is the energy corresponding to that level position in one isolated atom, E_{3d} (∞) is the level energy in bulk material. Value Δ describes surface contribution, determined by nanoobject type and dimensions. Δ is a product of γ_i factor ($i \leq 3$, layer number counted from the surface), corresponding to the ratio between the surface and the volume for different nanoobjects, depending on their dimensions multiplied by relative change of bond length ($c_i = d_i / d_0$), caused by reduction of CN z_i.

$$\begin{aligned}
&\Delta=\sum_i \gamma_i\left(c_i^{-m}-1\right),\\
&c_i=d_i/d_0=2/\left[1+\exp[(12-z_i)/8z_i]\right]\\
&\gamma_i=\frac{N_i}{N}=\frac{V_i}{V}=\left[4\pi\left((L-(i-1))d\right)^2 d_i\right]/\left(\frac{4\pi}{3}(Ld)^3\right)=\frac{3}{L}\left(1-\frac{i-1}{L}\right)^2\frac{d_i}{d}
\end{aligned} \tag{3}$$

In these equations, L is the number of atoms fitting along radius $L = D/2d_0$ (D is cluster diameter or cross-section of quantum wire, or thickness of continuous layer). Generally speaking, for small size objects it is necessary to take an average d value which will differ from d0. Total atom CN z_0 for Ge lattice (just like for Si and C - diamond) is 12, its value for the top cluster layer depending on radius, and it is determined by the formula: $z_1 = 4\times(1 - 0.75 / L)$.

Notably, fall in CN z from 12 to 8, 6 and 4 leads to relative reduction of bond length by 3 %, 4 % and 12 %, respectively. For ZnSe/Ge system, the difference in lattice constants about $\approx$ 0.2% in heterointerface zone should not considerably influence the characteristics of the layer material. Reduction of coordination number leads to reduction of relative length of ci bond in i-th layer according to (2) and to energy increase on CN z_i. Such change can be calculated, using the adjustable parameter m, which depends on material and is usually 1 for metals and about 4 for semiconductors. It was shown in (Sun, 2004), that good coordination between experimental data and calculations for clusters is obtained when z_1 for the top layer is defined according to equation (3), and for continuous layer with a flat surface $z_1 = 4$, and CN for the second and the third layers in both cases is taken as $z_2 = 6$ and $z_3 = 12$.

Relation between shifts of core Ge 3d level at the beginning and at the end of recrystallization process will be:

$$\frac{E_{3d}(\Delta_1)-E_{3d}(1)}{E_{3d}(\Delta_2)-E_{3d}(1)}=\frac{1+\Delta_1}{1+\Delta_2} \tag{4}$$

In our case, as the initial state at temperature of 293 K we considered amorphous layer containing clusters 6 atoms in diameter, which corresponds to the average thickness of the initial layer. The final condition at 623K is a continuous film 3 monolayers thick, taking into account consolidation of the structure during recrystallization and assuming that some large clusters do not melt. Values γ_i for both cluster and layer and d_i / d_0 relation was calculated in a way similar to (Pan et al., 2002; Sun, 2004), for silicon in substitution of Ge bond length is d_0 = 0.1366 nm. Proceeding from the selected initial and final states using adjustment method, we found parameter $m = 3.2$, provided that $\Delta E_{3d}(\infty) = [E_{3d}(\infty) - E_{3d}(1)] = 1.76$ eV, i.e. it corresponds to the energy of 3d level in single Ge atom $E_{3d}(1) = 27.2 \pm 0.1$ eV, with reference value of its position in bulk material $E_{3d}(\infty) = 28.95$ eV.

If we compare the obtained values for Ge with the data for silicon from (Sun, 2004), then for Si $m = 4.88$, and energy difference for Si 2p core level is $\Delta E_{2p}(\infty) = -2.46$ eV. It is obvious, that the value of adjustable parameter for germanium is approximately one and a half times smaller than that for silicon.

Since the recrystallization process of an amorphous layer goes through a polycrystalline state, the cluster size must grow with temperature. This feature was used to check the value of adjustable parameter m. The table below shows $1+\Delta_L$ values calculated for each temperature of the specimen for matching radii L and respective experimental relative change of Ge 3d peak position. Reference value of Se 3d peak position in ZnSe is 54.2 eV, and values of $E_{3d}(1)$ and $\Delta E_{3d}(\infty)$ for Ge were found above from the initial and the final state of the layer. Table 2 shows that we may observe a gradual increase in the average cluster size in the course of recrystallization (first column).

L D/2d	T [K]	ΔE_{3d} (T) = = E(Se 3d)-E(Ge 3d) [eV]	$\frac{E(Se3d)-\Delta E3d(T)-E3d(1)}{-\Delta E_{3d}(\infty)}$	$1+\Delta_L$
3	293	23.90	1.88	1.881
4	323	24.31	1.64	1.582
6	373	24.82	1.35	1.349
7.7	423	24.98	1.26	1.261
8	453	25.0	1.25	1.250

Table 2.

It is safe to assume, that cluster growth passes through coalescence of small nucleating seeds, melting with a rise in substrate temperature. In the framework of the same BOLS model, it is possible to determine the cluster melting temperature, knowing its size and m parameter by the following equation:

$$\frac{T_m(L) - T_{m0}}{T_{m0}} = \sum_{3} \gamma_i \left(\frac{z_i}{z_0} c_i^{-m} - 1 \right) \quad (5)$$

where $T_m(L)$ is the melting temperature for a cluster of radius L, T_{m0} is the melting temperature of bulk material, and other designations are the same. Now, proceeding from the cluster size calculated from 3d level shift, determined using XPS method, it is possible to calculate the melting temperature according to eq (5) and to compare it with the experimental data. As we can see from Fig. 14, the calculated curve 1 is situated approximately 400 degrees higher in the temperature scale, than the obtained points. Closer to room temperature, there is strong deviation from calculations even taking into account the above mentioned shift by 400 degrees (curve 2).

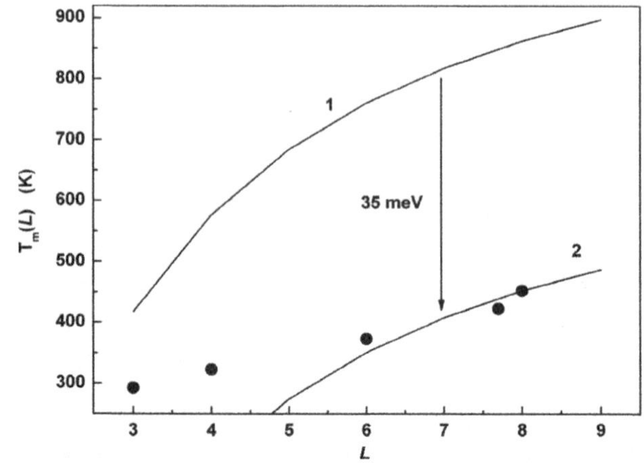

Fig. 19. Dependence between cluster melting temperature and its radius, curve 1 - calculation, 2 - the same dependence shifted in energy by 35 meV, black circles - experimental points.

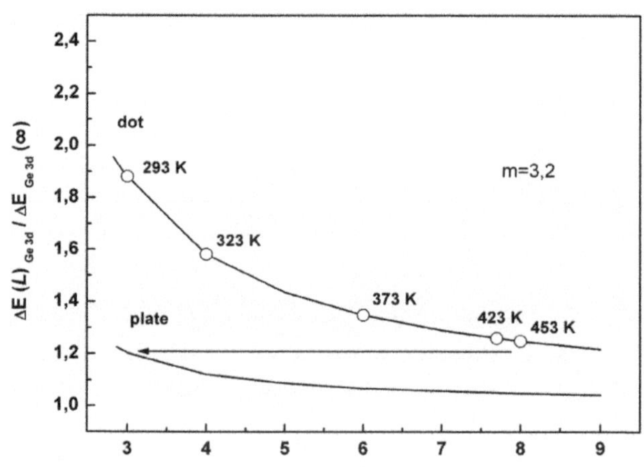

Fig. 20. Calculated dependence of the value of relative shift $[E_{3d}(L) - E_{3d}(1)]/\Delta E_{3d}(\infty)$ of Ge 3d level for cluster (continuous curve) and continuous layer (bottom dashed curve) on characteristic L parameter: radius for a cluster and thickness for a layer. Transition to continuous monocrystal layer during recrystallization is shown with an arrow; specimen temperature is shown over the experimental points.

Such discrepancy needs further explanation. In a previous work (Goldstein, 1995), the method of transmission electron microscopy allowed to observe *in situ* the melting process of Si nanoclusters, prepared according to a special procedure on a thin aluminium foil. Direct observation of changes in cluster forms and their crystal structure at different specimen temperatures via electronic microscope, made it possible to determine their melting temperature. It turned out, that melting temperature of Si clusters about 3nm in diameter was in the range between 50 and 100 °C. This value is lower than that calculated using BOLS model by about 700° (Sun et al., 2001). Experimental data are once again significantly understated in comparison with theoretical predictions. Presumably, the observed deviations in both cases may be explained by physical influence of the "interrogating" radiation. It can be assumed, that part of X-ray quantum energy during external photoemission is spent not only on electron emission, but also to excite crystal lattice vibrations (optical phonon energy is 35 meV in Ge and 63 meV in Si). Such additional heating of the cluster will reduce the observed melting temperature, and deviations will be the higher the smaller the cluster size and the higher the phonon energy are.

In the same work (Goldstein, 1995) coalescence of clusters and absorption of smaller clusters by larger ones was observed during melting. Similar processes should be taking place in our case. Using the described collection of experimental data, the recrystallization of a nanoscale Ge layer may be conceived as follows. With a rise in specimen temperature the influence of the electron (RHEED) or X-ray (XPS) beam causes melting of the smallest clusters, accompanied by their coalescence and enlargement. This process is favoured by a weak energy of Ge bond, with the epitaxial ZnSe layer and respectively low activation energy of adatom diffusion on the surface. After reaching a temperature about 620 K, a continuous epitaxial intermediate Ge layer is formed, in which theoretically there could be Ge clusters with a higher melting temperature than the specified one. This sequence of events is shown in Fig. 22, and is indirectly confirmed by Ge 3d peak dispersion behaviour with temperature. Decrease in peaks dispersion in Fig. 2 indicates increase of the system homogeneity.

Thus, using the XPS method it is possible to observe fairly delicate effects related to a change of physical properties of the material on nanolevel. At the same time, it must always be kept in mind that any measurements are always accompanied with interaction between the device and the object, which introduces a certain error into the research results.

6. Epitaxial layers of CaF_2 and BaF_2 on Si

Epitaxial films of fluorides of alkaline-earth metals with fluorite structure are extensively studied because of their dielectric properties, allowing their application as isolating layers in various electronic structures. For example, such materials as CaF_2 and BaF_2 are used as isolating buffer on Si for subsequent epitaxy of structures with quantum dots and solid solutions of lead chalcogenides. In the latter case, it is particularly topical in production of monolithic photodetectors using silicon technology. In (Hirose et al., 1992) the problem is studied of using CaF_2-SrF_2 fluorides as an electronic monocrystal resist on GaAs. That was possible because the influence of electronic beam causes breakdown of the compound accompanied by evaporation of fluorine. Subsequent metal oxidation and its dissolution make it possible to form the required pattern, and to preserve epitaxial isolating layer in other parts of the structure.

In our work, we studied the mechanism of decomposition of CaF_2 and BaF_2 layers during the surface excitation with a diffractometer electron beam during MBE on Si substrate (Suprun & Shcheglov, 2008). It was shown, that such influence causes formation of such macrodefects as pores. The degree of their development depends on doze, i.e. it is defined by the amount of received charge. It is accompanied by dissociation of the material under the influence of high-energy electrons, fluorine desorption, and transport of positive metal ions from the electron charge drain area. This phenomenon must be considered to exclude formation of defects in the working area when trying to obtain isolating layers. Moreover, this effect can be used in electronic lithography excluding contact between the produced structure and the atmosphere on intermediate stages. In this case, epitaxy combined with electronic scanning will allow immediate formation of the required pattern on the substrate, and preservation of epitaxial protective dielectric in other places. The next step could involve growth of the required layer in preliminary irradiated areas, doping, etc. without specimen exposing to the atmosphere.

Si-CaF_2 and-BaF_2 heterosystems have a few more peculiarities, related to the interface formation and which have not been satisfactory explained so far. It is clear, that some defects - stoichiometry irregularities, formed in the zone of transition from crystal lattice of one material to another, become electrically active centres. They can develop during observation of different electrophysical characteristics. The statement looks quite obvious: the more defects, the worse. More specifically, the greater is the mismatch in lattice constant in the heterojunction, the higher is the density of electrically active states in the interface. Let us consider an example which contradicts such simplified approach to the problems related to physical and chemical processes of heterointerface formation and its electronic properties.

Si-CaF_2 heterosystem has been quite extensively studied, which is true both for its formation using the MBE method and its electrophysical properties. Existing literature describes results of studying the behaviour of capacitance-voltage C-V characteristics of MIS-structures with CaF_2 films grown at different temperatures (Cho et al., 1992). It was shown, that the value of capacity modulation of the obtained structures depends on the temperature of CaF_2 deposition. The higher was the substrate temperature during the growth of epitaxial layers in the interval (300 - 700)°C, the narrower was the range of capacity change from voltage, and no modulation at all was observed in specimens obtained at $T \approx 700°C$. Analysis of electrophysical data and XPS results allowed authors of (Ioannov-Sougleridis et al., 1998; Olmstead et al., 1986) to draw a conclusion, that the reasons of the observed effect are physical and chemical changes in Si-CaF_2 heterointerface. Partial dissociation of CaF_2 molecules occurs during deposition, and the intensity of the process grows with a rise in the substrate temperature. This is accompanied by a reduction of Si-F bonds density at the interface, which can be observed as decrease in the intensity of high-energy component of F 2p doublet in XPS spectra. Thus, different types of bonds can be formed at the heterointerface - Si-F ionic type bonds, mainly at low temperature, and relatively weaker metal bond Si-Ca at high temperature and lack of fluorine. The authors of the aforementioned works offered a detailed analysis of Si 2p peak components. It was shown, that its chemical shift by 0.83 eV towards higher binding energies in relation to bulk component (Si-Si bond) takes place during formation of Si-F bond, and shift by 0.36 eV towards lower binding energies occurs in the presence of bond with calcium.

The results of our experiments (Suprun & Shcheglov, 2008) with Si-CaF_2 system are in good agreement with other well-known data. Besides, we noted improvement in layers adhesion with a decline in deposition temperature, which is consistent with the assumption that predominant type of bonds in the heterointerface zone changes with temperature. As for Si-BaF_2 heterosystem, there is a rather small number of works devoted to its analysis. This is caused, first of all, by a very large mismatch in lattice constants of these materials ≈ 14% (for comparison Si-CaF_2 -≈ 0.6%), therefore it seems that there might be problems to obtain good quality interface in this case. Authors of (Belenchuk et al., 2000) analyzed structure and morphology of the surface of BaF_2 films grown with the help of MBE method on Si substrates (001). One of the growth peculiarities here is the formation of an oriented BaF_2 layer (111), which means that there is a change of crystallographic orientation in relation to the substrate. The atomic force microscopy AFM showed the presence of well-developed block borders several microns in size. Electrophysical properties of such structures were never studied. According to our estimations, the density of dangling bonds at the interface from the side of Si, having a smaller lattice constant, will reach values of 3×10^{14} cm^{-2}, and mismatch dislocations form a grid with a step about 3 nm.

The results of our experiments, studying the system Si-BaF_2, allowed us to assume that BaF_2 films on silicon were grown in MBE installation according to the procedure described in (Suprun & Shcheglov, 2008). We used a thermal evaporator as a BaF_2 source, loaded with nutrient prepared from a monocrystal. Deposition rate was 2 – 6 nm per minute. Thickness of the obtained layers was in the range of (0.2 - 0.4) micron. We used Si (001) plates 76 mm in diameter as substrates, initially covered with oxide and undergoing standard chemical preparation before they were loaded into the installation. The substrates were cleaned by heating in vacuum at 850°C until the appearance of the surface superstructure 7x7, then BaF_2 deposition began at temperature of (750 – 780)°C. Diffraction picture in the process of epitaxy corresponded to formation of a block monocrystal. To protect the surface of the obtained film from the atmosphere influence, in some cases we used CaF_2 deposition at 700°C.

To study the chemical structure of the interface, with the help of XPS method, a layer of BaF_2 ≈ 1 nm thick was grown at a substrate temperature 780°C. The spectra of Si-BaF_2 specimen were obtained using Al K_α line as the source of radiation; spectra were acquired with 0.6 eV resolution with analyzer inclination 20° and 70° off the normal to the sample surface in order to vary the effective depth of analysis.

Fig. 23 shows Si 2p peak. Table 1 below shows the spin-orbital splitting values of this peak (components ratio in the doublet is 2:1), as well as values of full width at half-height of Gauss and Lorentz functions used in experimental spectra decomposition, energy shifts for surface component of silicon, and its compounds in relation to its bulk component. The values shown in the table 3 were taken from (Olmstead et al., 1986). The shift value for Si-Ba bond was determined in the course of spectrum decomposition, and slightly differed from similar energy shift for Si-Ca bond.

Si 2p spectra in Fig. 23 (energy position of bulk component ***V*** - 99.01 eV), were acquired at specimen inclination angle 20 (above) and 70 degrees to the normal. Relative contribution of each component (according to the area under the peak) is shown in table 4.Experimental data are shown in fig. 23 by circles, dashed lines corresponding to decomposition components, continuous line being the sum of all components. Designations on the spectra: ***V*** is a component, corresponding to Si bulk bond, ***S*** is a surface component, Si (Ba) is silicon-barium bond. Note, that presence of fluorine in the heterointerface area is within the error of the fitting procedure for the initial experimental spectrum. Comparison of peaks in the top and bottom Fig. 23 indicates that Si-BaF_2 interface is formed owing to Si-Ba bond: intensity of this component sharply increases with reduction of the effective depth of analysis. This conclusion is consistent with a type of spectrum corresponding to F 2p doublet (similar to the one described in (Olmstead et al., 1986)), where the intensity of high-energy components is smaller, indicating a lack of fluorine in the interface zone.

Si 2p	Energy [eV]
Spin-orbital splitting (eV)	0.600
Gaussian width (eV)	0.6
Lorentzian width (eV)	0.7
Surface component S (eV)	0.44
Chemical shift- Si (eV) in bond	
Si-F	0.83
Si-Ba	-0.35
Si-Ca	-0.36

Table 3.

Component	S_k / S_0 (20°)	S_k / S_0 (70°)
Bulk (V)	0.874	0.393
Si-Ba	0.099	0.603
Surface component S	0.026	0.003
Si-F	0.001	0.001

Table 4.

Thus, high-temperature growth of both calcium fluoride, according to the data in the research literature, and barium fluoride (in our experiments) results in the predominant formation of silicon bond with alkaline-earth metal in the interface zone. Of particular interest are the obtained measurement results of the electrophysical characteristics of Si-BaF_2 specimens. It is known, that no capacity modulation was observed in Si-CaF_2 MIS-structures obtained at temperature 700°C; moreover, the mismatch in lattice constant for Si-BaF_2 system is more than 10 times higher. Judging by the conventional opinion, there can hardly be something good to expect in this situation.

We performed measurements of C-V and G-V characteristics using immitance bridge with error in capacity measurement ±0.1%, and of parallel conductivity with error ±1%. Two types of contacts were used in specimens measurement: mercury contact S = 1.6 10^{-3} cm^2 by area, and mask deposited in vacuum contact (Ti + Ag) S = 5 10-3 cm^2 in area. Properties of these structures were measured at frequencies 1.66 kHz, 16.6 kHz and 100 kHz.

Before the formation of a metal electrode with the help of LEF 3 ellipsometer with operating wave length of laser radiation 0.63 micron, we defined refraction index (*n*) and thickness (*d*) of the BaF_2 layer. The spread in refraction index values at different points on the surface was in the range from 1.465 to 1.473, and the spread in layer thickness was from 373 nm to 388 nm. According to reference data (Grigorjeva, 1991) at wavelength of 0.5893 microns refraction index of BaF_2 volume material is 1.4775, and at wavelength of 1 micron it equals 1.4686.Thus, the refraction index of deposited films is consistent with the previously known results. Electric strength of the layers was measured at let-through current 10^{-5} A, and was in the range of $(0.6 - 1.5) \times 10^{6}$ V×cm^{-1}.

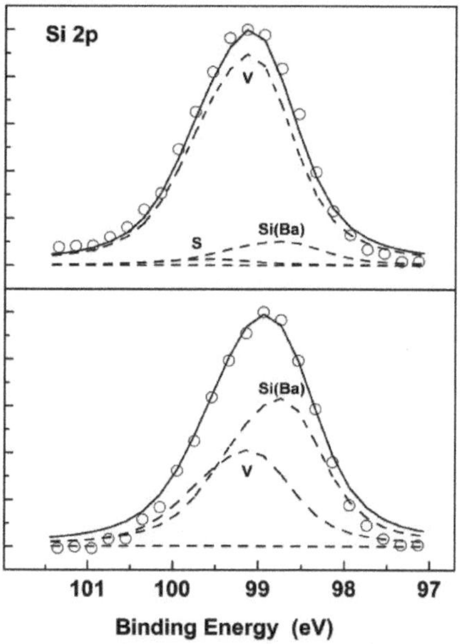

Fig. 21. Si 2p spectra acquired for the specimen inclination angle 20° (above) and 70° (below) to the normal. Components designations are described in the text.

Fig. 24 and 25 show the results of our analysis of Si(001)-BaF_2 specimens (substrate temperature during BaF_2 growth was 780°C) and Si(001)-BaF_2-CaF_2 (substrate temperature during BaF_2 growth was 780°C, temperature during CaF_2 deposition was 700°C and its thickness was 40 nm). High-resistivity silicon with n-type conductivity was used as a substrate.

The characteristics shown were measured during the change of the polarizing voltage from a strong inversion mode of MIS-structures to a strong enhancement mode in the range (–15; +15)V. In the enhancement mode the capacity of MIS- structure, as wellknown, equals the dielectric layer capacity (C_d). Using the ellipsometric measurements data of the thickness of BaF_2 layer and the measured value C_d with the known area of metal contact S, we obtained the dielectric constant for grown BaF_2 layers.

Calculated value of $\varepsilon_{\text{BaF2}} = 6.9$ differs from the value $\varepsilon_{\text{BaF2}} = 7.4$, obtained in [7] for a monocrystal. The difference may be caused by an error in the estimation of the contact area, owing to considerable relief of the surface or peculiar properties of the film itself.

Fig. 24 and 25 illustrate frequency dispersion of C-V dependences connected with states existing at the interface. These states are also characterized by the presence of maxima on G/ω-V dependences. Note, that no peak is observed on G/ω-V characteristics at the frequency of 100 kHz. It proves that density of the states at the interface with the time constant τ, corresponding to that frequency and with smaller values is very low. Thus, C-V characteristics for the studied specimens, measured at 100 kHz can be considered as high-frequency.

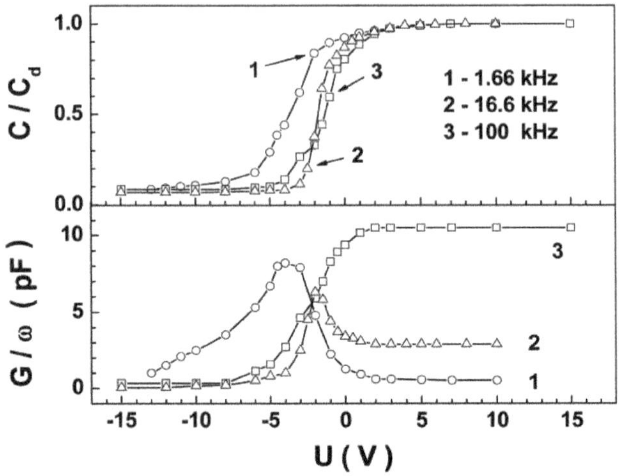

Fig. 22. C-V (top) and G-V (bottom) characteristics of Si(001)-BaF_2-Hg structures, measured for three different frequency values.

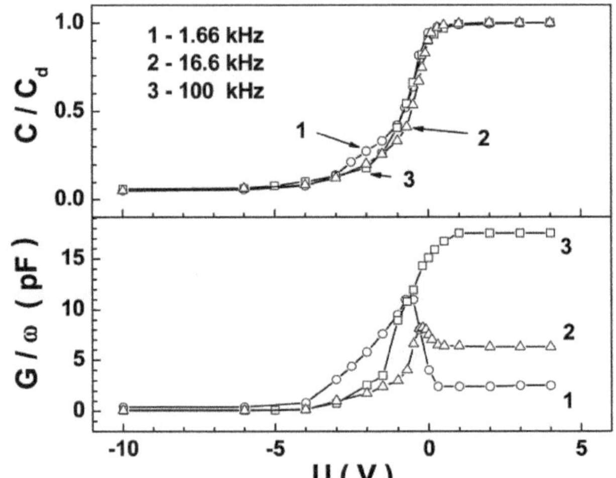

Fig. 23. C-V (top) and G-V (bottom) characteristics of Si(001)-BaF_2-CaF_2-(Ti +Ag) structures, measured for three different frequency values.

From C-V and C-V и G/ω-V characteristics we obtained the density of states N_{SS} at Si/BaF_2 interface. The calculated values of density of state for Si(001)/BaF_2 specimen are as follows: at τ = 100 μsec - N_{SS} = 1.84×10^{11} $eV^{-1} \times cm^{-2}$, at at τ = 10 μsec - N_{SS} = 9.8×10^{10} $eV^{-1} \times cm^{-2}$. For the second specimen at τ = 100 μsec - N_{SS} = 1.1×10^{11} $eV^{-1} \times cm^{-2}$, and at τ = 10 μsec - N_{SS} = 1.07×10^{11} $eV^{-1} \times cm^{-2}$.

The experimental data given above bring to the conclusion, that the capacitance modulation of Si-BaF_2 structure is observed if the predominant bond in the heterointerface is Si-Ba. This case essentially differs from the results obtained for Si-CaF_2 system. Density of dangling bonds at the interface Si-BaF_2, calculated from the difference between lattice constants approximately by three orders of magnitude, exceeds the density of states obtained experimentally on the basis of electrophysical characteristics. Analysis of the interface with the help of XPS method, and the method used to obtain the structure guarantees the absence of any foreign elements which could neutralize dangling Si bonds. It is safe to assume, that high density of spatially ordered defects itself forms a quantum subsystem, but this problem requires additional studies.

7. Conclusion

XPS technique allows to study various processes on a level of the electronic shell structure. Such approach breaks the familiar sight of many concepts, so we can see the "virtual" reference point, which relates to the energy of a pure isolated atom. We seldom realize that in experiments we only can observe reliably transitions between various chemical states of materials. In such transitions, for example, at the boundary between two stoichiometric monocrystals, such a macro-level characteristic as composition is determined by matching dipole moments, i.e. by a condition that has to be met on a micro-level. Application of XPS method also brings to the conclusion that obtaining any experimental data is unavoidably accompanied by an interaction with the analyzed object, which changes its energy state.

To summarize, it is necessary to point out that the discussed experimental results, obtained on many diverse heterostructures, show that XPS in situ technique can be effectively applied to study various phenomena at interfaces and chemical reactions in solid phase.

8. Acknowledgments

The work was supported by the Russian Foundation for Basic Research under Project No. 11-07-12067. The authors wish to thank V.G. Erkov for providing data on electrophysical properties of Si--BaF_2 MIS-structures.

9. References

Ahrenkel, R.K.; Kazmerski, L.L.; Ireland, P.J.; Jamjoum, O.; Russell, P.E.; Dunlavy, D.; Wagner, R.S.; Pattillo, S. & Jervis, T. (1982). Reduction of surface states on GaAs by the plasma growth of oxyfluorides. J. Vac. Sci. Technol., Vol. 21(2), pp. 434-437, ISSN 0022-5355

Bauer, R.S.; (1983). The role of metastable surfaces in determining MBE heterojunction structure: GaAs/Ge interfaces. J. Vac. Sci, Technol. B, 1, p. 314, ISSN 0734-211

Belenchuk, A.; Fedorov A., Huhtinen H., Kantser V., Laiho R., Shapoval O. & Zakhvalinskii V., (2000). Growth of (111)-oriented PbTe films on Si(001) using a BaF2 buffer Thin Solid Films, pp. 358, 277-282. ISSN 0040-6090

Briggs, D. & Seah M.P. (Eds.) (1990). Auger and X-ray Photoelectron Spectroscopy, Practical Surface Analysis, Vol. 1, ISBN-13: 978-0471953401, Wiley, Chichester

Carli, R. & Bianchi C.L. (1994). XPS analysis of gallium oxides. Appl. Sur. Sci., 74, 1, pp. 99-102, ISSN 0169-4332

Chang, C.C.; Chang, R.P.H. & Murarka, S.P. (1978). Plasma-grown oxide on GaAs. J. Electrochem. Soc., Vol. 125(3), pp. 481-487, ISSN 0013-4651

Cho, C.-C. ; Kim T. S., Gnade B. E., Liu H. & Y. Nishioka Y. (1992). Electrical properties of low-temperature-grown CaF_2 on Si(111). Appl. Phys. Lett., 60, pp. 338-340, ISSN 0003-6951

Davey, J.E.; Tierman R.J., & Pankey T. & Montgomery M.D. (1963). The effect of vacuum-evaporation parameters on the structural, electrical and optical properties of thin germanium films. Solid-State Electr., 6, 3, 205-212, ISSN 0038-1101

Dultsev, F.N. & Kesler, V.G. (2009). Etching and oxidation of InAs in a planar inductively coupled plasma. Appl. Surf. Sci., Vol. 256, pp. 246-250, ISSN 0169-4332

Franciosi, A. & Van de Walle C.G. (1996). Heterojunction band offset engineering. Surf. Sci. Reports, 25, pp.1-140, ISSN 0167-5729

Gleim, Th.; Heske C., Umbach E., Schumacher C., Gundel S., Faschinger W., Fleszar A., Ammon Ch., Probst M., & Steinruck H.-P. (2003). Formation of the ZnSe/(Te/)GaAs(100) heterojunction. Sur. Sci., 531, 77-85, ISSN 00396028

Goldstein, A.N. ; (1995). The melting of silicon nanocrystals: Submicron thin-film structures derived from nanocrystal precursors. Appl. Phys. A 62, 1, pp. 33-37, ISSN 0947-8396

Grigorjeva, I.S. & Mejlihov E.Z. (Eds.), (1991). *Fizicheskie velichiny*. Energoatomizdat, ISBN: 5-283-04013-5, Moskva

Grigorovich, S.L.; Volkov, A.S. & Lotsman, A.P. (1981). Study of the structure of anode oxide films based on indium arsenide using the method of Auger electron spectroscopy. Electronic technology. Serial materials, Issue 3 (152), pp. 39-42, ISSN

Grunthaner, P.J.; Grunthaner F.J., Fathauer R.W., Lin T.L., Hecht M.H., Bell L.D., Kaiser W.J., Schowengerdt F.D. & Mazur J.H. (1989). Hydrogen-terminated silicon substrates for low-temperature molecular beam epitaxy. J.Thin Solid Films, 183, pp. 197-212, ISSN 0040-6090

Gurtov, V.A.; Zolotov, M.V.; Kovchavtsev, A.P. & Kuryshev, G.L. (1986). Volume charge in MIS-structures based on indium arsenide. Microelectronics, Volume 15, issue 2, pp. 142-145, ISSN 0544-1269

Hirose, Y. ; Horng S., Kahn A., Wrenn C. & Pfeffer R. (1992). Electron beam patterning of epitaxial CaF_2 and $Ca_{0.5}Sr_{0.5}F_2$/(100)GaAs J.Vac.Sci.Technol. A 10, 4, pp. 960-964, ISSN 0734-2101

Holloway, P. H. & McGuire G.E. (Eds.), (1995). Handbook of Compound Semiconductors, William Andrew Inc., New Jersey, ISBN 0-8155-1374-7

Howard, A. ; Clark D.N.S., Mitchell C.E.J., Egdell R.G. & Dhanak V.R. (2002). Initial and final state effects in photoemission from Au nanoclusters on TiO2(110). Surf. Sci. 518, 3, pp. 210-224, ISSN 00396028

Ioannov-Sougleridis, V. ; Tsakiri V., Nassiopoulou A. G., Photopoulos P., Bassani F. & d` Avitaya F.A. (1998). Electroluminescence from Si/CaF2 Multilayers Grown by Molecular Beam Epitaxy. Phis. Stat. Sol. (a) 165, pp. 97-103,. ISSN 1862-6319

Isomura, N. ; Tsukamoto S., Iizuka K. & Arakawa Y. (2007). Investigation on GaAs(001) surface treated by As-free high temperature surface cleaning method. J. Cryst. Growth, 301-302, pp. 26-29, ISSN 0022-0248

Kawanaka, M. & Sone J. (1990). Current-voltage characteristics of p-Ge/n-GaAs heterojunction diodes grown by molecular beam epitaxy. J. Electronic Materials, 19, 6, pp. 575-580, ISSN 0361-5235

Kesler, V.G.; Seleznev, V.A.; Kovchavtsev, A.P. & Guzev, A.A. (2010). Composition, morphology and surface recombination rate of HCl-isopropanol treated and vacuum annealed InAs(111)A surfaces. Appl. Surf. Sci., Vol. 256, pp. 4626-4632, ISSN 0169-4332

Kesler, V.G.; Kovchavtsev, A.P.; Guzev, A.A.; Panova, Z.V. & Kuryshev, G.L. (2010a). Anodic oxidation of indium arsenide in glow discharge oxygen plasma. Fundamental problems of modern materials engineering, Vol. 7, No. 4, pp. 82-88, ISSN 1811-1416

Kesler, V.G. (2011). Study of InAs oxidation process in glow discharge plasma using the XPS method in situ. Journal of structural chemistry, 52, , No. 8, pp. S156-S164, ISSN 0136-7463

Kornyushkin, N.A.; Valisheva, N.A.; Kovchavtsev, A.P. & Kuryshev, G.L. (1996). Influence of properties of the interface and deep levels in low-energy gap on voltage-capacitance characteristic of MIS-structures on indium arsenide. FTP, Vol.30, issue 5, pp. 914-917, ISSN 0015-3222

Kovchavtsev, A.P.; Kuryshev, G. L.; Kanter, Yu.O.; Olshanetsky, B.Z.; Demyanov E.A.; Krytsyn, S.M. & Stenin, S.I. (1986). Influence of the composition of transitive layer on the value of low-voltage hysteresis of voltage-capacitance characteristics of MIS-structures based on indium arsenide. Surface. Physics, chemistry, mechanics, Vol. 10, pp. 132-137, ISSN

Kovchavtsev, A.P. et al. (2001). Photodetectors with charge injection on indium arsenide. In: Matrix photodetectors of infra-red range, V.N.Ovsyuk, G.L.Kuryshev, J.G.Sidorov et al., pp. 10–113, Nauka, ISBN 5-02-031717-9, Novosibirsk.

Lay, G.L. ; Mao D., Kahn A. & Hwu Y. (1991). Margaritondo G. High-resolution synchrotron-radiation core-level spectroscopy of decapped GaAs(100) surfaces. Phys. Rev. B, 43, pp. 14301-14304, ISSN 1098-0121

Leandersson, S., K.; Gothelid, M.; Tjernberg, O. & Karlsson, U.O. (2003). Interaction between oxygen and InAs(111) surfaces, influence of the electron accumulation layer. Appl. Surf. Sci, Vol. 212-213, pp. 589-594, ISSN 0169-4332

Li N., Harmon, E. S.; Hyland, James; Salzman, David B.; Ma, T. P.; Xuan, Yi & Ye, P.D. (2008). Properties of InAs metal-oxide-semiconductor structures with atomic-layer-deposited Al_2O_3 dielectric. Appl. Phys. Lett., Vol. 92, p. 143507, ISSN 0003-6951

Monch, W. & Grant H. (1980). Combined LEED, AES, and work function studies during the formation of Ge : GaAs(110) heterostructures. J.Vac. Sci, Technol. 17, 5, pp. 1094-1100, ISSN 0022-5355

Nanda, J. ; Kuruvilla B.A. & Sarma D.D. (1999). Photoelectron spectroscopic study of CdS nanocrystallites. Phys. Rev. B 59, pp. 7473-7479, ISSN 1098-0121

Neizvestny, I.G. ; Suprun S.P., Shumsky V.N., Talochkin A.B., Fedosenko E.V., Burbaev T.M. & Kurbatov V.A. (2001). Quantum dots of Ge in a GaAs/ZnSe/Ge unstrained heterosystem: fabrication and properties. Nanotech. 12, 12, pp. 437-440, ISSN 0957-4484

Olmstead, M.A. ; Uhrberg I.G., Bringans R.D. & Bachrach R.G. (1986). Photoemission study of bonding at the CaF_2-on-Si(111) interface. Phys. Rev. B 35, pp.7526-7532, ISSN 1098-0121

Pan, L.K. ; Sun C.Q., Tay B.K., Chen T.P. & Li S. (2002). Photoluminescence of Si Nanosolids near the Lower End of the Size Limit. J. Phys. Chem. B, 106, 45, pp. 11725-11727, ISSN 1089-5647

Pashley, M.D. ; (1989). Electron counting model and its application to island structures on molecular-beam epitaxy grown GaAs(001) and ZnSe(001). Phys. Rev. B, 40, 10481-10487, ISSN 1098-0121

Schmeißer, D. ; Böhme O., Yfantis A., Heller T., Batchelor D.R., Lundstrom I. & Spetz A.L. (1999). Dipole Moment of Nanoparticles at Interfaces. Phys. Rev. Lett. 83, pp. 380-383, ISSN 0031-9007

Schwartz, R.J.; Dockerty, R.C. & Thompson, H.W. (1971). Capacitance voltage measurements on n-type InAs MOS diodes. Sol. St. Electron., Vol.14, pp. 115-124, ISSN 0038-1101

Seah, M.P., Gilmore I.S. & Spencer S.J. (2001). Quantitative XPS. Journal of Electron Spectroscopy and Related Phenomena, 120, pp. 93–111. ISSN: 0368-2048

Seah, M.P. & Spenser, S.J. (2002). Ultrathin SiO_2 on Si. II Issues in quantification of the oxide thickness. Surf. Interface Anal., Vol. 33, p. 640-652, ISSN 0142-2421

Sorokin, I.N. & Gatko, L.E. (1985). Influence of fluorine on growth and properties of anode oxide layers of indium arsenide. Inorganic materials, Vol. 21, No. 4, pp. 537-540, ISSN 0002-337X

Stucki, F., Lapeyre G.J., Bauer R.S., Zurcher P. & Mikkelsen J.C., Jr. (1983). Photoemission study of GeAs(01): A model for the As-stabilized Ge surface on GaAs/Ge heterojunctions. J. Vac. Sci, Technol. B, 1, pp. 865-870, ISSN 0734-211

Sun, C. Q. (2007). Size dependence of nanostructures: Impact of bond order deficiency. Progress in Solid State Chemistry, 35, pp. 1-159, ISSN: 0079-6786

Sun, Y.; Pianetta Po-Ta Chen, P.; Kobayashi, M.; Nishi, Y.; Goel, N.; Garner, M. & Tsai, W. (2008). Arsenic-dominated chemistry in the acid cleaning of InGaAs and InAlAs surfaces. Appl. Phys. Lett., Vol. 93, p. 194103, ISSN 0003-6951

Sun, C.Q. ; (2004). Surface and nanosolid core-level shift: Impact of atomic coordination-number imperfection. Phys. Rev. B 69, 045105, ISSN 1098-0121

Sun, C.Q. ; Chen T.P., Tay B.K., Li S., Huang H., Zhang Y.B., Pan L.K., Lau S.P. & Sun X.W. (2001). An extended `quantum confinement' theory: surface-coordination imperfection modifies the entire band structure of a nanosolid. J. Phys. D: Appl. Phys. 34, 24, pp. 3470-3479, ISSN 0022-3727

Suprun, S.P.; Gutakovskiy A.K. Torlin M.A., Fedosenko E.V., Sherstyakova V.N., & Shumsky V.N. (1995). Structural and electrophysical properties of PtSi_Si the Shottky barriers J.Phys. Low-Dim. Struct., 10/11, pp. 125-132, ISSN 0204-3467

Suprun, S.P.; Talochkin A.B., Gutakovskiy A.K. & Shumsky V.N. (1995a). Determination Of Thickness Of Heteroepitaxial Ultra Thin Layers By RAMAN Spectroscopy. J.Phys. Low-Dim. Struct., 1, pp. 59-66, ISSN 0204-3467

Suprun, S.P. ; Torlin M.A., Fedosenko E.V., Sherstyakova V.N., Shumsky V.N. (1997). Characterization of surface beam roughness during molecular epitaxy on GaAs substrates by mesurements of scattered laser radiation intensity, pp. 119-122, Proceedings of ISDRS-97, Sharlottesville, USA, December

Suprun, S.P. & Fedosenko E.V. (2009). Formation of the GaAs-Ge heterointerface in the presence of oxide. JETP Letters, 89, 2, pp. 94–97, ISSN 0044-4510

Suprun, S.P. & Fedosenko E.V. (2007). Low-temperature recrystallization of Ge nanolayers on ZnSe. Semiconductors, 41, pp. 590-595, ISSN 1063-7826

Suprun, S.P. ; Sherstyakova V.N. & Fedosenko E.V. (2009). Epitaxial growth of ZnSe on GaAs with the use of the ZnSe compound as the source. Semiconductors, 43, pp. 1526-1531, ISSN 1063-7826

Suprun, S. P. & Shcheglov D. V. (2008). Effect of an Electron Beam on CaF2 and BaF2 Epitaxial Layers on Si. JETP Letters, Vol. 88, No. 6, pp. 365–369. ISSN 0044-4510

Talochkin, A.B. ; Teys S.A. & Suprun S. P. (2005). Resonance Raman scattering by optical phonons in unstrained germanium quantum dots. Phys. Rev. B 72, pp. 115416, ISSN 1098-0121

Tereshchenko, O.E.; Alperovich, V.L. & Terekhov, A.S. (2006). Composition and structure of chemically prepared GaAs(111)A and (111)B surfaces. Surf. Sci., Vol. 600, p. 577, ISSN 0039-6028

Tereshchenko, O.E.; D. Paget, P. Chiradia, J.E. Bonnet, F. Wiame, and A. Naleb-Ibrahimi, Preparation of clean reconstructed InAs(001) surfaces using HCl/isopropanol wet treatments, Appl. Phys. Lett. 82 (24) (2003) 4280

Tereshchenko, O.E.; Paget, D.; Chiradia, P.; Bonnet, J.E.; Wiame, F. & Naleb-Ibrahimi, A. (2003). Preparation of clean reconstructed InAs(001) surfaces using HCl/isopropanol wet treatments. Appl. Phys. Lett., Vol. 82 (24), p. 4280, ISSN 0003-6951

Torkhov, N.A. ; (2003). Formation of a native-oxide structure on the surface of n-GaAs under natural oxidation in air. Semiconductors, 37, 10, pp. 1177-1184, ISSN 1063-7826

Valisheva, N.A.; Levtsova, T.A.; Logvinsky, L.M.; Kovchavtsev, A.P.; Kuryshev, G.L.; Petrenko, A.Z. & Petrenko I.P. (1999). Influence of fluorine on the interface anode oxide-indium arsenide. Journal of Surface Investigation. X-Ray, Synchrotron and Neutron Techniques, No. 11, pp. 53-58, ISSN 0207-3528

Van Buuren, T. ; Weilmeier M.K., Athwal I., Colbow K.M., Mackenzie J.A., & Tiedje T. (1991). Oxide thickness effect and surface roughening in the desorption of the oxide from GaAs. Appl. Phys. Lett., 59, 4, pp. 464-466, ISSN 0003-6951

Yamada, M. & Ide Y. (1995). Anomalous behaviors observed in the isothermal desorption of GaAs surface oxides. Sur. Sci., 339, pp. L914-L918, ISSN 00396028

Yu E.T., McCaldin J.O. & McCaldin T.C., Band offset in semiconductor heterojunctions. Solid St. Phys., 46 (1992), pp. 1-146, ISSN 0-12-607746-0

Part 7

The Role of Stoichiometry in Energy Production

14

Chemical Transformations in Inhibited Flames over Range of Stoichiometry

O.P. Korobeinichev, A.G. Shmakov and V.M. Shvartsberg
Institute of Chemical Kinetics & Combustion,
Siberian Brunch of Russian Academy of Sciences, Novosibirsk,
Russia

1. Introduction

Addition of chemically active compounds to flame, which are able to change the flame velocity, the flame propagation limits and the other macrokinetic parameters, seems to be the most effective way to control combustion. Of special interest are chemically active inhibitors producing a noticeable effect on flame at low concentrations, which do not change the flame stoichiometry. Thousands of elementary reactions involving hundreds of species proceed in hydrocarbon flame. However, the key reactions are those involving atoms and free radicals; with their reaction rates being much faster than those of the other reactions. The inhibitors mainly interact and affect the above processes.

2. Methods and approaches

2.1 Flame structure

Knowledge of the chemistry and mechanism of combustion at the molecular level makes it possible to create combustion models capable of predicting many combustion characteristics important for practice, such as the rate and completeness of combustion, temperature and composition of the products under various conditions, and also to control the process of combustion by means of selecting conditions that would ensure the required characteristics of combustion. The flame structure is the main source of information on the chemistry of combustion. One of the most effective method for studying the flame structure is probe mass spectrometry, a versatile method capable of (1) identifying the components present in the flame, (2) determining their quantitative composition (concentrations), and (3) measuring these concentrations in the combustion zone (examine the structure of these flames). An efficient approach to studying combustion chemistry is to combine experimental methods with numerical simulations within the framework of a detailed kinetic mechanism. This combination makes it possible not only to understand the chemical mechanism of the process, but also to develop a kinetic model and constantly refine it by comparing its predictions with experimental data. At present, this approach is widely used in combustion research, including its application to studying the chemistry of combustion of organophosphorus compounds (OPCs). The most efficient method for determining the chemical structure of flames is the probing molecular-beam mass spectrometry (MBMS) with soft ionization, a technique that makes it possible to detect atoms, radicals, and labile components in the flame. It was successfully applied to studying the combustion of OPCs at the Cornel University (USA) (Werner & Cool, 1999) and at the Institute of Chemical Kinetics and Combustion of the Siberian Division of the Russian Academy of Sciences (Korobeinichev et al., 1996).

2.1.1 Molecular beam mass spectrometry

The best way to analyze the combustion products in situ is to use molecular beam sampling from the flame with the help of a sonic probe that forms a supersonic outflow of products into vacuum, which, passing through a skimmer, transforms into a molecular beam. The expansion of the products quenches chemical reactions in the sampled gases. The skimmer cuts out the central part of the flow, free from the products of possible heterogeneous catalytic reactions on the internal walls of the probe. The composition of the molecular beam is analyzed by a soft-ionization mass spectrometry. The molecular beam method ensures the preservation of the sample during extraction and transport to the analyzer. Mass spectrometry makes it possible to simultaneously detect *in situ* all flame components, a capability that is beyond the reach of any other method. Figure 1 shows the setup with molecular beam mass spectrometric sampling; for details, see (Korobeinichev et al., 1996).

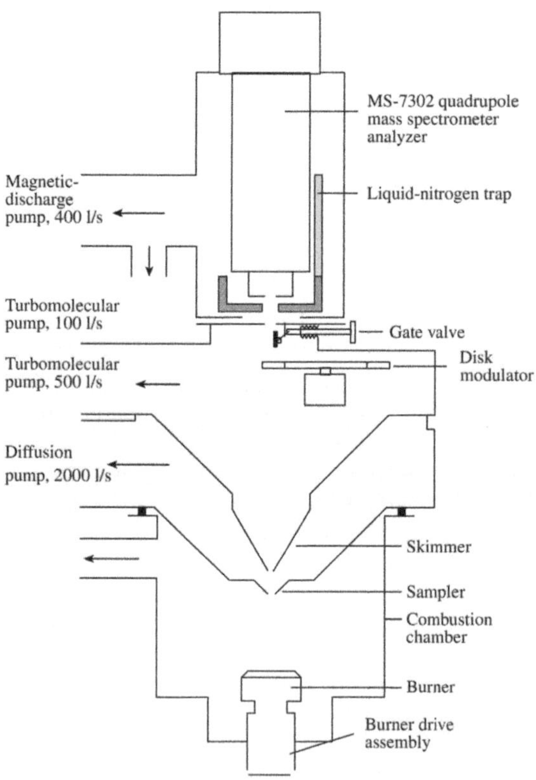

Fig. 1. Schematic of the MBMS setup for sampling gaseous flames.

The setup at the Institute of Chemical Kinetics and Combustion employed soft electron-impact ionization (7-20 eV). The low scatter of electron energies was achieved by using cathode voltage-drop compensation (Dodonov et al., 1990). This technique makes it possible to decrease the fragmentation of ions, an effect that interferes with measurements of the concentrations of atoms, radicals, and other labile species. The ionization potentials of PO, PO_2, HOPO, $HOPO_2$, and $(HO)_3PO$ were determined by measuring the ionization efficiency curves during direct sampling of flames seeded with OPCs. The accuracy of measuring the ionization potentials is determined by the ionizing electron energy scatter, signal-to-noise ratio, and intensity of the signal itself.

2.1.2 Microthermocouple measurements of the flame temperature

The temperature profiles in flames are usually measured with Pt/Pt+10% Rh thermocouples 0.05-0.02 mm in diameter. The ends of the thermocouple are welded to a 0.2-mm-diameter wire fabricated from the same material. Springs provide a steady stretching of the thermocouple and made it possible to prevent it from being deformed in the flame (Korobeinichev et al., 1996). Upon welding the surface of the thermocouple is coated with SiO_2 or Ceramobond 569 (Burton et al., 1992) to eliminate catalytic processes on its surface. The corrections for thermal emission are estimated using the formula from (Kaskan, 1957). To take into account the thermal disturbances introduced into the flame by the probe, the spatial variation of the temperature is measured with a thermocouple positioned at a distance of 0.25–0.30 mm from the probe tip.

2.2 Methods for measuring the laminar flame speed

The laminar flame speed was measured on a Mache - Hebra burner (a modification of the Bunsen burner). A quartz tube with a converging nozzle at the end appears as the burner. Such a nozzle is needed to make the visible flame contour take the shape of a regular cone. The laminar flame speed was calculated from the measured flow rate of the combustible mixture and the surface area of the flame cone. The size of the flame cone was identified with its luminescent contour or by shadow photography method. The error in determination of the flame speed by this method is 5% (for stoichiometric methane - air flame). The laminar flame speed was also measured by using the heat flux method (De Goey et al., 1993; Van Maaren et al., 1994), which makes it possible to determine this parameter with a high accuracy (±1% for a stoichiometric methane - air flame) over a wide range of compositions of the combustible mixture. The flat burner was a copper disk with small orifices; thermocouples were welded into orifices at various distances from the burner axis. The temperature of the disk was 60°C, whereas the combustible mixture temperature was 35°C. While passing through the orifices, the mixture was heated. By varying the flow rate of the combustible mixture, it is possible to achieve a uniform radial distribution of temperature over the disk surface, a situation that corresponds to the equality of the heat flux from the flame to the burner surface. In this case, the conditions of combustion are close to adiabatic. This means that the velocity of the combustible mixture equals the laminar flame speed.

2.3 Simulation of the structure and speed of a laminar flame

The simulation of the flame structure was performed using the PREMIX and CHEMKIN-II computer codes (Kee et al., 1989a 1989b), which make it possible to calculate the concentration profiles of species in a flame stabilized over a flat burner and the laminar flame speed by using a detailed mechanism composed of elementary chemical reactions and databases of thermodynamic and transport properties. Due to the existence of the heat fluxes from the flame to the sampler and burner, notably at atmospheric pressure, the flame structure was calculated employing the experimentally measured temperature profile. For this purpose, as in (Biordi et al., 1974), we used the data obtained with the help of a thermocouple located near the inlet orifice of the probe. The simplest mechanism of methane oxidation included 58 reactions and 20 species. Of these reactions, 23 belong to the hydrogen oxidation mechanism. This set of reactions was successfully used in modelling the structure of a stoichiometric hydrogen - oxygen flame stabilized over a flat burner at a pressure of 47 Torr (Korobeinichev et al., 1999a, 1999b, 2000) and of lean methane - oxygen flame stabilized at 76 Torr (Korobeinichev et al., 1999c, 2001). The structure and laminar flame speed of methane - oxygen flames at atmospheric pressure was calculated using the GRI 3.0 mechanism (Smith et al., 1999), a more complex kinetic mechanism composed of 325 reactions involving 53 species. The structure of propane - oxygen flames was simulated using 469 reactions involving 77 species (Curran et al., 2003, 2004). The structure of hydrogen - oxygen flames seeded with trimethylphosphate (TMP) and Dimethylmethylphosphonate (DMMP) was determined using a kinetic model of OPCs destruction in flames involved 35 steps. This model was initially developed on the basis of experimental data on the structure of flames seeded with DMMP (Werner & Cool, 1999; Korobeinichev et al., 1996), thermochemical data obtained by Melius, and the mechanism proposed by Twarowski (Twarowski, 1993a, 1993b, 1995); it was tested by comparing the experimental and theoretical results on the structure of H_2 - O_2 - Ar flames seeded with TMP and DMMP (Korobeinichev et al., 1999b, 2000, 2001) at a pressure of 47 Torr and methane - oxygen flames seeded with TMP at 76 Torr (Korobeinichev et al., 2001). This model includes the mechanism developed by Twarowski (Twarowski, 1993a, 1993b, 1995) with modified rate constants from (Korobeinichev et al., 2000). Later, the model was refined by altering the rate constants of six key steps and was successfully applied to calculating the structure of flames at atmospheric pressure. The enthalpies of formation of phosphorus oxyacids were calculated by various quantum-chemical methods (Glaude et al., 2000; Mackie et al., 2002; Korobeinichev et al., 2005). At the same time, the enthalpies of formation of some activated complexes were calculated by varying their structures (intermediate states) for different pathways of the key reactions. As a result, it was found that some of the steps in the mechanism of the destruction of OPCs are nonelementary, consisting of a sequence of elementary transformations (Korobeinichev et al., 2005, Jayaweera et al., 2005). Based on the calculations performed, the authors of (Korobeinichev et al., 2005, Jayaweera et al., 2005) developed a more detailed mechanism of the destruction of OPCs, which, in addition, was capable of describing propane - air mixtures of various compositions at atmospheric pressure. This mechanism consists of 210 reactions and 41 phosphorus-containing species.

3. Phosphorus containing compounds

3.1 Inhibition and promotion. Low-pressure hydrogen flames

One of the most interesting finding, made during investigation of OPCs combustion chemistry, is the promotion effect of phosphorus-containing compounds on hydrogen low-pressure flames.

Hastie and Bonnell (Hastie & Bonnell, 1980) studied the effect of TMP on atmospheric-pressure methane–oxygen and hydrogen–oxygen flames of various types and compositions by using MBMS and optical and spectroscopic methods. Although these authors have not observed flame promotion but inhibition only, they proposed reactions for flame inhibition via H and OH recombination catalyzed by phosphorus-containing compounds. Analyzing the proposed mechanism Hastie and Bonnell suggested that this kinetic scheme can provide promotion effect.

The first experimental observations of a decrease of the flame zone and a rise of the temperature of low-pressure $H_2/O_2/Ar$ flame as TMP was added were made by Korobeinichev et al. (Korobeinichev et al., 1994, 1996) (see Fig. 2). This phenomenon was termed promotion. The singularity of the phenomenon consisted in the following. A decrease in the width of the flame zone and a rise in the flame temperature occurred simultaneously with a decrease of radicals concentration in the flame (Korobeinichev et al., 1999b, 2001). Later it was shown that the promotion effectiveness for a stoichiometric flame at 47 Torr increases with OPCs loading and riches its maximum at concentration of 0.6% by volume (Korobeinichev et al., 2001). A further increase of OPCs loading results in a drop of the flame speed (Fig. 3).

It is noteworthy that Korobeinichev and co-authors (Korobeinichev et al., 1999d) measured concentration profiles of phosphorus oxides and acids PO, PO_2, HOPO, $HOPO_2$, and $(HO)_3PO$ in the flame that are the key flame species responsible for H and OH catalytic recombination. In the same work the authors demonstrated that the same chemical processes are responsible for both promotion and inhibition.

Twarowski (Twarowski, 1993a 1993b, 1995) made a significant progress in understanding the chemistry of catalytic recombination of H and OH with participation of phosphorus-containing flame species. Using a flash photolysis method, Twarowski studied the rate of H and OH recombination in the sampled combustion products (T=2000 K) of $H_2/O_2/Ar/P_2H_6$ flame. A phosphine additive was demonstrated to accelerate H and OH recombination. Twarowski assumed this process to be a recombination of H and OH radicals catalyzed by phosphorus oxides and acids (PO, PO_2, HOPO, and $HOPO_2$). He developed a reaction mechanism for phosphorus-catalyzed recombination. Later Twarowski refined his model (Twarowski, 1996).

A more detailed explanation of the nature of the promotion effect was given by Bolshova and Korobeinichev (Bolshova & Korobeinichev, 2006). The authors simulated the structure and speed of various low-pressure and atmospheric-pressure flames with various equivalence ratios. The authors revealed the influence of equivalence ratio on the inhibition and promotion effectiveness. In particular, the lean $H_2/O_2/Ar$ flames were shown to be promoted most effectively and the promotion efficiency decreased as the equivalence ratio rose. As to inhibition, an inverse dependence was observed: higher inhibition efficiency was observed for richer atmospheric-pressure $H_2/O_2/Ar$ flames.

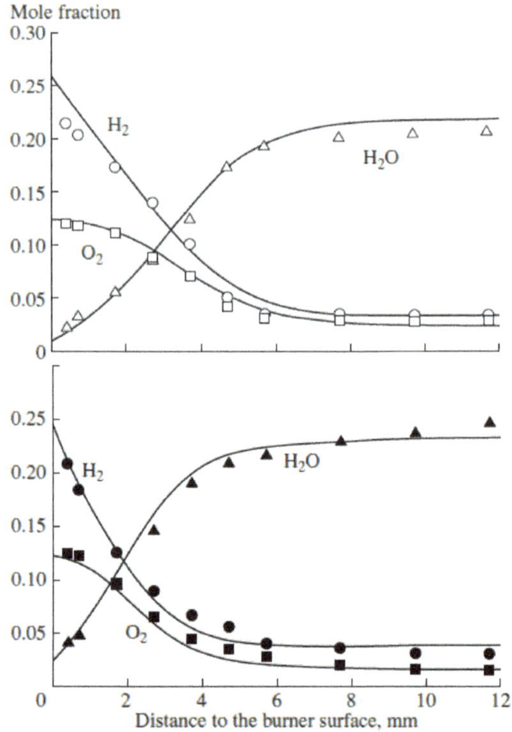

Fig. 2. Effect of a 0.2% TMP additive (lower plot) on the concentration profiles of the stable components of a hydrogen–oxygen flame stabilized over a flat burner at a pressure of 47 Torr; the points and lines are the experimental and simulation results, respectively.

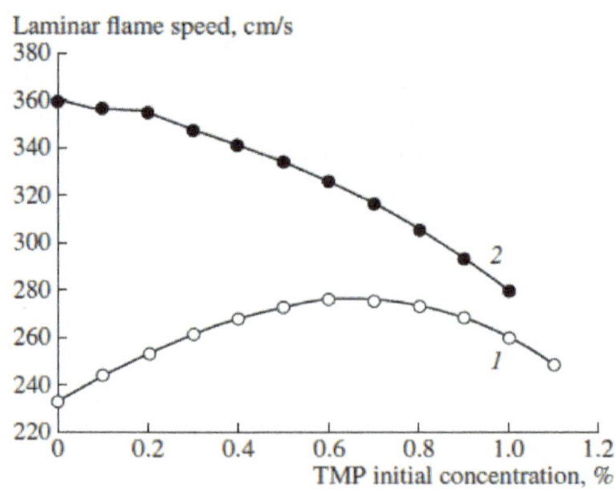

Fig. 3. Dependence of the speed of hydrogen–oxygen–TMP flame on the initial concentration of TMP at (*1*) 47 and (*2*) 760 Torr.

Bolshova and Korobeinichev performed the sensitivity analysis and revealed that a rise of low-pressure $H_2/O_2/Ar$ flame speed, as TMP loading is increased, is accompanied by a rise of the net rate of the chain-branching reaction $H + O_2 = O + OH$. It occurs due to a rise of the flame temperature throughout the flame zone. The flame temperature rise is explained by a release of additional heat due to catalytic pathway for water formation from H and OH. Increasing the additive loading increases both the branching rate and the chain termination rate. However, because of the high concentration of the chain carriers in low-pressure $H_2/O_2/Ar$ flames, the increase in the branching reaction rate dominates over the chain termination, resulting in the growth of the flame speed. In the case of atmospheric-pressure flames, where the radicals concentration is much lower and the catalytic recombination does not result in the flame temperature rise, the OPC addition to flame leads to a decrease in the flame speed.

It is noteworthy that OPCs additive increases the chain-termination rate, thus accelerating approach of the system to thermodynamic equilibrium. Therefore, at certain concentration, e.g. 0.6% by volume for the stoichiometric flame at 47 Torr, the promotion efficiency reaches its maximum.

3.2 Mechanism for inhibition of hydrogen flames at atmospheric pressure

3.2.1 Effect of the equivalence ratio and the degree of dilution with an inert on the speed of a $H_2/O_2/N_2$ flame doped with TMP

At present time a number of kinetic models for flame inhibition and promotion by OPCs are available. The last and the most justified version of the mechanism, which was developed on the basis of experimental results on speed of TMP-doped C_3H_8/air flames and quantum-mechanical calculations (Korobeinichev et al., 2005; Jayaweera et al., 2005) was used for predicting many experimental data including flame suppression (Shmakov et al., 2006), chemical structure of diffusive counterflow (Knyazkov et al., 2007) and premixed (Korobeinichev et al., 2007) hydrocarbon/air flames doped with OPCs. In spite of a satisfactory prediction of speed and structure of lean and stoichiometric flames, the mechanism predicted concentration profiles of labile species in rich flames with lower accuracy (Korobeinichev et al., 2007). To explain a sharp decrease of the inhibition effectiveness of hydrocarbon flames at equivalence ratio >1.2–1.3 and a disagreement between modeling and experiment (Korobeinichev et al., 2007), a formation of inactive P- and C-containing species in rich flames, which are not considered by the model was proposed. Therefore, it was important to check on this suggestion and to validate the mechanism by comparing the measured and simulated speed and structure of atmospheric-pressure $H_2/O_2/N_2$ flames of various fuel/oxidizer ratios.

Figure 4 presents the measured speeds of $H_2/O_2/N_2$ flames versus ϕ (equivalence ratio $\phi=([H_2]/[O_2])/([H_2]_{st}/[O_2]_{st})$, where $[H_2]$, $[O_2]$, $[H_2]_{st}$, $[O_2]_{st}$ - concentration of hydrogen and oxygen in studied and stoichiometric combustible mixture, respectively) in the range from 0.32 to 2.8 with the dilution ratio D ($D=[O_2]/([O_2]+[N_2])$, where $[O_2]$ and $[N_2]$ concentration of oxygen and nitrogen in combustible mixture) in the range from 0.209 to 0.077. The same figure shows data (Egolfopoulos, 1990) obtained using a counterflow burner for a combustible mixture at room temperature and converted to the conditions of our experiments ($t_0=35^0C$). Good agreement between our and literature data indicates the correctness of our measurements. The flame speeds predicted by the model are slightly lower than the experimental data.

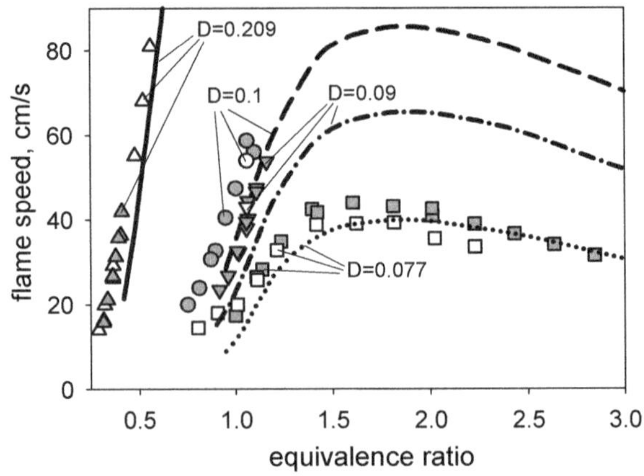

Fig. 4. Speed of $H_2/O_2/N_2$ flames with different dilution ratios (D) versus equivalence ratio; symbols - experiment (gray symbols - (Korobeinichev et al., 2009), open symbols - (Egolfopoulos & Law,1990)), lines - modeling.

The addition of 0.04% TMP (by volume) to the flames leads to a significant decrease in their speed. Figure 5 gives the measured flame speeds (symbols) and those calculated (dashed curves) using the model of (Jayaweera, 2005) versus ϕ for a flame doped with TMP. It is evident from Fig. 5 that the experimental and calculated results differ by a factor of 1.3 to 2. This difference, however, is not due to systematic measurement errors. The observed disagreements between the measured and calculated speeds are different for flames with and without TMP additives: for undoped flames, the flame speed predicted by the model is slightly underestimated, and for flames with the additive, it is overestimated.

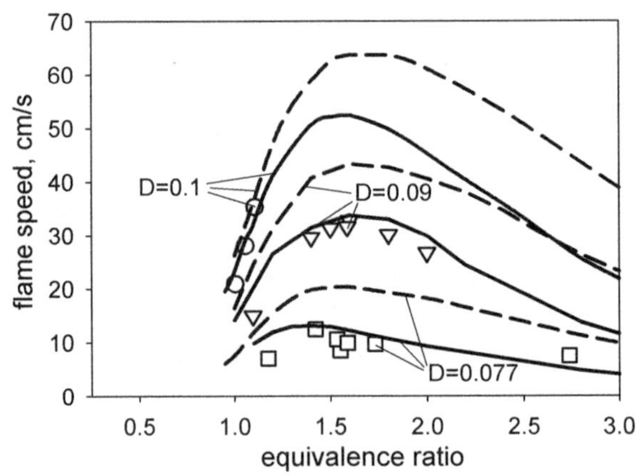

Fig. 5. Speed of $H_2/O_2/N_2$ flames doped with 0.04% TMP with different D versus equivalence ratio; symbols - experiment, dashed lines - modeling using mechanism (Jayaweera et al., 2005), solid lines - modeling using the updated mechanism.

An analysis of the speed sensitivity of a $H_2/O_2/N_2$ flame doped with TMP with respect to the rate constants of the reactions involving P-containing species shows that, in contrast to hydrocarbon–air flames, the $H_2/O_2/N_2$ flame speed is the most significantly sensitive to the primary stages involving TMP and its primary destruction products, which proceed in the low-temperature flame zone. These reactions are given in Table 1. It should be noted that in methane–air and propane–air flames, the flame speed was most significantly affected by the atom and radical recombination reactions involving phosphorus oxyacids, which proceed in the high-temperature flame zone (Korobeinichev et al., 1999b). Because the rate constants of the stages presented in Table 1 have previously been estimated only approximately (Glaude et al., 2000), we changed the pre-exponential factors of their rate constants as is shown in Table 1 in order to obtain agreement between the calculated and measured speeds of TMP-doped flames.

Reaction	A^a, (Jayaweera et al., 2005)	$A_{modified}^a$
$(CH_3O)_3PO + H = (CH_3O)_2PO(OCH_2) + H_2$	2.2×10^9	4.4×10^9
$(CH_3O)_2PO(OCH_2) + O = OP(OCH_3)_2O + CH_2O$	5.0×10^{13}	1.0×10^{13}
$(CH_3O)_2PO(OCH_2) = OP(OCH_3)_2 + CH_2O$	2.0×10^{13}	2.0×10^{12}

[a]Units are cm^3, mole, s.

Table 1. Three important reactions and pre-exponential factors of their rate constants before and after modification.

The results of flame speed calculations using the changed reaction rate constants are given in Fig. 5. It is evident that the changed mechanism provides an adequate description of the experimental data on the effect of TMP additives on the $H_2/O_2/N_2$ flame speed. We note that the sensitivity coefficients for the same reactions in CH_4/air flames doped with TMP are negligibly small. For example, for a CH_4/air flame (ϕ=1.1, 0.06% TMP), the flame speeds calculated for the original mechanism and the mechanism with the changed rate constants of the three reactions given above differ by only 0.2 %. Thus, the changed mechanism describes the propagation speed of both hydrocarbon–oxygen and hydrogen–oxygen flames doped with TMP. The calculations showed that the change in the rate constants of the reactions did not lead to appreciable changes in the flame structure for ϕ=1.6, including the concentration profiles of the final products of TMP conversion – PO, PO_2, HOPO and $HOPO_2$.

The changed rate constants of the three reaction were used to calculate the dependence of the inhibition effectiveness F ($F=(U_0-U)/U_0$, where U_0 and U are the speeds of the undoped flames and flames doped with 0.04 % TMP, respectively) on ϕ. The calculated dependences are presented in Fig. 6. It can be seen that F increases as ϕ increases from 1 to 3 and as D decreases from 0.209 to 0.077. Modeling data with updated mechanism for D=0.077 are in a good agreement with experiment.

It is important to note that the dependence of F on ϕ for $H_2/O_2/N_2$ differs from that for hydrocarbon flames (Korobeinichev et al., 2007; Rybitskaya et al., 2007). In hydrocarbon flames doped with 0.06% TMP, the inhibition effectiveness F increases slightly as ϕ increases from 0.7 to 1.2–1.3, and a further increase in ϕ from 1.3 to 1.5 leads to an abrupt decrease in F by a factor of 1.5 to 2 (Korobeinichev et al., 2007).

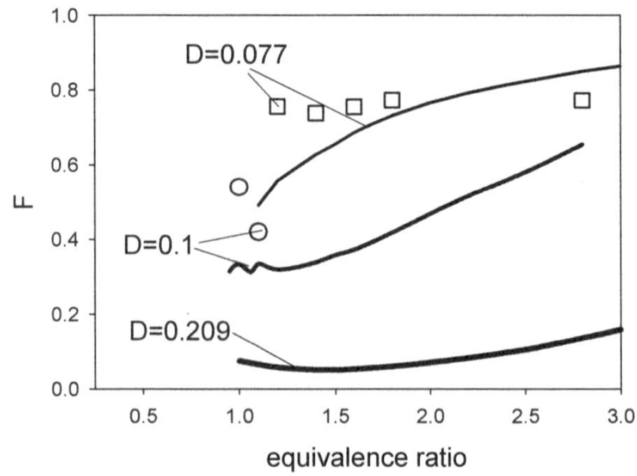

Fig. 6. Inhibition effectiveness F for $H_2/O_2/N_2$ flames doped with 0.04% TMP with different dilution ratios (D) versus equivalence ratio; symbols - experiment, lines - modeling using the updated mechanism.

The addition of TMP to flames reduces the maximum concentration of H atoms in the chemical reaction zone of a flame. Figure 7 shows the relative reduction in the maximum concentration of H atoms $\Delta[H]_{max}$ ($\Delta[H]_{max}=1-[H]^{d}_{max}/[H]^{0}_{max}$) and OH radicals $\Delta[OH]_{max}$ ($\Delta[OH]_{max}=1-[OH]^{d}_{max}/[OH]^{0}_{max}$) due to the addition of TMP versus ϕ, obtained from structure simulations for freely-propagating flames without additives and with 0.04% TMP added. Where $[H]_{max}$ and $[OH]_{max}$ - maximal concentrations of H and OH in the flame zone, subscripts "d" and "0" are related to doped and clear flames respectively. It is evident from Figs. 6 and 7, that there is a correlation between the dependences of F and $\Delta[H]_{max}$ and $\Delta[OH]_{max}$ on ϕ and D.

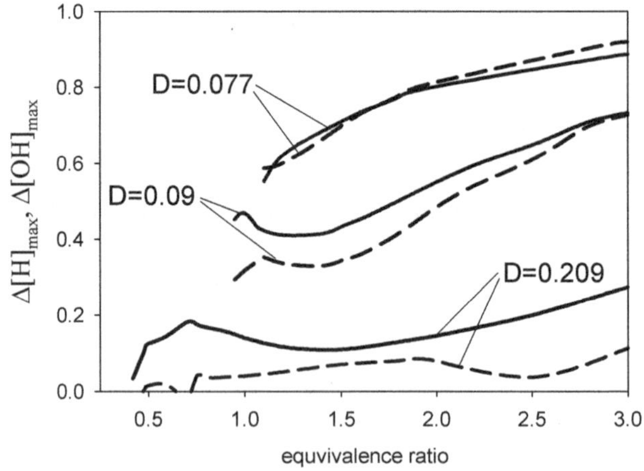

Fig. 7. Relative decrease of H (solid lines) and OH (dashed lines) maximal concentration in flames doped with 0.04% TMP with different D versus equivalence ratio; modeling data using the updated mechanism.

An analysis of the flame speed sensitivity coefficients with respect to the reaction rate constants for a flame with D=0.09 shows (Fig. 8) that an increase in ϕ from 1.1 to 1.9 primarily enhances the role of the reactions of hydrogen atoms with TMP and its primary destruction product:

$$(CH_3O)_3PO + H = (CH_3O)_2PO(OCH_2) + H_2 \qquad (1)$$

$$(CH_3O)_2PO(OCH_2) + H = (CH_3O)_3PO \qquad (2)$$

and the reactions with the HOPO species:

$$HOPO + H = H_2 + PO_2 \qquad (3)$$

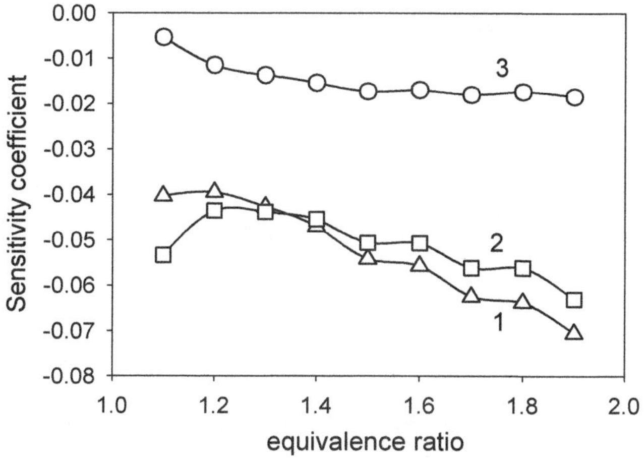

Fig. 8. Sensitivity coefficients of speed of $H_2/O_2/N_2$ flames (ϕ = 1.6, D = 0.077, 0.09, and 0.209) doped with 0.04% TMP to rate constants of reactions (1)-(3) involving P-containing species; modeling is based on mechanism (Jayaweera et al., 2005)

These reactions are involved in the catalytic recombination cycles of H atoms with the formation of H_2. It is these reactions that are responsible for the rise in the inhibition effectiveness with increasing of ϕ. Under the catalytic recombination responsible for scavenging of radicals in OPC-doped flames, all the authors meant reactions involving phosphorus oxides and acids (PO, PO_2, HOPO, and $HOPO_2$). It was assumed that organophosphorus combustion intermediates play a negligible role in the inhibition processes. However, in these specific flames reactions of OPCs with active species are of importance.

An analysis of the speed sensitivity coefficients of TMP-doped $H_2/O_2/N_2$ flames with respect to the rate constants of the main hydrogen combustion reactions and the reactions involving P-containing species revealed the main stages responsible for an increase in F with decreasing D. From the data given in Fig. 9 for the main hydrogen combustion reactions, it is evident that a decrease in D leads primarily to a growth in the role of the $H+O_2=O+OH$ branching reaction. In flames with ϕ=1.6, the flame speed sensitivity coefficient with respect to the rate constant of this reaction increased by a factor of 4 as D decreased from 0.209 to 0.077. In addition, the sensitivity coefficient for this reaction is much larger than that of the remaining reactions. Furthermore, dilution with N_2 changes the role of the $H+O_2(+M)=HO_2(+M)$ recombination reaction, for which the sensitivity coefficient is negative for flames with dilution coefficients D=0.077 and D=0.09 and is positive for D=0.209.

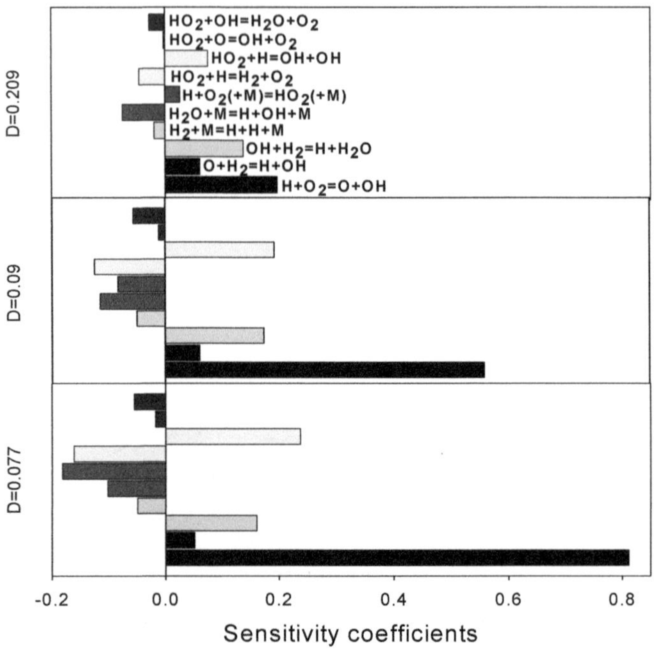

Fig. 9. Sensitivity coefficients of speed of $H_2/O_2/N_2$ flame (φ = 1.6, D = 0.077, 0.09, 0.209) doped with 0.04% TMP to rate constants of 10 key reactions of hydrogen combustion.

However, an analysis of the flame speed sensitivity to the rate constants of chain termination reactions - radical recombination upon interaction with TMP and its destruction products (Fig. 10) - shows that decreasing D increases the sensitivity to these reactions much more strongly than the sensitivity to the branching reaction constant rate. For example, as D decreases from 0.209 to 0.077, the sensitivity coefficient with respect to the rate constants of reactions (1) and (2) increases by a factor of 8 and 20 times, respectively. This is much greater than the increase in the flame speed sensitivity coefficient with respect to the branching reaction. An analysis of the simulation results shows that, a decrease in D from 0.209 to 0.077 also results in a factor of 2-5 increase in the ratio of the maximum rate of recombination of H atoms by reactions (1) and (2) to the maximum rate of the branching reaction. It is this factor that is responsible for the increase in the inhibition effectiveness F with decreasing D, and, possibly, for the increase in F with increasing of ϕ. Thus, the ratio of the chain termination rate to the chain branching rate is an important parameter that determines the hydrogen-oxygen flame speed.

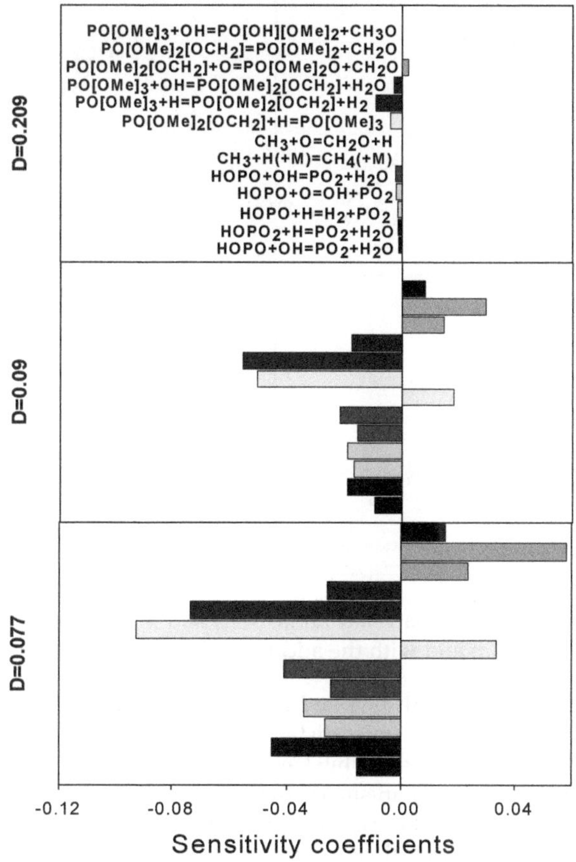

Fig. 10. Sensitivity coefficients of speed of $H_2/O_2/N_2$ flame (ϕ = 1.6, D = 0.077, 0.09, 0.209) doped with 0.04% TMP to rate constants of 13 key reactions of the inhibition mechanism (Jayaweera et al., 2005) involving P-containing species.

3.3 The mechanism for inhibition of hydrocarbon flames at atmospheric pressure

The inhibition effectiveness of hydrocarbon flames by phosphorus-containing inhibitors also increases with the rise of fuel excess in unburnt gases.

The burning velocities of propane-air and methane-air flames were measured on a flat burner using the heat flux method (De Goey et al., 1993; Van Maaren et al.,1994). This technique allows burning velocities to be measured with much higher accuracy (3% for stoichiometric flames and 5-10% for lean and rich flames) over a wide range of equivalence ratios.

Experimental data (Rybitskaya et al., 2007) on the burning velocities of propane-air and methane-air flames without additives (Fig. 11.) are in good agreement with literature data obtained using the same method (Bosschaart et al., 2004).

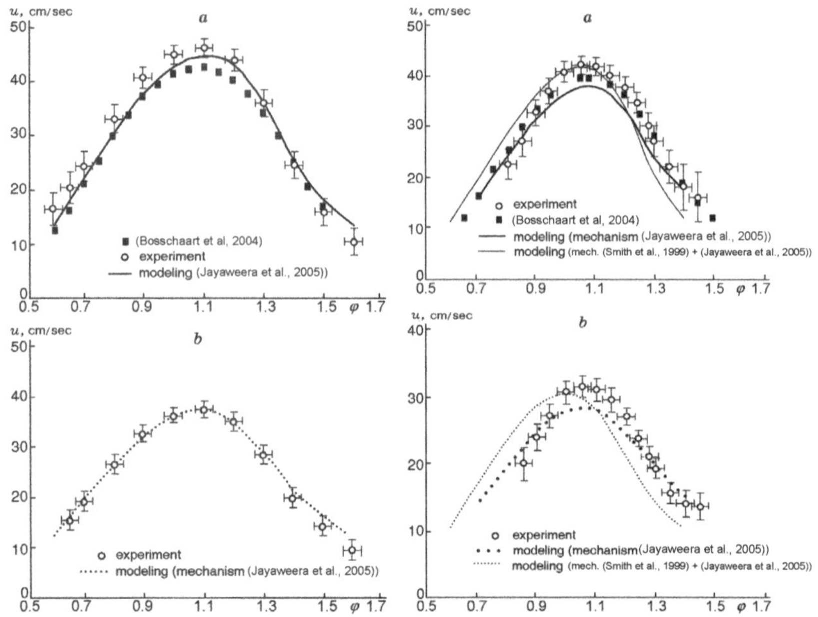

Fig. 11. Burning velocity of a propane–air flame (left) and methane-air flame (right) versus equivalence ratio without (a) and with the addition of 600 ppm TMP (b).

A comparison of the experimental data and modeling results for flames without additives and with 0.06% TMP showed to be in a satisfactory agreement for propane flames and differ somewhat for methane–air flames (Rybitskaya et al., 2007). For methane–air flames, the predictions using the GRI 3.0 mechanism are in better agreement with the experimental data for near stoichiometric flames, whereas the mechanism (Jayaweera, et al., 2005) provides a better agreement for lean and rich flames. Figure 12 presents the inhibition effectiveness of propane–air and methane–air flames by TMP, which is defined as $F = (u_0 - u)/u_0$, where u_0 and u are the burning velocities of the undoped and doped flames, respectively. The simulation predicts that the effectiveness increases with φ up to 1.3 and 1.2 for C_3H_8 and CH_4 flames, respectively, and then decreases in the region of rich flames. In the experiments, the inhibition effectiveness is also observed to decrease sharply in rich flames, whereas in lean and near-stoichiometric flames, the experimental dependence $F(\phi)$ is weakly expressed. The discrepancies between the simulation and measurement data may be due to drawbacks of both the mechanism for the reactions involving OPCs and the hydrocarbon combustion mechanisms. This is supported by the fact that two different kinetic schemes of hydrocarbon combustion (Jayaweera, et al., 2005; Smith et al., 1999) used in burning velocity calculations for TMP-doped flames with the same set of TMP reactions yield somewhat differing results. It noteworthy that the hydrocarbon combustion mechanisms used in the calculations have previously been tested by various researchers by comparing simulation and experimental results flame speed, structure, ignition delay in shock waves, and oxidation in a flow reactor. Therefore, the application of the two different hydrocarbon combustion mechanisms (Jayaweera, et al., 2005; Smith et al., 1999) allows estimating how these models influence the predicted speed of TMP-doped flames.

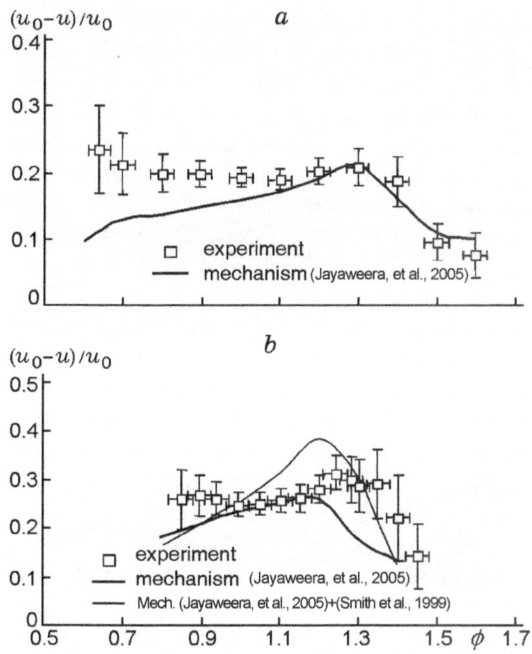

Fig. 12. Inhibition effectiveness versus equivalence ratio in propane–air (a) and methane–air (b) flames with the addition of TMP.

As the modeling results depend on a mechanism for hydrocarbon combustion, it is reasonable to performed modeling using several kinetic schemes. As can be seen from Fig. 11b, although these two different models give differing absolute values of F for $\phi \approx 1.2$, the resulting dependences $F(\phi)$ are qualitatively similar. The burning velocity simulation for a hydrogen–air flame (at p = 1 bar and T_0 = 298 K) without additives and with 0.1% TMP predicts the inverse dependence of the inhibition effectiveness. Figure 13 shows flame velocity sensitivity coefficients versus ϕ for changes in the rate constants of the most important reactions in TMP-doped propane–air and hydrogen–air flames. The sensitivity coefficients were determined by the formula $[(u - u_{5A})/u]\times100\%$, where u and u_{5A} are the burning velocity calculated using the mechanism [(Jayaweera, et al., 2005) and the flame speed calculated using the same mechanism modified the pre-exponential factor of the reaction rate constant increased by a factor of five. Thus, the reactions PO_2+ H+M = HOPO+M and $HOPO_2$+H = PO_2+H_2O are the most significant in lean propane–air flames because these flames are dominated by compounds of phosphorus with a greater degree of oxidation, such as PO_2 and $HOPO_2$. The same reaction $HOPO_2$+H = PO_2+H_2O makes a significant contribution to the inhibition of lean hydrogen–air flames. For rich propane–air flames, the major contribution to the inhibition comes from the reaction HOPO + OH = PO_2 + H_2O, which is responsible for the sharp decrease in the effectiveness at ϕ = 1.3. The sensitivity coefficient dependence of this reaction (see Fig. 13a) correlates with the dependence of the inhibition effectiveness of TMP (see Fig. 12a): both have a maximum at $\phi \approx 1.3$.

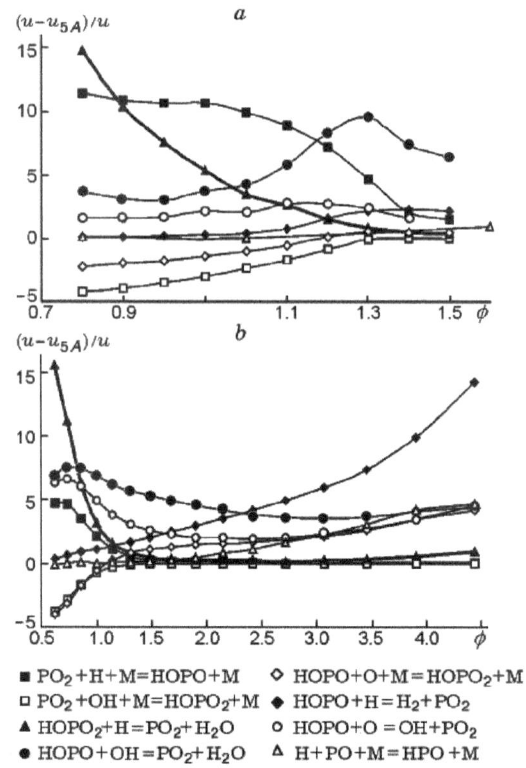

Fig. 13. Burning velocity sensitivity coefficients versus reaction rate constants in propane–air (a) and hydrogen–air (b) flames doped with TMP.

For rich hydrogen-air flames at $\phi > 2.5$, the highest contribution to the inhibition comes form the reaction HOPO+H = H_2 + PO_2, which, together with the other reactions, is responsible for an increase in the inhibition effectiveness as ϕ rise. Thus, the differences in the behavior of the inhibition effect of hydrocarbon-air and hydrogen- air flames occurs due to changes in the importance of the reactions of radicals with the phosphorus-containing products of TMP combustion. The curves of the inhibition effectiveness versus φ correlate with the those of the concentrations of H atoms and OH radicals versus in the chemical reaction zone ϕ of TMP- doped flames. Figure 14 presents the maximum concentrations of H and OH radicals in propane–air flames without additives and doped with TMP and their relative decrease versus equivalence ratio. As can be seen from the figure, the curve of the relative decrease in the radical concentration ($\Delta C_i/C_i$) and the curve of the inhibition effectiveness (see Fig. 12) have a break at ϕ = 1.3, which indicates a direct relationship between the decrease in the concentration of the active radicals and the effect of the TMP additive. Similar curves for hydrogen–air flames are presented in Fig. 15. The reduction in the inhibition effectiveness in rich hydrocarbon flames may be due to an increase in the concentration of organophosphorus products of incomplete TMP destruction in the postflame zone at high equivalence ratios.

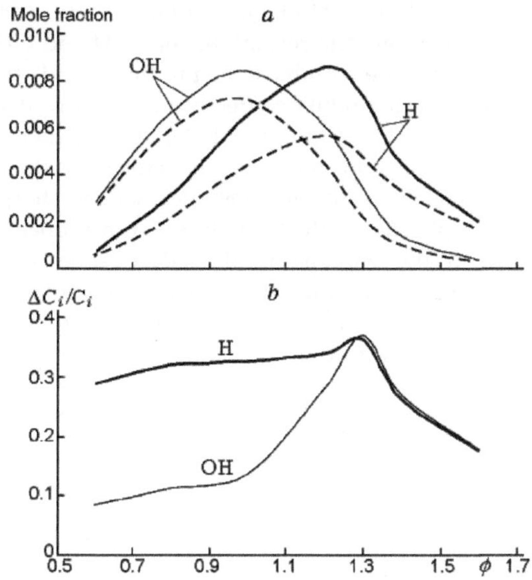

Fig. 14. Maximum concentrations of active radicals in propane–air flames (a) without additives (solid curves) and doped with 0.06% TMP (dashed curves) and the relative decrease in the concentrations (b).

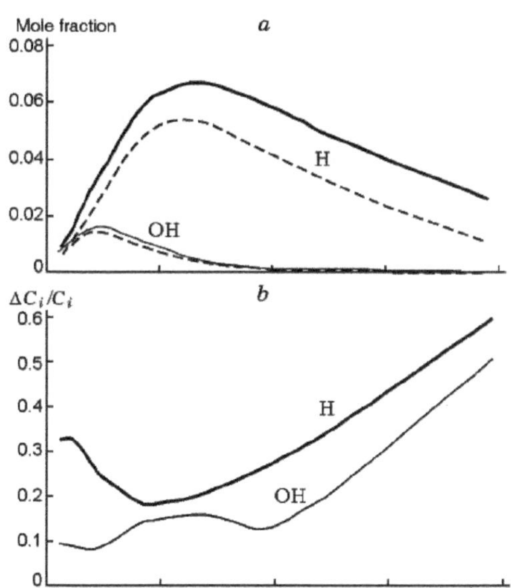

Fig. 15. Maximum concentrations of active radicals in hydrogen–air flames (a) without additives (solid curves) and with 0.1% TMP (dashed curves) and the relative decrease in the concentrations (b).

Species such as CH_3PO, CH_3PO_2, CH_3OPO, CH_3OPO_2, and CH_2OPO_2 are probably ineffective or incapable of catalyzing the recombination of H and OH radicals; therefore, a rise of their concentration in the combustion products leads to a decrease in the concentration of active species and inhibition effectiveness. Figure 16 demonstrates the ratio of the total concentration of CH_3PO, CH_3PO_2, CH_3OPO, CH_3OPO_2, and CH_2OPO_2 in the post-flame zone of propane–air and methane–air flames to the TMP loading versus equivalence ratio. These data obviously demonstrate that the sharp increase in the total concentration of organophosphorus products is observed in propane–air flames at $\phi = 1.4$, this almost coinciding with the maxima in the effectiveness curves in Fig. 12. It is noteworthy that, for the methane–air flames, the similar inhibition effectiveness curve has a maximum at $\phi = 1.2$–1.3 and a sharp increase in the total concentration of the organophosphorus products is observed at the same ϕ. Therefore, the accumulation of the catalytically ineffective organophosphorus intermediates in the rich OPCs-doped flame reduces the inhibition effectiveness.

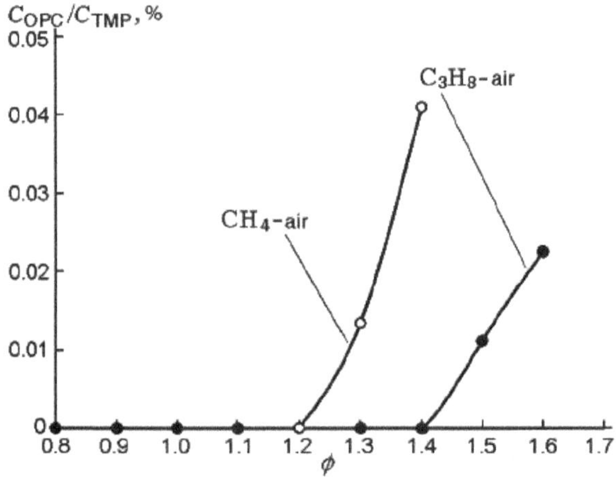

Fig. 16. Total concentration of CH_3PO, CH_3PO_2, CH_3OPO, CH_3OPO_2, and CH_2OPO_2 normalized to the initial TMP concentration in the flame (C_{OPC}/C_{TMP}) versus equivalence ratio.

One of the important combustion characteristics of premixed mixtures are flammability concentration limits (FCLs). The effect of various fire suppressants and inhibitors (chlorofluorohydrocarbons, bromine-containing hydrocarbons) on the FCLs of hydrocarbon–oxygen mixtures has been studied earlier (Saito et al., 1995; Shebeko et al., 2000) but the effect of OPCs have not been examined. Investigation of the inhibitors (including OPCs) influence on the FCLs allows, on the one hand, to evaluate the possibilities of their use as fire suppressants, and, on the other hand, to validate the inhibition mechanism by comparing experimental and modeling results. At present, there are a number of methods for determining FCLs, e.g. the method proposed by Coward and Jones (Coward & Jones, 1952); the standard method (State Standard No. 12.1.044-89; Baratov et al., 1990); a method using a cylindrical burner (Ishizuka, 1991; Hichens et al., 1999). In practice, these and other methods give different values for the FCLs of the same combustible mixture, the results being greatly dependent on the design features of the experimental setups. In addition, there are a number of factors that are difficult to take into account in modeling but which have an appreciable influence on FCLs. The most important of them are the heat losses due to radiation and convection, the influence of the source and method of flame ignition, and the flame stretching effects (due to the presence of velocity gradients) (Ishizuka, 1991; Hertzberg, 1976). The opposed-flow burner method proposed by Law et al. (Law et al. 1986) to determine FCLs allows one to minimize the effect of heat losses on test results and to control the flame stretching. Law et al. (Law et al. 1986) used this method for determining FCLs of methane- and propane/air mixtures. It was also employed in (Womeldorf et al., 1995; Womeldorf & Grosshandler, 1996; Grosshandler et al., 1998) to determine the flammability limits of mixtures of fluorochlorohydrocarbons and air. Knyazkov et al (Knyazkov et al., 2008) used this method to study the effect of TMP additives on FCLs of methane-air mixtures. A significant advantage of the opposed-flow burner method for determining FCLs is that processes in this system can be modeled using detailed kinetic schemes. This allows one, by comparing calculated and experimental results, to obtain new data on the kinetic inhibition mechanism and to improve the employed kinetic scheme in order to increase the calculation accuracy.

The experimental setup used to determine FCLs of premixed combustible mixtures doped with OPCs is an opposed-flow burner equipped with a gas supply system and a system for supplying additives of the substances studied. TMP was added to both flows of the combustible mixture by two saturators. The required TMP vapor concentration was achieved by placing the saturators in a water bath at a thermostatically controlled temperature. It is known that the effect of any inhibitor on a flame is due to thermal-physics and chemical factors.

Experimental curves of the extinction velocity gradient K_{ext} for methane-air flames with and without TMP additives (in the TMP concentration range of 0–0.5% by volume) versus volumetric concentration of methane in the mixture are given in Fig. 17. For comparison, the figure also gives data for methane-air mixtures doped with CF_3Br. The obtained experimental points K_{ext} (values at various concentrations of CH_4 and a specified concentration of TMP) are approximated by lines which extrapolate the data to the value of $K_{ext} = 0$. Figure 17 gives data for the methane concentrations near the lower (Fig. 17a) and upper (Fig. 17b) flammability limits of the gas mixture studied. The methane concentrations in the mixture corresponding to the upper or lower flammability limits of the combustible mixture with and without additives were determined by the points of intersection of the straight lines with the abscissa. The data presented in Fig. 17 were used to obtain dependences of the upper and lower FCLs of methane- air mixtures on the volumetric concentrations of TMP and CF_3Br.

These dependences are given in Fig. 18; for comparison, the figure also shows the data for the concentration limits of CH_4/air/CF_3Br mixtures at a temperature of ≈353 K obtained by Saito et al. (Saito et al., 1995) using a cylindrical burner. As can be seen from Fig. 18, the flammability limits of CH_4/air mixtures are 4±0.2% CH_4 (the lower limit) and 16.2 ± 0.2% CH_4 (the upper limit).

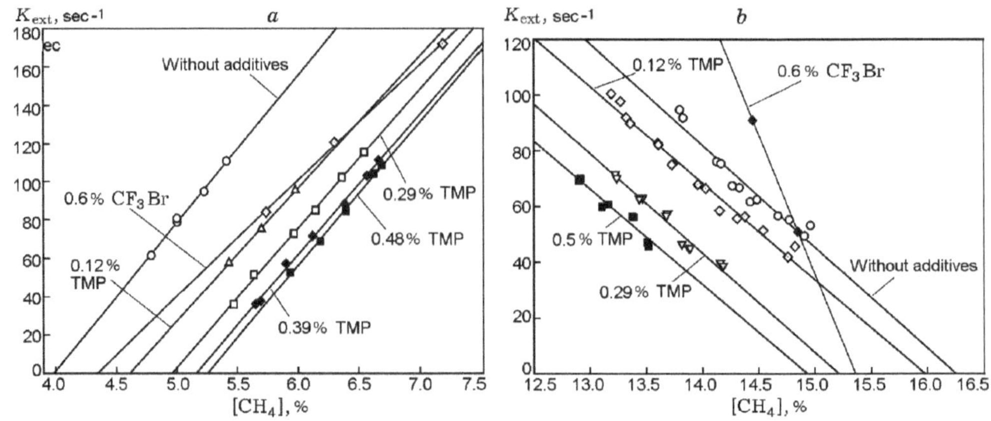

Fig. 17. Dependence of the extinction velocity gradient on methane concentration extrapolated to the value of $K_{ext} = 0$ for various TMP contents in the mixture: (a) lean flame; (b) rich flame.

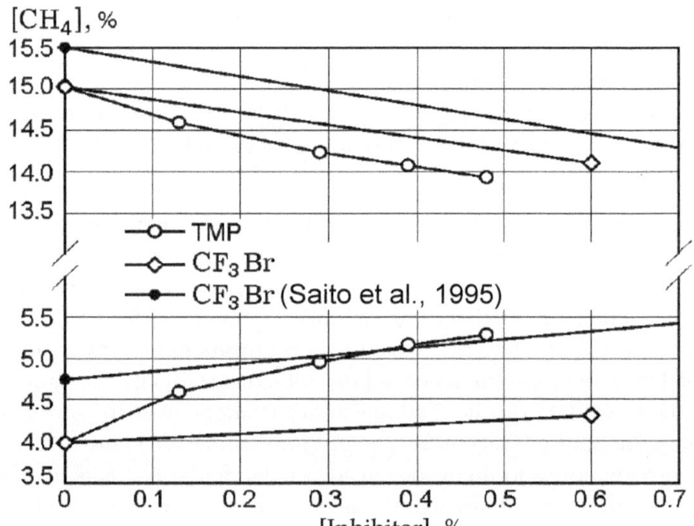

Fig. 18. Upper and lower flammability concentration limits of methane–air mixtures versus volumetric concentration of the inhibitor (CF_3Br and TMP).

These data differ from the results of Saito et al. (Saito et al., 1995), according to which the flammability limits are in the methane concentration range of 4.75–15.5%. The observed difference is obviously not due to the difference between the temperatures of the initial combustible mixtures in the experiments because it is only ≈15 K, but it is due, as noted above, by the difference between the techniques used in the studies. It should also be noted that if the experiments described in the present paper were performed at a temperature of the initial mixture of ≈298 K, the flammability concentration limits would be somewhat narrower. Hichens et al. showed (Hichens et al., 1999) that changing the temperature of the initial methane–air mixture by 70 K led to an approximately 0.5% change in each limit. In view of this, the concentration limit for these conditions should be 4.5–15.7 % CH_4. An analysis of the literature (Saito et al., 1995; Baratov et al., 1990; Ishizuka, 1991; Hichens et al., 1999; Womeldorf & Grosshandler, 1996), shows that, for methane- air mixtures under normal conditions, various methods give lean flammability limits in the [CH_4] range of 4.0 to 5.3% and rich limits in the range of 13.8–15.6%. Thus, the values of the lower and upper limits obtained in the work fall in these ranges, i.e., the employed technique yields reasonable results in agreement with literature data. Experiments on determining the effect of CF_3Br additives on the flammability limits made it possible, on the one hand, to reproduce the data of (Saito et al., 1995) and thus to test the procedure of determining the limits, and, on the other hand, to compare the effects of TMP and CF_3Br on the FCLs. The experiments were performed at a 0.6% concentration of CF_3Br in the combustible mixture. Because the flammability limits of the combustible mixture without additives do not coincide with those obtained in (Saito et al., 1995), we compare data on changes in the limit (upper or lower) at specified concentrations of CF_3Br. From Fig. 18 it is evident that, according to (Saito et al., 1995), the addition of 0.6% CF_3Br leads to a 1% decrease in the upper concentration limit for methane and a ≈0.5% increase in the lower limit. This agrees with the result obtained in our work using an opposed-flow burner. The addition of TMP to the combustible mixture, as is seen from Fig. 18, narrows the flammability concentration limits. The relative effect of TMP reduces as its concentration is increased; in particular, the addition of 0.12% TMP changes both limits for [CH_4] by ≈0.5%, and to change the limits by 1%, it is necessary to add ≈0.4% TMP. It is worth noting that, in contrast to CF_3Br, the addition of TMP has the same effect on both the lower and the upper concentration limit within the experimental error. For example, the addition of 0.48% TMP reduces the upper limit for [CH_4] by ≈1.1% and increases the lower limit by ≈1.3%. As noted above, the effect of CF_3Br on the upper limit is twice that on the lower limit. The addition of 0.3% TMP increases the lower limit four times more effectively than the addition of CF_3Br of the same concentration. Furthermore, the addition of TMP reduces the upper limit two times more effectively than the addition of CF_3Br. It is known that many flame inhibitors, such as fluorinated hydrocarbons (for example, CF_3Br), influence the upper and lower concentration limits differently: they reduce the upper limit more strongly and increase the lower limit less strongly (Saito et al., 1995; Shebeko et al., 2000). This effect is due mainly to the fact that the addition of an inhibitor to a combustible mixture leads to a change in the equivalence ratio, and, thus, to a reduction in both concentration limits for methane, and, as result, to different effectiveness of the additive for the upper and lower limits. In the experiments with the addition of TMP to the combustible mixture, this effect was not observed. Figure 19 gives experimental and calculated dependences $K_{ext} = f([CH_4])$ for flames without additives and doped with 0.12% TMP. It is evident that, in both cases, the experimental dependences $K_{ext} = f([CH_4])$ agree better with the predicted dependences for lean mixtures than for rich mixtures. Figure 20 gives experimental and calculated dependences of the upper and lower flammability concentration limits on the volumetric concentration of TMP in methane–air mixtures. It is evident that, for both the doped undoped flames, the predicted lean concentration limits are also in better agreement with the corresponding experimental values than the values of the rich limits. In particular, the value of 4.2% predicted for the of the flame without TMP is in good agreement with the experimental result of 4%. According to the modeling data, the increase in the lower concentration limit due to the addition of 0.12% TMP to the combustible mixture was 0.7%, which also agrees with the experimental data (within the measurement error). The predicted rich concentration limit for the undoped flame is seen to fit well to the experimental value. However, there is a difference between the experimental data and modeling results for the reduction in the rich limit due to the addition of 0.12% TMP to the mixture. This is most likely due to the drawbacks of the kinetic mechanism. For example, the mechanism does not take into account heavy hydrocarbons which are known to form in rich flames. The experimental finding that TMP has the same effect on both the lower and upper concentration limits (in contrast to, for example, CF_3Br, which, as noted above, changes the upper limit more strongly than the lower limit) agrees qualitatively with data (Korobeinichev et al., 2007) on the effect of the addition of TMP on the velocity of premixed propane–air flames with various equivalence ratios. It was found that the inhibition effectiveness determined from the decrease in the flame propagation velocity drops with increasing ϕ in the range of $\phi > 1.3$ (see explanation above).

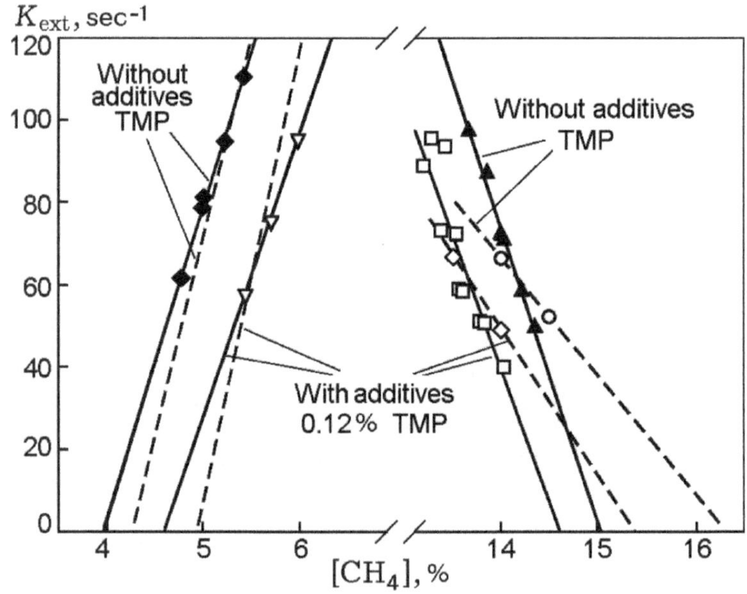

Fig. 19. Experimental and calculated extinction velocity gradients versus volumetric concentration of CH_4 in methane–air mixtures (without additives and doped with 0.12% TMP): the solid curves are extrapolations of the experimental data, and the dashed curves are extrapolations of the calculation data.

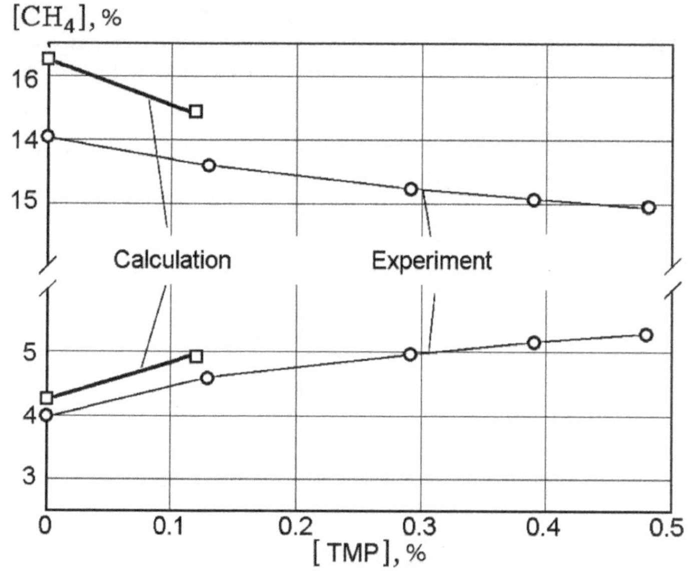

Fig. 20. Experimental and calculated upper and lower flammability concentration limits of methane–air mixtures versus volumetric concentration of TMP.

4. Iron-containing compounds

4.1 The mechanism for inhibiting hydrogen flames at atmospheric pressure

The first studies of iron-containing species (ICS) as flame inhibitors were performed in the early 1960s. Lask and Wagner (Lask & Wagner, 1962) were the first to measure the effect of iron pentacarbonyl $Fe(CO)_5$ on the speed of H_2/air flame and to show that it was 2 orders of magnitude greater than that of CF_3Br. Miller et al. (Miller et al., 1963) studied the effect of certain inhibitors, including $Fe(CO)_5$, on the speed of H_2/air flames for various stoichiometry and the inhibitor concentration of 0.5% by volume. Unfortunately, addition of $Fe(CO)_5$ at this concentration greatly changed the fuel equivalence ratio of the initial flame because 2.5% by volume of carbon monoxide was formed in the flame and as a result the effect of iron alone was difficult to assess.

An increased interest in metal-containing compounds, including iron pentacarbonyl and ferrocene $Fe(C_6H_6)_2$ as flame inhibitors, has arisen after the ban on the production of halogenated fire-extinguishing agents (such as CF_3Br) with a high ozone depletion potential. Searches for new inhibitors and fire suppressants have led scientists to pay attention to ICS compounds. Reinelt and Linteris (Reinelt & Linteris, 1996) studied the effect of $Fe(CO)_5$ and ferrocene on premixed laminar flame speeds and extinction strain rates of opposed-jet diffusion flames. The inhibition effectiveness was shown to strongly depend upon the dopant loading; the maximum inhibition was achieved at a concentration as low as 100 μL/L (μL/L is equivalent to parts per million by volume) and was then little affected by a further increase in the concentration. Rumminger et al. (Rumminger et al., 1999) attribute this to the formation of iron-containing particles in the flame, resulting in a reduction in the gas-phase concentration of iron species (which are responsible for flame inhibition). According to data (Reinelt & Linteris, 1996), an increase in the oxygen volume fraction in a O_2/N_2 mixture reduces the inhibition effect. This may be due to the combustibility of iron pentacarbonyl and ferrocene.

Linteris et al. (Linteris et al., 2000a) performed experimental and modeling studies of the effect of $Fe(CO)_5$ and $Fe(C_6H_6)_2$ additives on the speed of a $CH_4/O_2/N_2$ flame. The numerical study was based on a kinetic model (Rumminger et al., 1999), which provided a good fit to experimental results. Ferrocene, whose molecule contains more fuel than that of iron pentacarbonyl, showed a lower inhibition effect than $Fe(CO)_5$.

Rumminger et al. (Rumminger et al., 1999) determined the key chemical processes responsible for flame inhibition

$$FeOH+H = FeO+H_2$$

$$FeO+H_2O = Fe(OH)_2$$

$$Fe(OH)_2+H=FeOH+H_2O$$

This is equivalent to the recombination of hydrogen atoms H + $H=H_2$. The catalytic recombination of O atoms also involves three elementary reactions

$$Fe+O_2 + M = FeO_2 + M$$

$$FeO_2 +O = FeO + O_2$$

$$FeO + O = Fe + O_2$$

The role of the formation of iron-containing particles in the inhibition of hydrogen and methane flames by a $Fe(CO)_5$ additive was studied by Rumminger & Linteris (Rumminger & Linteris, 2000a, 2002). The results showed that the formation of iron particles reduced the inhibition effect of the iron pentacarbonyl additive. This supports the hypothesis of a gas-phase mechanism for flame inhibition by ICS.

Experimental studies of cup-burner flame suppression by metallic compounds (Linteris et al., 2004) have provided further evidence for the gas-phase mechanism of inhibition by ICS. Models ignoring solid-phase formation in flame predict the greater inhibition effect than experimental data.

Spatial variations in the temperature and concentration of atomic iron in a low-pressure lean (φ = 0.37) $H_2/O_2/Ar/Fe(CO)_5$ flame were measured using laser-induced fluorescence (Wlokas et al., 2009; Staude et al., 2009a). Iron pentacarbonyl was found to decompose in the flame to produce atomic iron, which was then transformed to iron oxides and hydroxides. On the basis of a previously developed mechanism (Rumminger et al., 1999), a reduced 12-step mechanism for flame inhibition by $Fe(CO)_5$ was developed and validated by comparing measured and simulated concentration profiles of atomic iron.

A low-pressure, rich (φ=2.3), laminar, premixed propene/oxygen/argon flame doped with ferrocene was studied experimentally using molecular beam mass spectrometry (MBMS) and laser-induced fluorescence (LIF) and by numerical simulations (Tian et al., 2009). The flame temperature was obtained by two-line OH LIF measurements, and the additive was found to increase the postflame temperature by 40 K. MBMS analysis of the species profiles of important intermediates in flames with and without ferrocene doping showed a slight increase in the maximum concentration of species, such as CH_2O, C_5H_5, and C_6H_6. At the same time, the dopant slightly decreased the maximum concentration of the propargyl radical C_3H_3, which is known to be an intermediate in the formation of soot precursors. The MBMS measurements showed that the flame velocity decreased with the addition of ferrocene, which was not predicted by the model.

Staude and Atakan (Staude & Atakan, 2009b) carried out equilibrium calculations for iron-doped hydrogen/oxygen/argon and propene/oxygen/argon gas mixtures under combustion-relevant conditions. It is noteworthy that condensed Fe-containing compounds were considered in the calculations. The focus was on iron intermediates and the conditions under which condensed phases of iron or iron species could be expected in the flame. The stoichiometry (ϕ= 0.37, 1, and 2.3), temperature (1000-2500 K), and pressure (0.03-1 bar) were varied, allowing for a prediction of which gas-phase iron species might be expected in measurable concentrations under the flame conditions used. The effect of the sampling probe on the composition of the combustion products, which are cooled during probing, was discussed.

The developed model for flame inhibition by iron compounds was continually validated by comparing the measured and modeled burning velocity of near-stoichiometric flames doped with $Fe(CO)_5$ and ferrocene. A significant progress in understanding the features of the inhibition chemistry under lean and rich conditions was made by measuring and simulating the H_2/air flame speed over a wide range of equivalence ratios (Fig. 21) (Gerasimov et al., 2011). Furthermore, Gerasimov et al. (Gerasimov et al., 2011) performed unique measurements of the spatial variation of concentration of some of $Fe(CO)_5$ destruction products (Fe, FeO_2, FeOH, and $Fe(OH)_2$) in premixed atmospheric-pressure $H_2/O_2/N_2$ flame using probing molecular beam mass spectrometry with soft ionization by electron impact. A comparison between experimental and numerical results revealed that the mechanism used in the study (Rumminger et al., 1999) satisfactorily predicts speeds of atmospheric-pressure H_2/air flames over a wide range of equivalence ratios and concentration profiles FeO_2 and $Fe(OH)_2$ in the flame.

Shvartsberg et al. (Shvartsberg et al., 2010) showed that the inhibition effectiveness (expressed as the relative decrease of the flame speed as the dopant is added) substantially depends upon the equivalence ratio of unburnt gases: the minimum effectiveness is observed at $\phi \approx 2$, and the maximum effectiveness is observed in lean flames (Fig. 22.). At the same time the same authors demonstrated that the maximum total rate of the active species (H, O, and OH)) consumption is observed in stoichiometric flame that at first glance seems contrary to the effectiveness curves (Fig. 23). This illusory contradiction was adjusted by suggesting that the inhibition effectiveness over a wide range of equivalence ratios of hydrogen flames depends not only by the rate of chain termination in Fe-involving reactions but also by the rate of chain branching ($H + O_2 = O + OH$). This suggestion was supported by comparing curves of the inhibition effectiveness and the ratio of the integrated production rates of chain carriers (H + O + OH) to the chain-branching rate (Shvartsberg et al., 2010). The products of atomic iron oxidation in H_2/air flames mainly catalyze H atom recombination, and the recombination rate is largely determined by the concentration of iron hydroxides FeOH and $Fe(OH)_2$. It happens because the rate of consumption of O atoms is much lower than that of H atoms and its sum with rate of OH production gives a value close to zero. OH radicals are produced in Fe-involving reaction in atmospheric-pressure H_2/air flame all over range of equivalence ratios (Shvartsberg et al., 2010).

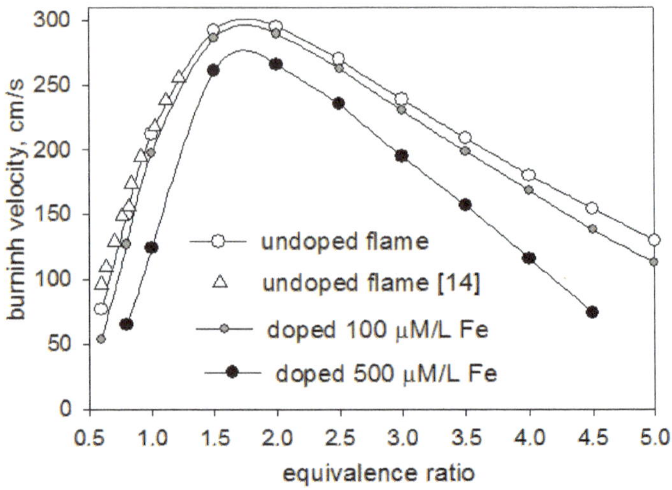

Fig. 21. Speed of H_2/air flames without additives and doped with 100 and 500 μL/L atomic iron versus the equivalence ratio at a pressure of 0.1 MPa. Lines, modeling results; symbols, literature data.

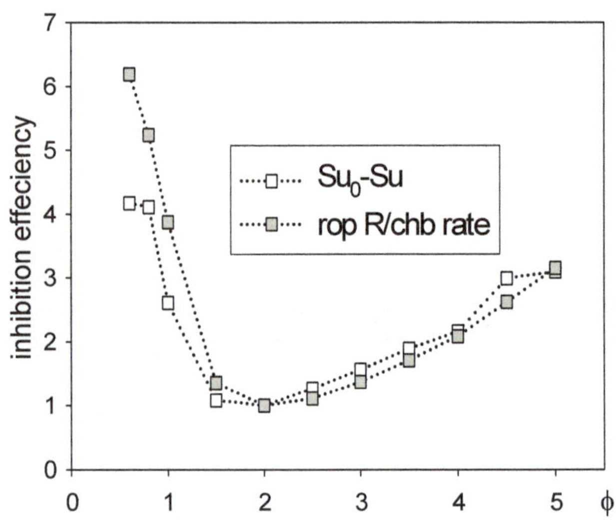

Fig. 22. Inhibition effectiveness of atmospheric-pressure H_2/air flames doped with 100 μL/L atomic iron, expressed as (1) the decrease in flame speed because of doping (open symbols) and (2) the ratio of the integrated production rates of chain carriers (H + O + OH) to the chain-branching rate (gray symbols).

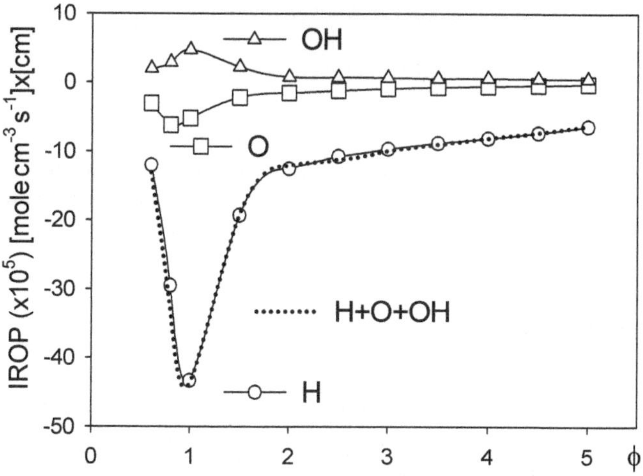

Fig. 23. Integrated production rate (mol cm^{-3} s^{-1})x(cm) of H, O, and OH and the total production rate of all of these species in reactions involving ICS in atmospheric-pressure H_2/air flames doped with 100 µL/L atomic iron versus the equivalence ratio.

According to (Shvartsberg et al., 2010) the variation of net rate of production/consumption of the chain carriers in the Fe-involving reactions versus ϕ is mainly associated with a variation of ICS composition and variation of concentration of the chain carriers in the reaction zone of the flames with flame stoichiometry. The change of the flame temperature with ϕ plays minor role in variation of the rate of production of the active flame species. It may be explained by relatively low activation energies of the Fe-involving reactions from the mechanism (Linteris et al., 2000a), which do not exceed 6500 J/mol.

The chemical processes responsible for changes in the production rates of H, O, and OH as ϕ changes were identified. Three flames were chosen for the analysis: a lean flame with ϕ=0.6, a stoichiometric flame with the maximum rate of radical production, and the richest flame with ϕ=5. The rate of H atom production in separate reactions was calculated, and the key reactions for each of the chosen flames were determined. In addition, to evaluate the role of each reaction qualitatively the authors (Shvartsberg et al., 2010) calculated the contribution of each reaction to the total rate of H consumption, considering only the reactions involving ICS. It was found out that, regardless of the flame equivalence ratio, the key reactions responsible for H atom removal are as follows:

$$FeOH+H = FeO+H_2$$

$$Fe(OH)_2+H = FeOH+H_2O$$

This was also shown previously for a stoichiometric $CH_4/O_2/N_2$ flame (Rumminger et al., 1999). In the richest flame (ϕ=5), the reaction $FeO_2 + H + M = FeOOH + M$ also makes a noticeable contribution to the removal of H atoms from the flame.

Using analogues approach, the key Fe-involving reactions of O atoms consumption were found (Shvartsberg et al., 2010):

$$FeOH + O = FeO + OH$$

$$FeO_2 + O = FeO + O_2$$

$$Fe + O_2 = FeO + O$$

Although reaction of FeOH with O atom is a key reaction of O atoms consumption, it is unlikely to contribute to the inhibition effectiveness because it transforms one chain carrier to another ($O \rightarrow OH$). This fact largely determines the symmetry of the O and OH curves of rate of production of these species.

5. Super-effective inhibitors

Among the chemically active inhibitors (such as phosphorus-, bromine-, iodine- containing compounds, etc.), compounds of alkali metals are the least studied experimentally, which is explained by their very low volatility.

The minimum extinguishing concentrations (MECs) are one of the most important characteristics of flame suppressants, proving estimates of prospects for their future application. Shmakov et al. (Shmakov et al., 2006) applied the cup-burner technique to study a number of recently synthesized organophosphorus compounds and inorganic and organic potassium salts (K_3PO_4, $KOOCH_3$, KOOCCOOK, and $K_4[Fe(CN)_6]$). Aqueous solutions of the examined salts were fed through a nebulizer to the heated air flow in the same way as was done for OPCs. The mass median particle diameter of the salt solution aerosols was 10–20 μm, and after water evaporation, it decreased to 2-5 μm. The mass-median diameter of the aerosol particles was determined with the aid of a five stage impactor (Korobeinichev et al., 2003). Concentrations near the cup were found by sampling the air–aerosol flow through the aerosol filter and determining the mass of the aerosol deposited on it. The MECs of the examined compounds were calculated taking into account the effect of water contained in the solution. Most of the experiments were performed at a constant temperature of the air flow of 75°C. Under these conditions, the aerosol particles of the liquid substances completely vaporized.

Among the OPCs tested, the most effective flame suppressant (in terms of the volume concentration of vapor) is $[(CF_3)_2CHO]_2P(O)CF_3$; it is followed (in decreasing order of the MEC) by $[(CF_3)_2CHO]_2P(O)C_2H_5$; $[(CF_3)_2CHO]_3P$; $(CF_3CH_2O)_2P(O)CF_3$; $(CF_3CH_2O)_3P$; $[(CF_3)_2CHO]_2P(O)CH_3$. The minimum extinguishing concentration of such effective fire suppressant as $(CF_3CH_2O)_3P$ varies from 1 to 3% (by volume). The results of experiments (Table 2) on extinguishing a diffusion *n*-heptane/air flame by aqueous salt solutions show that potassium salts are an order of magnitude more effective than OPCs and halons (Linteris, 2001).

Salt	minimal extinguishing concentration	
	mole fraction (×100)	g/m³
K_3PO_4	No extinguishing at 1%	
$KOOCH_3$	0.25	10.9
KOOCCOOK	0.13	9.6
$K_4[Fe(CN)_6]$	0.035	6.6

Table 2. The studied salts and their minimal extinguishing concentration

The effectiveness of flame suppression by an organic salt per a molecule is directly proportional to the number of potassium atoms in it. The possible mechanism for flame inhibition by potassium salts is following:

$$\text{Potassium salt} \rightarrow K_2O, KOH, \text{etc.}$$

$$KOH + H \rightarrow K + H_2O$$

$$K + OH + M \rightarrow KOH + M$$

$$\text{Total: } H + OH \rightarrow H_2O$$

A distinction is observed for $K_4[Fe(CN)_6]$ because this salt contains not only potassium but also iron. Some iron compounds, for example, $Fe(CO)_5$, are known to be effective flame inhibitors. In the case of $K_4[Fe(CN)_6]$, potassium and iron act jointly in extinguishing flames but the obtained data do not provide a quantitative estimate of the synergetic effect of their joint action.

The expected effectiveness of K_3PO_4 should be much higher than the effectiveness of the OPC but the experiments disprove this assumption. In flame, thermally stable potassium phosphate K_3PO_4 does not dissociate into reactive inhibitors – potassium oxides and phosphorus oxyacids. Thus, the use of combined flame suppressants based on OPCs and MCCs is not promising.

The obtained data suggest very intense chemical reactions of inhibition in the flames doped with alkali metal compounds. The inhibition chemistry of the alkali metals has been studied insufficiently and may become an object of future research.

6. Conclusion

Summarizing the obtained data on inhibition chemistry, we came to following conclusions.

1. The effectiveness of a certain inhibitor depends on features of its combustion chemistry, flame stoichiometry and conditions (pressure). These developments should be necessarily taken into consideration in the case of practical application of an inhibitor.
2. In fact, it is not valid to expect an effective flame inhibitor to be also effective for auto-ignition. Moreover, the same compound can promote the auto-ignition. As an example we can mention iron pentacarbonyl that inhibits atmospheric-pressure H_2/air flames but can reduce the ignition delay of the same mixtures.
3. The inhibition effectiveness of flames with appreciably different equivalence ratio is explained by not only different rate of the chain carriers consumption. A flame with maximum burning velocity (or near-stoichiometric flame) all other things being equal, are inhibited less effectively than rich or lean flames.
4. The inhibition chemistry of essentially rich hydrocarbon flames with equivalence ratio close to the flammability limit is not adequately investigated. It is explained by very complex combustion chemistry of soot and its precursors formation, which interaction with inhibitor was not explored.
5. A search for novel alternative chemically active inhibitors should be continued and it must be based on fundamental understanding of combustion chemistry of certain

compounds. Simultaneously the search should involve various flames over a wide range of equivalence ratios, auto-ignition of various mixtures, and other objects of study.

7. References

Baratov, A.; Korol'chenko, A. & Kravchuk, G. (1990). *Fire and Explosion Hazard of Substances and Materials and Means of Fire Suppression: Handbook,* Khimiya, Moscow [in Russian]

Biordi, J.; Lazarra, C. & Papp, J. (1974). Molecular beam mass spectrometry applied to determining the kinetics of reactions in flames. I. Empirical characterization of flame perturbation by molecular beam sampling probes. *Combustion and Flame,* Vol.23, No. 1, pp. 73-82, ISSN 0010-2180

Bolshova, T. & Korobeinichev, O. (2006). Promotion and Inhibition of a Hydrogen–Oxygen Flame by the Addition of Trimethyl Phosphate, *Combustion, Explosion, and Shock Waves,* Vol. 42, No. 5, pp. 493–502, ISSN 0010-5082

Bonne, U.; Jost, W. & Wagner, H. (1962). Iron Pentacarbonyl in CH_4-O_2 and Air Flames, *Fire Res. Abstr. Rev.,* Vol. 4, pp. 6–18

Bosschaart, K.;de Goey, L. & Burgers, J. (2004). The laminar burning velocity of flames propagation in mixtures of hydrocarbons and air measured with the heat flux method, *Combustion and Flame,* Vol. 136, No. 3, pp. 261–269, ISSN 0010-2180

Burton, K.; Ladouceur, H. & Fleming J. (1992). An Improved Noncatalytic Coating for Thermocouples. *Combust. Sci. Technol.,* Vol. 81, pp. 141-145, ISSN 0010-2202

Coward, H. & Jones, G. (1952). Limits of flammability of gases and vapors, *Bureau of Mines Bull. No. 503,* Washington, D.C., USA

Curran, H.; Gallagher, S.; Conaire, M. & Simmie J. (2003). Comprehensive Modelling Study of Methane Oxidation. *Proceedings of 3rd European Combustion Meeting,* Orleans, France, October 2003

Curran, H.; Jayaweera, T.; Pitz, W. & Westbrook, C. (2004). A Detailed Modeling Study of Propane Oxidation, *Proceedings of Western States Section meeting of The Combustion Institute,* Davis, CA, USA, March 2004

De Goey, L.; Van Maaren, A. & Quax, R. (1993). Stabilization of Adiabatic Premixed Laminar Flames on a Flat Flame Burner. *Combust. Sci. Technol.,* Vol. 92, p.201-207, ISSN 0010-2202

Dodonov, S.; Karpov, A. & Pevzner, A. (1990) Information-measuring system for research of streams are charged particles. USSR Inventor's Certificate No. 1457716, *Byull. Izobret.,* No. 47. p.272

Egolfopoulos, F. & Law. C. (1990).An experimental and computational study of the burning rates of ultra-lean to moderately-rich $H_2/O_2/N_2$ laminar flames with pressure variations, *Proc. Combust. Inst.,* Vol. 23, pp. 333-340, ISSN 1540-7489

Gerasimov, I.; Knyazkov, D.;Shmakov, A.; Paletsky, A.; Shvartsberg, V.; Bolshova, T. &Korobeinichev, O. (2011).Inhibition of Hydrogen-Oxygen Flames by Iron Pentacarbonyl at Atmospheric Pressure, *Proc. Combust. Inst.,* Vol. 33, No.2, pp. 2523-2529, ISSN 1540-7489

Glaude, P.; Curran H.; Pitz, J. &Westbrook, C. (2000). Kinetic Study of the Combustion of Organophosphorus Compounds, *Proc. Combust. Inst.,* Vol. 28, pp. 1749-1756, ISSN 1540-7489

Grosshandler, W.; Donnelly, M. & Womeldorf, C. (1998). Lean flammability limit as a fundamental refrigerant property: Phase III, *NIST Interim Tech. Report*, September 2011, Available from: http://www.fire.nist.gov/bfrlpubs/fire98/PDF/f98061.pdf

Hastie, J. & Bonnell, D. (1980).Molecular Chemistry of Inhibited Combustion Systems, *National Bureau of Standards*, Washington, D.C., 1980, NBSIR 80–2169

Hertzberg, M. (1976). The theory of flammability limits: natural convection, *Bureau of Mines.* Rep. of Investigation No. RI-8127

Hichens, R.; Dlugogorski, B. & Kennedy, E. (1999). Advantages and drawbacks of tubular flow burner for testing flammability limits,*Proceedings of Halon Options Technical Working Conference HOTWC-1999,* Albuquerque, NM, USA, April 1999.

Hughes, K.; Turanyi, T. & Pilling, M. (2001). The Leeds methane oxidation mechanism, ver 1.5, In: *Combustion Simulation,* September 2011, Available from: http://garfield.chem.elte.hu/Combustion/methane.htm

Ishizuka, S. (1991). Determination of flammability limits using a tubular flame geometry, *J. Loss Prev. Process. Ind.*, Vol. 4, pp. 185–193, ISSN 0950-4230

Jayaweera, T.; Melius, C.; Pitz, W.; Westbrook, C.; Korobeinichev, O.; Shvartsberg, V.; Shmakov, A.; Rybitskaya, I. & Curran, H. (2005). Flame Inhibition by Phosphorus-Containing Compounds over a Range of Equivalence Ratios, *Combustion and Flame,* Vol. 140, No.1-2 pp. 103-115, ISSN 0010-2180

Kaskan, W. (1957). The Dependence of Flame Temperature on Mass Burning Velocity. *Proc. Combust. Inst.*, Vol. 6, pp. 134-143, ISSN 1540-7489

Kee, R.; Crcar, J.; Smooke, M. & Miller J. (1989a). A Fortran Program for Modeling Steady Laminar One-Dimensional Premixed Flames, *Sandia National Laboratories,* 1989, Report No. SAND85–8240

Kee, R.; Rupley, F. & Miller, J. (1989b). CHEMKIN-II:A Fortran Chemical Kinetics Package for the Analysis of the Gas Phase Chemical Kinetics, *Sandia National Laboratories,* 1989, Report No. SAND89–8009B

Knyazkov, D.; Shmakov, A. & Korobeinichev, O. (2007). Application of molecular beam mass spectrometry in studying the structure of a diffusive counterflow flame of CH_4/N_2 and O_2/N_2 doped with trimethylphosphate, *Combustion and Flame,* Vol. 151, No.1-2 pp. 37-45, ISSN 0010-2180

Knyazkov, D.; Yakimov, S. Korobeinichev, O. & Shmakov, A. (2008). Effect of Trimethylphosphate Additives on the Flammability Concentration Limits of Premixed Methane–Air Mixtures,*Combustion, Explosion, and Shock Waves,* Vol. 44, No. 1, pp. 9–17, ISSN 0010-5082

Korobeinichev, O.; Chernov, A. & Shvartsberg, V. (1994). Mass-spectrometric Investigation of the Structure of a Stoichiometric $H_2/O_2/Ar$ Flame Doped with Trimethylphosphate and N-Tributylphosphate, *Preprint of Paper - American Chemical Society, Division of Fuel Chemistry,* Vol. 39, No. 1, pp. 139-197, Washington, D.C., USA

Korobeinichev, O.; Ilyin, S.; Mokrushin, V. & Shmakov, A. (1996). Destruction Chemistry of Dimethyl Methylphosphonate in $H_2/O_2/Ar$ Flame Studied by Molecular Beam Mass Spectrometry. *Combust. Sci. Technol.,* Vol. 116–117, pp. 51-67, ISSN 0010-2202

Korobeinichev, O.; Shvartsberg, V.; Il'in, S.; Chernov, A. & Bolshova, T. (1999a). Laminar Flame Structure in a Low-Pressure Premixed $H_2/O_2/Ar$ Mixture. *Combustion, Explosion, and Shock Waves,* Vol. 35, No. 3, pp.239-244, ISSN 0010-5082

Korobeinichev, O.; Shvartsberg, V. & Chernov, A. (1999b). The Destruction Chemistry of Organophosphorus Compounds in Flames—II: Structure of a Hydrogen-Oxygen Flame Doped with Trimethyl Phosphate. *Combustion and Flame,* Vol. 118, pp.727-732, ISSN 0010-2180

Korobeinichev, O.; Bolshova, T.; Shvartsberg, V.; Chernov, A. & Mokrushin, V. (1999c). Inhibition Effect of TMP on $CH_4/O_2/Ar$ and $H_2/O_2/Ar$ Flames, *Proceedings of Halon Options Technical Working Conference HOTWC-1999,* Albuquerque, NM, USA, April 1999.

Korobeinichev, O.;Ilyin, S.; Shvartsberg, V. & Chernov, A. (1999d). The Destruction Chemistry of Organophosphorus Compounds in Flames—I: Quantitative Determination of Final Phosphorus-Containing Species in Hydrogen-Oxygen Flames, *Combustion and Flame,* Vol. 118, pp. 718-726, ISSN 0010-2180

Korobeinichev, O.; Ilyin, S.; Bolshova, T.; Shvartsberg V. & Chernov A. (2000). The Chemistry of the Destruction of Organophosphorus Compounds in Flames—III: The Destruction of DMMP and TMP in a Flame of Hydrogen and Oxygen. *Combustion and Flame,* Vol. 121, pp. 593-609, ISSN 0010-2180

Korobeinichev, O.; Bolshova, T.; Shvartsberg V. & Chernov A. (2001). Inhibition and Promotion of Combustion by Organophosphorus Compounds Added to Flames of CH_4 or H_2 in O_2 and Ar, *Combustion and Flame,* Vol. 125, pp. 744-751, ISSN 0010-2180

Korobeinichev, O.; Shmakov, A.; Shvartsberg, V.; Knyazkov, D.; Makarov, V.; Koutsenogii, K.; Samsonov, Yu.; Nifantev, E.; Kudryavtsev I.; Goryunov, E.; Nikolin, V. & Kaledin,V. (2003). Study of Effect of Aerosol and Vapors of Organophosphorus Fire Suppressants on Diffusion Heptane and Premixed C_3H_8/Air Flames, *Proceedings of Halon Options Technical Working Conference HOTWC-2003,* Albuquerque, NM, USA, May 2003.

Korobeinichev, O.; Shvartsberg, V.; Shmakov, A.; Bolshova, T.; Jayaweera, T.; Melius, C.; Pitz, W.; Westbrook, C. & Curran, H. (2005). Flame Inhibition by Phosphorus-Containing Compounds in Lean and Rich Propane Flames, *Proc. Combust. Inst.,* Vol. 30, pp. 2353-2360, ISSN 1540-7489

Korobeinichev, O.; Shvartsberg, V.;Shmakov, A.; Knyazkov, D. & Rybitskaya, I. (2007). Inhibition of Atmospheric Lean and Rich $CH_4/O_2/Ar$ Flames by Phosphorus-Containing Compound, *Proc. Combust. Inst.,* Vol. 31, pp. 2741-2748, ISSN 1540-7489

Korobeinichev, O.; Rybitskaya, I.; Shmakov, A.; Chernov, A.; Bolshova, T. & Shvartsberg, V.(2009). Inhibition of atmospheric-pressure $H_2/O_2/N_2$ flames by trimethylphosphate over range of equivalence ratio, *Proc. Combust. Inst.,* Vol. 32, pp. 2591-2597, ISSN 1540-7489

Lask, G. &Wagner, H. (1962). Influence of additives on the velocity of laminar flames, *Proc. Combust. Inst.,* Vol. 8, pp. 432-438, ISSN 1540-7489

Law, C.; Zhu, D. & Yu, G. (1986). Propagation and extinction of stretched premixed flames, *Proc. Combust. Inst.,* Vol. 21, pp. 1419–1426, ISSN 1540-7489

Li, J.; Zhao, Zh.; Kazakov, A. & Dryer, F. (2004). An Updated Model and Discussion of Modeling Challenges in High-Pressure H_2/O_2 Flames, *Int. J. Chem. Kinetics,* Vol.36, No. 10, pp. 566-575, ISSN 1097-4601

Linteris, G.; Rumminger, M.; Babushok ,V. & Tsang, W. (2000a). Flame inhibition by ferrocene and blends of inert and catalytic agents, *Proc. Combust. Inst.,* Vol. 28, pp. 2965-2972, ISSN 1540-7489

Linteris, G. (2001). Suppression of cup-burner diffusion flames by supereffective chemical inhibitors and inert compounds, *Proceedings of Halon Options Technical Working Conference HOTWC-2001,* Albuquerque, NM, USA, April 2001.

Linteris, G.; Katta, V. & Takahashi, F. (2004). Experimental and numerical evaluation of metallic compounds for suppressing cup-burner flames, *Combustion and Flame,* Vol. 138, pp. 78-96, ISSN 0010-2180

MacDonald, M.; Gouldin, F. &Fisher, E. (2001). Temperature dependence of phosphorus-based flame inhibition, *Combustion and Flame,* Vol. 124, No. 4, pp. 668-683, ISSN 0010-2180

Mackie, J.; Bacskay, G.; & Haworth, N. (2002). Reactions of Phosphorus-Containing Species of Importance in the Catalytic Recombination of H + OH: Quantum Chemical and Kinetic Studies, *J. Phys. Chem. A,* Vol. 106, No. 45, pp. 10825-10830, ISSN 1089-5639

Miller, D.; Evers, R. & Skinner, G. (1963). Effects of various inhibitors on hydrogen-air flame speeds, *Combustion and Flame,* Vol. 7, pp. 137-142, ISSN 0010-2180

Reinelt, D. & Linteris, G. (1996). Experimental study of the inhibition of premixed and diffusion flames by iron pentacarbonyl, *Proc. Combust. Inst.,* Vol. 26, pp. 1421-1428, ISSN 1540-7489

Rumminger, M.; Reinelt, D.; Babushok, V. & Linteris, G. (1999). Numerical study of the inhibition of premixed and diffusion flames by iron pentacarbonyl, *Combustion and Flame,* Vol. 116, pp. 207-219, ISSN 0010-2180

Rumminger, M. & Linteris, G. (2000). The role of particles in the inhibition of premixed flames by iron pentacarbonyl, *Combustion and Flame,* Vol. 123, pp. 82-94, ISSN 0010-2180

Rumminger, M. & Linteris, G. (2002). The role of particles in the inhibition of counterflow diffusion flames by iron pentacarbonyl, *Combustion and Flame,* Vol. 128, pp. 145-164, ISSN 0010-2180

Rybitskaya, I.; Shmakov, A. & Korobeinichev, O. (2007). Propagation Velocity of Hydrocarbon–Air Flames Containing Organophosphorus Compounds at Atmospheric Pressure, *Combustion, Explosion, and Shock Waves,* Vol. 43, No. 3, pp. 253–257, ISSN 0010-5082

Saito, N.; Saso, Y.; Liao, C.; Ogawa, Y. & Jnoue, Y. (1995). Flammability peak concentrations of halon replacements and their function as fire suppressant, *Halon Replacements: Technology and Science, ACS Symp. Ser., Amer. Chem. Soc.,* pp. 243–257

Shebeko, Yu.; Azatyan, V.; Bolodian, I.; Navzenya, V.; Kopylov, S. Shebeko, D. & Zamishevski, E. (2000). The influence of fluorinated hydrocarbons on the combustion of gaseous mixtures in a closed vessel, *Combustion and Flame,* Vol. 121,pp. 542–547, ISSN 0010-2180

Shmakov, A.; Korobeinichev, O.; Shvartsberg, V.; Knyazkov, D.; Bolshova, T. & Rybitskaya, I. (2004). Inhibition of Premixed and Non-Premixed Flames with Phosphorus-Containing Compounds, *Proc. Combust. Inst.,* Vol. 30, No. 2, pp. 2342–2352, ISSN 1540-7489

Shmakov, A.; Korobeinichev, O.; Shvartsberg, V.; Yakimov, S.; Knyazkov, D.; Komarov, V. & Sakovich, G. (2006). Testing ogranophosphorus, organofluorine, and metal-containing compounds and solid-propellant gas-generating compositions doped with phosphorus-containing additives as effective fire suppressants, *Combust. Explos. Shock Waves,* Vol. 42, No. 6, pp.678–687, ISSN 0010-5082

Shvartsberg, V.; Bolshova, T. &Korobeinichev O.(2010). Numerical Study of Inhibition of Hydrogen/Air Flames by Atomic Iron, *Energy and Fuels,* Vol.24, No. 3, pp 1552–1558, ISSN 0887-0624

Smith, G.; Golden, D.; Frenklach, M.; Moriarty, N.; Eiteneer, B.; Goldenberg, M.; Bowman, T.; Hanson, R.; Song, S.; Gardiner, W.; Lissianski, V. & and Qin, Z. (1999).GRIMec3.0, In: *GRI-Mech Home Page,* September 2011, Available from: http://www.me.berkeley.edu/gri_mech/

Staude, S. & Atakan, B. (2009). An Investigation of Equilibrium Iron Thermochemistry in Flames,*The Open Thermodynamics Journal,* Vol. 3, pp. 42-46, ISSN 1874-396X

Staude, S.; Hecht, C.; Wlokas, I.; Schulz, C.; Atakan, B. (2009). Experimental and Numerical Investigation of $Fe(CO)_5$ Addition to a Laminar Premixed Hydrogen/Oxygen/Argon Flame, *Z. Phys. Chem.,* Vol. 223, pp. 639-649, ISSN 0942-9352

Tian, K.; Li, Z.; Staude, S.; Li, B.; Sun, Z.; Lantz, A.; Alden, M. & Atakan B. (2009). Influence of ferrocene addition to a laminar premixed propene flame: Laser diagnostics, mass spectrometry and numerical simulations, *Proc. Combust. Inst.,* Vol. 32, pp. 445–452, ISSN 1540-7489

Twarowski, A. (1993a). The influence of phosphorus oxides and acids on the rate of H + OH recombination. *Combustion and Flame,* Vol. 94, pp. 91-107; ISSN 0010-2180

Twarowski, A. (1993b). Photometric determination of the rate of H2O formation from H and OH in the presence of phosphine combustion products, *Combustion and Flame,* Vol. 94, pp. 341-348, ISSN 0010-2180

Twarowski, A. (1995). Reduction of a phosphorus oxide and acid reaction set, *Combustion and Flame,* Vol. 102, pp. 41-54, ISSN 0010-2180

Twarowski, A. (1996). The Temperature Dependence of H+OH Recombination in Phosphorus Oxide Containing Combustion Gases, *Combustion and Flame,* Vol. 105, pp. 407-413, ISSN 0010-2180

Van Maaren, A.; Thung, D. & De Goey, L. (1994). Measurement of Flame Temperature and Adiabatic Burning Velocity of Methane/Air Mixtures. *Combust. Sci. Technol.,* Vol. 96, pp. 327-344, ISSN 0010-2202

Werner, J. & Cool, T. (1999). Kinetic Model for the Decomposition of DMMP in a Hydrogen/Oxygen Flame. *Combustion and Flame,* Vol. 117, pp. 78-98, ISSN 0010-2180

Wlokas, I.; Staude, S.; Hecht, C.; Atakan, B. & Schulz, C. (2009). Measurement and simulation of Fe-atom concentration in premixed $Fe(CO)_5$-doped low-pressure H_2/O_2 flames, *Proceedings of the European Combustion Meeting ECM-2009,* Vienna, Austria, April 2009.

Womeldorf, C.; King, M.; & Grosshandler, W. (1995). Lean flammability limit as a fundamental refrigerant property: Phase I, In: *NIST Interim Tech. Report,* September 2011,Available from: http://www.fire.nist.gov/bfrlpubs/fire95/PDF/f95083.pdf

Womeldorf, C. & Grosshandler, W. (1996). Lean flammability limit as a fundamental refrigerant property: Phase II, *NIST Interim Tech. Report,* September 2011, Available from: http://www.fire.nist.gov/bfrlpubs/fire96/PDF/f96072.pdf

Zachariah, M. & Smith, O. (1987). Experimental and numerical studies of sulfur chemistry in $H_2/O_2/SO_2$ flames, *Combustion and Flame,* Vol. 69, No. 2, pp. 125-139, ISSN 0010-2180

15

Stoichiometric Approach to the Analysis of Coal Gasification Process

Mamoru Kaiho and Osamu Yamada
National Institute of Advanced Industrial Science and Technology
Japan

1. Introduction

Coal is a solid fuel and less convenient for storage and transportation than petroleum and natural gas. In addition, it usually holds undesirable compounds containing S, N, and so on. Gasification converts coal into H_2, CO, and CH_4 by the reaction with gasifying agents such as O_2 and H_2O. There are three types of commercialized processes, 1) fixed bed gasifier: lump coal is gasified in a shaft reactor at 900~1000℃, 2) fluidized bed gasifier: crashed coal is gasified in a fluidized reactor at around 900℃, and 3) entrained bed gasifier: pulverized coal is gasified by burner system at 1350~1600℃. In order to improve the performance of 500~2000t/d plants, operating conditions should be appropriately determined and controlled based on the understanding of chemical reaction process that occurs in gasifier. It has been believed that coal supplied to gasifier is decomposed thermally to produce gases such as H_2, CO, CO_2, H_2O, and CH_4, tar, and char. Tar and char react with O_2 and H_2O supplied to form H_2, CO, CO_2 and CH_4. However, it is quite difficult to understand gasification mechanism only based on the kinetics of above reactions.

An equilibrium constant has been often used to understand the composition of gasification products. It is, however, only available to the stable state, and therefore inapplicable to the analysis of transient composition of gases in operation, which is constantly fluctuating.

As mentioned above, the chemical process in a large scale gasifier cannot be perfectly explained by kinetics and equilibrium. Composition of gases produced by gasification is a very clear indicator of the chemical state in the gasifier. We propose a stoichiometric method to analyze the reaction process of coal gasification.

2. Derivation of chemical reaction formula

A gasification reaction is composed of various kinds of chemical processes such as pyrolysis of coal, decomposition of tar, oxidation of char, combustion of gas, shift reaction, and formation of various organic compounds. In order to elucidate the reaction process, the method to delve into the composition of gas for information on the reaction state in the reactor needs to be established. As has been mentioned, kinetic and equilibrium theories are not available for this purpose.

2.1 Molecular formula of coal

Coal scientists imagine a coal to be a complex of heterogeneous macromolecular compounds composed of aromatic rings, aliphatic chains, and various kinds of functional groups containing heteroatoms such as O, N, and S. One of the problems to express the reaction formula of gasification is how we express the molecular formula of coal. Even if a relatively accurate reaction formula could be obtained, it would not be available for investigation of practical analysis. Since the molecular structure of most of product is very simple, the chemical process may be discussed sufficiently based on the CH_mO_n regardless of the detailed structure of coal molecule.

2.2 Estimation of reaction formula

We express the gasification reaction by formula (1), where CH_mO_n and $CH_{m'}O_{n'}$ are coal and tar respectively.

$$CH_mO_n + \alpha O_2 + \beta H_2O \rightarrow \gamma H_2 + \delta CO + \varepsilon CO_2 + \eta CH_4 + \zeta C_2H_4 + \theta C_2H_6 + \lambda CH_{m'}O_{n'} \quad (1)$$

Equations concerning elemental balance for C, H and O for formula (1) are described as (2), (3) and (4) respectively.

$$1 = \delta + \varepsilon + \eta + 2\zeta + 2\theta + \lambda \quad (2)$$

$$m + 2\beta = 2\gamma + 4\eta + 4\zeta + 6\theta + m'\lambda \quad (3)$$

$$n + 2\alpha + \beta = \delta + 2\varepsilon + n'\lambda \quad (4)$$

Let the total moles of product gases in formula (1) be equal to Σ.

$$\Sigma = \gamma + \delta + \varepsilon + \eta + \zeta + \theta \quad (5)$$

When the concentrations of H_2, CO, CO_2, CH_4, C_2H_4, and C_2H_6 (dry and N_2 free) are represented by p, q, r, s, t, and u respectively, the mole number of each gases is described as follows.

$$H_2 \quad ; \quad \gamma = p\Sigma \quad (6)$$

$$CO \quad ; \quad \delta = q\Sigma \quad (7)$$

$$CO_2 \quad ; \quad \varepsilon = r\Sigma \quad (8)$$

$$CH_4 \quad ; \quad \eta = s\Sigma \quad (9)$$

$$C_2H_4 \quad ; \quad \zeta = t\Sigma \quad (10)$$

$$C_2H_6 \quad ; \quad \theta = u\Sigma \quad (11)$$

A sampling gas for analysis has usually been drawn out from the main stream of product gas, passed into cooler to remove condensable H_2O and tar, and then measured its volume by gas meter and analyzed its composition by various kinds of analyzer. Yield of tar has been evaluated as its moles per unit volume of gas produced. The molar yields of tar is presented by v, λ in formula (1) is described as equation (12) that is written in the same form as that of equation (6) ~ (11).

$$\lambda = v\Sigma \tag{12}$$

Consequently, we obtain ten solutions, α ~ η and Σ, mathematically because we could prepare eleven equations from (2) to (12).

Equation (2) is rewritten as (13) by employing (7) ~ (12).

$$1 = q\Sigma + r\Sigma + s\Sigma + 2t\Sigma + 2u\Sigma + v\Sigma \tag{13}$$

(13) is rearranged to equation (14).

$$\Sigma = \frac{1}{q + r + s + 2t + 2u + v} \tag{14}$$

Yields of gases and tar, γ ~ λ in (1) are expressed by equation (15) ~ (21).

$$H_2 \quad ; \quad \gamma = \frac{p}{q + r + s + 2t + 2u + v} \tag{15}$$

$$CO \quad ; \quad \delta = \frac{q}{q + r + s + 2t + 2u + v} \tag{16}$$

$$CO_2 \quad ; \quad \varepsilon = \frac{r}{q + r + s + 2t + 2u + v} \tag{17}$$

$$CH_4 \quad ; \quad \eta = \frac{s}{q + r + s + 2t + 2u + v} \tag{18}$$

$$C_2H_4 \quad ; \quad \zeta = \frac{t}{q + r + s + 2t + 2u + v} \tag{19}$$

$$C_2H_6 \quad ; \quad \theta = \frac{u}{q + r + s + 2t + 2u + v} \tag{20}$$

$$\text{Tar} \quad ; \quad \lambda = \frac{v}{q + r + s + 2t + 2u + v} \tag{21}$$

The expression for β in formula (1) is written from (3) as follows;

$$\beta=\gamma+2\eta+2\zeta+3\theta+0.5m'\lambda-0.5m$$

When γ, η, ζ, θ, and λ of above equation are substituted by the expression of (15), (18), (19), (20), and (21) respectively, equation (22) is obtained.

$$H_2O \quad ; \quad \beta = \frac{p+2s+2t+3u+0.5m'}{q+r+s+2t+2u+v} - 0.5m \tag{22}$$

Expression of α is written by following equation from (4).

$$\alpha = \frac{\delta+2\varepsilon+\beta+n'\lambda-n}{2}$$

When δ, ε, β, and λ are substituted by (16), (17), (21), and (22) respectively, equation (23) is obtained.

$$\alpha = \frac{(-p+q+2r-2s-2t-3u)-05m'v+n'v)}{2(q+r+s+2t+2u+v)} + 2.5m - 0.5n \tag{23}$$

3. Feature of this method

Since every equation used to evaluate α ~ λ has been derived from (2), (3), and (4) without any arbitrary assumption and approximation, it may be applicable to any practical process. Since the left side of (1) is the reactant of gasification, and the right side is product, although (1) is written in the simplest form, it expresses exactly the material balance of gasification. Therefore, we can readily estimate the molar amounts of H_2, CO, and CH_4 from γ, δ, and η to judge whether the molar ratio of O_2 or H_2O to coal is appropriate or not from α or β. In addition, each value of α ~ λ is estimated simultaneously by arithmetic calculations using the values of concentration of each gas and H/C and O/C ratios of coal and tar. Formula (1) seems to be also obtained from the flow rates of coal, gasifying agents, and each gas produced. The carbon conversion for practical process has been usually found to be less than 100%, however, the accuracy of industrial instruments to control each flow rates of raw materials (coal, O_2, and H_2O) or to measure that of products (gas, tar, drain, ash, and residual char) are insufficient to get a formula satisfying the law of conservation of mass.

4. Analysis of reaction formula obtained

We consider that a reference standard should be necessary to elucidate the reaction process concealed in formula (1). We tried to prepare the reaction formula expressed in mathematical form by a theoretical approach for this purpose. The quantitative details of formula (1) obtained from composition of gas becomes apparent by the comparison with a theoretical formula which is derived from (24) by mathematical means.

$$CH_mO_n+0.5(1-n)O_2 \rightarrow 0.5mH_2+CO \tag{24}$$

4.1 Hypotheses for the chemical process

We assumed that the reaction process of gasification has been divided for convenience into two categories, namely partial oxidation and secondary reaction. Coal is converted into H_2, CO, CO_2, and H_2O by the reaction with O_2 and H_2O. Most solid-gas reactions except for hydrogenation described later are classified in it. In secondary reaction step, the gas produced by partial oxidation changes its composition by shift reaction and formation of organic compounds. We transform standard reaction formula (24) mathematically according to both steps mentioned above and derived a formula having the same form of formula (1).

4.2 Partial oxidation step

The amount of O_2 in (24), i.e. 0.5(1- n), is regarded as a standard amount of O_2 for gasification. α in (1) is written as bellow.

$$\alpha = 0.5(1-n) + O_{ex} \tag{25}$$

In the case of $O_{ex}>0$, it can be assumed that coal is firstly gasified with $0.5(1 - n)O_2$ and converted into product gas, i.e., $0.5mH_2$ + CO. Then $2O_{ex}$ mole of the product gas is burned with O_{ex} mole of O_2, $2O_{ex}$ mole of H_2O and CO_2 in total was generated. When moles of H_2 burned is taken as variable x, that of CO is presented as $(2O_{ex}-x)$ and reaction formula after partial oxidation step was written as follows;

$$CH_mO_n + \{0.5(1-n) + O_{ex}\}O_2 \rightarrow (0.5m-x)H_2 + (1-2O_{ex}+x)CO + (2O_{ex}-x)CO_2 - xH_2O \tag{26}$$

In the case of $O_{ex} < 0$, since α is not enough to complete (24), $-2O_{ex}$ of residual carbon is produced intermediately. The reaction is expressed by following formula.

$$CH_mO_n + \{0.5(1-n) + O_{ex}\}O_2 \rightarrow 0.5mH_2 + (1+2O_{ex})CO - 2O_{ex}C$$

$-2O_{ex}C$ in above formula should be gasified with $-2O_{ex}H_2O$ as follows;

$$-2O_{ex}C - 2O_{ex}H_2O \rightarrow -2O_{ex}CO - 2O_{ex}H_2$$

The final formula after partial oxidation step is estimated as follows;

$$CH_mO_n + \{0.5(1-n) + O_{ex}\}O_2 - 2O_{ex}H_2O \rightarrow (0.5m - 2O_{ex})H_2 + CO \tag{27}$$

On the other hand, we can understand the partial oxidation step in further detail from a broader standpoint of view. Since it is stipulated that organic constituents are not produced in the partial oxidation, formula can be expressed as follows.

$$CH_mO_n + \alpha O_2 + \beta H_2O \rightarrow \gamma H_2 + \delta CO + \varepsilon CO_2$$

The elemental balance of this formula is given below.

$$1 = \delta + \varepsilon \tag{C}$$

$$m+2\beta=2\gamma \tag{H}$$

$$n+2\alpha+\beta=\delta+2\varepsilon \tag{O}$$

The calculation, 2(C)+0.5(H)-(O) was performed.

$$\begin{array}{rcl} 2(C) & : & 2=2\delta+2\varepsilon \\ 0.5(H) & : & 0.5m+\beta=\gamma \\ +\big)\quad -(O) & : & -n-2\alpha-\beta=-\delta-2\varepsilon \\ \hline \multicolumn{3}{c}{2(1+0.25m-0.5n)-2\alpha=\delta+\gamma} \end{array}$$

As theoretical moles of O_2 for complete combustion of coal is defined by μ (μ= 1+0.25m-0.5n), the equation that shows the relationship between the sum of H_2 and CO produced and amount of O_2 reacted is obtained.

$$(\gamma+\delta)=2\mu-2\alpha$$

When the both side is divided with μ and let α/μ be replaced by χ, (28) is obtained. χ is called oxygen ratio and generally used in the analysis of combustion.

$$(\gamma+\delta)/\mu=2-2\chi \tag{28}$$

(O) - (C) is performed, equation (29) is obtained.

$$\begin{array}{rcl} (O) & : & n+2\alpha+\beta=\delta+2\varepsilon \\ +\big)\quad -(C) & : & -1=-\delta-\varepsilon \\ \hline \multicolumn{3}{c}{\varepsilon-\beta=2\alpha-(1-n)} \end{array}$$

$$(\varepsilon-\beta)/\mu=2\chi-(1-n)/\mu \tag{29}$$

Since β is moles of H_2O decomposed, –β means that produced. Therefofe, (ε–β) in equation (29) indicates the sum of CO_2 and H_2O produced in the case of $\chi > 0.5(1-n)/\mu$.

In the case of $\chi \leq 0.5(1-n)/\mu$, the following equation is valid where –ε means moles of CO_2 decomposed.

$$(\beta-\varepsilon)/\mu=(1-n)/\mu-2\chi \tag{30}$$

The variations of (γ + δ)/μ, (ε – β)/μ, and (β – ε)/μ are exhibited graphically against χ in Fig. 1.

(28) is represented by black solid line with slope -2, (29) is done by red solid one with slope 2, and (30) is done by blue dotted lines with slope -2. Since μ and χ are the parameters that have been commonly used in the theoretical analysis of combustion, Fig. 1 allows us to investigate the partial oxidation step from the macroscopic point of view that covers not only gasification but also combustion. The partial oxidation step can be realized as general comprehension of oxidation process of coal based on Fig. 1.

Partial oxidation step can be further subdivided into various elemental reaction such as pyrolysis, combustion of C and H in coal, water-gas reaction, and Boudouard reaction. Taking all these into account, however, the reaction model may complicate analysis of gasification too much and may be far from practical application.

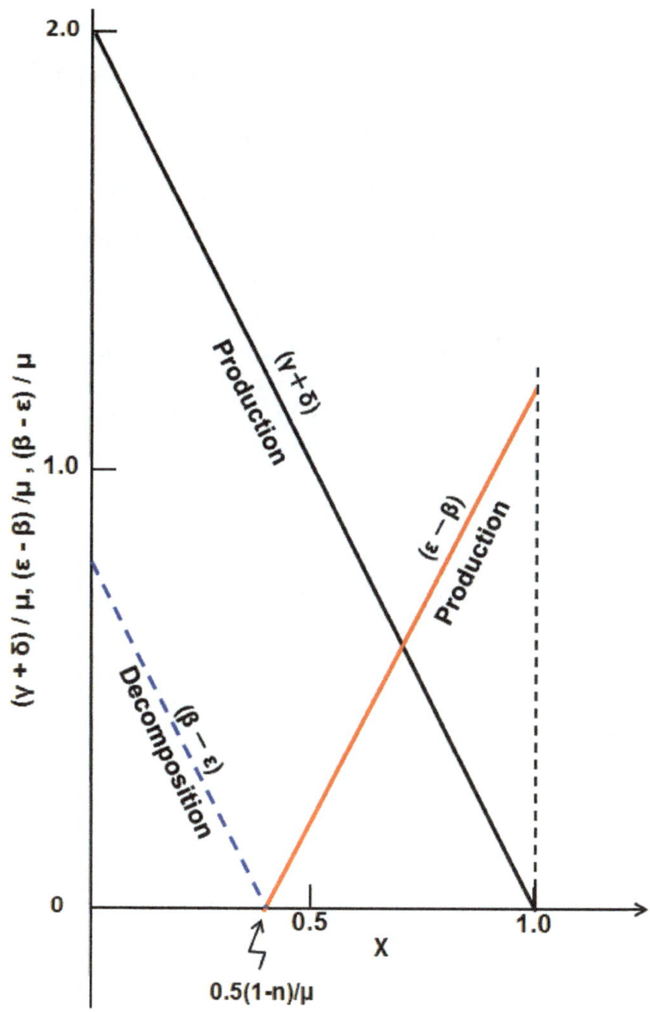

Fig. 1. Stoichiometry of partial oxidation step.

4.3 Shift reaction

In the case of shift reaction, we can easily express the variation of yields of CO, H_2O, H_2, and CO_2 numerically based on its reaction formula.

$$CO + H_2O \rightarrow H_2 + CO_2$$

In the case of $O_{ex}>0$, let y be moles of CO caused shift reaction, formula (26) is modified as follows;

$$\begin{aligned} &CH_mO_n + \{0.5(1-n) + O_{ex}\}O_2 + yH_2O \rightarrow \\ &\quad (0.5m + y - x)H_2 + (1 - 2O_{ex} - y + x)CO + (2O_{ex} + y - x)CO_2 + xH_2O \end{aligned} \tag{31}$$

In the case of $O_{ex} \leq 0$, formula (28) is modified as follows;

$$\begin{aligned} &CH_mO_n + \{0.5(1-n) + O_{ex}\}O_2 + (-2O_{ex} + y)H_2O \rightarrow \\ &\quad (0.5m - 2O_{ex} + y)H_2 + (1-y)CO + yCO_2 \end{aligned} \tag{32}$$

4.4 Formation of organic compounds

Organic constituent such as CH_4, C_2H_4, and C_2H_6 are produced by the synthetic reaction as well as pyrolysis of coal and hydrogenation reaction. There is no question that the synthetic process is classified as secondary reaction step. However, the idea that pyrolysis and hydrogenation are also classified in the same step appears doubtful.

The formulae of pyrolysis, hydrogenation, and synthetic reaction are descrived as follows;

Pyrolysis:

$$l'C + m'H + n'O + l'H_2O \rightarrow C_{l'}H_{m'}O_{n'} + l'H_2O \tag{33}$$

Hydrogenation reaction:

$$l'C + 0.5m'H_2 + n'O + l'H_2O \rightarrow C_{l'}H_{m'}O_{n'} + l'H_2O \tag{34}$$

Synthetic reaction:

$$l'CO + (l' + 0.5m')H_2 + n'O \rightarrow C_{l'}H_{m'}O_{n'} + l'H_2O \tag{35}$$

Where $C_{l'}H_{m'}O_{n'}$ is organic compound and $l'H_2O$ is moisture in an atmosphere. When $m'H$ in the left side of (33) cause $0.5m'H_2$, it changes into (34). When water-gas reaction is occurred between $l'C$ and $l'H_2O$ in the left side of (34), it turns into formula (35). In other words, it means that the effect of the formation of organic compound on the yields of other inorganic components can be appreciated mathematically according to formula (35). Since O atom in the product gas is very reactive, however, it readily reacts with H_2. Therefore, formula (35) is finally altered as follows;

$$l'CO + (l' + 0.5m' - n')H_2 + n'H_2O \rightarrow C_{l'}H_{m'}O_{n'} + l'H_2O \tag{36}$$

The difference in formation processes of organic compounds in (1) is shown as follows;

$$CH_4 \begin{cases} \text{Pyrolysis} & ; \quad C+4H+H_2O \rightarrow CH_4+H_2O & (37) \\ \text{Hydrogenation} & ; \quad C+2H_2+H_2O \rightarrow CH_4+H_2O & (38) \\ \text{Synthesis} & ; \quad CO+3H_2 \rightarrow CH_4+H_2O & (39) \end{cases}$$

$$C_2H_4 \begin{cases} \text{Pyrolysis} & ; \quad 2C+4H+2H_2O \rightarrow C_2H_4+2H_2O & (40) \\ \text{Hydrogenation} & ; \quad 2C+2H_2+2H_2O \rightarrow C_2H_4+2H_2O & (41) \\ \text{Synthesis} & ; \quad 2CO+4H_2 \rightarrow C_2H_4+2H_2O & (42) \end{cases}$$

$$C_2H_6 \begin{cases} \text{Pyrolysis} & ; \quad 2C+6H+2H_2O \rightarrow C_2H_6+2H_2O & (43) \\ \text{Hydrogenation} & ; \quad 2C+3H_2+2H_2O \rightarrow C_2H_6+2H_2O & (44) \\ \text{Synthesis} & ; \quad 2CO+5H_2 \rightarrow C_2H_6+2H_2O & (45) \end{cases}$$

$$CH_{m'}O_{n'} \begin{cases} \text{Pyrolysis} & ; \quad C+m'H+n'O+H_2O \rightarrow CH_{m'}O_{n'}+H_2O & (46) \\ \text{Hydrogenation} & ; \quad C+0.5m'H+n'O+H_2O \rightarrow CH_{m'}O_{n'}+H_2O & (47) \\ \text{Synthesis} & ; \quad CO+(1+0.5m'-n')H_2+n'H_2O \rightarrow CH_{m'}O_{n'}+H_2O & (48) \end{cases}$$

In the case of $O_{ex}>0$, when z mole of CH_4 is formed, formula (31) is modified.

$$\begin{aligned} &CH_mO_n+\{0.5(1-n)+O_{ex}\}O_2+(y-x-z)H_2O \rightarrow \\ &\quad (0.5m+y-x-3z)H_2+(1-2O_{ex}-y+x-z)CO+(2O_{ex}+y-x)CO_2+zCH_4 \end{aligned} \quad (49)$$

When w mole of C_2H_4 is formed, (49) is transformed into (50) according to (42).

$$\begin{aligned} &CH_mO_n+\{0.5(1-n)+O_{ex}\}O_2+(y-x-z-2w)H_2O \rightarrow \\ &\quad (0.5m+y-x-3z-4w)H_2+(1-2O_{ex}-y+x-z-2w)CO \\ &\quad\quad +(2O_{ex}+y-x)CO_2+zCH_4+wC_2H_4 \end{aligned} \quad (50)$$

When u mole of C_2H_6 is formed, (50) is modified to (51) according to (45).

$$\begin{aligned} &CH_mO_n+\{0.5(1-n)+O_{ex}\}O_2+(y-x-z-2w-2u)H_2O \rightarrow \\ &\quad (0.5m+y-x-3z-4w-5u)H_2+(1-2O_{ex}-y+x-z-2w-2u)CO \\ &\quad\quad +(2O_{ex}+y-x)CO_2+zCH_4+wC_2H_4+uC_2H_6 \end{aligned} \quad (51)$$

When v mole of $CH_{m'}O_{n'}$ is formed, (51) is transformed into (52) according to (48).

$$\begin{aligned} &CH_mO_n+\{0.5(1-n)+O_{ex}\}O_2+\{y-x-z-2w-2u-(1-n')v\}H_2O \rightarrow \\ &\quad [0.5m+y-x-3z-4w-5u\{(1-n')+0.5m'\}v]H_2 \\ &\quad\quad +(1-2O_{ex}-y+x-z-2w-2u-v)CO+(2O_{ex}+y-x)CO_2 \\ &\quad\quad\quad +zCH_4+wC_2H_4+uC_2H_6+vCH_{m'}2O_{n'} \end{aligned} \quad (52)$$

In the case of $O_{ex} < 0$, formula (32) is transformed step by step along the reaction formulae in which CH_4, C_2H_4, C_2H_6, and $CH_{m'}O_{n'}$ are synthesized.

$$CH_mO_n + \{0.5(1-n)+O_{ex}\}O_2 + (-2O_{ex}+y-z)H_2O \rightarrow (0.5m-2O_{ex}+y-3z)H_2 + (1-y-z)CO + yCO_2 + zCH_4 \quad (53)$$

When w mole of C_2H_4 is formed according to formula (42), the formula is as follows.

$$CH_mO_n + \{0.5(1-n)+O_{ex}\}O_2 + (-2O_{ex}+y-z-2w)H_2O \rightarrow (0.5m-2O_{ex}+y-4w)H_2 + (1-y-z-2w)CO + yCO_2 + zCH_4 + wC_2H_4 \quad (54)$$

When u mole of C_2H_6 is formed according to formula (45), the formula is as follows.

$$CH_mO_n + \{0.5(1-n)+O_{ex}\}O_2 + (-2O_{ex}+y-z-2w-2u)H_2O \rightarrow (0.5m-2O_{ex}+y-4w-5u)H_2 + (1-y-z-2w-2u)CO + yCO_2 + zCH_4 + wC_2H_4 + uC_2H_6 \quad (55)$$

When v mole of $CH_{m'}O_{n'}$ is formed according to formula (48), the formula is obtained as (41).

$$CH_mO_n + \{0.5(1-n)+O_{ex}\}O_2 + \{-2O_{ex}+y-z-2w-2u-(1-n')v\}H_2O \rightarrow [0.5m-2O_{ex}+y-4w-5u-\{(1-n')+0.5m'\}v]H_2 + (1-y-z-2w-2u-v)CO + yCO_2 + zCH_4 + wC_2H_4 + uC_2H_6 + vCH_{m'}O_{n'} \quad (56)$$

(52) and (56) show the stoichiometric structure of coal gasification. We can investigate the chemical process of gasification from a stoichiometric point of view that is different from conventional kinetic or chemical equilibrium viewpoint.

4.5 Evaluation of practical value of each variable

The numerical expression of each component is shown in Table 1.

(1)	(52) : $O_{ex} \geqq 0$	(56) : $O_{ex} < 0$
α	$0.5(1-n)+O_{ex}$	$0.5(1-n)+O_{ex}$
β	$y-x-z-2w-2u-(1-n')v$	$-2O_{ex}+y-z-2w-2u-(1-n')v$
γ	$0.5m+y-x-3z-4w-5u-\{(1-n')+0.5m'\}v$	$0.5m-2O_{ex}+y-4w-5u-\{(1-n')+0.5m'\}v$
δ	$1-2O_{ex}-y+x-z-2w-2u-v$	$1-y-z-2w-2u-v$
ε	$2O_{ex}+y-x$	y
η	z	z
ζ	w	w
θ	u	u
λ	v	v

Table 1. Comparison of each coefficient in (1) with (52) and (56)

The functions that express the inorganic components or β, γ, δ, and ε differ appreciably with the sign of O_{ex}. In the case of $O_{ex}>0$, the value of O_{ex} is calculated by equation (25) and that of z, w, u, and v are found to be equal to η, ζ, θ, and λ respectively. Following equation is obtained relating to the value of x and y.

$$\varepsilon = 2O_{ex} + y - x \tag{57}$$

Since the value of x, i.e. quantity of H_2 burned with $O_{ex}O_2$, cannot be determined only from the results of ultimate analysis and gas analysis, we assume reasonable value of x based on the rate of combustion of H_2 and CO. When gasifying temperature is too low to maintain the combustion rate of CO, only H_2 seems to be virtually burned. In this case, x become $2O_{ex}$ and equation (58) is valid.

$$x = 2O_{ex} \tag{58}$$

When (58) is substituted in (57), (59) is obtained.

$$y = \varepsilon \tag{59}$$

In usual gasification, the product gas in (24) is burned with $O_{ex}O_2$ as it is, x is calculated by $2O_{ex}\{0.5m/(1+0.5m)\}$ and equation (60) is obtained.

$$y = \varepsilon - 2O_{ex} + \{mO_{ex}/(1+0.5m)\} \tag{60}$$

In the case of $O_{ex}<0$, we can find the relation, $y = \varepsilon$, from Table 1. Since the quantity of O_2 is not enough to satisfy the standard reaction formula, CO_2 cannot be produced by the combustion of CO.

The quantity of H_2 or CO_2 formed by shift reaction has never been evaluated. Our approach can account for the contribution of shift reaction to the composition of gas based on the rational assumption of the quantity of H_2 burned.

5. Application to underground coal gasification

The principle of underground coal gasification (UCG) is shown schematically in Fig. 2.

UCG is a process that gasifies coal in seam with O_2 or air injected through a borehole drilled from the surface. The gas produced is withdrawn to the surface through another borehole. In recent years, UCG technique is advanced drastically by application of horizontal digging technique developed in oil excavation, and several commercial processes have been scheduled.

It is difficult to insert various sensors to measure temperature, pressure, flow rate of gas from ground surface into reacting spots. Chemical phenomenon occurred in UCG has not been well understood compared with surface gasification processes because of the lack of information described above. Therefore, the reaction formula, heat of gasification, and adiabatic temperature of UCG should be helpful to understand the reaction process. We applied our method to the data of gas composition and ultimate analysis of coal obtained by UCG tests carried out at five China mines, and investigated the feature of each chemical processes.

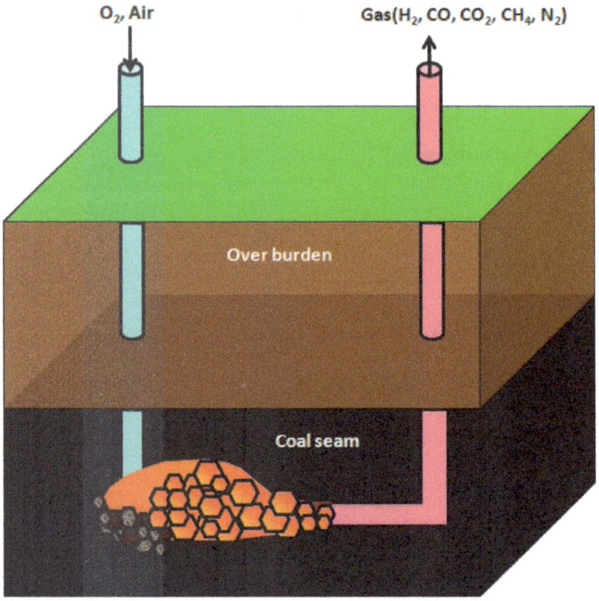

Fig. 2. The scheme of UCG

5.1 Reaction formulla and heat of reaction

Composition of gas in the data reported by China University of Mining and Technology consists of H_2, CO, CO_2, CH_4, and N_2, and none of the other hydrocarbons nor tar were measured.

$$CH_mO_n + \alpha O_2 + \beta H_2O \rightarrow \gamma H_2 + \delta CO + \varepsilon CO_2 + \eta CH_4 \quad (1')$$

The value of α, β, γ, δ, ε, and η in (1′) is estimated by equation (23), (22), (15), (16), (17), and (18) respectively.

Heat of reaction of gasification, h_r (kcal/mol-coal) is calculated by (61).

$$hr = \gamma h_{H2} + \delta h_{CO} + \eta + h_{CH4} - h_{Coal} \quad (61)$$

Here, h_{Coal}, h_{H2}, h_{CO}, and h_{CH4} are molar combustion heats (kcal/mol) of coal, H_2, CO, and CH_4 respectively. The values of h_{H2}, h_{CO}, and h_{CH4} are -68.32, -67.64, and -212.80 kcal/mol respectively.

5.2 Estimation of adiabatic temperature

In the case of UCG, it is presumed that h_r generated is partly transferred to the wall made of coal or char. As the heat conducted to the wall is utilized in water gas reaction, pyrolysis, drying, and preheating of coal effectively. Consequently, the reactor of UCG can be thought as adiabatic one, and most of h_r turns to the sensible heat of gas in the reactor.

Gas in the reactor consists of the products presented in right side of formula (1′), N_2, and H_2O. The total moles of product gas, i.e. Σ in (1′), is expressed by $\Sigma = 1/(q + r + s)$ as shown in equation (13), the yield of N_2 per 1 mol of coal gasified is defined by ζ.

$$\zeta = \frac{(\text{Concentration of } N_2)}{(q + r + s)}$$

The amount of residual H_2O remained in the reactor is expressed by θ. In the case of UCG, it is difficult to estimate θ precisely based on the result of gas analysis. We estimate θ based on the equilibrium relationship of shift reaction.

$$K = \frac{[H_2][CO_2]}{[CO][H_2O]}$$

The relationship between equilibrium constant of shift reaction and temperature is shown in Table 2.

T(℃)	K	T(℃)	K	T(℃)	K
350	2.075×10^{1}	700	1.549×10^{0}	1050	5.162×10^{-1}
400	1.177×10^{1}	750	1.257×10^{0}	1100	4.647×10^{-1}
450	7.319×10^{0}	800	1.042×10^{0}	1150	4.222×10^{-1}
500	4.887×10^{0}	850	8.801×10^{-1}	1200	3.870×10^{-1}
550	3.453×10^{0}	900	7.553×10^{-1}	1250	3.576×10^{-1}
600	2.553×10^{0}	950	6.574×10^{-1}	1300	3.329×10^{-1}
650	1.959×10^{0}	1000	5.793×10^{-1}		

Table 2. Equilibrium constant of shift reaction

Heat of product gas, Q_{sh} is calculated by the integration of thermal capacity of each component gas, which is given by following eq., $Cp = a+bT+cT^2+dT^3$ (kcal/kg-mol-deg), from 298K to gasification temperature T. The values of a ~ d of each gas are shown in Table 3.

Gas	a	$b\times10^{2}$	$c\times10^{5}$	$d\times10^{9}$
H_2	6.952	-0.04576	0.09563	-0.2079
CO	6.726	0.04001	0.1283	-0.5307
CO_2	5.316	1.4285	-0.8362	1.784
CH_4	4.75	1.2	0.303	-2.630
C_2H_4	0.944	3.735	-1.993	4.22
C_2H_6	1.648	4.124	-1.530	1.74
N_2	6.903	-0.03753	0.193	-0.6861
H_2O	7.7	0.04594	0.2521	-0.8387

Table 3. The values of a ~ d to estimated Cp of gas

Taking account of residual water equilibrated with $[H_2]$, $[CO]$, and $[CO_2]$, adiabatic gasification temperature or T_{ad} that satisfies $Q_{sh} = h_r$ was estimated by the rule of trial and error.

5.3 Analysis of chemical process

We estimated reaction formula of gasification, heat of reaction, and gasification temperature of UCG carried out at Fuxin mine, Xinghe mine, Liuzhuang mine, Ezhuang mine, and Xiyang mine in China. The results of ultimate analysis and molecular formulae estimated for five coals were shown in Table 4 with their heating values.

Coal mine	Ultimate analysis (%-daf)					Molecular formula	Heating Value
CH_mO_n	C	H	O	N	S	CH_mO_n	(MJ/mol)
Fuxin coal	79.7	5.35	12.91	1.36	0.68	$CH_{0.806}O_{0.121}$	0.486
Xinghe coal	81.67	5.57	9.08	1.39	1.7	$CH_{0.818}O_{0.083}$	0.532
Liuzhuang coal	82.66	5.63	8.85	1.47	1.09	$CH_{0.817}O_{0.080}$	0.493
Ezhuang coal	82.87	5.68	9.22	1.4	0.83	$CH_{0.822}O_{0.083}$	0.498
Xiyang coal	92.34	3.22	2.48	1.17	0.79	$CH_{0.418}O_{0.020}$	0.512

Table 4. Ultimate analysis, molecular formula and heating values of coals

As an example of analysis of chemical process of UCG, results for Ezhuang mine are summarized below. The daily variation of the concentration of each gas components is shown in Fig. 3.

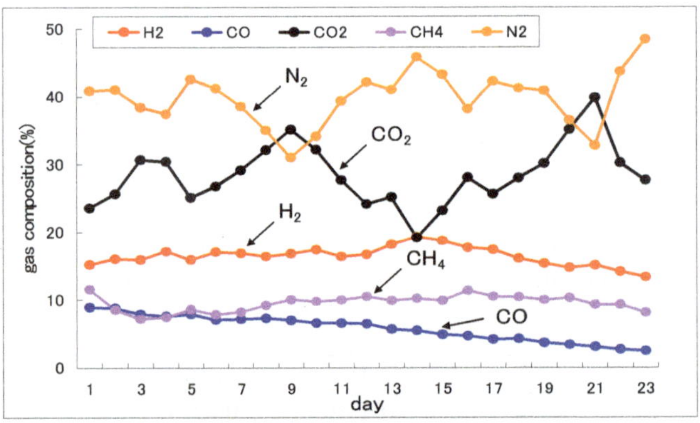

Fig. 3. The daily variation of the concentration of each gas

The change in the each coefficient of reaction formula (1′) with elapse of time was shown in Fig. 4. Value of α may be applicable to estimate the amount of coal gasified. Therefore, total volume of the cave formed in coal seam can be estimated by the integrated value of α.

Value of β is important information to understand the reaction mechanism quantitatively.

The linear relationship was found in the plots of the values of other coefficients against α as shown in Fig. 5 and 6.

Figs. 5 and 6 show that yields of combustible gas such as H_2, CO, and CH_4 decreased in proportion to increase of oxygen reacted. The quantity of H_2O reacted decreased with α in the same way. On the other hand, the yield of CO_2 increased with α. It was thus proved that α was responsible for the formation of every component gas in the case of UCG carried out in Ezhuang mine.

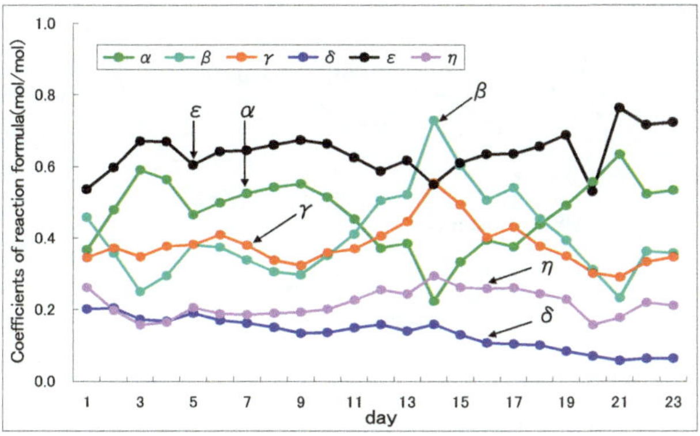

Fig. 4. Change in the coefficients with the elapse of time

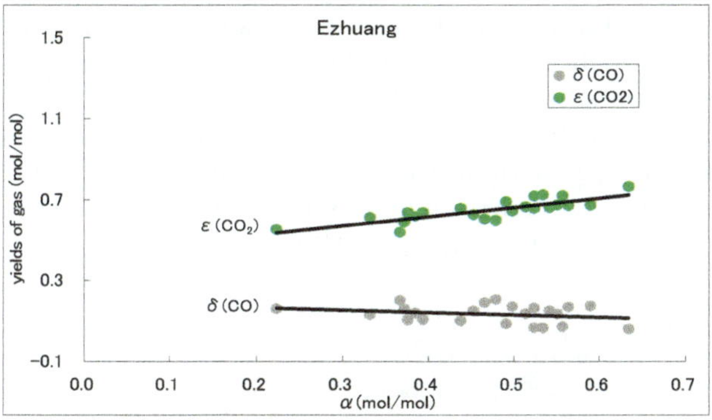

Fig. 5. Relationship between α and other coefficient

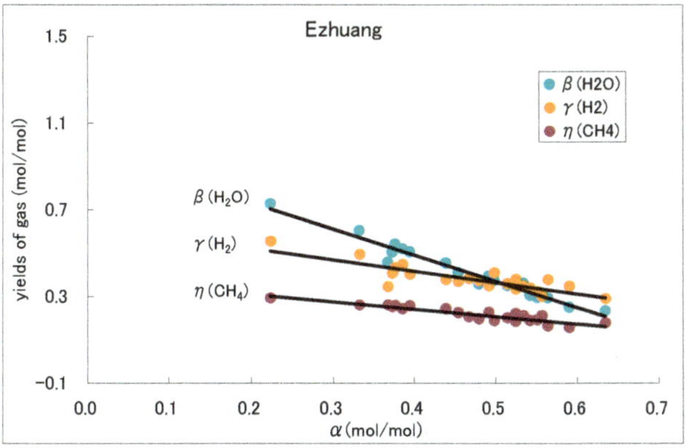

Fig. 6. Relationship between α and other coefficient

The change in h_r and T_{ad} with elapse of time was shown in Fig. 7.

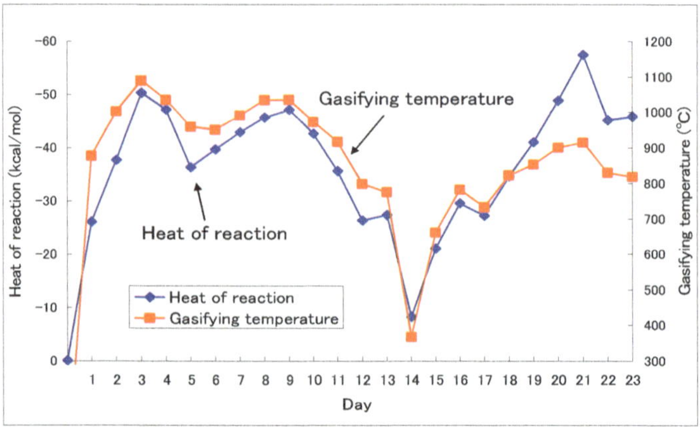

Fig. 7. Change in h_r and T_{ad} with elapse of time

Fig. 8 indicates the plot of h_r vs. α, and T_{ad} vs. α. A plot of h_r vs. α gave a straight line, but that of T_{ad} vs. α was widely scattered. The distribution of plots may be attributed to the uncertainty in the estimation of residual H_2O in product gas, which was calculated from equilibrium of shift reaction. It is presumed that the gas composition in UCG reactor has not been attained to chemical equilibrium. The approach to improve the reliability of estimation for T_{ad} without depending upon the equilibrium theory has not been reported. Our attempt might be the first to predict the temperature of UCG reactor.

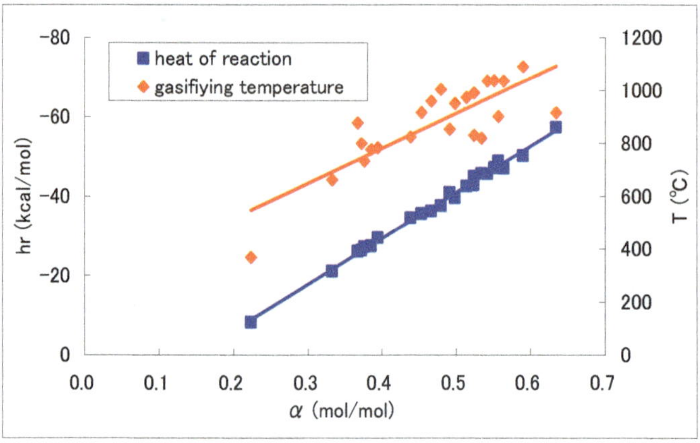

Fig. 8. indicates the plots of h_r vs. α, and T_{ad} vs. α.

5.4 Comparison of reaction process in five coal mines

We compared the result of analysis of the data of UCG carried out at five coal mines.

Average compositions of gases produced by UCG processes are listed in Table 5.

	Average compositions of gases (%)					
Coal mine	O_2	H_2	CO	CO_2	CH_4	N_2
Fuxin	0.71	8.28	16.5	7.01	5.18	57.68
Xinghe	-	21.72	8.82	13.59	2.64	53.36
Liuzhuang	1.49	14.77	13.98	10.73	3.69	55.31
Ezhuang	0	16.42	5.83	28.47	9.48	39.8
Xiyang	0.2	9.3	9.2	9.2	7.2	64.7

Table 5. Average compositions of gases produced by UCG

The mean value of each coefficient of reaction formula of UCG is shown in Table 6.

The average coefficients of reaction formula and indicators of gasification (mol/mol)								
Coal mine	α	β	γ	δ	ε	η	O_{ex}	y
Fuxin	0.347	0.247	0.287	0.573	0.245	0.182	-0.093	0.244
Xinghe	0.333	0.688	0.882	0.348	0.545	0.108	-0.126	0.520
Liuzhuang	0.376	0.399	0.537	0.499	0.365	0.135	-0.084	0.319
Ezhuang	0.477	0.397	0.375	0.133	0.650	0.217	0.019	0.583
Xiyang	0.171	0.717	0.363	0.359	0.359	0.281	-0.319	0.354

Table 6. The average coefficients of reaction formula of gasification

The result of analysis of the data obtained by fixed bed gasifier which is a conventional surface process is summarized in Table 7 as a reference.

The average coefficients of reaction formula and indicators of gasification(mol/mol)								
Prosess	α	β	γ	δ	ε	η	O_{ex}	y
Wilputte	0.337	0.316	0.515	0.705	0.183	0.112	-0.122	0.183
Riley-Morgan	0.428	0.370	0.517	0.707	0.284	0.009	-0.048	0.284
Riley-Morgan	0.407	0.256	0.560	0.658	0.279	0.062	-0.020	0.279
Riley-Morgan	0.433	0.215	0.503	0.723	0.224	0.053	-0.022	0.224
Lurgi	0.331	0.224	0.623	0.466	0.396	0.136	0.018	0.018

(Chemistry of Coal Utilization, Elliott.M.A.Editor, p1615)

Table 7. The coefficients of reaction formula of surface fixed bed gasification

Value of each coefficient in Table 7 distributed in narrower range compared with Table 6. Comparing to the fixed bed process, the value of α of four mines except for Xiyang were around the same as well as those of fixed bed. The relatively small α of Xiyang may be due to its different chemical composition and heating value as shown in Table 4.

The partial oxidation process is considered to be practically governed by the value of O_{ex}. On the other hand, the secondary reaction proceeded more actively compared to surface fixed bed gasification.

The change in h_r with elapse of time for five mines is shown in Fig. 9.

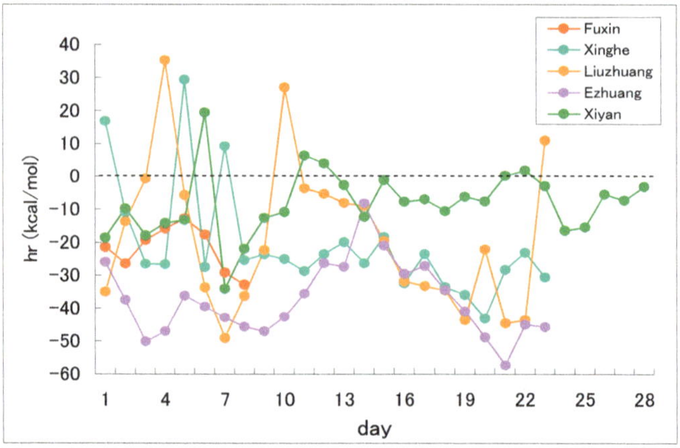

Fig. 9. The change in h_r with elapse of time

Although the rate of air supply is nevertheless kept constant during every tests, h_r repeated increase and decrease. It occasionally rose into positive value indicating that the endothermic reaction process proceeded.

The relationship between h_r and α is shown in Fig. 10. As plots in the region of $h_r > 0$ stays the same straight line characteristic for each coal mine, it is indicated that $h_r > 0$ has not accidentally obtained. Positive value of h_r is assumed to be attributed to excessive water gas reaction promoted by the heat accumulated at the wall of gasifier. The periodic fluctuation of h_r may indicate an essential feature of actual UCG reactions.

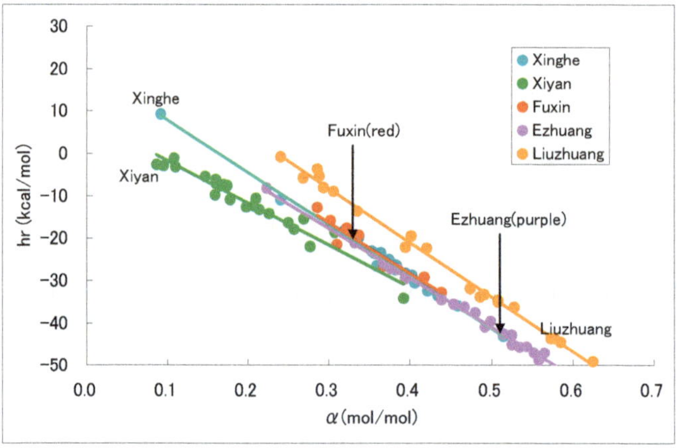

Fig. 10. Relationship between α and h_r

Relationship between α and β was shown in Fig. 11.

The plot of α vs. β gave a straight line in all coal mines. O_{ex} in UCG is almost zero or negative as shown in Table 9. In the case of $O_{ex} < 0$, $-2O_{ex}H_2O$ was consumed in water gas reaction to compensate the lack of O_2. Therefore, the amount of water reacted basically increased in proportion to the reduction of oxygen reacted. Besides, it is considered that progress of shift reaction and formation of CH_4 influenced on the slope of each straight line.

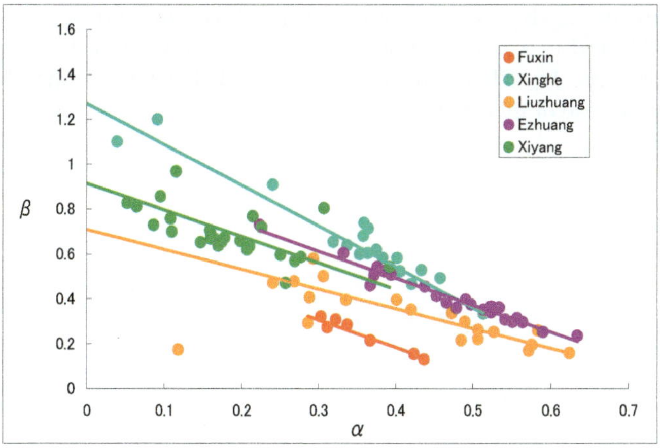

Fig. 11. Relationship between α and β

Plot of y vs. α for each mine is shown in Fig. 12.

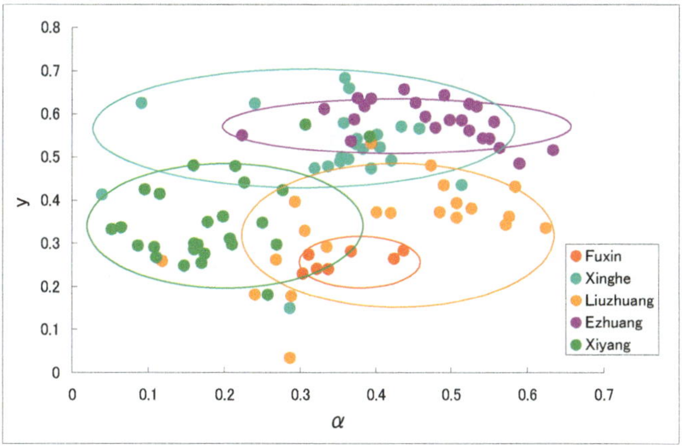

Fig. 12. Relationship between α and y

Since the plots scattered in the range indicated by ellipse and the linearity was hardly found, it is considered that shift reaction was not affected by partial oxidation.

The plot of η vs. α is shown in Fig. 13.

A relatively linear relationship is found to exist between η and α showing decrease in η along with increase in α. This is presumably a result of thermal effect on the stability of CH_4

determined by the chemical equilibrium. Average η of Xiyang mine is found to be 0.281mol/mol, which is much larger than results of other mines as shown in Table 6. As carbon content of Xiyang coal is 92.34%, it should be classified as anthracite. It is generally accepted that most of H atoms in molecular structure of anthracite are combined with C atoms directly at the rim of aromatic rings, and very limited numbers of methyl group or hydrocarbon chains exist. This means that CH_4 is produced by hydrogenation or synthesis reaction between H_2 and CO, not by pyrolysis. It is well known that hydrogenation occurs only with high pressure H_2. Therefore, it is considered that CH_4 is mainly produced by synthetic route in the case of this coal.

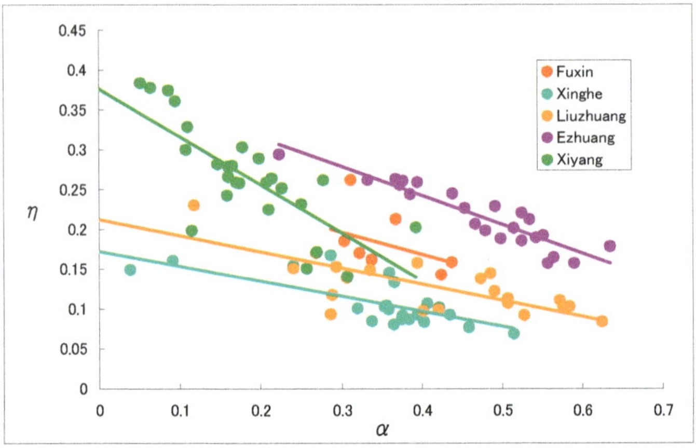

Fig. 13. Relationship between α and η

Plots of γ, δ, and ε vs. α are shown in Figs. 14, 15, and 16.

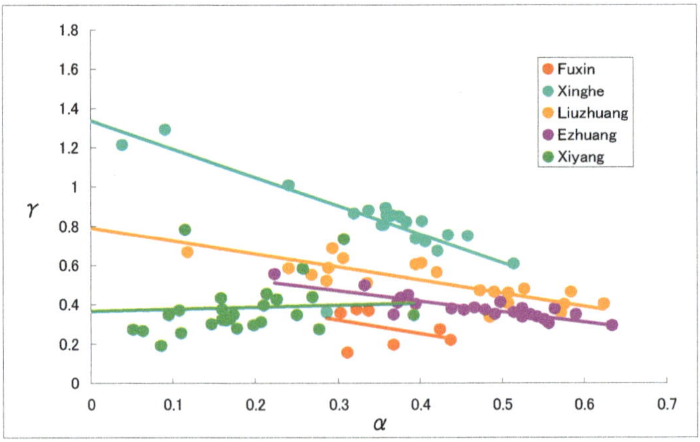

Fig. 14. Relationship between α and γ

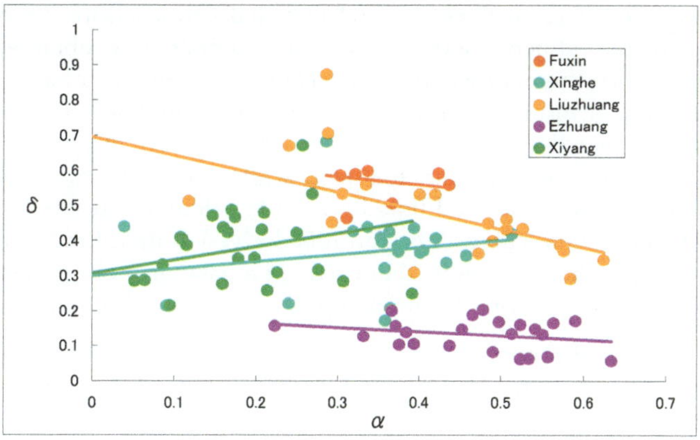

Fig. 15. Relationship between α and δ

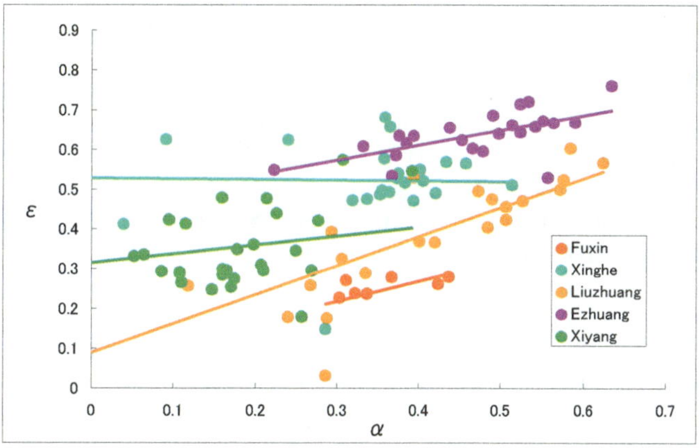

Fig. 16. Relationship between α and ε

Amounts of H_2 and CO have been used as indicator to estimate chemical process of surface gasification. For example, the yields of H_2 and CO are expected to increase with decrease in the amount of O_2 reacted. Such tendency, however, was not found in the plots of γ vs. α and δ vs. α presumably due to the appreciable progress of secondary reaction. In the case of UCG, we need to evaluate the amounts of H_2, CO, and CO_2 after estimation of the effect of secondary reaction.

6. Conclusion

It is generally accepted that gasification consists of more than five chemical processes such as pyrolysis, partial oxidation of char, further decomposition of tar, secondary reactions, and combustion of char or gas. It is obviously difficult to simulate actual coal gasification precisely by applying reliable scientific analysis of fundamental experiment. Since coal gasification is a very complicated both from experimental and theoretical points of view, its chemical process cannot be completely understood merely by the accumulation of kinetic data.

Our method described here may be the first one that can scientifically elucidate reaction process based on the stoichiometry using gas composition obtained at a practical gasification plant. Since this method is constructed based on stoichiometry of the reaction formula without any arbitrary assumption and approximation, it is applicable to any gasification process regardless of the type of gasifier or a rank of coal used. The feature of this study is to elucidate gasification mathematically based on material balance of coal gasification reactions which was traditionally used to calculate carbon conversion and cold gas efficiency. The mathematical reaction formula derived in this study offered a novel point of view to estimate practical reactions that occur in a gasifier more precisely and it should help to attain optimum operation condition in practical gasification plant.

As a good example of the application of our method, results of the analysis of UCG is introduced. Actual operation data of UCG carried out at five mines in China was investigated by our method. We have consequently succeeded to obtain the reaction formula of gasification, progress of shift reaction, reaction heat of gasification, and adiabatic gasification temperature. These results allow us to understand the partial oxidation step and secondary reaction step of UCG.

7. References

Elliott, M, ed. Chemistry of Coal Utilization, 2nd Suppl. Chapter 24 Coal Gasification Process , pp. 1615, Wiley (1981)

Kaiho, M., Yamada, O., Yasuda, H, and Shimada, S. (2009). International Conference on Coal Science & Technology, proceedings of 2009 ICCS&T, in CD-ROM(14-4), Cape Town, South Africa, October 2009.

Kaiho, M., Yasuda, H and Yamada, O., Shimada, S. and Fujioka, M. International Conference on Coal Science & Technology, proceedings of 2011 ICCS&T, in electronic storage device(P2_29), Oviedo, Spain, October 2011

Kaiho, M.; Yamada, O. and Shimada, S. (2008). 25th International Pittsburgh Coal Conference, proceedings of PCC2008, in CD-ROM(32-5), Pittsburgh, USA, September 2008.

Yamada, O., Kaiho, M., Shimada, and Fujioka, M. and Liang, J. (2010). 27th International Pittsburgh Coal Conference, proceedings of PCC2010, in CD-ROM(14-5), Istanbul Turkey, October 2010

Improved Combustion Control in Diesel Engines Through Active Oxygen Concentration Compensation

Jason Meyer and Stephen Yurkovich
The Ohio State University and University of Texas at Dallas
USA

1. Introduction

Like any other chemical reaction, the process of burning fuel depends on the quantities of the reactants involved; for a combustion reaction, those are fuel and oxygen. The unsteady nature of a diesel engine both inside and outside of the cylinders, however, makes the problem of combustion control much more complex than simply regulating the trapped oxygen mass and the fuel injection mass. Because liquid fuel is injected directly into the combustion chamber, several processes must occur before combustion can commence. First, the liquid fuel stream must be atomized. Next, the fuel droplets must be vaporized. Finally, the fuel must mix with the oxygen already residing in the cylinder to form a combustible mixture. Even though the ratio of the total oxygen mass to the total mass of fuel is often significantly lean relative to stoichiometry, the local ratios can be much richer than stoichiometry. As the fuel spray mixes with the trapped charge gases, the local oxygen to fuel ratios at the boundaries of the fuel spray approach stoichiometry and the fuel begins to ignite. Once the combustion process has started, the burn rate is primarily limited by the rate at which the unburned fuel can mix with the remaining oxygen.

Modern diesel engines have sophisticated fuel injection systems that allow for multiple fuel injection pulses, thus providing the ability to flexibility shape the combustion process. The timing of these injections and the fraction of the fuel delivered by each injection strongly affects the combustion process. Injecting fuel prior to the main injection (termed a pilot injection) adds energy and turbulence to the combustion mixture thereby reducing the ignition delay of the main injection, especially at low temperatures (MacMillan et al., 2009; Osuka et al., 1994). This allows for the majority of the heat release to occur when the piston is close to top-dead-center which results in a higher thermodynamic efficiency. However, the improvement in fuel consumption can also be accompanied by an unwanted increase in NO_x emissions (Benajes et al., 2001).

Injecting fuel after the main injection (termed a post injection) adds energy to the system after the primary combustion process has occurred. This energy increases the temperature of the combustion gases at the end of the combustion cycle without increasing the peak combustion temperature, which dictates the production of NO_x. A higher end-of-cycle temperature promotes higher soot oxidation rates and an overall decrease in the engine-out soot levels (Benajes et al., 2001). Higher end-of-cycle temperatures are also very important to the efficiency of aftertreatment components such as selective catalytic reduction systems.

Controlling a diesel engine is particularly challenging because the evolution of the combustion process depends on numerous factors including: the air charge density, the mass of oxygen gas, the mass of dilutent gases, the mixture homogeneity, the fuel injection mass, the fuel injection timing, the fuel injection pressure and the speed of the engine. Further complicating the control problem, combustion performance is often defined with respect to multiple metrics including torque response, noise, fuel consumption and emissions production. To achieve the best trade-off between these factors, the fueling injection parameters must be precisely matched to the in-cylinder conditions. Typically, a diesel engine is controlled using the type of control structure shown in Figure 1. Such a control system consists of a supervisory controller, a setpoint determination system including a feed-forward fueling controller and an air path controller (Guzzella & Amstutz, 1998). At the highest level, the supervisory controller selects the system operating mode depending on the dynamics of the system (i.e. steady-state versus transient operation), the barometric pressure, the states of the aftertreatment systems (e.g. the soot level within a diesel particulate filter) and other off-nominal considerations (e.g. turbocharger surge prevention).

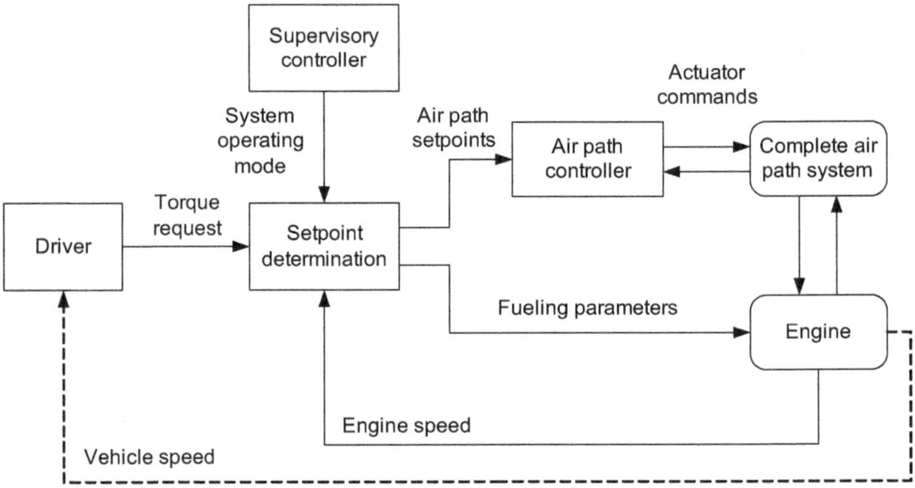

Fig. 1. Current approach to diesel engine control

The setpoint determination system first calculates the mass of fuel required to achieve the driver's torque request. Then based on the system operating mode, the setpoint determination system simultaneously identifies the desired air path response and the desired fuel injection strategy for the current engine speed and calculated fuel mass. Typically, this is achieved using different sets of three-dimensional air path setpoint and fueling parameter setpoint tables. The air path setpoints are passed to an air path controller, whereas the fueling parameters directly drive the fuel injection system. Because of the mass accumulation and mass transport dynamics within the air path system, the response time of a closed loop air path control system is much slower than the response time of a fueling controller. While it may take many engine cycles for an air path controller to change the cylinder contents from one set of desired conditions to another, the fuel delivery method can be arbitrarily changed each injection event.

Selecting the fueling parameters based on the engine speed, desired torque request and system operating mode inherently ignores the dynamics of the air path system. As a result, the injected fuel may combust in a manner which produces less torque, more emission, more noise and/or undesirable engine-out gas temperatures. Additionally, the current method of fueling control lacks robustness. Any variation in the behavior of the air path system can lead to combustion performance degradation.

This chapter develops a new approach to diesel engine fueling control which controls the fuel injections process based on the in-cylinder contents. An integral part of this design is a new air path oxygen dynamics model. By directly predicting the transport delay within the air path system, this model achieves nearly the same accuracy as a one-dimensional model but has the complexity of a zero-dimensional model. The performance of both the delay based oxygen dynamics model and the proposed fueling control structure are demonstrated in simulation.

2. In-cylinder contents based diesel fueling control

Although diesel engine combustion depends on numerous factors, a production diesel engine cannot account for all of them. Firstly, identifying each factor and its exact impact is far too time consuming. Engine calibration processes must be kept as short as possible to minimize both cost and the time to market. Secondly, an engine control unit (ECU) has limited processing power and on-board memory. Lastly, many of the factors that affect combustion such as in-cylinder motion (e.g. swirl and tumble) are extremely difficult to measure or estimate. Rather than predicting each of the factors influencing combustion and optimizing the fuel injection parameters online, the optimal fuel parameters are identified *a priori* and stored in lookup tables.

During the engine calibration process, the fueling parameters and air path setpoints which produce the best trade-off between torque response, noise, fuel consumption and emissions production are established for a wide range of engine operating conditions. In a conventional diesel engine control structure, the air path setpoints and fueling parameters identified as optimal are then fit to sets of three-dimensional tables indexed by engine speed, the fuel injection mass and the system operating mode. The fueling parameters selected by this mapping structure do not directly account for the instantaneous cylinder conditions but rather indirectly account for the nominal set of cylinder conditions that correspond to ideal air path performance. During transients or in the presence of system faults, however, the instantaneous in-cylinder conditions deviate from their nominal values. As a result the typical control structure can significantly misestimate the fueling parameters.

A more robust alternative is to schedule the fueling parameters using a variable which represents the achieved in-cylinder contents rather than the expected conditions. Figure 2 illustrates how such a selection scheme could be integrated into the overall diesel engine control structure. As with a traditional controller, the fueling parameters are still selected based on sets of lookup tables, but in this architecture one of the indexing variables is related to the properties of the cylinder contents. With this structure, the fueling controller is no longer autonomous with respect to the air path system. The interconnection between the fueling controller and the air path system through the cylinder contents predictor significantly improves the robustness of the overall engine control system. It should be noted that the air path setpoints can still be selected based on the system operating mode.

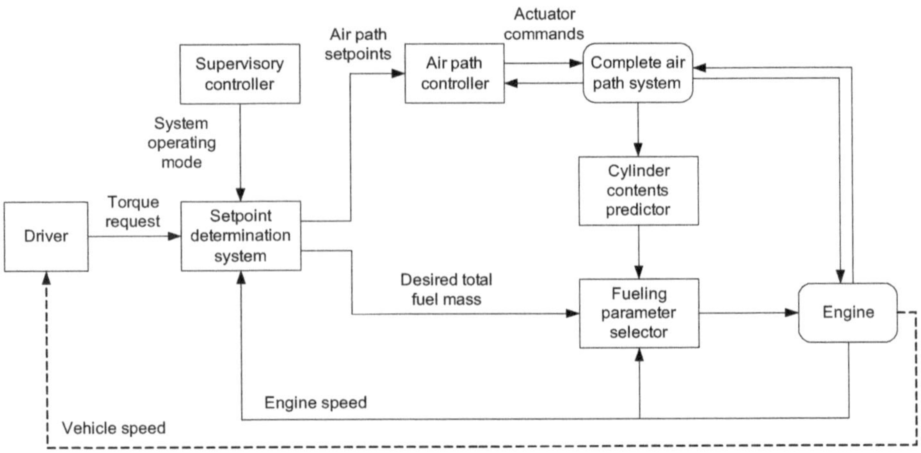

Fig. 2. Proposed air path and fueling control approach

From an engineering standpoint, the objective of a lookup table based fueling parameter selection system should be to predict the fueling parameters as accurately as possible. Although any of the numerous in-cylinder properties which affect combustion could be used as a table indexing variable, not every variable is equally viable. Logically, the factor that best predicts the variations in the optimal fueling parameters should be chosen. In this work, to identify the most appropriate variable and test the performance of the in-cylinder contents based fueling controller, a sophisticated engine model of a a six cylinder heavy-duty diesel engine was developed in the commercially available engine modeling software GT-Power.

3. GT-Power Model Description

For many in-cylinder properties such as the oxygen concentration, experimental measurements are very difficult to obtain and infeasible in a production setting. Using a high fidelity GT-Power engine model, however, the in-cylinder properties including the oxygen concentration can be predicted with reasonable confidence. A GT-Power model can accurately predict both the one-dimensional gas dynamics within an engine air path system as well as the combustion phenomena. In GT-Power, the process by which injected fuel mixes with air and combusts is based on the modeling work by Morel & Wahiduzzaman (1996). As will be shown next, a GT-Power model can predict the two most important engine performance metrics, fuel consumption and NOx emissions, quite well.

The results presented in this chapter are based on a GT-Power model of a six cylinder heavy-duty diesel engine with a variable geometry turbocharger (VGT) and an exhaust gas recirculation (EGR) system. This model was calibrated using steady-state data collected experimentally and then validated using transient data. For the steady-state calibration, each of the actuator models were tuned to ensure that the VGT and EGR actuators affect the air path in a manner consistent with the experimental data. Quantities such as the fresh air flow, EGR flow, turbine speed, exhaust pressure and intake pressure were all analyzed. Because the specific fuel consumption and NO_X production predictions are so important to the selection of the fueling parameters, particular attention was given to the predictive combustion model. The correlations between the experimental and predicted brake specific fuel consumption

(BSFC) and NO_x production are shown in Figure 3. The R^2 correlation defined by

$$R^2 = 1 - \frac{\sum_{k=1}^{n_{data}} (y_{pred} - y_{meas})^2}{\sum_{i=1}^{n_{data}} (y_{meas} - \overline{y_{meas}})^2}. \tag{1}$$

was 94.2% for the BSFC prediction and 82.9% for the NO_x prediction, both reasonably good for this type of model.

To validate the GT-Power model of the engine, the response of the model over a heavy duty FTP (Federal Test Procedure) drive cycle was compared to experimental data. In the simulation, the measured trajectories for the engine speed, the EGR valve position and the VGT position along with the command trajectories for the fueling parameters for the experimental drive cycle were imposed as inputs to the GT-Power model. For each of the relevant quantities, the predicted and experimentally measured values had very high correlations. The R^2 correlations were 91.1% for the fresh air mass flow rate, 91.3% for the EGR mass flow rate, 90.9% for the oxygen concentration of the exhaust stream and 81.1% for the NO_x production. To account for the dynamics of the NO_x sensor, the GT-Power predictions were filtered with a first order filter and then compared to the measurement data. Figure 4 depicts how well the dynamic variation in each of these four quantities is captured by the GT-Power model for the first 100 seconds of the drive cycle.

3.1 Table based fueling parameter representation comparisons

Once the GT-Power model was calibrated, a series of sensitivity studies were performed to identify the in-cylinder contents variable that has the most significant impact on the combustion process. For this study, the EGR valve position, VGT position, total fuel injection mass and engine speed were considered to be independent control variables. Approximately 20,000 different combinations of these inputs were simulated. The engine under investigation uses a pilot, a main and a post injection. Each injection is defined by two parameters, a start of injection (SOI) timing and a fuel injection quantity; however, the sum of the injection quantities must add to the desired value. At each of the 20,000 points, a total of five fueling parameters (main SOI, pilot SOI, post SOI, pilot quantity and post quantity) were optimized. This data represented steady-state operation at various altitudes as well as transient operation. It should be noted that an experimental study of this magnitude would take many months or

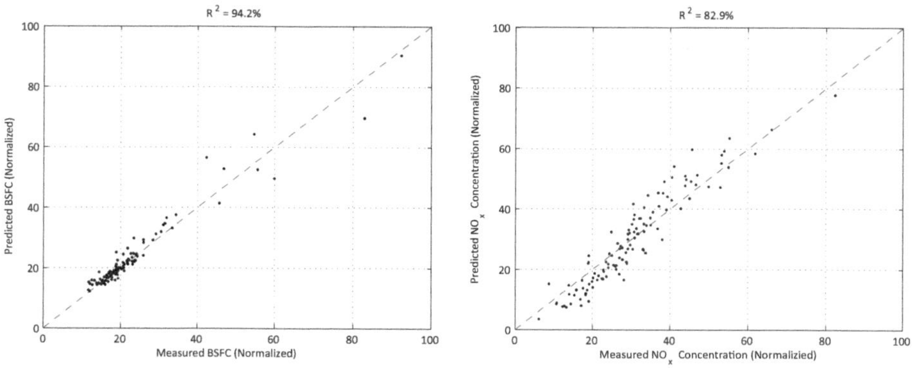

Fig. 3. Steady-state BSFC and NO_x comparisons: Experimental versus GT-Power

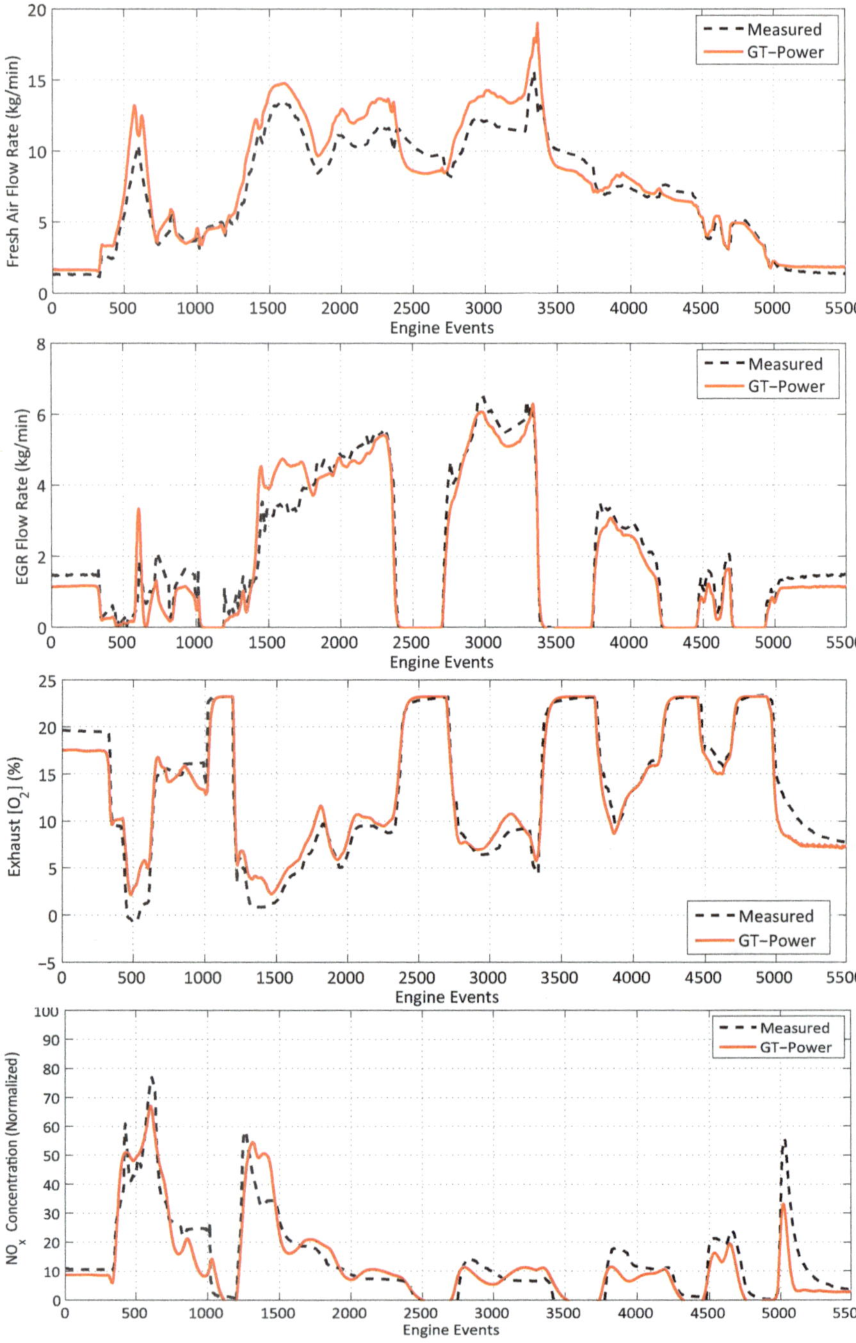

Fig. 4. Transient GT-Power model validation results

years. In a GT-Power simulation environment, however, these tests were completed in just a few weeks.

For simplicity, only two factors were considered when evaluating the engine performance, the indicated specific fuel consumption (ISFC) and the brake specific NO_x ($BSNO_x$). The calibration goal was the minimization of ISFC while limiting the $BSNO_x$ to less than 1.0 g/hp-hr. Figure 5 compares the $BSNO_x$ versus ISFC trade-off produced by the optimal set of fueling parameters to the trade-offs of alternative sets of fueling parameters for a single operating condition (1000 RPM, 65 mg fuel). Because of the proprietary nature of the data, all of the ISFC values presented in this chapter have been normalized with respect to the full range of achievable ISFC values (i.e. percentage of full scale). The 20,000 sets

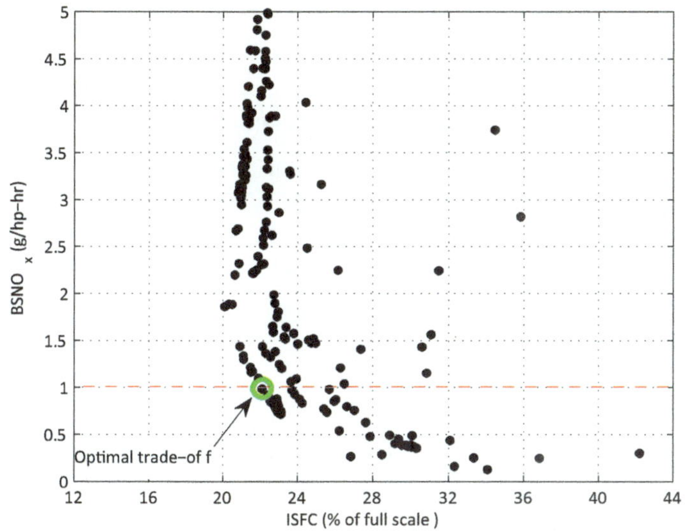

Fig. 5. Fueling parameter calibration example (1000 RPM, 65 mg fuel)

of optimal fueling parameters produced by this study were then used to calibrate sets of three-dimensional fueling parameter prediction tables, each of which used different indexing variables. Mirroring the typical approach currently in production, two of the indexing variables were select to be engine speed and fuel injection mass. For the third, the following variables were tested: total trapped charge mass, trapped oxygen mass, trapped burned gas mass, in-cylinder oxygen concentration, mass ratio of fuel to oxygen and charge density. At a minimum, three different system operating modes (steady-state, transient and altitude) are required to characterize the different modes of normal engine operation. Fundamentally, the only difference between the transient, steady-state and altitude fueling tables within a conventional controller is that they correspond to different air path setpoints. In industry, air path and fueling tables use a very fine grid spacing for the engine speed and fuel mass variables, on the order of 20 values each. Because the proposed in-cylinder contents based fueling tables cover at least three different system operating modes, the size of the three-dimensional tables was selected to be 10 x 10 x 10 to keep the total number of entries comparable to current practice.

To quantify the effectiveness of each candidate table indexing variable, the R^2 fit correlations between the table predicted fueling parameters and the optimal values of the fueling parameters were calculated. All of the correlations are presented in Table 1. The tables indexed by in-cylinder oxygen concentration (ratio of the oxygen mass to the total gas mass) best predicted the optimal fueling parameters. It is well known in the diesel engine research community that the in-cylinder oxygen concentration has a strong effect on the combustion process. As the oxygen concentration decreases due to the addition of a dilutant such as recirculated exhaust gas, the heat release rate decreases. To offset this effect, the start of combustion must be advanced by advancing the SOI timings. It is therefore not surprising that the optimal fueling parameters are strongly correlated to the in-cylinder oxygen concentration. As long as the in-cylinder oxygen concentration can be dynamically estimated, these results indicate that lookup tables indexed by engine speed, fuel mass and in-cylinder oxygen concentration can closely predict the optimal fueling parameters.

Table 1. Fueling table R^2 fit comparison (all values in %)

	Main SOI	Pilot SOI	Post SOI	Pilot Qty	Post Qty
$\frac{\text{Oxygen Mass}}{\text{Total Mass}}$	95.0	94.3	96.9	90.0	90.8
Burned Mass	94.7	93.7	94.5	84.2	85.0
Total Mass	88.4	83.5	95.9	87.9	86.4
$\frac{\text{Fuel Mass}}{\text{Oxygen Mass}}$	88.1	83.2	95.5	87.1	85.2
Oxygen mass	76.6	69.2	89.1	77.7	79.6
Charge Density	73.2	69.6	89.1	80.8	68.0

4. Delay based oxygen dynamics modeling

For an in-cylinder oxygen concentration based fueling controller to be practical, the in-cylinder oxygen concentration must be estimated reliably with a model simple enough for real time implementation in a production ECU. A wide range of real-time oxygen concentration prediction models have been studied, including models derived from the first law of thermodynamics (Ammann et al., 2003; Canova et al., 2009; van Nieuwstadt et al., 2000), linear parameter varying models (Jung & Glover, 2006; Wang, 2008; Yan & Wang, 2010), steady-state relationships (Chen & Yanakiev, 2005) and empirical models (Alberer & Del Re, 2009; Langthaler & Del Re, 2007; 2008). All of these approaches are based on zero-dimensional system descriptions and therefore do not account for the time-varying transport delays which characterize the air path system. By ignoring these delays, the predicted in-cylinder oxygen concentration tends to lead the actual oxygen concentration. Although most of these models are simple enough to be implemented in a production ECU, the oxygen concentration prediction errors can be significant, especially during rapid transients. To directly account for the time varying transport delays, a delay based oxygen dynamics model has been derived from a one-dimensional description of the air path oxygen dynamics.

Assuming a uniform radial distribution, the engine air path system can be partitioned into three one-dimensional control volumes corresponding to the exhaust, EGR and intake systems. The longitudinal oxygen dynamics within each of these systems can be described by the following one-dimensional oxygen diffusion-convention equation:

$$\frac{\partial[O_2]}{\partial t} - D\frac{\partial^2[O_2]}{\partial z^2} + v\frac{\partial[O_2]}{\partial z} + \dot{r}_{O_2} = 0, \tag{2}$$

where z is the characteristic length variable, v is the fluid velocity, $[O_2]$ is the oxygen concentration on a per mass basis, D is the diffusivity of oxygen in the gaseous mixture and $\dot{r}_{O_2}$ is the rate of oxygen is produced by chemical reactions. If all of the chemical reactions are assumed to occur inside the combustion chamber, then the oxygen production term can be eliminated. Next, the one-dimensional control volumes can be spatially discretized into many cells. Using a discretization length of Δz, this gives rise to

$$\frac{d[O_2]_i(t)}{dt} = D\frac{[O_2]_{i-1}(t) - 2[O_2]_i(t) + [O_2]_{i+1}(t)}{\Delta z^2} - v\frac{[O_2]_i(t) - [O_2]_{i-1}(t)}{\Delta z}$$

where $[O_2]_i$ is oxygen concentration of the ith control volume cell. Assuming constant cross sectional area, the oxygen dynamics equation can be rewritten in terms of the volumetric flow rate in the manner

$$\frac{d[O_2]_i(t)}{dt} = \Big(D_1 + \frac{q_{in}(t)}{V_{cell}}\Big)[O_2]_{i-1}(t) + \Big(-2D_1 - \frac{q_{in}(t)}{V_{cell}}\Big)[O_2]_i(t) + D_1[O_2]_{i+1}(t) \tag{3}$$

where

$$D_1 = \frac{D}{\Delta z^2}, \tag{4}$$

q_{in} is the volumetric flow rate entering the control volume and V_{cell} is the volume of a unit cell.

Next, this time domain oxygen concentration model can be converted into the engine event domain in which a combustion occurrence is defined as an event. In the event domain, the number of events per engine cycle is equal to the number of cylinders. One can convert between the event domain defined by variable e and the time domain defined by variable t using

$$t = \frac{120}{n_{cyl}N}e \tag{5}$$

where n_{cyl} is the number of cylinders and N is the engine speed. The derivative of this equation can be approximated as

$$dt = \frac{120}{n_{cyl}N}de, \tag{6}$$

because the rate of change of engine speed is insignificant in this setting. The volumetric flow rates can also be converted from the time domain to the event domain. Define $q^{\star}_{in}(e)$ as the volumetric flow rate entering the system on per event which satisfies

$$q^{\star}_{in}(e) = \frac{120}{n_{cyl}N}q_{in}(t). \tag{7}$$

With these relationships, the continuous time domain oxygen concentration model described by (3) can be converted to the continuous event domain model

$$\frac{n_{cyl}N}{120}\frac{d[O_2]_i(e)}{de} = \Big(D_1 + q_{in}^{\star}\frac{n_{cyl}N}{120V_{cell}}\Big)[O_2]_{i-1}(e) + \Big(-2D_1 - q_{in}^{\star}\frac{n_{cyl}N}{120V_{cell}}\Big)[O_2]_i(e) + D_1[O_2]_{i+1}(e). \tag{8}$$

By defining D_2 as

$$D_2 = \frac{120}{n_{cyl}N}D_1, \tag{9}$$

the event domain model can be reduced to

$$\frac{d[O_2]_i(e)}{de} = \Big(D_2 + \frac{q_{in}^{\star}}{V_{cell}}\Big)[O_2]_{i-1}(e) + \Big(-2D_2 - \frac{q_{in}^{\star}}{V_{cell}}\Big)[O_2]_i(e) + D_2[O_2]_{i+1}(e).$$

Applying the discrete Euler approximation with an integration step (in event space) of Δe, the changing rate of the oxygen concentration within a unit control volume can be approximated as

$$\frac{d[O_2]_i(e)}{de} \approx \frac{[O_2]_i(e+\Delta e) - [O_2]_i(e)}{\Delta e}. \tag{10}$$

With this approximation the oxygen dynamics equation becomes

$$[O_2]_i(e+\Delta e) = \Delta e\Big(D_2 + \frac{q_{in}^{\star}}{V_{cell}}\Big)[O_2]_{i-1}(e) + \Big(1 - 2\Delta e D_2 - \Delta e\frac{q_{in}^{\star}}{V_{cell}}\Big)[O_2]_i(e) + \Delta e D_2[O_2]_{i+1}(e). \tag{11}$$

When the gas velocity is sufficiently large, the diffusion effects can be ignored. Under these circumstances, the oxygen dynamics reduce to

$$[O_2]_i(e+\Delta e) = \Delta e\frac{q_{in}^{\star}}{V_{cell}}[O_2]_{i-1}(e) + \Big(1 - \Delta e\frac{q_{in}^{\star}}{V_{cell}}\Big)[O_2]_i(e). \tag{12}$$

If additionally the volumetric flow rate were constant, then the convection dynamics could be reduced to a pure transport delay by choosing a special time step Δe_1 defined as

$$\Delta e_1 = \frac{V_{cell}}{q_{in}^{\star}}. \tag{13}$$

With this choice, the oxygen concentration of the ith control volume cell can be predicted with

$$[O_2]_i(e+\Delta e_1) = [O_2]_{i-1}(e). \tag{14}$$

This result can then be extended to predict the oxygen concentration at the outlet of the subsystem $[O_2]_{outlet}$ as the following function of the oxygen concentration at the inlet of the subsystem $[O_2]_{inlet}$:

$$[O_2]_{outlet}(e+\Delta e_2) = [O_2]_{inlet}(e). \tag{15}$$

where

$$\Delta e_2 = \frac{V_{sys}}{q^{\star}_{in}}. \quad (16)$$

and V_{sys} is the total volume of the subsystem.

Using (15) and (16) to approximate the transport delays, a simple discrete time air path model can be generated. Define k as a discrete time index in engine events (i.e. $\Delta e = 1$). Nominal transport delay estimates of the exhaust system (${d_{exh}}^{\dagger}$), the EGR system (${d_{egr}}^{\dagger}$) and intake system (${d_{int}}^{\dagger}$) can be produced using

$$ {d_{exh}}^{\dagger}(k) = \frac{V_{exh} p_{exh}(k)}{\dot{m}^{\star}_{exh}(k) T_{exh}(k) R}, \quad (17)$$

$$ {d_{egr}}^{\dagger}(k) = \frac{V_{egr} p_{egr}(k)}{\dot{m}^{\star}_{egr}(k) T_{egr}(k) R}, \quad (18)$$

$$ {d_{int}}^{\dagger}(k) = \frac{V_{int} p_{int}(k)}{(\dot{m}^{\star}_{egr}(k) + \dot{m}^{\star}_{fresh}(k)) T_{int}(k) R}. \quad (19)$$

where V_{exh}, V_{egr} and V_{int} represent the volumes of the exhaust, EGR and intake systems; p_{exh}, p_{egr} and p_{int} represent the pressures within the exhaust, EGR and intake systems; T_{exh}, T_{egr} and T_{int} represent the temperatures within the exhaust, EGR and intake systems; $\dot{m}^{\star}_{egr}$ represents the EGR mass flow rate on a per event basis and $\dot{m}^{\star}_{fresh}$ represents the mass flow rate of fresh air on a per event basis. Although these estimates assume the gas velocities within each system are constant (which may not always be the case), this method of estimating the transport delays is much more accurate than current methods which ignore the delays altogether. Through the use of memory buffers, these nominal delays can be used to predict the oxygen concentrations leaving the exhaust, EGR and intake systems based on the oxygen concentrations entering these systems. Before these delay estimates can be used, they must be converted to integers and bounded to account for the fixed allocation of memory. With the following three equations, the nominal exhaust, EGR and intake delays can be converted into integer delays:

$$ d_{exh}(k) = \begin{cases} \lfloor {d_{exh}}^{\dagger}(k) \rfloor & \text{if } {d_{exh}}^{\dagger}(k) < d_{exh,max} \\ d_{exh,max} & \text{otherwise} \end{cases}, \quad (20)$$

$$ d_{egr}(k) = \begin{cases} \lfloor {d_{egr}}^{\dagger}(k) \rfloor & \text{if } {d_{egr}}^{\dagger}(k) < d_{egr,max} \\ d_{egr,max} & \text{otherwise} \end{cases} \quad (21)$$

and

$$ d_{int}(k) = \begin{cases} \lfloor {d_{int}}^{\dagger}(k) \rfloor & \text{if } {d_{int}}^{\dagger}(k) < d_{int,max} \\ d_{int,max} & \text{otherwise} \end{cases} \quad (22)$$

where $d_{exh,max}$ is the largest allowable exhaust system transport delay, $d_{egr,max}$ is the largest allowable EGR system transport delay and $d_{int,max}$ is the largest allowable intake system transport delay.

During each event new exhaust gas enters the EGR system. The oxygen concentration of the gases leaving the EGR system depends on the oxygen concentration of the gases currently in the EGR system and the oxygen concentration of the exhaust gases that enter the EGR system. Because of transport delay, the oxygen concentration of the gases entering the system depends on the oxygen concentration of the exhaust gas that was expelled several events prior. Accordingly, the nominal oxygen concentration at the confluence function where the EGR system meets the intake system $[O_2]_{egr}{}^{\dagger}$ can be approximated with

$$[O_2]_{egr}{}^{\dagger}(k) = [O_2]_{exh}{}^{\dagger}(k - d_{exh}(k) - d_{egr}(k))\frac{\dot{m}^{\star}_{egr}(k)}{\epsilon\, \dot{m}^{\star}_{egr,max}} + [O_2]_{egr}{}^{\dagger}(k-1)\frac{\epsilon\, \dot{m}^{\star}_{egr,max} - \dot{m}^{\star}_{egr}(k)}{\epsilon\, \dot{m}^{\star}_{egr,max}} \qquad (23)$$

where $[O_2]_{exh}{}^{\dagger}$ is the nominal oxygen concentration at the exhaust ports, $\dot{m}^{\star}_{egr,max}$ is the maximum possible mass flow rate of EGR and $\epsilon > 1$ captures the diffusion phenomena. In the same manner, the oxygen concentration at the intake ports depends on the oxygen concentrations of the ambient air and the gases leaving the EGR system, time-shifted to account for the transport delay of the intake system. Using their respective mass flow rates as weighting factors, the nominal oxygen concentration at the intake ports, $[O_2]_{int}{}^{\dagger}$, can be modeled according to

$$[O_2]_{int}{}^{\dagger}(k) = \frac{[O_2]_{egr}{}^{\dagger}(k - d_{int}(k))\dot{m}^{\star}_{egr}(k - d_{int}(k)) + [O_2]_{amb}\dot{m}^{\star}_{fresh}(k - d_{int}(k))}{\dot{m}^{\star}_{egr}(k - d_{int}(k)) + \dot{m}^{\star}_{fresh}(k - d_{int}(k))} \qquad (24)$$

where $[O_2]_{amb}$ is the oxygen concentration of the ambient air.

The nominal in-cylinder oxygen concentration depends on the oxygen concentration of the air charge entering the cylinders and the residual gas left in the cylinders. Define the trapped residual fraction γ_{res} as the ratio of the trapped residual mass $\dot{m}^{\star}_{res}$ to the total trapped mass $\dot{m}^{\star}_{trapped}$ as in

$$\gamma_{res}(k) = \frac{\dot{m}^{\star}_{res}(k)}{\dot{m}^{\star}_{trapped}(k)}. \qquad (25)$$

Using this definition, the total trapped mass $\dot{m}^{\star}_{trapped}$ can be related to the mass flow rate into the engine with

$$m^{\star}_{trapped}(k) = \frac{\dot{m}^{\star}_{egr}(k - d_{int}(k)) + \dot{m}^{\star}_{fresh}(k - d_{int}(k))}{1 - \gamma_{res}(k)}. \qquad (26)$$

By assuming the trapped residual gases are equivalent to the gases that were just exhausted, the nominal in-cylinder oxygen concentration, $[O_2]_{cyl}{}^{\dagger}$, can be approximated by

$$[O_2]_{cyl}{}^{\dagger}(k) = (1 - \gamma_{res}(k))[O_2]_{int}{}^{\dagger}(k) + [O_2]_{exh}{}^{\dagger}(k)\gamma_{res}(k). \qquad (27)$$

Define $d_{int,exh}$ as the fixed delay between the induction and exhaust strokes. After $d_{int,exh}$ events, the inducted cylinder contents are burned and expelled into the exhaust system; therefore, the oxygen concentration at the exhaust ports can be predicted with

$$[O_2]_{exh}{}^{\dagger}(k + d_{int,exh}) = \frac{[O_2]^{\dagger}_{cyl}(k)m^{\star}_{trapped}(k) - OFR_s m^{\star}_{fuel}(k)}{m^{\star}_{trapped}(k) + m^{\star}_{fuel}(k)} \qquad (28)$$

where OFR_s is the stoichiometric mass ratio of oxygen to fuel. When estimating the transport delays, the nominal values are always rounded down. As a result, the in-cylinder oxygen concentration predicted in (27) tends to underestimate the total transport delay of the exhaust gas recirculation loop. To compensate for this effect and to remove any artificial jaggedness, a final in-cylinder oxygen concentration estimate $[O_2]_{cyl}$ can be predicted by filtering the nominal in-cylinder oxygen concentration according to

$$[O_2]_{cyl}(k) = \beta[O_2]_{cyl}{}^{\dagger}(k) + (1-\beta)[O_2]_{cyl}(k-1) \tag{29}$$

where β is the filter constant. To test the performance of this methodology, a delay based oxygen dynamics model was implemented into the GT-Power model of the six cylinder diesel engine. The oxygen dynamics model was then used to predict the in-cylinder oxygen concentration and the exhaust oxygen concentration over the first 100 seconds of a heavy duty FTP drive cycle. The delay based oxygen dynamics model produces excellent performance as shown in Figure 6. The top plot in Figure 6 compares the in-cylinder oxygen concentration predicted by GT-Power to the in-cylinder oxygen concentration predicted by the delay based oxygen dynamics model, whereas the bottom plot compares the predicted oxygen concentration at the exhaust ports. By directly predicting the transport delay, the oxygen concentrations predicted by a delay based oxygen dynamics model are always in phase with the actual oxygen concentrations. The root mean squared (RMS) errors in predicting the oxygen concentrations inside the cylinders and at the exhaust ports were 0.112% and 0.424% respectively. A delay based oxygen dynamics model achieves remarkable accuracy but is simple enough to be implemented in a production ECU.

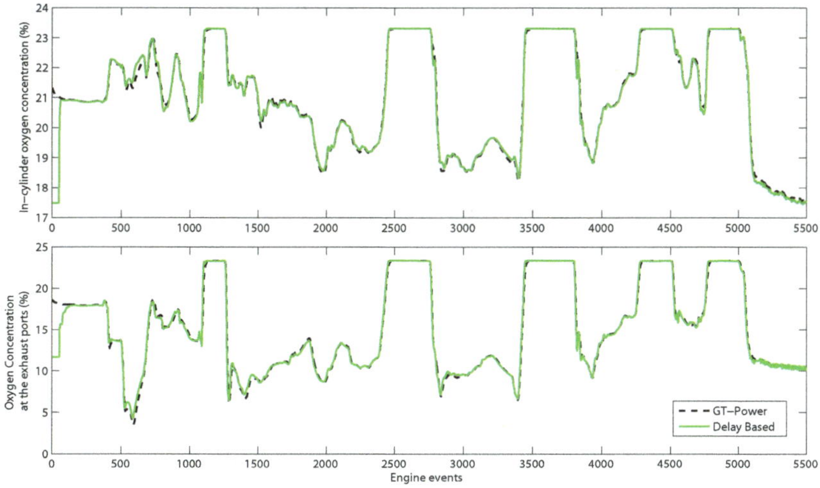

Fig. 6. Oxygen concentration prediction comparison: GT-Power versus delay based oxygen dynamics model

5. Transient performance comparison

Following the control structure shown in Figure 2, an in-cylinder oxygen concentration based fueling controller was implemented in GT-Power. Within this controller, the five fueling parameters (main SOI, pilot SOI, post SOI, pilot quantity and post quantity) are predicted with five sets of 10 x 10 x 10 tables indexed by the engine speed, the total fueling mass and the in-cylinder oxygen concentration. The in-cylinder oxygen concentration is dynamically predicted using a delay based oxygen dynamics model, whereas the exact values for the total fueling and engine speed are used directly. To provide a benchmark, a conventional fueling controller based on five sets of 20 x 20 tables indexed by the total fueling mass and the engine speed was also implemented into a separate GT-Power model.

Ideally, an air path controller would perfectly regulate the air path variables to their desired setpoints. To represent this scenario, the optimal air path actuator trajectories were directly specified and both fueling controllers were simulated in GT-Power for the first 100 seconds of a heavy duty FTP drive cycle. Under these conditions, both of the fueling control structures produced similar results. Figure 7 compares the instantaneous ISFC produced by the optimal fueling parameters to the instantaneous ISFC produced by the fueling parameters selected by the oxygen concentration based fueling controller and the conventional fueling controller. Similarly, Figure 8 compares the instantaneous $BSNO_X$ performance. It should be noted that for some sections of this drive cycle it was not possible to simultaneously meet the driver's torque request and produce a $BSNO_X$ level below 1.0 g/hp-hr. In these situations, the $BSNO_X$ level was minimized as much as possible while still achieving the desired torque. The spans of events without data correspond to decelerations where no fuel was injected. Neither controller was able to perfectly predict the optimal fueling parameters, thus resulting in small differences in the instantaneous values of ISFC and $BSNO_X$. However, these differences do not significantly affect the overall drive cycle performance.

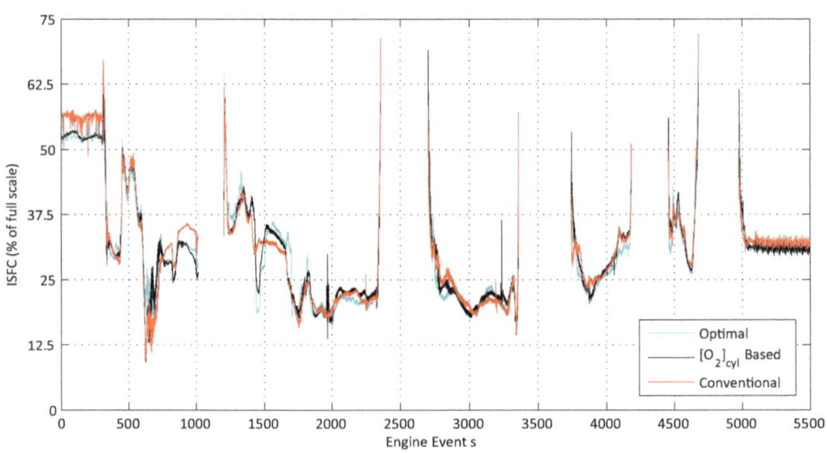

Fig. 7. ISFC Comparison: Ideal air path response

The drive cycle averaged ISFC and $BSNO_X$ results are reported in Table 2. With respect to the benchmark set by the optimal fueling parameters, the average ISFC in percentage of full scale produced by the oxygen concentration based fueling controller was 0.44% larger and the

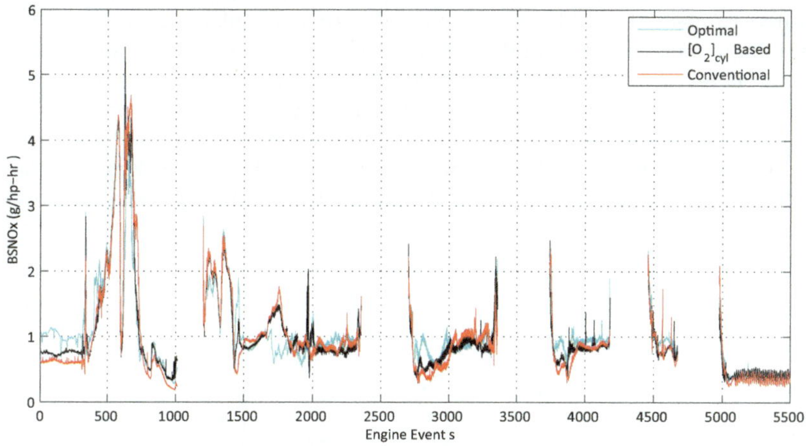

Fig. 8. BSNOx Comparison: Ideal air path response

average ISFC produced by the conventional fueling controller was 0.69% larger. Because of the inverse relationship between ISFC and $BSNO_X$, both the oxygen concentration based fueling controller and the conventional fueling controller produced on average less NO_X than the optimal fueling parameters. Recall that the optimization objective of the fueling parameter calibration was to minimize the ISFC under the constraint that the $BSNO_X$ is less than 1.0 g/hp-hr. Even though the two fueling controllers produced less NO_X on average, the optimal fueling parameters met the $BSNO_X$ target more consistently. For the oxygen concentration based fueling controller, 1.07% fewer combustion events achieved a $BSNO_X$ value less than 1.2 g/hp-hr, as compared to the optimal fueling parameters. For the conventional controller, 1.59% fewer combustion events met this criteria.

	ISFC (% of full scale)	BSNOx (g/hp-hr)	% of data with BSNOx < 1.2 g/hp-hr
Optimal	29.75	0.970	83.6%
$[O_2]_{cyl}$ Based	30.19	0.942	82.5%
Conventional	30.44	0.939	82.0%

Table 2. Drive cycle averaged combustion performance results: Ideal air path response

Although both the oxygen concentration based fueling controller and the conventional fueling controller produced on average higher values of ISFC and exceeded a $BSNO_X$ value of 1.2 g/hp-hr more frequently, the magnitude of these differences was very small for both controllers. For the in-cylinder oxygen concentration based fueling controller, these results demonstrate that the sets of three-dimensional tables indexed by engine speed, total fuel mass and in-cylinder oxygen concentration effectively predict the optimal fueling parameters and that the in-cylinder oxygen concentration prediction errors are practically negligible. In these types of ideal conditions where the trajectories of the air path setpoint variables are perfectly achieved, a conventional fueling controller with two-dimensional tables also effectively predicts the fueling parameters.

In practice, the instantaneous values of the air path setpoint variables will often differ from their desired values. Although a closed loop air path controller is designed to minimize this difference, many factors can limit the convergence rate or altogether prevent a desired value from being achieved. These factors include component failure, turbocharger surge/choke behavior, imposed air to fuel ratio limits and, most importantly, the dynamics of the air path system. To compare the robustness of an in-cylinder oxygen concentration based fueling controller to a conventional fueling controller with respect to these types of errors, the same drive cycle was simulated but with slightly different air path responses. Uniform bias shifts

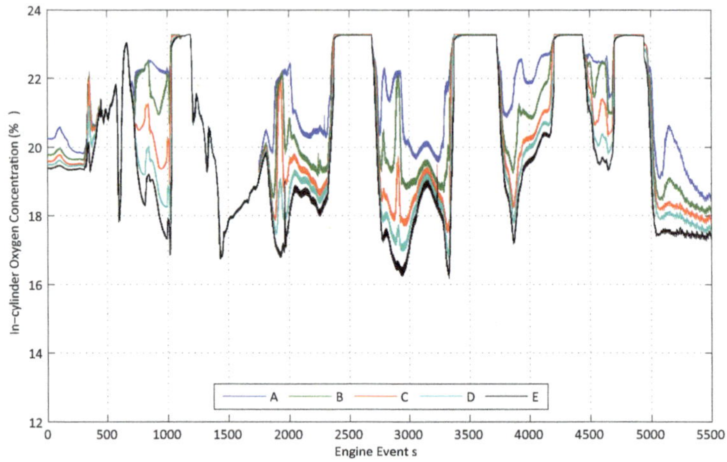

Fig. 9. In-cylinder oxygen concentration variations: Cases A - E

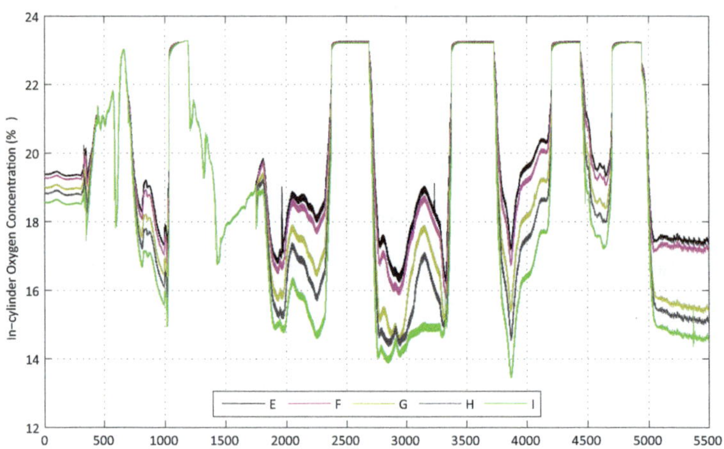

Fig. 10. In-cylinder oxygen concentration variations: Cases E - I

		$[O_2]_{cyl}$ Based		Conventional	
Case	$[O_2]_{cyl}$ (%)	ISFC (% of full scale)	$BSNO_x$ (g/hp-hr)	ISFC (% of full scale)	$BSNO_x$ (g/hp-hr)
A	20.57	41.38	1.758	32.31	2.788
B	19.95	36.06	1.367	31.00	1.907
C	19.43	32.56	1.138	30.44	1.369
D	19.07	31.06	1.020	30.31	1.102
E	18.74	30.19	0.942	30.44	0.939
F	18.53	30.32	0.875	31.13	0.839
G	17.72	30.94	0.826	33.75	0.631
H	17.36	32.06	0.778	36.06	0.562
I	16.82	34.75	0.711	41.06	0.493

Table 3. Drive cycle averaged combustion performance results: Shifted air path actuator commands

were applied to the actuator position trajectories to produces these differences. A total of eight additional drive cycles were simulated for each controller.

The actuator shifts applied to the first four drive cycles labeled "A", "B", "C" and "D" decreased the EGR flow rate and increased the fresh air flow rate. The resulting effects on the in-cylinder oxygen concentration are shown in Figure 9. The ideal case tested previously is labeled as case "E". For the remaining four tests labeled "F", "G", "H" and "I," the shifts applied to the actuator commands caused a relative increase in the EGR flow rate. Figure 10 depicts the in-cylinder oxygen concentrations resulting from these shifts. The types of uniform bias errors simulated in these tests are not meant to represent typical air path errors; rather, they were selected to help identify the robustness of each controller.

All of the drive cycle averaged results are compiled in Table 3. This same data is also visually depicted in Figure 11. These results clearly show that the two fueling controllers respond to actuator bias errors differently. Because a conventional fueling controller does not consider the state of the air path system, the conventional fueling controller responded exactly the same for each drive cycle. Although this response produced excellent performance when the air path setpoint trajectories were perfectly tracked, this same response produced poor performance when the air path setpoints were not achieved.

In general, as the in-cylinder oxygen concentration decreases (i.e. dilutent concentration increases), the heat release rate decreases. To offset this effect, the fuel injection timing should be advanced. Conversely, as the oxygen concentration increases (i.e. diluent concentration decreases), the combustion temperature tends to increase, thus increasing the NO_x production rate. To prevent this increase, the fuel injection timings should be retarded. These types of corrective actions are not applied by a conventional fueling controller. For drive cycles (A, B, C, D) with actuator biases which cause an increase in the in-cylinder oxygen concentration, the conventional fueling controller consistently injected fuel sooner than it should, resulting in drastically higher $BSNO_x$ values. On the other hand, for drive cycles (F, G, H, I) with actuator biases which cause a decrease in the in-cylinder oxygen concentration, the conventional fueling controller consistently injected fuel later than it should, resulting in larger ISFC values.

With an in-cylinder oxygen concentration based fueling controller, the changes to the oxygen concentration caused by the air path actuator biases are correctly compensated. Based on the

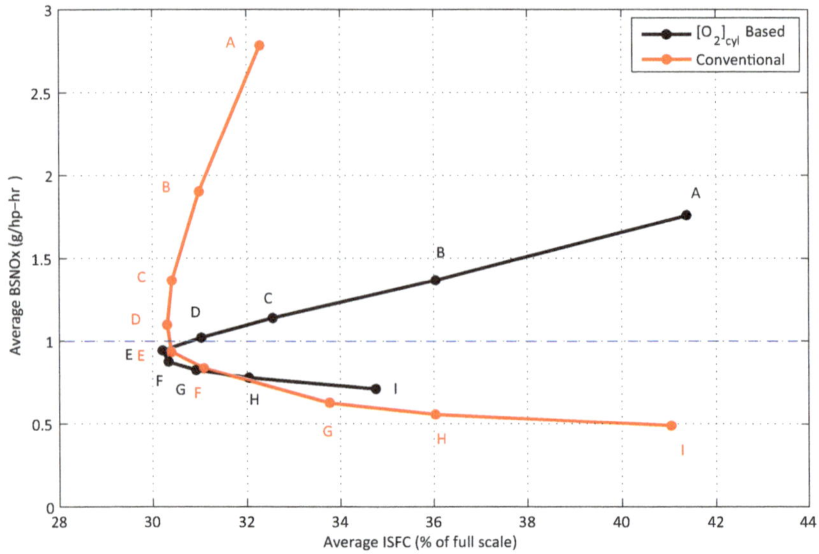

Fig. 11. Trade-off between cycle averaged $BSNO_X$ and cycle average ISFC

objective function used to calibrate the controller, the in-cylinder oxygen concentration based fueling controller produces a nearly constant NO_X production rate of about 1 g/hp-hr. Under certain conditions, however, this target NO_X production rate cannot be achieved through fuel injection adjustments alone. In these cases, the oxygen concentration based fueling controller selects the fueling parameters that minimize the $BSNO_X$ while meeting the torque request. For the drive cycles in which the EGR flow rate was decreased (A, B, C, D), these types of conditions were encountered more frequently.

To further illustrate these effects, the ISFC and $BSNO_X$ values produced by the two controllers have been directly compared on an event-by-event basis to generate families of probability density functions. The magnitude of the difference in the ISFC values achieved by the two fueling controllers depends on the magnitude of the air path setpoint errors. This is also true for the $BSNO_X$ difference. Figure 12 depicts how the distribution of the ISFC difference between the conventional fueling controller and the oxygen concentration based fueling controller varied as a function of the in-cylinder oxygen concentration difference (achieved value minus the value corresponding to the optimal air path setpoints). The median ISFC difference is also included in this figure on the bottom plane (i.e. 0% probability). Each of the probability density functions corresponds to a particular oxygen concentration difference. The random variable responsible for these variations corresponds to the combined effects of engine speed, the fuel injection mass and the cylinder conditions.

When the oxygen concentration difference for a given combustion event was approximately -2%, for example, the ISFC produced by the conventional controller was between 0% and 15% larger than the ISFC produced by the oxygen concentration based fueling controller for 91.9% of the occurrences. This same information can be extracted from Figure 12 by integrating the

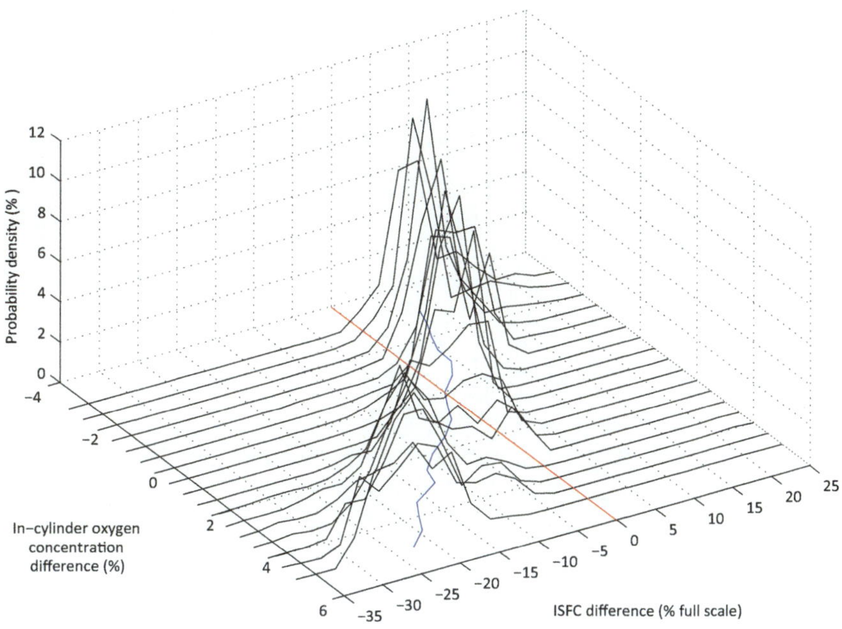

Fig. 12. Instantaneous ISFC difference (Conventional - $[O_2]_{cyl}$ based) probability density functions

probability density function curve corresponding to an oxygen concentration difference of -2% from 0% to 15%. Each of the distributions has approximately the same degree of narrowness indicating that the compensatory adjustments made by an in-cylinder oxygen concentration based fueling controller influence the combustion performance in a consistent manner.

A similar set of probability density functions shown in Figure 13 were created to illustrate how the $BSNO_x$ difference between the conventional fueling controller and the oxygen concentration based fueling controller varies as a function of the in-cylinder oxygen concentration difference. These probability density functions are also characterized by very narrow distributions. For negative oxygen concentration differences, the two controllers tended to produce approximately the same $BSNO_x$. When the oxygen concentration difference is positive, an in-cylinder oxygen concentration based fueling control produced significantly less NO_x. In fact, the magnitude of these changes was frequently larger than 1.0 g/hp-hr. These conditions correspond to combustion events where the oxygen concentration based fueling controller remained near the $BSNO_x$ target of 1.0 g/hp-hr while the conventional fueling controller produced more than double that same target.

The occurrence frequency of each (ISFC, $BSNO_x$) pair encountered across all of the drive cycle simulations has also been analyzed for each control structure. Figures 14 and 15 depicts the result on a logarithmic scale. To put this data in perspective, the drive cycle averaged (ISFC, $BSNO_x$) pairs originally presented in Figure 5 have also been superimposed

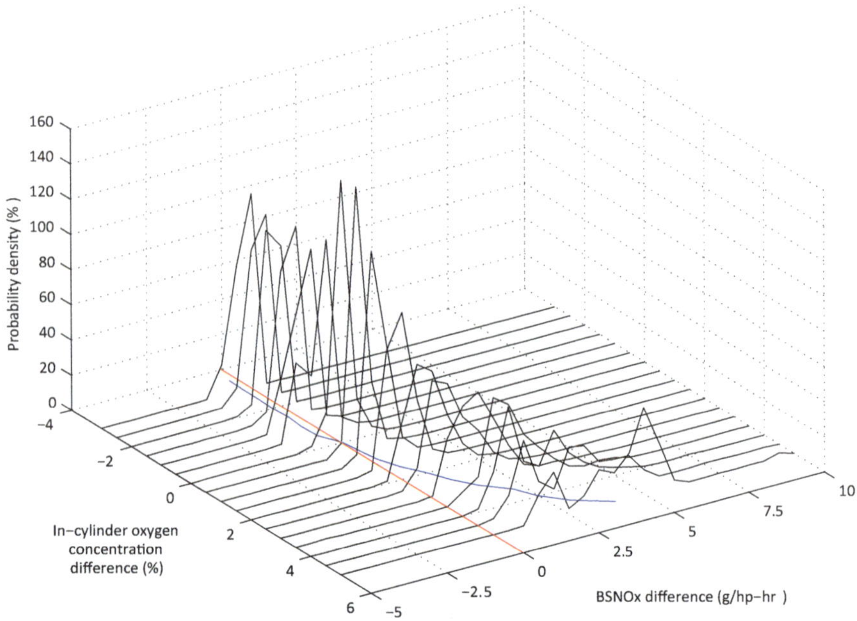

Fig. 13. Instantaneous $BSNO_x$ difference (Conventional - $[O_2]_{cyl}$ based) probability density functions

into these distribution charts as white curves. The combustion produced by the in-cylinder oxygen concentration based controller most frequently resulted in (ISFC, $BSNO_x$) pairs within the range [18, 35] % × [0.4, 1.4] g/hp-hr. Conversely, the combustion produced by the conventional controller most frequently resulted in a wider range of (ISFC, $BSNO_x$) pairs contained within the box outlined by [18, 40] % × [0.1, 1.7] g/hp-hr. Moreover, the $BSNO_x$ produced by the oxygen concentration based fueling controller for almost all of the combustion events was below 4.5 g/hp-hr, whereas this $BSNO_x$ exceeded 4.5 g/hp-hr on numerous occasions with the conventional fueling controller.

		$[O_2]_{cyl}$ Based		Conventional	
	$[O_2]_{cyl}$	ISFC	$BSNO_x$	ISFC	$BSNO_x$
Case	(%)	(% of full scale)	(g/hp-hr)	(% of full scale)	(g/hp-hr)
J	18.55	32.75	0.925	33.38	1.007
K	18.67	33.44	0.972	33.50	1.115
L	18.12	34.38	0.982	36.88	1.040
M	18.70	35.69	1.133	36.06	1.745

Table 4. Drive cycle averaged combustion performance results: Dynamic air path actuator errors

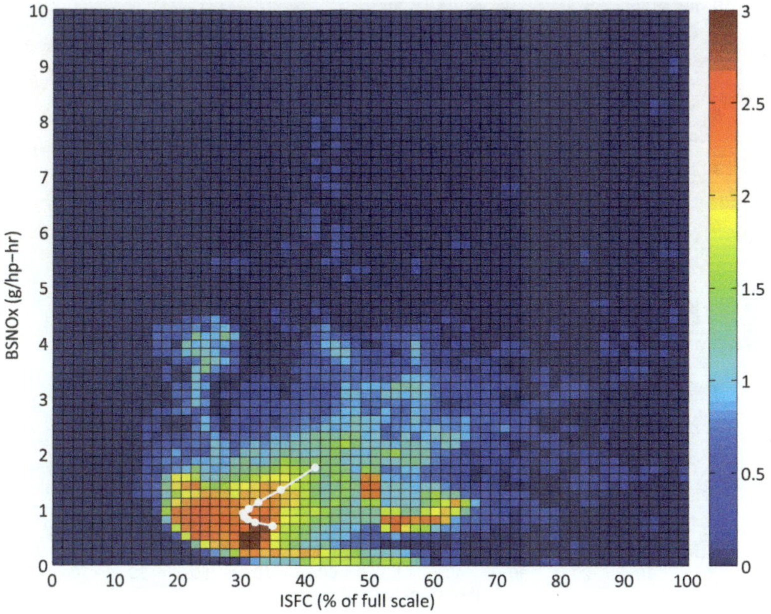

Fig. 14. Distribution of (ISFC, $BSNO_X$) pairs for the oxygen concentration based fueling controller

In contrast to the constant bias errors tested thus far, most air path errors are transient in nature. The relative performance of the two fueling controllers depends on the air path trajectory, so an infinite number of air path error trajectories are possible. To demonstrate some of the possible outcomes, a few hypothetical dynamic air path error trajectories have been simulated. The first two cases labeled "J" and "K" had dynamic errors which are approximately of the same order as those seen experimentally. The second two cases labeled "L" and "M" contained larger dynamic errors (both in amplitude and duration). The drive cycle averaged performance results for these drive cycles are assembled in Table 4. The cycle averaged (ISFC, $BSNO_X$) pairs are also plotted in Figure 16. Included for reference are the cycle averaged data pairs for drive cycles A through I.

For each of the four drive cycles, the in-cylinder oxygen concentration based fueling controller produced more desirable combustion performance than the conventional fueling controller in terms of both fuel consumption and NO_X production. Between the two cases which simulated moderate dynamic air path errors, the cycle averaged ISFC achieved by the oxygen concentration fueling controller in percentage of full scale was as much as 0.63% lower than the conventional fueling controller. The oxygen concentration based fueling controller also reduced the cycle averaged $BSNO_X$ by as much as 0.143 g/hp-hr. The larger dynamic errors simulated in cases L and M produced even larger relative changes. Under these conditions, the cycle averaged ISFC produced by the oxygen concentration based fueling controller in percentage of full scale was as much as 2.50% lower than the conventional fueling controller and the cycle averaged $BSNO_X$ was decreased by much as 0.612 g/hp-hr.

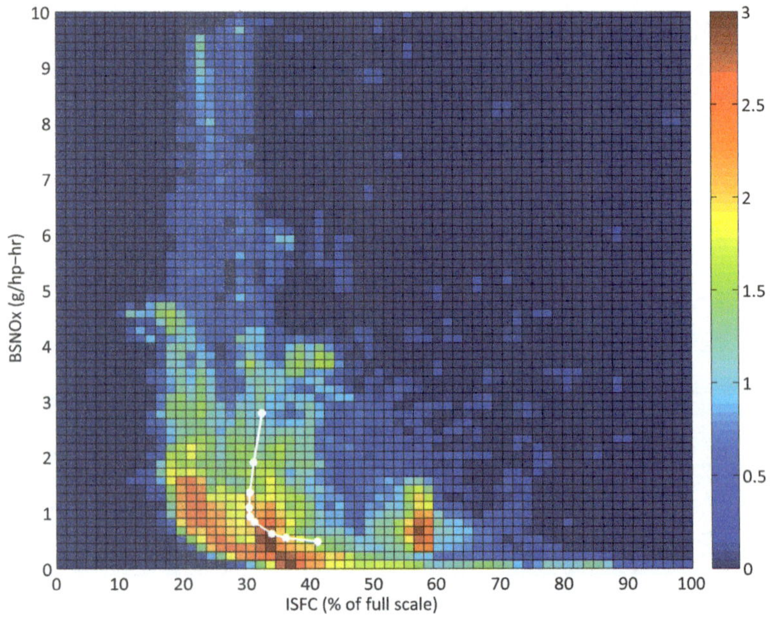

Fig. 15. Distribution of (ISFC, $BSNO_X$) pairs for the conventional fueling controller

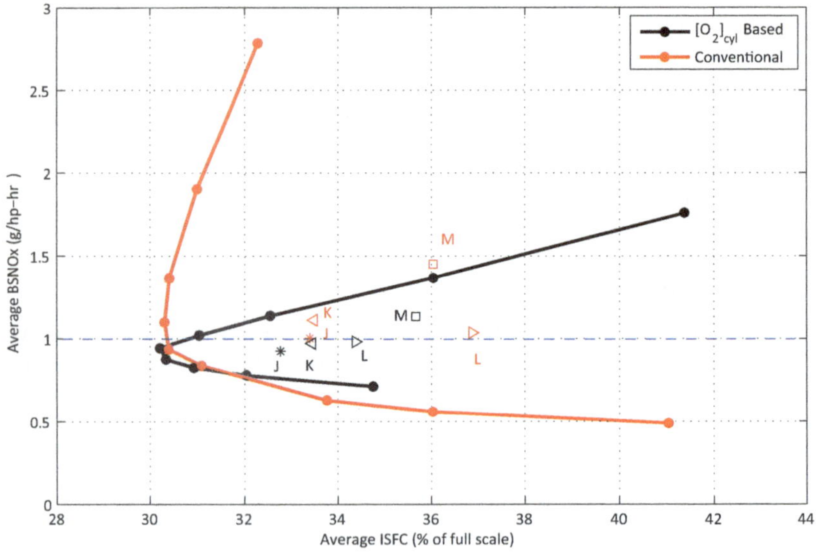

Fig. 16. Trade-off between cycle averaged $BSNO_X$ and cycle averaged ISFC: Dynamic air path errors

6. Conclusions

Modern diesel engines have sophisticated fuel injection systems that allow for multiple fuel injection pulses, thus providing flexibility in shaping the combustion process. The timing of each of these injections and fraction of the total fuel delivered by each injection can be modified to ensure that the combustion produces a desired trade-off between performance metrics. A typical diesel engine controller determines both the desired air path setpoints and the fueling parameters based on a feed-forward mapping of the engine speed, the fuel injection mass and the system operating mode. The main difference between different system operating modes such as transient, steady-state and altitude is that they require different air path setpoints to achieve the desired level of performance. The combustion process depends on quantities within the cylinder after the intake valve closes, and not on how those quantities got there. Provided they can be estimated, scheduling the fuel injection parameters on the in-cylinder conditions rather than the system operating mode is a more robust and effective solution.

The air path of a diesel engine, including the intake manifold, exhaust manifold and the exhaust gas recirculation loop is characterized by mass accumulation and mass transport dynamics. Without a thorough understanding of these dynamics, the in-cylinder content cannot be estimated accurately. Beginning from the one-dimensional oxygen diffusion-convention equation, a delay based oxygen dynamics model capable of dynamically predicting the in-cylinder oxygen concentration was designed and demonstrated in this chapter. This model is quite simple and could be run in real time in a production ECU. Enabled by the delay based oxygen dynamics model, an in-cylinder oxygen concentration based fueling controller was also developed. The resulting in-cylinder oxygen concentration based fueling controller demonstrated in simulation the ability to consistently limit the $BSNO_x$ to below a desired level while minimizing the ISFC and smoothly meeting the torque request. In steady-state, the performance of an oxygen concentration based fueling controller is equivalent to a conventional controller. During a transient, however, the robustness provided by scheduling the fueling parameters based on the in-cylinder oxygen concentration allows the fueling controller to maintain any desired set of performance trade-offs. Even when a component within the air path system fails or when exceptional control action is taken to prevent the turbocharger system from reaching its surge or choke limits, an oxygen concentration based fueling controller achieves very good performance.

7. References

Alberer, D. & Del Re, L. (2009). Fast oxygen based transient diesel engine operation, *SAE International Journal of Engines* 2(1): 405–413. No. 2009-01-0622.

Ammann, M., Fekete, N. P., Guzzella, L. & Glattfelder, A. H. (2003). Model-based control of the VGT and EGR in a turbocharged common-rail diesel engine: Theory and passenger car implementation, *SAE*. No. 2003-01-0357.

Benajes, J., Molina, S. & Garcia, J. (2001). Influence of pre- and post-injection on the performance and pollutant emissions in a HD diesel engine, *SAE*. No. 2001-01-0526.

Canova, M., Midlam-Mohler, S., Guezennec, Y. & Rizzoni, G. (2009). Mean value modeling and analysis of HCCI diesel engines with external mixture formation, *Journal of Dynamic Systems, Measurements, and Control* 131: 011002–1 – 011002–14.

Chen, S. K. & Yanakiev, O. (2005). Transient NOx emissions reduction using exhaust oxygen concentration based control for a diesel engine, *SAE*. No. 2005-01-0372.

Guzzella, L. & Amstutz, A. (1998). Control of diesel engines, *IEEE Control Systems Magazine* pp. 53–71.

Jung, M. & Glover, K. (2006). Calibratable linear parameter-varying control of a turbocharged diesel engine, *IEEE Transactions on Control Systems Technology* 14(1): 45–62.

Langthaler, P. & Del Re, L. (2007). Fast predictive oxygen charge control of a diesel engine, *Proceedings of the 2007 American Control Conference*.

Langthaler, P. & Del Re, L. (2008). Robust model predictive control of a diesel airpath, *Proceedings of the 2008 International Federation of Automatic Control*.

MacMillan, D., La Rocca, A., Shayler, P. J., Morris, T., Murphy, M. & Pegg, I. (2009). Investigating the effects of multiple injections on stability at cold idle for DI diesel engines, *SAE*. No. 2009-01-0612.

Morel, T. & Wahiduzzaman, S. (1996). Modeling of diesel combustion and emissions, *XXVI FISITA Congress*.

Osuka, I., Nishimura, M. & Tanaka, Y. (1994). Benefits of new fuel injection system technology on cold startability of diesel enginess, *SAE*. No. 940586.

van Nieuwstadt, M. J., Kolmanovsky, I. V. & Moraal, P. E. (2000). Coordinated EGR-VGT control for diesel engines: an experimental comparison, *SAE*. No. 2000-01-0266.

Wang, J. (2008). Air fraction estimation for multiple combustion mode diesel engines with dual-loop EGR systems, *Control Engineering Practice* 16: 1479–1486.

Yan, F. & Wang, J. (2010). In-cylinder oxygen mass fraction cycle-by-cycle estimation via a Lyapunov-based observer design, *Proceedings of the 2010 American Control Conference*.

Permissions

The contributors of this book come from diverse backgrounds, making this book a truly international effort. This book will bring forth new frontiers with its revolutionizing research information and detailed analysis of the nascent developments around the world.

We would like to thank Dr. Alessio Innocenti, for lending his expertise to make the book truly unique. He has played a crucial role in the development of this book. Without his invaluable contribution this book wouldn't have been possible. He has made vital efforts to compile up to date information on the varied aspects of this subject to make this book a valuable addition to the collection of many professionals and students.

This book was conceptualized with the vision of imparting up-to-date information and advanced data in this field. To ensure the same, a matchless editorial board was set up. Every individual on the board went through rigorous rounds of assessment to prove their worth. After which they invested a large part of their time researching and compiling the most relevant data for our readers. Conferences and sessions were held from time to time between the editorial board and the contributing authors to present the data in the most comprehensible form. The editorial team has worked tirelessly to provide valuable and valid information to help people across the globe.

Every chapter published in this book has been scrutinized by our experts. Their significance has been extensively debated. The topics covered herein carry significant findings which will fuel the growth of the discipline. They may even be implemented as practical applications or may be referred to as a beginning point for another development. Chapters in this book were first published by InTech; hereby published with permission under the Creative Commons Attribution License or equivalent.

The editorial board has been involved in producing this book since its inception. They have spent rigorous hours researching and exploring the diverse topics which have resulted in the successful publishing of this book. They have passed on their knowledge of decades through this book. To expedite this challenging task, the publisher supported the team at every step. A small team of assistant editors was also appointed to further simplify the editing procedure and attain best results for the readers.

Our editorial team has been hand-picked from every corner of the world. Their multi-ethnicity adds dynamic inputs to the discussions which result in innovative outcomes. These outcomes are then further discussed with the researchers and contributors who give their valuable feedback and opinion regarding the same. The feedback is then collaborated with the researches and they are edited in a comprehensive manner to aid the understanding of the subject.

Apart from the editorial board, the designing team has also invested a significant amount of their time in understanding the subject and creating the most relevant covers. They scrutinized every image to scout for the most suitable representation of the subject and create an appropriate cover for the book.

The publishing team has been involved in this book since its early stages. They were actively engaged in every process, be it collecting the data, connecting with the contributors or procuring relevant information. The team has been an ardent support to the editorial, designing and production team. Their endless efforts to recruit the best for this project, has resulted in the accomplishment of this book. They are a veteran in the field of academics and their pool of knowledge is as vast as their experience in printing. Their expertise and guidance has proved useful at every step. Their uncompromising quality standards have made this book an exceptional effort. Their encouragement from time to time has been an inspiration for everyone.

The publisher and the editorial board hope that this book will prove to be a valuable piece of knowledge for researchers, students, practitioners and scholars across the globe.

List of Contributors

Limin Cao, Xiangyi Zhang and Wenkui Wang
State Key Laboratory of Metastable Materials, Yanshan University, China

Min Feng
Department of Physics and Astronomy, University of Pittsburgh, USA

Hélio A. Duarte and Maicon P. Lourenço
Department of Chemistry, ICEx, Universidade Federal de Minas Gerais, Belo Horizonte, MG, Brazil

Thomas Heine
School of Engineering and Science, Jacobs University Bremen, Bremen, Germany

Luciana Guimarães
Department of Natural Science, Universidade Federal de São João Del Rei, São João Del Rei, MG, Brazil

Azusa N. Hattori
Nanoscience and Nanotechnology Center, The Institute of Scientific and Industrial Research, Osaka University, Japan

Katsuyoshi Endo
Research Center for Ultra-Precision Science and Technology, Graduate School of Engineering, Osaka University, Japan

Elena Rogacheva
National Technical University "Kharkov Polytechnic Institute", Ukraine

Yu-Xiang Zheng, Rong-Jun Zhang and Liang-Yao Chen
Department of Optical Science and Engineering, School of Information Science and Engineering, Fudan University, Shanghai, China

Kwo Young
Ovonic Battery Company, Rochester Hills, Michigan, USA

Mauvy Fabrice
ICMC Bordeaux, CNRS, Université de Bordeaux, UPR 9048, 33608 Pessac, Cedex, France

Fouletier Jacques
LEPMI, UMR 5279, CNRS, Grenoble INP, Université de Savoie, Université Joseph Fourier, BP75. 38402 Saint Martin d'Hères, France

Andrzej Stokłosa
Cracow University of Technology, Faculty of Chemical Engineering and Technology, Cracow, Poland

Masato Kato
Japan Atomic Energy Agency, Japan

Norlida Kamarulzaman
Centre For Nanomaterials Research, Institute of Science, Universiti Teknologi MARA, Shah Alam, Selangor, Malaysia

Mohd Hilmi Jaafar
Centre For Foundation Study In Sciences, University of Malaya, Kuala Lumpur, Malaysia

W.M. Tang, C.H. Leung and P.T. Lai
Department of Electrical and Electronic Engineering, The University of Hong Kong, Hong Kong, China

Naira B. Yeritsyan and Lida A. Khachatryan
Name Institute of General and Inorganic Chemistry, National Academy of Science, Republic of Armenia

Sergey P. Suprun, Valeriy G. Kesler and Evgeniy V. Fedosenko
Institute of Semiconductor Physics, Siberian Branch of Russian Academy of Sciences, Russia

O.P. Korobeinichev, A.G. Shmakov and V.M. Shvartsberg
Institute of Chemical Kinetics & Combustion, Siberian Brunch of Russian Academy of Sciences, Novosibirsk, Russia

Mamoru Kaiho and Osamu Yamada
National Institute of Advanced Industrial Science and Technology, Japan

Jason Meyer and Stephen Yurkovich
The Ohio State University and University of Texas at Dallas, USA